Number Theory
with Applications

James A. Anderson

University of South Carolina at Spartanburg

James M. Bell

Milliken & Company

PRENTICE HALL
Upper Saddle River, NJ 07458

Library of Congress Cataloging-in-Publication Data

Anderson, James A. (James Andrew),
  Number theory with applications / James A. Anderson & James M.
Bell

      p. cm.
  Includes bibliographical references (p. –).
  ISBN 0-13-190190-7 (alk. paper)
    1. Number theory.   I. Bell, James M. (James Milton),
II. Title.
QA241.A496   1996
512'.72—dc20   96-25134

                                                    CIP

Acquisition Editor: George Lobell
Editorial Assistant: Gale Epps
Editorial Director: Tim Bozik
Editor-in-Chief: Jerome Grant
Assistant Vice President of Production and Manufacturing: David W. Riccardi
Editorial/Production Supervision: Jennifer Fischer
Managing Editor: Linda Mihatov Behrens
Executive Managing Editor: Kathleen Schiaparelli
Manufacturing Buyer: Alan Fischer
Manufacturing Manager: Trudy Pisciotti
Marketing Manager: John Tweeddale
Marketing Assistant: Diana Penha
Creative Director: Paula Maylahn
Art Director: Jayne Conte
Cover Art/Design: Kiwi Design

Printed in the United States of America

10   9   8   7   6   5   4   3   2   1

ISBN   0-13-190190-7

Prentice-Hall International (UK) Limited, London
Prentice-Hall of Australia Pty. Limited, Sydney
Prentice-Hall Canada Inc., Toronto
Prentice-Hall Hispanoamericana, S.A., Mexico
Prentice-Hall of India Private Limited, New Delhi
Prentice-Hall of Japan, Inc., Tokyo
Simon & Schuster Asia Pte. Ltd., Singapore
Editora Prentice-Hall do Brasil, Ltda., Rio de Janeiro

To our wives
Marilyn and Pat
and to our children
Kristin and Andy Anderson
and
Bruce, Keith, and Sarah Bell

Preface

This undergraduate text is intended to be flexible enough so that it may be used by anyone requiring a beginning course in number theory, including mathematics, secondary education, computer science, engineering, and science majors. Thus, the purpose of this book is to provide a basic core of topics useful for any number theory course and to supplement this core with numerous applications and advanced topics to provide maximum flexibility for the instructor. We have included a fair amount of history since it not only makes the number theory more interesting and relevant but also helps unify one of the oldest branches of mathematics.

Number theory is an excellent area in which to learn to do proofs. For this reason we have tried to present a self-contained logical development of basic number theory. Some abstract algebra is introduced in Chapter 2 but it requires no previous knowledge of the subject. For those who want to begin the axiomatic development of number theory with Peano's Postulates, we have included them in Appendix B. We have also given an extended discussion of logic and proofs as well as a formal development of polynomials for those who wish to use them. None of the appendices is essential for the main body of the book. Chapter 0 has been included for students who have no background in sets and may be skipped if they are already familiar with the material.

The applications are a very significant part of the text. Most are drawn from computer science, engineering, and science; however, we feel that it is also helpful for all students to see practical uses of number theory and hope that they will find them interesting. The applications are fully developed so that they may be used without reference to outside sources.

The core consists of Sections 1 to 6 of Chapter 1, Sections 1 to 4 of Chapter 2, and Sections 1 to 9 of Chapter 3. Chapters 4 to 7 depend only upon the core and hence may be covered in any order, so that any of these chapters may be included or omitted as desired by the instructor.

Matrices are used in Section 3 of Chapter 3 and more extensively in Sections 4 and 5 of Chapter 7. Although the use of matrices is self contained and requires little previous knowledge of matrices, all of these sections may be skipped without affecting the continuity of the book.

For a course intended primarily for science or computer science majors, we suggest covering Chapters 1 to 5 including as many of the applications as possible. For a course intended primarily for secondary education majors, in addition to the core, the course could begin with the logic included in Appendix A and/or the set theory of Chapter 0 if needed or desired. As time allows, Sections 1 to 4 of Chapter 5 or Sections 1 to 3 and 6 to 8 of Chapter 7 could also be covered as well as selected applications. In a course for mathematics majors having a strong background, material in all eight chapters and Appendix B would be valuable. There is sufficient material that the instructor may select the topics and applications of most interest.

Acknowledgments

We wish to thank Jerome Lewis for suggesting several of the applications included in this book, Linda and Jimmie Gilbert for their encouragement and practical help in completing this project, and James D. Spencer for proofreading the manuscript. We also thank the following reviewers for their valuable suggestions and comments:

John R. Burke	Gonzaga University
Joel Freedman	University of British Columbia
Jack Hughes	Clearwater Christian College
Wells Johnson	Bowdoin College
Libby Krussel	University of California, Santa Barbara
William Pardon	Duke University
Charles Parry	Virginia Tech
Don H. Tucker	University of Utah, Salt Lake City
Ti Wang	University of Toronto
Kenneth Williams	Carleton University

Contents

Chapter 0
SETS

0.1 Sets and Relations

In this book a set is understood to be a well-defined collection of objects or elements. In reality this statement is not a definition but the avoidance of a definition. This "definition" is similar to the one given by Georg Cantor, who was instrumental in the early development of set theory. The inadequacy of this definition became apparent when paradoxes or contradictions were developed by the Italian logician Burali-Forti in 1879 and later by Bertrand Russell. Axiomatic systems have been developed for set theory in hope of avoiding further contradictions and paradoxes. These systems include the Zermelo-Fraenkel-von Neumann system, the Gödel-Hilbert-Bernays system and the Russell-Whitehead system. These systems are beyond the scope of this book, so we will essentially leave sets as undefined objects and assume that sets have been selected so that we can determine their elements and so that no contradictions are produced.

Finite sets may be described by listing their elements. This is done notationally by listing the elements between two braces. For example, $\{1, 2, 3, 4\}$ is the set containing the elements 1, 2, 3, and 4. The set of positive integers may be denoted by $\{1, 2, 3, 4, \ldots\}$ and the first n positive integers by $\{1, 2, 3, \ldots, n\}$, where three dots are used to indicate the continuation of a pattern. Set builder notation is also used to describe sets. Braces are again used to denote the set; and within these braces, the characteristic describing the set is given. Thus, if T is a set, then the set $\{x : x$ is in T and x has property $P\}$ is intended to contain exactly those objects of the set T having the property P. Only objects of T possessing the property P are in the set, and all objects of T having the property P are in the set.

We formalize some of the ideas discussed above in the following definition.

Definition 0.1 *If a is one of the objects of the set A, we say that a is an element of A or a belongs to A. This condition is denoted by $a \in A$. If it is not true that a is an element of A, we write $a \notin A$.*

1

Thus $3 \in \{1,2,3,4\}$ but $5 \notin \{1,2,3,4\}$.

If T is the set $\{1,3,5,7\}$, then the set builder notation $\{x : x \in T$ and $2x + 1 < 8\}$ describes the set $\{1,3\}$. If T is the set from which elements are taken, we write $\{x : x \in T$ and $P(x)\}$ or equivalently, $\{x \in T : P(x)\}$, where $P(x)$ is a property of x such as "$2x + 1 < 8$." If the set T is understood, for brevity we may write $\{x : P(x)\}$.

Definition 0.2 *A set A is a* subset *of a set B (denoted $A \subseteq B$) if every element of A is an element of B; that is, if $x \in A$, then $x \in B$. In particular, every set is a subset of itself. If it is not true that A is a subset of B, this condition is denoted by $A \nsubseteq B$. Thus $A \nsubseteq B$ if and only if there is an x in A that is not in B.*

Thus $\{1,2,3\} \subseteq \{1,2,3,4\}$; but $\{1,2,3,4\} \nsubseteq \{1,2,3\}$.

Sets are equal provided that they contain exactly the same elements. If A is the set $\{2,4,6\}$ and B is the set $\{x : x$ is an even positive integer less than $7\}$, then we have different names for the same set. This discussion is formalized in the following definition.

Definition 0.3 *If A and B are sets, then we say that A equals B, written $A = B$, whenever, for any x, $x \in A$ if and only if $x \in B$. If $A \subseteq B$ and $A \neq B$, we write $A \subset B$ and say that A is a* proper subset *of B.*

These definitions immediately provide a way to establish the equality of two sets.

Theorem 0.4 *For sets A and B, $A = B$ if and only if A is a subset of B and B is a subset of A; that is, $A = B$ if and only if $A \subseteq B$ and $B \subseteq A$.*

Proof.   Assume that $A = B$. If $x \in A$, then $x \in B$ so that $A \subseteq B$; and if $x \in B$, then $x \in A$ so that $B \subseteq A$.

Conversely, assume that $A \subseteq B$ and $B \subseteq A$. Thus, if $y \in A$, then $y \in B$; and if $y \in B$, then $y \in A$. This last sentence is equivalent to stating that $y \in A$ if and only if $y \in B$. Thus $A = B$. ■

Thus, proving that two sets are equal is a two-step process:

> Prove that the first set is a subset of the second.
> Prove that the second set is a subset of the first.

At this point introduce two new sets, the universal set and the empty set. In a sense they are opposites since the empty set contains no elements and the universal set contains "all" elements.

Definition 0.5 *The* empty set, *denoted by $\varnothing$ or by $\{\}$, is the set which contains no elements. The* universal set $\mathbf{U}$ *is a set which has the property that all sets under consideration are subsets of it.*

The universal set is the collection of all of the objects that are the subject of discussion, that is, the context. In discussions about number theory, the universal set is usually the set of all integers. In calculus, the universal set may be the set of all real numbers or the set of all points in n-dimensional space. The existence of a set with no elements can be made plausible by considering set builder notation and using a property that nothing satisfies. Let $A = \{x : x$ is an integer and $2x = 5\}$. Since A is a set and since no x satisfies the conditions to belong to A, A has no elements.

By definition, every set is a subset of the universal set and the empty set is a subset of every set. Note that although called the universal set, $\mathbf{U}$ is not uniquely defined.

The next concepts use existing sets to form new sets.

Definition 0.6 *The* intersection *of sets A and B is the set of all elements that are in both A and B. The intersection of sets A and B is denoted $A \cap B$. More succinctly, $A \cap B = \{x : x \in A \text{ and } x \in B\}$.*

For example, if $A = \{1, 2, 3, 4, 5\}$ and $B = \{1, 3, 5, 7, 9\}$, then $A \cap B = \{1, 3, 5\}$.

Definition 0.7 *The* union *of sets A and B is the set of all elements that are in either A or B. The union of sets A and B is denoted $A \cup B$. Equivalently, $A \cup B = \{x : x \in A \text{ or } x \in B\}$.*

For example if $A = \{1, 2, 4, 6, 7\}$ and $B = \{2, 3, 4, 5, 6\}$, then $A \cup B = \{1, 2, 3, 4, 5, 6, 7\}$. The union $A \cup B$ is a new set formed from A and B by combining the elements of A and B.

Definition 0.8 *Let A and B be sets. The* difference *between A and B, denoted by $A - B$, is the set of all elements which are in A but are not in B. Equivalently, $A - B = \{x : x \in A \text{ and } x \notin B\}$. The* symmetric difference *of A and B, denoted by $A \triangle B$, is the set $(A - B) \cup (B - A)$.*

For example, if $A = \{1, 2, 4, 6, 7\}$ and $B = \{2, 3, 4, 5, 6\}$, then $A - B = \{1, 7\}$ and $A \triangle B$ is the set $\{1, 3, 5, 7\}$. The symmetric difference of A and B is comprised of those elements that are in one and only one of the two sets A and B.

Definition 0.9 *The* complement *of a set A, denoted by A', is the set of all elements of the universe which are not in A. Hence*

$$A' = \mathbf{U} - A = \{x : x \in \mathbf{U} \text{ and } x \notin A\}$$

If $\mathbf{U}$ is the set of positive integers and $A = \{2, 4, 6, 8, \ldots\} = \{x \in \mathbf{U} : x = 2n$ for some positive integer $n\}$ is the set of all even positive integers, then $A' = \{1, 3, 5, 7, \ldots\}$ is the set of all odd positive integers.

Definition 0.10 *The* power set *of a set A, denoted by $\mathcal{P}(A)$, is the set consisting of all subsets of A.*

Hence the power set of the set $A = \{1, 2, 3\}$ is the set

$$\mathcal{P}(A) = \{\varnothing,\ \{1\},\ \{2\},\ \{3\},\ \{1,2\},\ \{2,3\},\ \{1,3\},\ \{1,2,3\}\}$$

When A has 3 elements, $\mathcal{P}(A)$ has $2^3 = 8$ elements; or equivalently, A has $2^3 = 8$ subsets. In general, if A has n elements, the set $\mathcal{P}(A)$ has 2^n elements and A has 2^n subsets.

Another operation between sets which is commonly used is the Cartesian product, which is defined as follows:

Definition 0.11 *Let a and b be elements of a set. Let (a, b) be the set $\{a, \{a, b\}\}$. The object (a, b) is called an* ordered pair *with first component a and second component b. The* Cartesian product *of A and B, denoted by $A \times B$, is the set $\{(a, b) : a \in A \text{ and } b \in B\}$. The set $A \times B$ consists of all ordered pairs having the first component an element of A and the second component an element of B.*

We are justified in referring to "first" and "second" components because $(a, b) = (c, d)$ if and only if $a = c$ and $b = d$. Let $A = \{1, 2, 3\}$ and $B = \{r, s\}$; then

$$A \times B = \{(1, r),\ (1, s),\ (2, r),\ (2, s),\ (3, r),\ (3, s)\}$$

These are the same type of ordered pairs that are used to graph functions in algebra and calculus.

Definition 0.12 *A* relation R *between sets A and B is a subset of $A \times B$. If $(a, b) \in R$, we sometimes denote this by aRb and say that a is* related to b; *or that a is related to b by means of R.*

If $A = \{1, 2, 3\}$ and $B = \{r, s\}$ so that

$$A \times B = \{(1, r),\ (1, s),\ (2, r),\ (2, s),\ (3, r),\ (3, s)\}$$

then $R = \{(1, r), (1, s), (3, s)\}$ is a relation between A and B. We may also write $3Rs$ because $(3, s) \in R$. $A \times B$ has 6 elements so that there are $2^6 = 64$ subsets of $A \times B$. Therefore, there are 64 different relations on $A \times B$.

Associated with each relation R on $A \times B$ is a relation R^{-1} on $B \times A$.

Definition 0.13 *Let $R \subseteq A \times B$ be a relation on $A \times B$. Then the relation R^{-1} on $B \times A$ is defined by*

$$R^{-1} = \{(b, a) : (a, b) \in R\}$$

That is, $(b, a) \in R^{-1}$ if and only if $(a, b) \in R$; or, equivalently, $bR^{-1}a$ if and only if aRb.

In the example following Definition 0.12, $R = \{(1, r), (1, s), (3, s)\}$ so that $R^{-1} = \{(r, 1), (s, 1), (s, 3)\}$.

Two given relations may be combined to generate a new relation. One method of combining relations is called composition.

Definition 0.14 *Let $R \subseteq A \times B$ be a relation on $A \times B$ and $S \subseteq B \times C$ be a relation on $B \times C$. The* composition *of S and R is the relation $T \subseteq A \times C$ defined by*

$$T = \{(a,c): \text{ there is an element } b \text{ such that } (a,b) \in R \text{ and } (b,c) \in S\}$$

This set is denoted by $T = S \circ R$.

Example 0.15 *Let $A = \{1,2,3\}$, $B = \{x,y\}$, and $C = \{\square, \triangle, \bigcirc, \bigstar\}$, and let the relations R on $A \times B$ and S on $B \times C$ be given by*

$$\begin{aligned} R &= \{(1,x),(1,y),(3,x)\} \\ S &= \{(x,\square),(x,\triangle),(y,\bigcirc),(y,\bigstar)\} \end{aligned}$$

Then

$$S \circ R = \{(1,\square), (1,\triangle), (1,\bigcirc), (1,\bigstar), (3,\square), (3,\triangle)\}$$

because, if we test all 12 possibilities, only the following produce elements of $S \circ R$:

$$\begin{aligned} (1,x) \in R \text{ and } (x,\square) \in S &\quad \text{imply that} \quad (1,\square) \in S \circ R \\ (1,x) \in R \text{ and } (x,\triangle) \in S &\quad \text{imply that} \quad (1,\triangle) \in S \circ R \\ (1,y) \in R \text{ and } (y,\bigcirc) \in S &\quad \text{imply that} \quad (1,\bigcirc) \in S \circ R \end{aligned}$$

$$\vdots$$

$$(3,x) \in R \text{ and } (x,\triangle) \in S \quad \text{imply that} \quad (3,\triangle) \in S \circ R$$

We now consider the special case of relations on $A \times B$ when $A = B$, giving relations on $A \times A$.

Definition 0.16 *A relation R on $A \times A$ is* reflexive *if for all a in A, (a,a) is in R. R is* symmetric *if for all a and b in A, (a,b) in R implies that (b,a) is in R. R is* transitive *if for all a, b, and c in A, whenever (a,b) and (b,c) are in R, (a,c) is in R. R is* antisymmetric *if for all a and b in A, whenever (a,b) and (b,a) are in R, $a = b$.*

Example 0.17 *Let $A = \{1,2,3,4,5,6\}$ and the relation $R_1 \subseteq A \times A$ be the set $R_1 = \{(1,1),(2,2),(3,3),(4,4),(5,5),(6,6),(1,2),(1,4),(2,1),(2,4),(3,5),(5,3),(4,1),(4,2)\}$.*

The relation R_1 is reflexive by "inspection" since for each $a \in A$, we have $(a,a) \in R_1$. (For $a = 1$, $(1,1) \in R_1$. For $a = 2$, $(2,2) \in R_1$. ... For $a = 6$, $(6,6) \in R_1$.)

The relation R_1 is shown to be symmetric also by considering all possible cases.

Case	$(a,b) \in R_1$	(b,a)	$(b,a) \in R_1$?
1	$(1,2)$	$(2,1)$	Yes
2	$(1,4)$	$(4,1)$	Yes
3	$(2,1)$	$(1,2)$	Yes
$\vdots$	$\vdots$	$\vdots$	$\vdots$

Since every possible case of (a, b) in R_1 has been examined and it was found that the ordered pair (b, a) is in R_1, then R_1 is symmetric.

The relation R_1 is shown to be transitive by "inspection" or by exhaustion, as given in the following table, which examines all possible cases of $(a, b) \in R_1$ and $(b, c) \in R_1$.

Case	$(a, b) \in R_1$	$(b, c) \in R_1$	(a, c)	$(a, c) \in R_1$?
1	$(1, 2)$	$(2, 1)$	$(1, 1)$	Yes
2	$(1, 2)$	$(2, 2)$	$(1, 2)$	Yes
3	$(1, 2)$	$(2, 4)$	$(1, 4)$	Yes
4	$(1, 4)$	$(4, 1)$	$(1, 1)$	Yes
5	$(1, 4)$	$(4, 2)$	$(1, 2)$	Yes
$\vdots$	$\vdots$	$\vdots$	$\vdots$	$\vdots$

Since every possible case of $(a, b) \in R_1$ and $(b, c) \in R_1$ has been examined and it has been found that $(a, c) \in R_1$ in each such case, then R_1 is transitive.

R_1 is not antisymmetric because $(1, 2) \in R_1$ and $(2, 1) \in R_1$ but $1 \neq 2$.

Example 0.18 *Let $A = \{\square, \triangle, \bigcirc, \bigstar\}$ and let $R_2 \subseteq A \times A$ be defined by*

$$R_2 = \{(\square, \square), (\square, \triangle), (\square, \bigstar), (\triangle, \square), (\bigstar, \square), (\bigstar, \bigstar), (\bigcirc, \bigstar), (\bigcirc, \bigcirc)\}$$

R_2 is not reflexive because $\triangle \in A$ but $(\triangle, \triangle) \notin R_2$. R_2 is not symmetric because $(\bigcirc, \bigstar,) \in R_2$ but $(\bigstar, \bigcirc) \notin R_2$. R_2 is not antisymmetric because $(\triangle, \square) \in R_2$ and $(\square, \triangle) \in R_2$ but $\triangle \neq \square$. R_2 is not transitive because $(\triangle, \square) \in R_2$ and $(\square, \bigstar) \in R_2$ but $(\triangle, \bigstar) \notin R_2$.

If a relation R on A has certain properties just described, then it is further classified as follows:

Definition 0.19 *A relation R on $A \times A$ is an* equivalence relation *on A if it is reflexive, symmetric, and transitive. R is a* partial ordering *of A if it is reflexive, antisymmetric, and transitive.*

The relation R_1 in Example 0.17 was found to be reflexive, transitive, and symmetric; therefore, R_1 is an equivalence relation on the set $A = \{1, 2, 3, 4, 5, 6\}$. The relation R_2 in Example 0.18 was not reflexive and so is not an equivalence relation on the set $A = \{\square, \triangle, \bigcirc, \bigstar\}$.

Example 0.20 *Let A be the set of all integers. Define the relation $R_3 \subseteq A \times A$ by $R_3 = \{(a, b) : a - b = 5 \cdot k$ for some integer $k\}$. For example, $(0, 0) \in R_3$ because $0 - 0 = 0 = 5 \cdot 0 = 5 \cdot k$ for $k = 0$; and $(-11, 4) \in R_3$ because $(-11) - 4 = -15 = 5(-3) = 5 \cdot k$ for $k = -3$.*

The relation R_3 is reflexive. If a is an integer (i.e., $a \in A$), then $a - a = 0 = 5 \cdot 0 = 5 \cdot k$ for $k = 0$ so that $(a, a) \in R_3$.

The relation R_3 is symmetric. We need to show that if $(a, b) \in R_3$, then $(b, a) \in R_3$ by finding an integer k such that $b - a = 5 \cdot k$. But $(a, b) \in R_3$

implies there is an integer m so that $a - b = 5 \cdot m$ and

$$
\begin{aligned}
b - a &= -(a - b) \\
&= -(5 \cdot m) \\
&= 5 \cdot (-m) \\
&= 5 \cdot k
\end{aligned}
$$

where k is the integer $-m$. Thus $(b, a) \in R_3$.

The relation R_3 is transitive. Assume that a, b, and c are integers, $(a, b) \in R_3$, and $(b, c) \in R_3$. By definition,

$$(a, b) \in R_3 \text{ implies that } a - b = 5 \cdot k \text{ for some integer } k$$

and

$$(b, c) \in R_3 \text{ implies that } b - c = 5 \cdot m \text{ for some integer } m$$

Adding these two equalities gives

$$(a - b) + (b - c) = 5 \cdot k + 5 \cdot m$$

or

$$a - c = 5 \cdot (k + m) = 5 \cdot j$$

for the integer $j = k+m$. By definition of R_3, $(a, c) \in R_3$, and R_3 is transitive.

Since R_3 is reflexive, symmetric, and transitive, it is an equivalence relation.

Example 0.21 *Let $S = \{1, 2, 3\}$ and A be the power set of S.*

$$A = \mathcal{P}(S) = \{\varnothing, \{1\}, \{2\}, \{3\}, \{1, 2\}, \{1, 3\}, \{2, 3\}, \{1, 2, 3\}\}$$

Define the relation $R_4 \subseteq A \times A$ by $R_4 = \{(T, V) : T \subseteq V\}$. Thus, one has $(\{2\}, \{1, 2\}) \in R_4$ because $\{2\} \subseteq \{1, 2\}$ and $(\{2, 3\}, \{3\}) \notin R_4$ because $\{2, 3\} \not\subseteq \{3\}$. One can easily verify that R_4 is partial ordering of the subsets of S.

We shall see that an equivalence relation R on $A \times A$ is able to separate the elements of A into collections where the elements in a given collection are all similar to each other through the relation R, but the elements in different collections are not related. In the context of equivalence relations, these collections are called the equivalence classes of the relation R.

A physical analogy would be for A to be a collection of different-colored balls and R_1 to be the relation defined by $(a, b) \in R_1$ if and only if ball a has the same color as ball b. Since R_1 is an equivalence relation, each collection will consist of balls having the same color. If R_2 is the relation defined by $(a, b) \in R_2$ if and only if a and b have the same diameter, then each collection or equivalence class of R_2 will consist of balls having the same size.

Definition 0.22 *Let $a \in A$ and R be an equivalence relation on $A \times A$. Let $[a]$ denote the set $\{x : xRa\} = \{x : (x, a) \in R\}$, called the equivalence class*

*containing a. The symbol $[A]_R$ denotes the set of all equivalence classes of A
for the equivalence relation R.*

Example 0.23 *In Example 0.17,*

$$R_1 \quad = \quad \{(1,1), (2,2), (3,3), (4,4), (5,5), (6,6), (1,2),$$
$$(1,4), (2,1), (2,4), (3,5), (5,3), (4,1), (4,2)\}$$

*R_1 is an equivalence relation on the set $A = \{1,2,3,4,5,6\}$. We obtain the
equivalence classes of R_1 by calculating, exhaustively, the equivalence class of
each element of A.*

$$[1] \quad = \quad \{x : (x,1) \in R_1\} = \{x : xR_1 1\}$$
$$= \quad \{1,2,4\}$$

*where $1 \in [1]$ because $(1,1) \in R_1$, $2 \in [1]$ because $(2,1) \in R_1$, and $4 \in [1]$
because $(4,1) \in R_1$, and by direct inspection of the set R_1, there is no other
x in A such that $(x,1) \in R_1$.*

$$[2] \quad = \quad \{x : (x,2) \in R_1\}$$
$$= \quad \{2,1,4\}$$

$$[3] \quad = \quad \{x : (x,3) \in R_1\}$$
$$= \quad \{3,5\}$$

$$[4] \quad = \quad \{x : (x,4) \in R_1\}$$
$$= \quad \{4,1,2\}$$

$$[5] \quad = \quad \{x : (x,5) \in R_1\}$$
$$= \quad \{5,3\}$$

$$[6] \quad = \quad \{x : (x,6) \in R_1\}$$
$$= \quad \{6\}$$

There are only three distinct equivalence classes:

$$[1] = [2] = [4] = \{1,2,4\}$$

$$[3] = [5] = \{3,5\}$$

$$[6] = \{6\}$$

so that

$$[A]_{R_1} = \{[1], [3], [6]\} = \{\{1,2,4\}, \{3,5\}, \{6\}\}$$

Any member of an equivalence class will generate the equivalence class; that is, if $b \in [a]$, then $[a] = [b]$. Because of this property, we say that any member of an equivalence class represents the class. Every equivalence class contains at least one element because the relation is reflexive. No element is in two distinct equivalence classes (because the relation is transitive).

Example 0.24 *Consider the equivalence relation R_3 of Example 0.20. For the set A of all integers, $R_3 \subseteq A \times A$ was defined by $R_3 = \{(a, b) : a - b = 5 \cdot k \text{ for some integer } k\}$. The equivalence classes of R_3 will be calculated. Since*

$$
\begin{aligned}
[a] &= \{x : (x, a) \in R_3\} = \{x : xR_3a\} \\
&= \{x : x - a = 5 \cdot k \text{ for some integer } k\} \\
&= \{x : x = a + 5 \cdot k \text{ for some integer } k\}
\end{aligned}
$$

we see that

$$
\begin{aligned}
[0] &= \{\ldots, -15, -10, -5, 0, 5, 10, 15, 20, 25, \ldots\} \\
&= [5] = [10] = [-5] = \cdots
\end{aligned}
$$

$$
\begin{aligned}
[1] &= \{\ldots, -14, -9, -4, 1, 6, 11, \ldots\} \\
&= [6] = [11] = [-4] = \cdots
\end{aligned}
$$

$$
\begin{aligned}
[2] &= \{\ldots, -13, -8, -3, 2, 7, 12, \ldots\} \\
&= [7] = [12] = [-3] = \cdots
\end{aligned}
$$

$$
\begin{aligned}
[3] &= \{\ldots, -12, -7, -2, 3, 8, 13, \ldots\} \\
&= [8] = [13] = [-2] = \cdots
\end{aligned}
$$

$$
\begin{aligned}
[4] &= \{\ldots, -11, -6, -1, 4, 9, 14, \ldots\} \\
&[9] = [14] = [-1] = \cdots
\end{aligned}
$$

are the distinct equivalence classes for the equivalence relation R_3. Thus,

$$
[A]_{R_3} = \{[0], [1], [2], [3], [4]\}
$$

The elements of $[0]$ are "alike" in the sense that each is a multiple of five. The elements of $[3]$ are "alike" since when any one is divided by 5, the remainder is 3.

In both of Examples 0.23 and 0.24, the collection of equivalence classes divides up all of the elements of the set A into nonempty sets. These sets are *mutually exclusive* or *disjoint*, meaning that no two of them have an element in common. Such a division of a set is called a partition of the set.

Definition 0.25 *Let A be a set, and let $\langle A \rangle = \{A_1, A_2, A_3, \ldots\}$ be a set of nonempty subsets of A. The collection $\langle A \rangle$ is called a partition of A if both of the following are satisfied:*

(a) *$A_i \cap A_j = \varnothing$ for all $i \neq j$.*

(b) *A is the union of the A_i; that is, $a \in A$ if and only if $a \in A_i$ for some i.*

Theorem 0.26 *A nonempty set of subsets $\langle A \rangle$ of a set A is a partition of A if and only if $\langle A \rangle = [A]_R$ for some equivalence relation R.*

Proof. Let $\langle A \rangle = \{A_1, A_2, A_3, \ldots\}$ be a partition of A. Define a relation R on $A \times A$ by aRb if and only if a and b are in the same subset A_i for some i. Certainly for all a in A, aRa and R is reflexive. If a and b are in the same subset, then b and a are in the same subset and R is symmetric. Since the sets $A_i \cap A_j = \varnothing$ for $i \neq j$, if a and b are in the same subset and b and c are in the same subset, then a and c are in the same subset. Hence R is transitive and R is an equivalence relation.

Conversely, assume that R is an equivalence relation. We need to show that $[A]_R = \{[a] : a \in A\}$ is a partition of A. Certainly, for all a, $[a]$ is nonempty since $a \in [a]$. Obviously, A is the union of the $[a]$, $a \in A$. Assume that $[a] \cap [b]$ is nonempty and let $x \in [a] \cap [b]$. Then xRa and xRb, and by symmetry, aRx. But since aRx and xRb, by transitivity, aRb. Therefore, $a \in [b]$. If $y \in [a]$, then yRa and, by transitivity, yRb. Therefore, $[a] \subseteq [b]$. Similarly, $[b] \subseteq [a]$ so that $[a] = [b]$, and we have a partition of A. ∎

Example 0.27 *Let $A = \{\square, \triangle, \bigcirc, \bigstar\}$ and consider the partition $A_1 = \{\square\}$, $A_2 = \{\triangle, \bigcirc, \bigstar\}$. According to the proof of the preceding theorem, we should define R by $R = \{(a, b) : a \in A_i \text{ and } b \in A_i \text{ for some } i\}$. From the set A_1 we obtain the ordered pair $(\square, \square)$ in R. From A_2 we obtain these ordered pairs in R:*

	$\triangle$	$\bigcirc$	$\bigstar$
$\bigstar$	$(\triangle, \bigstar)$	$(\bigcirc, \bigstar)$	$(\bigstar, \bigstar)$
$\bigcirc$	$(\triangle, \bigcirc)$	$(\bigcirc, \bigcirc)$	$(\bigstar, \bigcirc)$
$\triangle$	$(\triangle, \triangle)$	$(\bigcirc, \triangle)$	$(\bigstar, \triangle)$

Thus,

$$R = \{(\square, \square), (\triangle, \bigstar), (\bigcirc, \bigstar), (\bigstar, \bigstar), (\triangle, \bigcirc),$$
$$(\bigcirc, \bigcirc), (\bigstar, \bigcirc), (\triangle, \triangle), (\bigcirc, \triangle), (\bigstar, \triangle)\}$$

is the relation generated by this partition.

Exercises

1. Let $A = \{1, 2, 3, 4, 5\}$
           $B = \{6, 7, 8, 9\}$
           $C = \{10, 11, 12, 13\}$
           $D = \{\square, \triangle, \bigcirc, *\}$

    Let $R \subseteq A \times B$, $S \subseteq B \times C$, and $T \subseteq C \times D$ be defined by

    $$R = \{(1, 7), (4, 6), (5, 6), (2, 8)\}$$
    $$S = \{(6, 10), (6, 11), (7, 10), (8, 13)\}$$
    $$T = \{(11, \triangle), (10, \triangle), (13, *), (12, \square), (13, \bigcirc)\}$$

Compute the sets:

(a) R^{-1} and S^{-1}

(b) $S \circ R$

(c) $S \circ S^{-1}$ and $S^{-1} \circ S$

(d) $R^{-1} \circ S^{-1}$

(e) $T \circ (S \circ R)$

(f) $T \circ S$

(g) $(T \circ S) \circ R$

2. Prove that composition of relations is associative. That is, assume that A, B, and C are sets and that $R \subseteq A \times B$, $S \subseteq B \times C$, and $T \subseteq C \times D$. Prove that $T \circ (S \circ R) = (T \circ S) \circ R$.

3. For sets A, B, and C, prove that

    (a) $(A \cup B) \times C = (A \times C) \cup (B \times C)$

    (b) $A \times (B \cap C) = (A \times B) \cap (A \times C)$

4. Let $A = \{\square, \triangle, \bigcirc, w, t, z, h, \bigstar\}$. The relation R on A is defined by $R = \{(\square, \square), (\triangle, \triangle), (\bigcirc, \bigcirc), (w, w), (t, t), (z, z), (h, h), (\bigstar, \bigstar), (\square, \bigcirc), (\bigcirc, z), (w, h), (t, \bigstar), (\square, z), (z, \square), (z, \bigcirc), (h, w), (\bigstar, t), (\bigcirc, \square)\}$.

    (a) Show that R is an equivalence relation on A.

    (b) Calculate all the equivalence classes of R.

5. Let $A = \{1, 2, 3\}$. Determine whether each of the following relations on A is reflexive, symmetric, transitive, or antisymmetric:

    (a) $R_1 = \{(2, 2), (1, 1)\}$

    (b) $R_2 = \{(1, 1), (2, 2), (3, 3)\}$

    (c) $R_3 = \{(1, 1), (2, 2), (3, 3), (1, 2), (2, 1), (3, 1), (1, 3)\}$

    (d) $R_4 = \{(1, 1), (2, 2), (3, 3), (1, 2), (3, 2), (2, 1)\}$

    (e) $R_5 = \{(1, 1), (2, 2), (3, 3), (1, 2), (2, 1), (2, 3), (3, 2), (1, 3), (3, 1)\}$

6. Let $f(x) = x^2 + 1$, where $x \in [-2, 4] = \{x : x \text{ is a real number and } -2 \leq$

$x \leq 4\} = A$. Define the relation R on $A \times A$ as follows:

$$(a, b) \in R \text{ if and only if } f(a) = f(b)$$

(a) Show that R is an equivalence relation on $A = [-2, 4]$.

(b) Calculate these equivalence classes:

$$[1], [2], [0], [3], [-1/2], \text{ and } [4]$$

(c) Sketch a graph of f and describe what elements of A are "alike" according to the equivalence relation R and how they are related to the graph of f. How many distinct equivalence classes does R have?

7. For $X \neq \emptyset$, define the relation I to be $I = \{x : x = (a, a)$ for some $a \in X\} = \{(a, a) \in X \times X : a \in X\}$. I is an equivalence relation on X and is called the identity relation on X. If R is a relation on X, prove that:

(a) R is reflexive if and only if $I \subseteq R$.

(b) R is symmetric if and only if $R = R^{-1}$.

(c) R is transitive if and only if $R \circ R^{-1} \subseteq R$.

8. Using Definition 0.1, prove that $(a, b) = (c, d)$ if and only if $a = c$ and $b = d$.

0.2   Functions

We have discussed two useful kinds of relations: equivalence relations and partial ordering relations. A third kind of relation is the function relation.

Definition 0.28 *A relation f on $A \times B$ is a function from A to B (denoted $f : A \rightarrow B$) if for every $a \in A$ there is one and only one $b \in B$ so that $(a, b) \in f$. If $f : A \rightarrow B$ is a function and $(a, b) \in f$, we say that $b = f(a)$. A is called the* domain *of the function f, and B is called the* codomain. *If $E \subseteq A$, then $f(E) = \{b : f(a) = b \text{ for some } a \text{ in } E\}$ is called the* image *of E. The image of A itself is called the* range *of f. If $F \subseteq B$, then $f^{-1}(F) = \{a : f(a) \in F\}$ is called the* preimage *of F.*

A function $f : A \rightarrow B$ is also called a mapping and we speak of the domain A being mapped into B by the mapping f. If $(a, b) \in f$ so that $b = f(a)$, then the element a is mapped to the element b. In the context above, the function f is referred to as a "function of one variable" and is often denoted using the symbolism $f(x)$, where x is called the "variable." If the domain of

f is the Cartesian product of two sets, say $C \times D$, then $f : C \times D \to B$ is said to be a "function of two variables" and is often denoted using the symbolism $f(x, y)$, where x and y are called the variables with x selected from C and y selected from D. Thus, if $(c, d) \in C \times D$, then we write $f((c, d))$ or simply $f(c, d)$, which gives rise to the notation above. Similarly, if the domain of f is $C \times D \times E$ and $f : C \times D \times E \to B$, then f is called a function of three variables.

Example 0.29 *Let $A = \{-2, -1, 0, 1, 2\}$ and $B = \{0, 1, 2, 3, 4, 5\}$. Define the relation $f \subseteq A \times B$ as $f = \{(-2, 5), (-1, 2), (0, 1), (1, 2), (2, 5)\}$. The relation f is a function from A to B because $f \subseteq A \times B$ and each of the five elements of A appears as the first component of an ordered pair in f exactly once.*

If $E = \{1, 2\}$, then

$$
\begin{aligned}
f(E) &= \{b : (a, b) \in f \text{ for some } a \text{ in } E\} \\
&= \{b : b = f(a) \text{ for some } a \text{ in } E\} \\
&= \{2, 5\}
\end{aligned}
$$

is the image of E under f.

If $F = \{0, 2, 3, 4, 5\} \subseteq B$, then

$$
\begin{aligned}
f^{-1}(F) &= \{b : \text{there is an } a \in A \text{ such that } f(a) = b\} \\
&= \{-1, 1, -2, 2\}
\end{aligned}
$$

is the preimage of F where $-1 \in f^{-1}(F)$ because $f(-1) = 2$, $1 \in f^{-1}(F)$ because $f(1) = 2$, $-2 \in f^{-1}(F)$ because $f(-2) = 5$, and $2 \in f^{-1}(F)$ because $f(2) = 5$. Note that the elements $0, 3,$ and 4 do not contribute any elements to $f^{-1}(F)$ because they are elements of B that are not "used" by the function f. The preimage can be empty as it would be for $W = \{0, 3\}$ since there is no $a \in A$ for which $f(a) = 0$ or $f(a) = 3$.

The image of the domain A is the range

$$
\begin{aligned}
f(A) &= \{b : f(a) = b \text{ for some } a \in A\} \\
&= \{1, 2, 5\}
\end{aligned}
$$

The elements of $f(A)$ are just those elements of B that are "used" by the function f.

The function f was defined in this example by explicitly listing the ordered pairs as $f = \{(-2, 5), (-1, 2), (0, 1), (1, 2), (2, 5)\}$. The function f of this example can equivalently be defined as $f = \{(a, b) : a \in A \text{ and } b = a^2 + 1\}$ so that $f(a) = a^2 + 1$. The equivalence of these two definitions of f is easily verified.

A function f from A to B can be further classified by whether there is more than one element of A associated with any of the elements of B and whether all of the elements of the codomain B are associated with corresponding members of the domain A.

Definition 0.30 *A function f is* one to one *or* injective *if $(a, b) \in f$ and $(a', b) \in f$ imply $a = a'$. A function f is* onto *or* surjective *if for every $b \in B$ there exist an $a \in A$ so that $(a, b) \in f$. An alternative statement meaning that f is one to one is that $f(a) = f(a')$ implies that $a = a'$. The function f is onto means that, for every $b \in B$, there exists an $a \in A$ such that $f(a) = b$.*

Let f be a function from A into the set B or $f : A \to B$. Certainly, $f \subseteq A \times B$ so that f is a relation on $A \times B$. Consider the *inverse relation* $f^{-1} \subseteq B \times A$ defined by $f^{-1} = \{(b, a) : (a, b) \in f\}$. The relation f^{-1} may not be a function from B to A even though f is a function from A to B. The proof of the following theorems are left to the reader.

Theorem 0.31 *If a function $f : A \to B$ is one to one and onto, then the inverse relation f^{-1} is a function from B into A that is also one to one and onto. Conversely, for $f : A \to B$, if f^{-1} is a function from B into A, then f is one to one and onto.*

Theorem 0.32 *If $f : A \to B$ is one to one and onto, then*

$$f(f^{-1}(b)) = b \text{ for each } b \text{ in } B$$

and

$$f^{-1}(f(a)) = a \text{ for each } a \text{ in } A.$$

Operations such as addition, subtraction, multiplication, and division are also examples of functions. In the case of subtraction of integers, $5 - 3$ means that an integer, namely 2, is assigned to 5 and 3 by the subtraction operation. We could write $-((5, 3)) = 2$ to emphasize the functional nature of the subtraction operation. However, $3 - 5 = -((3, 5)) = -2$. Subtraction assigns a unique integer to each pair (r, s) of integers. This special application of a function of two variables is called a binary operation and is defined next.

Definition 0.33 *A* binary operation *on the set A is a function $b : A \times A \to A$. The image of (r, s) under b is written as $b((r, s))$ using functional notation or as $r \, b \, s$ using what is called infix notation. Often, $b((r, s))$ is shortened to $b(r, s)$.*

Because the range of a binary operation on A is a subset of A, by definition, a binary operation exhibits the property of closure wherein the "result," $r \, b \, s$, of the operation on two members r and s of A is also a member of A. A ternary operation would be a function from $A \times A \times A$ into A.

Example 0.34 *Assume for the set Z of all integers that the function $\square$: $Z \times Z \to Z$ is defined by $\square(r, s) = r \, \square \, s = 3 \cdot r + s$, where we have assumed that addition and multiplication of integers have already been defined. For example, $5 \, \square \, 2 = 3 \cdot 5 + 2 = 17$ and $2 \, \square \, 5 = 3 \cdot 2 + 5 = 11$.*

We saw previously that if R is a relation on $A \times B$ and S is a relation on $B \times C$, then we could define a relation $S \circ R$ on $A \times C$, called the composition of S and R. If R and S are functions, then so is $S \circ R$. The next theorem gives some useful properties of composite functions.

Theorem 0.35 *Let $g : A \to B$ and $f : B \to C$; then:*

(a) *The composition $f \circ g$ is a function from A into C, denoted by*

$$f \circ g : A \to C$$

(b) *If $a \in A$, then $(f \circ g)(a) = f(g(a))$.*

(c) *If g and f are onto B and C, respectively, then $f \circ g$ is onto C.*

(d) *If g and f are both one to one, then $f \circ g$ is one to one.*

(e) *If g and f are both one to one and onto, then $f \circ g$ is one to one and onto.*

(f) $(f \circ g)^{-1} = g^{-1} \circ f^{-1}$.

Proof. **(a)** If $g : A \to B$ and $f : B \to C$, then $g \subseteq A \times B$ and $f \subseteq B \times C$. By definition of composition of relations, $f \circ g \subseteq A \times C$. Suppose that $(a, c) \in f \circ g$ and $(a, c') \subseteq f \circ g$. We need to show that $c = c'$.

Since $(a, c) \in f \circ g$, there is a member $b \in B$ with $(a, b) \in g$ and $(b, c) \in f$. Similarly, since $(a, c') \in f \circ g$, there is a member $b' \in B$ with $(a, b') \in g$ and $(b', c') \in f$. But g is a function so that $(a, b) \in g$ and $(a, b') \in g$ imply that $b = b'$. Now we have $(b', c') = (b, c') \in f$ and $(b, c) \in f$. Since f is also a function, $c = c'$. Therefore, $f \circ g$ is a function from A into C.

(b) If $g : A \to B$ and $a \in A$, then there is a b in B with $(a, b) \in g$ or, equivalently, with $b = g(a)$. Since $b \in B$ and $f : B \to C$, there is a $c \in C$ with $(b, c) \in f$, or equivalently, with $c = f(b)$. But $b = g(a)$ so that $f(g(a)) = f(b) = c = (f \circ g)(a)$ because $(a, b) \in g$ and $(b, c) \in f$ imply that $(a, c) \in f \circ g$.

The proofs of the remaining parts are left as exercises for the reader. ∎

As an application of functions we will define finite and infinite sets. For this purpose we will assume that certain subsets of integers and real numbers are known even though, for example, the logical development of integers occurs in Chapter 1. Let Z denote the set of integers, $\{\ldots, -2, -1, 0, 1, 2, \ldots\}$. Let N denote the set of positive integers so that $N = \{1, 2, 3, \ldots\}$. If n is a positive integer, then the notation $\{1, 2, \ldots, n\}$ is an abbreviation for the set $\{k : k$ is an integer and $1 \le k \le n\}$. In illustrations, R will denote the set of real numbers.

Definition 0.36 *A set A is* finite *if and only if*

(a) $A = \varnothing$,

or

(b) *$A \neq \varnothing$ and there is a positive integer n and a function*

$$f : \{1, 2, \ldots, n\} \to A$$

such that f is one to one and onto.

We say that A has zero elements and write $|A| = 0$ when $A = \varnothing$ and that A has n elements for some positive integer n and write $|A| = n$ if A satisfies Definition 0.36(b). The set A is infinite if and only if it is not finite. A set A is countably infinite if and only if there is a function $f : N \to A$ such that f is one to one and onto, where N is the set of positive integers. A set A is uncountable or uncountably infinite if and only if A is infinite but not countably infinite.

In effect, the definition says that the elements of a nonempty finite set A are "counted" by the integers $1, 2, 3, \ldots, n$. If A is countably infinite, then it is "counted" by the entire set of positive integers $1, 2, 3, \ldots$.

To establish that $A = \{\square, \triangle, \bigcirc\}$ is finite, we have only to find an appropriate function f. Let $n = 3$ and define $f : \{1, 2, 3\} \to A$ by $f(1) = \square$, $f(2) = \triangle$, and $f(3) = \bigcirc$. By inspection f is one to one and onto. Therefore, A is finite and has three elements.

The set $E = \{2, 4, 6, \ldots\} = \{k : k = 2j \text{ for some positive integer } j\}$, which consists of the even positive integers, is countably infinite because the function $f : N \to E$ defined by $f(j) = 2j$ for each $j \in N$ can be shown to be one to one and onto.

Definition 0.37 *Nonempty sets A and B are in* one-to-one correspondence *if and only if there is a function $f : A \to B$ that is one to one and onto.*

Evidently, if A and B are in one-to-one correspondence, so are B and A since $f^{-1} : B \to A$ is one to one and onto. Thus, sets that are in one-to-one correspondence are thought of as having the same "number" of elements. The next two theorems provide fundamental counting principles. The first theorem is often called the Addition Principle and the second is called the Multiplication Principle.

Theorem 0.38 *If A and B are finite disjoint sets with n and m elements, respectively, then $A \cup B$ is finite and has $n + m$ elements; that is, $|A \cup B| = |A| + |B|$.*

Proof. Assume that $A \neq \varnothing$, $B \neq \varnothing$, and that both $f : \{1, 2, \ldots, n\} \to A$ and $g : \{1, 2, \ldots, m\} \to B$ are one to one and onto. Define the function $h : \{1, 2, \ldots, n + m\} \to A \cup B$ as follows:

$$h(k) = \begin{cases} f(k) & \text{if } 1 \leq k \leq n \\ g(k - n) & \text{if } n < k \leq n + m \end{cases}$$

It is straightforward to show that h is one to one and onto. Therefore, $A \cup B$ is finite and has $n + m$ elements. The cases when $A = \varnothing$ or $B = \varnothing$ are immediate. ∎

Theorem 0.39 *If A and B are finite sets, then the number of ordered pairs in $A \times B$ is the number of elements of A times the number of elements of B, that is, $|A \times B| = |A| \cdot |B|$.*

Proof. Let $|A| = n$, $A = \{a_1, a_2, \ldots, a_n\}$, $|B| = m$, and $B = \{b_1, b_2, \ldots, b_m\}$. Define $f(a_i, b_j) = (i-1)m + j$; f is one to one and onto $\{1, 2, \ldots, mn\}$. ∎

Since the identity function $I : N \to N$ defined by $I(k) = k$ is one to one and onto, N is countably infinite. The definitions of infinite set and countably infinite set were given separately. We need to show that a countably infinite set is indeed infinite.

Theorem 0.40 *If A is countably infinite, then A is not finite.*

Proof. Assume that A is countably infinite, $f : N \to A$ is one to one and onto, and A is finite so that $g : \{1, 2, \ldots, n\} \to A$ is one to one and onto for some positive integer n. Then $h = f^{-1} \circ g : \{1, 2, \ldots, n\} \to N$ is one to one and onto by Theorem 0.35. Since $\{h(1), h(2), \ldots, h(n)\}$ is obviously a finite set of positive integers, let the largest one be $h(k)$. Since $1 + h(k)$ is a positive integer and $h(i) < 1 + h(k)$ for all i, $1 \le i \le n$, we have a contradiction of the fact that h is onto. Thus, A cannot be finite. ∎

Definition 0.41 *A* sequence *in a nonempty set A is a function f such that $f : N \to A$ or $f : \{1, 2, \ldots, n\} \to A$ for some positive integer n, where the former is called an* infinite sequence *and the latter a* finite sequence.

For example, let $f : \{1, 2, 3, \ldots\} \to \{9, 7\}$ be defined by $f(i) = 9$ if i is odd and $f(i) = 7$ if i is even. Then $f(1) = 9, f(2) = 7, f(3) = 9$, etc. Because sequences occur frequently, we often "contract" the notation and write $f_1 = f(1), f_2 = f(2), f_3 = f(3)$, and so on. We then equivalently refer to the sequence $f_1, f_2, f_3, \ldots = 9, 7, 9, 7, \ldots$ by saying that $f_i = 9$ if i is odd and $f_i = 7$ if i is even.

Again, if $A = \{9, 7\}$ and $n = 5$, define $f(i) = 9$ if $1 \le i \le 5$ and i is odd and $f(i) = 7$ if i is even. The finite sequence is given explicitly by $f(1), f(2), f(3), f(4), f(5)$, by f_1, f_2, f_3, f_4, f_5, or by $9, 7, 9, 7, 9$.

In most contexts, (infinite) sequences and finite sequences will be specified much less formally; however, occasionally it will be convenient to call upon the full definition and use the function associated with such sequences.

Exercises

1. Let $f : A \to B$ be one to one and onto and $g : B \to A$ be any function having the properties

$$(g \circ f)(a) = a \text{ for all } a \in A$$
$$(f \circ g)(b) = b \text{ for all } b \in B$$

Prove that $g = f^{-1}$. Note that this statement says that f^{-1} is the only function from B into A such that $f \circ f^{-1} = I_B$ and $f^{-1} \circ f = I_A$, where I_A and I_B are, respectively, the identity mappings (relations) on A and B (see Exercise 7 in Section 0.1).

2. If $f : A \to B$, then prove that f is one to one if and only if for all subsets X and Y of A, $f(X \cap Y) = f(X) \cap f(Y)$.

3. For $f : A \to B$, prove that:

    (a) The function f is one to one if and only if $(f^{-1} \circ f)(W) = W$ for every $W \subseteq A$, where f^{-1} indicates the preimage of a set.

    (b) The function f is onto if and only if $(f \circ f^{-1})(V) = V$ for every $V \subseteq B$.

4. Let $f : A \to B$, A_1 and A_2 be subsets of A, and B_1 and B_2 be subsets of B. Prove that:

    (a) If $A_1 \subseteq A_2$, then $f(A_1) \subseteq f(A_2)$.

    (b) If $B_1 \subseteq B_2$, then $f^{-1}(B_1) \subseteq f^{-1}(B_2)$.

    (c) $f(A_1 \cup A_2) = f(A_1) \cup f(A_2)$.

    (d) $f^{-1}(B_1 \cup B_2) = f^{-1}(B_1) \cup f^{-1}(B_2)$.

    (e) $f(A_1 \cap A_2) \subseteq f(A_1) \cap f(A_2)$.

    (f) $f^{-1}(B_1 \cap B_2) = f^{-1}(B_1) \cap f^{-1}(B_2)$.

    (g) $f^{-1}(B_1') = (f^{-1}(B_1))'$.

    (h) Give an example of a function f and sets A_1 and A_2 such that $f(A_1 \cap A_2) \neq f(A_1) \cap f(A_2)$.

5. If A is a nonempty set having n elements, prove that $P = \{(a, a') : a \in A, \ a' \in A, \text{ and } a \neq a'\} = A \times A - I$ has $n(n-1)$ elements, where $I = \{(a, a) : a \in A\}$ is the identity relation on A. Then P is the set of all ordered pairs formed of elements of A with unequal first and second components and can also be considered to be the set of permutations of objects from the set A taken two at a time.

6. Prove that every subset of a finite set is finite.

7. Prove that any set that contains an infinite subset is also infinite.

8. Prove that Z, the set of integers, is countably infinite.

9. If A and B are countably infinite, prove that:

   (a) $A \cup B$ is countably infinite if A and B are disjoint.

   (b) $A \cup B$ is countably infinite.

10. If A is finite and nonempty, prove that there is only one positive integer n such that there is a function $f : \{1, 2, \ldots, n\} \to A$ that is one to one and onto.

11. The Dirichlet Box Principle or Pigeonhole Principle is often stated as follows: If there are n boxes and m balls with $m > n$ and all the m balls are placed into the n boxes, then at least one box contains two balls. Establish the version of the Dirichlet Box Principle where every box contains at least one ball by proving the following: If n and m are positive integers with $n < m$ and $g : \{1, 2, \ldots, m\} \to \{1, 2, \ldots, n\}$ is onto, then g is not one to one.

0.3 Generalized Set Operations

In Section 0.1 we discussed set operations that dealt with finitely many sets. We now want to generalize those ideas in order to include unions and intersections of infinite collections of sets. For example, if A_1, A_2, and A_3 are subsets of $\mathbf{U}$, then we indicate their union by

$$B = A_1 \cup A_2 \cup A_3$$

Evidently, $x \in B$ if and only if $x \in A_1$, $x \in A_2$, or $x \in A_3$; that is, $x \in B$ if and only if x is in at least one of the three sets A_1, A_2, or A_3. Thus, $x \in B$ if and only if there is a $j \in \{1, 2, 3\}$ such that $x \in A_j$, or equivalently,

$$B = \{x : x \in \mathbf{U} \text{ and there is a } j \in \{1, 2, 3\} \text{ such that } x \in A_j\}$$

We define generalized unions and intersections as follows.

Definition 0.42 *If I is a set and A_i is a subset of the universal set $\mathbf{U}$ for each $i \in I$, then*

$$\bigcup_{i \in I} A_i = \{x : x \in \mathbf{U} \text{ and there is an } i \in I \text{ such that } x \in A_i\}$$

$$\bigcap_{i \in I} A_i = \{x : x \in \mathbf{U} \text{ and } x \in A_i \text{ for all } i \in I\}$$

The set $\bigcup_{i \in I} A_i$ is called the **generalized union** *and $\bigcap_{i \in I} A_i$ is called the* **gener-** *alized intersection of the collection of sets $\{A_i : i \in I\}$, which is sometimes* *written $\{A_i\}$ for short. The set I is called the* indexing set *and the collection $\{A_i\}$ is said to be* indexed by I.

Example 0.43 *Let* **U** *be the set* $N = \{1, 2, 3, \ldots\}$ *of positive integers and* $I = \{\square, \triangle, \bigcirc\}$. *Let*

$$
\begin{aligned}
A_\square &= \{x : x = 2k \text{ for some positive integer } k\} \\
A_\triangle &= \{x : x = 3k \text{ for some positive integer } k\} \\
A_\bigcirc &= \{x : x = 5k \text{ for some positive integer } k\}
\end{aligned}
$$

Then

$$
C = \bigcap_{i \in I} A_i = \{x : x \in A_i \text{ for all } i \in \{\square, \triangle, \bigcirc\}\}
$$

$$
= \{x : x \text{ is a multiple of } 30\}
$$

Example 0.44 *Let* $I = N = U = \{1, 2, 3, \ldots\}$ *and*

$$
A_i = \{x : x \in N \text{ and } x \geq i\} = \{i, i+1, i+2, \ldots\}
$$

Then

$$
\bigcup_{i \in I} A_i = N
$$

because $A_k \subseteq A_1$ *for all* k *and* $A_1 = N$. *Also,*

$$
\bigcap_{i \in I} A_i = \varnothing
$$

because if m *is a positive integer, then* $m \notin A_{m+1}$ *so that no positive integer is in every* A_i.

The special case of an empty index set is considered in the following theorem. The proof is left to the reader.

Theorem 0.45 *If, for each* $i \in I$, A_i *is a subset of the universal set* **U** *and* $I = \varnothing$, *then*

(a) $\displaystyle\bigcup_{i \in I} A_i = \varnothing$.

(b) $\displaystyle\bigcap_{i \in I} A_i = \mathbf{U}$.

If A_1 and A_2 are subsets of **U** and $I = \{1, 2\}$, then it is obvious that

$$
A_1 \cup A_2 = \bigcup_{i \in I} A_i
$$

and

$$
A_1 \cap A_2 = \bigcap_{i \in I} A_i
$$

so that the generalized set operations introduced in this section reduce to those introduced in Section 0.1 for finite nonempty indexing sets. Nearly all of the properties of finite unions and intersections have counterparts with generalized unions and intersections. Some are given in the following theorem.

Theorem 0.46 *Let $\{A_i : i \in I\}$ be a collection of subsets of the universal set $\mathbf{U}$ indexed by the nonempty set I and let $\{B_j : j \in J\}$ be a collection of subsets of the universal set $\mathbf{U}$ indexed by the nonempty set J. If S is a subset of $\mathbf{U}$, then*

(a) *For each $t \in I$, $\bigcap\limits_{i \in I} A_i \subseteq A_t \subseteq \bigcup\limits_{i \in I} A_i$.*

(b) $S \cup \left(\bigcap\limits_{i \in I} A_i \right) = \bigcap\limits_{i \in I} (S \cup A_i)$.

(c) $S \cap \left(\bigcup\limits_{i \in I} A_i \right) = \bigcup\limits_{i \in I} (S \cap A_i)$.

(d) $\left(\bigcap\limits_{i \in I} A_i \right) \cup \left(\bigcap\limits_{j \in J} B_j \right) = \bigcap\limits_{(i,j) \in I \times J} (A_i \cup B_j)$.

(e) $\left(\bigcup\limits_{i \in I} A_i \right) \cap \left(\bigcup\limits_{j \in J} B_j \right) = \bigcup\limits_{(i,j) \in I \times J} (A_i \cap B_j)$.

(f) $\left(\bigcap\limits_{i \in I} A_i \right)' = \bigcup\limits_{i \in I} A_i'$.

(g) $\left(\bigcup\limits_{i \in I} A_i \right)' = \bigcap\limits_{i \in I} A_i'$.

Proof. (d) If $x \in \left(\bigcap\limits_{i \in I} A_i \right) \cup \left(\bigcap\limits_{j \in J} B_j \right)$, then $x \in \bigcap\limits_{i \in I} A_i$ or $x \in \bigcap\limits_{j \in J} B_j$. If $x \in \bigcap\limits_{i \in I} A_i$, then for every $i \in I$, $x \in A_i$. Since $x \in A_i$ implies that $x \in A_i \cup B_j$ for any $j \in J$, then $x \in A_i \cup B_j$ for any $i \in I$ and $j \in J$. Thus, $x \in A_i \cup B_j$ for any $(i,j) \in I \times J$; and we obtain $x \in \bigcap\limits_{(i,j) \in I \times J} (A_i \cup B_j)$. We obtain similar results if $x \in \bigcap\limits_{j \in J} B_j$. Hence we have $\left(\bigcap\limits_{i \in I} A_i \right) \cup \left(\bigcap\limits_{j \in J} B_j \right) \subseteq \bigcap\limits_{(i,j) \in I \times J} (A_i \cup B_j)$.

Conversely, if $y \in \bigcap\limits_{(i,j) \in I \times J} (A_i \cup B_j)$, then for each ordered pair $(i,j) \in I \times J$, $y \in A_i \cup B_j$ so that either $y \in A_i$ or $y \in B_j$. We now show that either $y \in \bigcap\limits_{i \in I} A_i$ or $y \in \bigcap\limits_{j \in J} B_j$. If $y \notin \bigcap\limits_{i \in I} A_i$, then for some $k \in I$, $y \notin A_k$. But for any $j \in J$, $(k,j) \in I \times J$ and we know that $y \in A_k \cup B_j$, which implies that $y \in B_j$ for all j. We can then conclude that $y \in \bigcap\limits_{j \in J} B_j$. Similarly,

$y \notin \bigcap\limits_{j \in J} B_j$ implies that $y \in \bigcap\limits_{i \in I} A_i$. Thus $y \in \left(\bigcap\limits_{i \in I} A_i\right) \cup \left(\bigcap\limits_{j \in J} B_j\right)$ and

$\bigcap\limits_{(i,j) \in I \times J} (A_i \cup B_j) \subseteq \left(\bigcap\limits_{i \in I} A_i\right) \cup \left(\bigcap\limits_{j \in J} B_j\right)$.

The proofs of the other parts of this theorem are left to the reader. ∎

Exercises

1. Assume **U** is the set R of real numbers and for $i \in N$ we have
    $$A_i = \left(1 - \frac{1}{i}, 2 + \frac{1}{i}\right)$$
    where $(a, b) = \{x : x \in R \text{ and } a < x < b\}$. Determine:

    (a) $\bigcup\limits_{i \in N} A_i$

    (b) $\bigcap\limits_{i \in N} A_i$

2. Assume that $\mathbf{U} = N = \{1, 2, 3, \ldots\}$ and let $A_n = \{n + 1, n\}$ for each positive integer n. Determine:

    (a) $\bigcup\limits_{i \in I} A_i$ if $I = \{1, 2, 3, 4, 5\}$

    (b) $\bigcup\limits_{i \in I} A_i$ if $I = \{k : k \in N \text{ and } k \geq 5\}$

    (c) $\bigcup\limits_{i \in I} A_i$ if $I = N$

    (d) $\bigcap\limits_{i \in I} A_i$ if $I = \{1, 2, 3\}$

    (e) $\bigcap\limits_{i \in I} A_i'$ if $I = N$

    (f) $\bigcup\limits_{i \in I} A_i'$ if $I = N$

3. Let $\mathbf{U} = R$, the set of real numbers, and let $B_i = \left(2 + \frac{1}{i}, 8 - \frac{2}{i}\right)$ for each positive integer i. Determine:

    (a) $\bigcup\limits_{i \in N} B_i$ (b) $\bigcap\limits_{i \in N} B_i$ (c) $\bigcup\limits_{i \in N} B_i'$ (d) $\bigcap\limits_{i \in N} B_i'$

4. Let the collection $\{A_i : i \in I\}$ be a partition of the nonempty set X and

$S \subseteq X$. Prove that $S = \bigcup_{i \in I} (S \cap A_i)$.

5. Let $\{A_i : i \in I\}$ be a collection of subsets of $\mathbf{U}$ and $I = \bigcup_{j \in J} I_j$ so that $\{I_j : j \in J\}$ is indexed by J. Prove that:

(a) $\bigcup_{i \in I} A_i = \bigcup_{j \in J} \left(\bigcup_{i \in I_j} A_i \right)$ **(b)** $\bigcap_{i \in I} A_i = \bigcap_{j \in J} \left(\bigcap_{i \in I_j} A_i \right)$

Chapter 1
ELEMENTARY
PROPERTIES OF
INTEGERS

1.1   Introduction

When considering the integers, one might wonder whether they should be studied by a mathematician or a theologian. Indeed, as we shall see, they have been studied by both. Even accepting the existence of the set of integers requires an act of faith. After all, we cannot formally completely describe the integers and certainly no one has seen a complete set. To overcome this problem we shall follow the example of Humpty Dumpty in *Through the Looking-Glass and What Alice Found There* [29], who stated: "When I use a word, it means just what I choose it to mean — neither more nor less." We shall consider the integers to mean what we define them to be, nothing more, nothing less. The mathematician Leopold Kronecker stated that "God made the integers, all the rest is the work of man." To be on the safe side, most of this book is strictly about integers.

Throughout the centuries, the positive integers, especially certain ones, were shrouded in mystery. The Pythagoreans combined mathematics and religion, and in doing so, managed to produce important mathematics and to develop a mythology or superstition with regard to the positive integers. Led by Pythagoras, the Pythagorean school was founded at Croton in 532 B.C. Their contributions to mathematics involved considering numbers as an abstraction and trying to form clear proofs to help them understand the ways of God, whom they considered to be a mathematician. They were more interested in internal truths than in practical ones. Today we would call them pure mathematicians. They considered some numbers, such as six, to be perfect numbers because they are equal to the sum of their proper factors. They dealt with amicable pairs such as 284 and 220, where each is a sum

24

of the proper factors of the other. They were aware of triangular numbers such as 3, 6, 10, and 15, and square numbers. To understand why they were called triangular numbers, consider 10 bowling pins. They were credited with discovering that the intervals in music correspond to the relative lengths or ratio of the lengths of the vibrating strings. Obviously, the Pythagoreans were credited with a proof of the Pythagorean Theorem.

As mentioned before, the Pythagorean Brotherhood was also a religious cult. As did other religious groups of the time, Pythagoreans believed in the transmigration of the individual soul to another, even to an animal. Bodies served as tombs or prisons; and if one led a pure enough life, he was released from the body which held him captive. Their symbol was the pentagram, which still appears in religions and cults. They believed that positive integers comprised that which was "real" in the universe and that to understand the rules of the integers enabled one to control the universe.

There are many examples of Pythagorean numerology. The number one was the most revered number and the number of reason. Two was the first even number and represented diversity of opinion. It was also the first female number since even numbers were considered female, as well as infinite, unlimited, moving, the left side, crooked, darkness, and evil. Three was the first male number and represented harmony. Odd numbers in general represented the right side, masculine, limited, resting, straight, light, and goodness. The number four represented justice. Five represented marriage since it was the sum of the first male and female numbers. Six was the number of creation. The number ten was a "perfect number" since it was the sum of the first four integers.

The Pythagorean numerology affected many cultures. Plato believed that even numbers were human and inferior and odd numbers were heavenly and superior. Even numbers were an evil omen. Aristotle believed that four was the number of justice. During this period, a number mysticism developed where numbers were believed to have existed before objects they described and were believed to influence the objects. Numbers could be used to communicate with and influence the Divine.

As an example, the Judeo-Christian religion contains many references to special numbers. A partial listing includes:

The rain lasted forty days and forty nights.
Solomon and David ruled forty years.
Moses spent forty days on the mountain.
Christ was in the wilderness where he was tempted by Satan
    for forty days.
Christ was in the tomb for forty hours.
The earth was created in seven days, including a day of rest.
Noah took seven pairs of clean animals into the ark.
Jacob served seven years to get Leah and seven years to get Rachel.

Jacob and Leah had seven children.
There were twelve tribes of Israel.
There are twelve nights between Christmas and Epiphany.
Christ had twelve Disciples.
Heavenly Jerusalem has twelve gates.

Forty was an important number since it was considered the number of completeness. The Jewish alphabet had 22 letters, each associated with a number. The sums of these numbers in a word or sentence have been considered important, especially if the sum is 666, the sign of Satan. Many chapters in the Bible are acrostic; that is, they have 22 verses and each verse starts with the next letter of the alphabet. This helped in memorization but was also considered magical.

There are many other excellent examples. The Babylonians worshipped trinity (the sun, the moon, and Venus). They also worshipped four planet gods, Mercury, Mars, Jupiter, and Saturn, making a total of seven heavenly gods. Seven also denoted wholeness, and the number of gates to the netherworld. They had the eight-pointed star, which was the symbol of Ishtar, and the Deity resided on the eighth floor of the temple. The number twelve also had special meaning since the Babylonians recognized the twelve signs of the zodiac.

Although the natural numbers and integers are familiar to us and, indeed, are part of our culture in many ways, in this text we shall assume that the integers satisfy certain rules or axioms and use these rules to develop other properties of the integers. Hence we will be guided by our previous knowledge of integers, but will accept a statement about integers as true only if we can prove it using our axioms and the rules of logic. This axiomatic development of the integers is similar to the method used in high-school geometry.

Actually, we are building a model of the integers. Our "integers" are simply objects that obey our rules which we hope are also true statements about actual integers, whatever they are. A model of the integers is *complete* if it can be used to prove every true statement about the integers. It is *consistent* if no statement about the integers can be proved both true and false. Unfortunately, it has been shown that no formal axiomatic system about the integers is both complete and consistent. It is hoped that our model is consistent.

1.2   Axioms of the Integers

We will assume that we have a set of objects called integers. The set of integers is denoted by $Z = \{\ldots, -3, -2, -1, 0, 1, 2, \ldots\}$. The axiom system will specify the assumptions about these objects. Since the axioms below are intended to be about the integers, which are already familiar, we will occasionally use this previous familiarity to explain the axioms; however, such use in the illustrations does not presuppose their existence. None of the axioms or theorems depends in any way upon these "numerical" illustrations.

A relation and two operations will be defined. In addition, two integers 0 and 1 with special characteristics will be identified.

Equality of integers is defined by a relation $=$ that is a subset of $Z \times Z$, the set of all ordered pairs of integers. Thus, if a and b are integers and the ordered pair (a, b) is an element of the relation $=$, or, equivalently, $(a, b) \in =$, then we usually just write $a = b$, where the name of the relation is placed between the components of the ordered pair (infix placement).

First we shall assume the following axioms of equality.

(E1) For all integers a, $a = a$.

(E2) For all integers a and b, if $a = b$, then $b = a$.

(E3) For all integers a, b, and c, if $a = b$ and $b = c$, then $a = c$.

Axioms E1, E2, and E3 state that equality has the properties of an equivalence relation on Z. Refer to Section 0.1 for a fuller discussion of equivalence relations.

Two operations on integers are defined. An operation on the set Z is a function from $Z \times Z$ into Z that assigns to each ordered pair (a, b) of $Z \times Z$ an integer c of Z. The two functions are addition, denoted by $+$, and multiplication, denoted by $\cdot$ so that

$$+ : Z \times Z \to Z$$
$$\cdot : Z \times Z \to Z$$

Thus, $+(a, b) = c$ means that $a + b = c$ and $\cdot(a, b) = d$ means that $a \cdot b = d$.

That both $+$ and $\cdot$ are functions with domain $Z \times Z$ and with range Z simply means that each of $a + b$ and $a \cdot b$ specifies only one object and in each case that object is an integer. This last property (that $a + b$ and $a \cdot b$ are integers) is referred to as the closure property of the operations.

In the discussion above we have described $=$, $+$, and $\cdot$ only in the most general terms using set theoretic terminology; namely, $=$ is a relation on $Z \times Z$ and $+$ and $\cdot$ are functions from $Z \times Z$ into Z. The axioms themselves specify exactly what kind of relation $=$ is and what kind of functions $+$ and $\cdot$ are. We include these properties in the axioms below.

(I1) If a and b are integers, then $a + b$ and $a \cdot b$ are integers. The integers are *closed* under the operations of addition and multiplication.

(I2) If $a = b$ and $c = d$, then $a + c = b + d$ and $a \cdot c = b \cdot d$.

(I3) For all integers a and b, $a + b = b + a$ and $a \cdot b = b \cdot a$. The integers are *commutative* under the operations of addition and multiplication.

(I4) For all integers a, b, and c, $(a + b) + c = a + (b + c)$ and $a \cdot (b \cdot c) = (a \cdot b) \cdot c$. The integers are *associative* under the operations of addition and multiplication.

(I5) For all integers a, b, and c, $a \cdot (b + c) = (a \cdot b) + (a \cdot c)$. Multiplication of integers is *distributive* over addition.

(I6) There exist unique integers 0 and 1 such that for all integers a, $a + 0 = 0 + a = a$ and $a \cdot 1 = 1 \cdot a = a$. The integer 0 is called the *additive identity* of the integers and 1 is called the *multiplicative identity*.

(I7) For every integer a there is a unique integer, $-a$, called the *additive inverse* of a, such that $a + (-a) = (-a) + a = 0$.

(I8) If b and c are integers and $a \cdot b = a \cdot c$ for some nonzero integer a, then $b = c$. This is called the *multiplicative cancellation property*.

Axiom I2 relates equality to addition and multiplication stating that in an expression for an integer involving addition or multiplication, an integer within the expression may be replaced by another equal to it without affecting the integer described by the total expression. Axiom I6 specifies a particular integer, 0, called zero, whose effect, when added to any integer a, is to produce again the integer a. Likewise, Axiom I6 specifies a particular integer, 1, called one, whose effect, when multiplied by any integer a, is to produce again the integer a. The important idea here is that the single integer 0 "works" for every integer a with respect to addition and the single integer 1 "works" for every integer a. Axiom I6 further states that no other integers than 0 and 1 have these respective properties.

Axiom I7 guarantees that each integer a has a unique corresponding b with the property that $a + b = 0$ and $b + a = 0$. Since a uniquely specifies b, we can denote the unique b as $(-a)$ without any ambiguity. More formally, Axiom I7 states that there is a function $f : Z \rightarrow Z$ that is one to one so that $a + f(a) = f(a) + a = 0$. The important distinction we need to see here (as compared to I6) is that each integer a does require a different $-a$ with $a + (-a) = 0$. As an illustration, if $a = (-5)$, then $(-a)$ is 5 so that $a + (-a) = (-5) + 5 = 0$. Furthermore, $-a$ is $-(-5)$ since a is -5. We need to realize that despite the notation -5, we are not defining the subtraction operator, $-$, in Axiom I7.

We will use these axioms to prove the following theorem. Notice that in this proof we will mention each time the axioms above are used. In future proofs the properties described are used but may not be mentioned explicitly because most proofs would be too cumbersome if they were given in this much detail.

Theorem 1.1 *Let a, b, and c be integers.*

(a) *If $b + a = c + a$, then $b = c$.*

(b) *For each integer a, $a \cdot 0 = 0$.*

(c) *For each integer a, $-(-a) = a$.*

(d) $a \cdot (-b) = -(ab)$.

(e) $(-a) \cdot (-b) = a \cdot b$.

(f) $-(a+b) = (-a) + (-b)$.

Proof. (a) Assume that $b + a = c + a$. Then

$$
\begin{array}{rcll}
(b+a) + (-a) & = & (c+a) + (-a) & \text{by I2} \\
b + (a + (-a)) & = & c + (a + (-a)) & \text{by I4} \\
b + 0 & = & c + 0 & \text{by I7} \\
b & = & c & \text{by I6}
\end{array}
$$

(b)

$$
\begin{array}{rcll}
a \cdot 0 & = & a \cdot (0 + 0) & \text{by I6} \\
& = & a \cdot 0 + a \cdot 0 & \text{by I5}
\end{array}
$$

Hence

$$
\begin{array}{rcll}
0 + a \cdot 0 & = & a \cdot 0 & \text{by I6} \\
& = & a \cdot 0 + a \cdot 0 & \text{from part (a)}
\end{array}
$$

Thus,

$$
a \cdot 0 = 0 \quad \text{by part (a)}
$$

(c) The proof is left to the reader.

(d) By Axiom I7, $-(ab)$ is the unique element such that

$$
(-(a \cdot b)) + (a \cdot b) = 0
$$

Hence we need to show that

$$
(a \cdot b) + (a \cdot (-b)) = 0
$$

but

$$
\begin{array}{rcll}
(a \cdot b) + (a \cdot (-b)) & = & a \cdot (b + (-b)) & \text{by I5} \\
& = & a \cdot 0 & \text{by I7} \\
& = & 0 & \text{by part (b)}
\end{array}
$$

(e) Proof is left to the reader.

(f) By I7, we need to show that $((-a) + (-b)) + (a+b) = 0$. But

$$
\begin{array}{rcll}
((-a) + (-b)) + (a+b) & = & (((-a) + (-b)) + a) + b & \text{by I4} \\
& = & ((-a) + ((-b) + a)) + b & \text{by I4} \\
& = & ((-a) + (a + (-b))) + b & \text{by I3} \\
& = & (((-a) + a) + (-b)) + b & \text{by I4} \\
& = & (0 + (-b)) + b & \text{by I7} \\
& = & (-b) + b & \text{by I6} \\
& = & 0 & \text{by I7} \quad \blacksquare
\end{array}
$$

Theorem 1.1(a) is the cancellation property for addition (compare Axiom I8), which is shown to be a consequence of the axioms, whereas the cancellation property for multiplication appears as an axiom. Theorem 1.1(b) describes how the unique integer 0 behaves when it is used as a multiplier in a product of integers. Remember that the unique characteristic of 0 was with reference to addition only. Theorem 1.1(c) states that the additive inverse of the additive inverse of an integer a is the integer a. It also shows that the integers in Z are divided up into pairs (r, s) where each component is the additive inverse of the other. Theorem 1.1(d) to (f) give further properties of additive inverses. These are the familiar "sign" properties of integers.

A common convention or abbreviation in writing arithmetic expressions involving $+$ and $\cdot$ is to omit the multiplication operator and indicate multiplication by placing the integer symbols side by side. Thus, we write ab instead of $a \cdot b$, $a(b + c)$ instead of $a \cdot (b + c)$, and $ab + c$ instead of $a \cdot b + c$.

The associative properties of Axiom I4,

$$(a + b) + c = a + (b + c)$$

and

$$(a \cdot b) \cdot c = a \cdot (b \cdot c)$$

state that the parentheses may be removed without ambiguity.

We now define another operation, subtraction, on Z.

Definition 1.2 *If a and b are integers, then $a - b$ is the integer $a + (- b)$.*

This definition is justified because Axiom I7 guarantees that associated with b there exists a unique integer, namely $-b$, having the property that $b + (- b) = (- b) + b = 0$.

Within the model of the integers Z is a subset N having very important and powerful properties. Indeed, in Appendix B, the whole set of integers is constructed using only this subset. We define the *positive integers* or *natural numbers* N with the following axioms:

(N1) The integer 1 is a positive integer.

(N2) The positive integers are closed under addition and multiplication; that is, if a and b are positive integers, then $a + b$ and $a \cdot b$ are positive integers.

(N3) (Trichotomy Axiom) For every integer a, one and only one of the following is true:

(a) a is a positive integer.

(b) $a = 0$.

(c) $-a$ is a positive integer.

(N4) (Principle of Induction) If S is a subset of N with the following properties:

(a) 1 is in S.

(b) Whenever the integer n is in S, then $n + 1$ is in S.

Then $S = N$.

Trichotomy allows the integers to be partitioned into three disjoint sets — those integers that are positive, those that are zero, and those that are neither positive nor zero. Those integers that are neither positive nor zero are called *negative* integers. Thus, trichotomy justifies the following definition.

Definition 1.3 *An integer a is negative if and only if* $-a$ *is positive.*

The Principle of Induction is the main tool we have for proving that a statement is true for all of the positive integers. When considering whether a statement about an integer n is true for every positive integer, we obviously cannot test the truth of the statement for each individual integer n. It also cannot be assumed that because a statement is true for the first ten positive integers or even the first ten billion positive integers, the statement is true for all integers. The Principle of Induction has been compared to an infinite row of dominoes. Properly aligned dominoes have the property that if any one is pushed over, then the next one falls over. Let S be the set of dominoes that fall over. Thus, if (a) the first one is pushed over (the first domino is in S) and (b) each domino knocks over the next one (when domino n is in S, then domino $n + 1$ is in S), then they all fall down (S contains all the dominoes).

Example 1.4 *Assume that we want to calculate the sum of the first m positive integers, $1 + 2 + 3 + \cdots + m$, but want to find a method or formula for doing it which is more compact and requires less work. Such a formula would be useful if we needed to calculate the sum of the first $100,000$ positive integers but did not want to perform $99,999$ additions to obtain the answer. Assume we have reason to suspect that the sum of the first m positive integers is $\dfrac{m(m+1)}{2}$; so if we calculate the sum for several values of m and compare with $\dfrac{m(m+1)}{2}$, we obtain the table below.*

m	$1+2+3+\cdots+m$	$m(m+1)/2$
1	1	$1(1+1)/2 = 1$
2	3	$2(2+1)/2 = 3$
3	6	$3(3+1)/2 = 6$
4	10	$4(4+1)/2 = 10$
5	15	$5(5+1)/2 = 15$
6	21	$6(6+1)/2 = 21$
7	28	$7(7+1)/2 = 28$

The middle column gives the direct calculation of the sum of the first m posi-

tive integers. At least for the first 7 integers, the formula $\dfrac{m(m+1)}{2}$ *faithfully gives the correct sum of the first m positive integers. The question is whether* $\dfrac{m(m+1)}{2}$ *always gives the sum* $1+2+3+\cdots+m$. *The Principle of Induction is used to prove that it does.*

Let $S = \left\{ m : m \in N \text{ and } 1+2+3+\cdots+m = \dfrac{m(m+1)}{2} \right\}$; *that is, S is the set of only those positive integers for which the sum of the first m positive integers is* $\dfrac{m(m+1)}{2}$. *We first show that* 1 *is in S. The left side of the equality* $1+2+3+\cdots+m$ *gives* 1 *for* $m=1$. *On the right side we get*

$$\frac{m(m+1)}{2} = \frac{1(1+1)}{2} = 1$$

so the integer 1 *is in S. Next we assume that the positive integer n is in S or, equivalently, that the equation is true for* $m=n$. *That is, we assume that*

$$1+2+3+\cdots+n = \frac{n(n+1)}{2}$$

We now need to show that $n+1$ *is in S or, equivalently, that the equality is true for* $m = n+1$. *In other words, we need to show that*

$$1+2+3+\cdots+n+(n+1) = \frac{(n+1)((n+1)+1)}{2}$$

Using our assumption that

$$1+2+3+\cdots+n = \frac{n(n+1)}{2}$$

and comparing this equality to the one we need, we see that adding $n+1$ *to both sides of the assumed equality will produce the left side of the equality we are looking for:*

$$1+2+3+\cdots+n \;=\; \frac{n(n+1)}{2}$$

$$1+2+3+\cdots+n+(n+1) \;=\; \frac{n(n+1)}{2} + (n+1)$$

$$\;=\; \frac{(n+1)((n+1)+1)}{2}$$

We have proved that

$$1+2+3+\cdots+n+(n+1) = \frac{(n+1)((n+1)+1)}{2}$$

which means that the equality holds for $m = n+1$. *This result ensures that* $n+1$ *is in the set S. Hence, by Axiom N4 (that is, by induction), every positive integer is in S or, equivalently, the equation is true for all positive integers.*

Example 1.5 *Show that $m^3 - m$ is divisible by 3 for every positive integer m. (Note that a positive integer t is divisible by 3 provided that there is a positive integer k so that $t = 3k$.) The proof will proceed by induction. Let $S = \{m : m^3 - m \text{ is divisible by } 3\}$.*

Show that $1 \in S$. For $n = 1$ we calculate $n^3 - n = 1^3 - 1 = 1 - 1 = 0$, which is divisible by 3 because $0 = 3 \cdot 0$. Thus $1 \in S$.

Show that whenever n is a positive integer in S, then $n+1$ is in S. Assume that $n \in S$. This means that $n^3 - n$ is divisible by 3 and that $n^3 - n = 3k$ for some positive integer k. (Note: This assumption is called the "induction hypothesis" or the "inductive assumption.") We must show that $n+1$ is in S or, equivalently, that $(n+1)^3 - (n+1)$ is divisible by 3 or that there is a positive integer w so that $(n+1)^3 - (n+1) = 3w$. We now focus on the positive integer $(n+1)^3 - (n+1)$. The strategy here is to manipulate $(n+1)^3 - (n+1)$ so that we will be able to use our knowledge about n (namely, that $n^3 - n = 3k$). Thus,

$$
\begin{aligned}
(n+1)^3 - (n+1) &= (n^3 + 3n^2 + 3n + 1) - (n+1) \\
&= (n^3 - n) + (3n^2 + 3n) \\
&= 3 \cdot k + 3(n^2 + n) \\
&= 3 \cdot w
\end{aligned}
$$

where $w = k + (n^2 + n)$. This result shows that $(n+1)^3 - (n+1)$ is divisible by 3 and consequently, that $n + 1 \in S$. Therefore, by induction, $S = N$ and $m^3 - m$ is divisible by 3 for every positive integer m.

Example 1.6 *Show that a plane divided up into regions by any number of distinct straight lines can be painted with black and white paint in such a way that any two regions having a common boundary will be painted in different colors as depicted in Figure 1.*

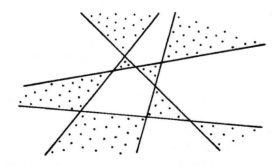

Figure 1: Painted regions.

Since the statement of the problem asks about all possible numbers of regions and lines, we have infinitely many possibilities. This type of problem is one that the Principle of Induction is designed to handle. We have a choice of whether to show that any number m of such regions can be properly painted or that any number of regions generated by m lines can be properly painted. In either case, if we can show the property for all occurrences of m regions or all occurrences or m lines, we will have shown that any such drawing could be painted. In the first case, we would say that we are proceeding by induction upon the number m of regions. In the second case, we would say that we are proceeding by induction upon the number m of lines. We will use the number of lines involved in a drawing.

Let $P(m)$ be the statement that a drawing made with m lines can be properly painted (i.e., with two colors black and white with no adjacent regions that share a common boundary segment having the same color).

We choose

$$M = \{m : P(m) \text{ is true}\}$$
$$= \{m : \text{a drawing made with } m \text{ lines can be colored}\}$$

Thus, if we show that $M = N$, then a drawing created with any number of lines can be properly painted. $M \subseteq N$ by definition.

(a) *Show that $1 \in M$. $1 \in M$ provided that $P(1)$ is true or provided that a drawing made with only 1 line can be painted. It is known from geometry that a line drawn in the plane divides the plane into two regions. Choose one to paint black and paint the other one white as shown in Figure 2. Then $1 \in M$.*

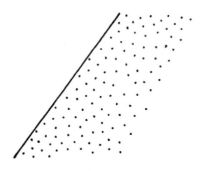

Figure 2: Painted with one line.

(b) *Show that whenever $m \in M$, then $m+1 \in M$. Assume that m is a positive integer in M, which means that $P(m)$ is true and any drawing made with m lines can be properly painted. Consider the positive integer $m + 1$. We need to show that $P(m + 1)$ is true, or, equivalently, that any drawing made with $m + 1$ lines can be properly painted. In doing so, we are allowed to use "the induction hypothesis," namely, the knowledge that any drawing made with m lines can be properly painted. So consider a drawing made with $m + 1$ lines* *as depicted in Figure 3.*

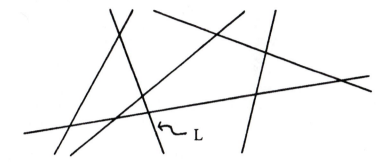

Figure 3: m + 1 lines.

To be able to proceed, we must conceive of, or create, a situation for which we can use the knowledge that $m \in M$ or that any drawing with m lines can be properly painted. Toward this end, select one of the lines, say line L, and remove it.

The drawing that remains is made of only m lines; so since we assumed that any drawing made with m lines can be properly painted, we can properly paint the drawing in Figure 4.

Now replace the line L in its original position and change the color of each region on one side of the line L to the opposite color (change whites to black and change blacks to white) as shown in Figure 5. This drawing is properly colored because each boundary that occurred on the chosen side of L where the changes were made originally had different colors on either side so that if the colors were switched, the colors on either side of any such boundary would still be different. The only other boundaries affected were those along L; but if L divided a region, then the color was changed on one side of L, making different colors on either side of such a boundary. Thus, this drawing with $m + 1$ lines can be properly painted. This result means that $P(m + 1)$ is true and that $m + 1 \in M$.

In summary, we have shown parts (a) and (b) of the Induction Principle. Thus, $M = N$. We often say at this point that "by induction" we have shown that $M = N$. Thus, a drawing with any (finite) number of lines can be properly painted.

The properties of the equality ($=$) relation are given in Axioms E1 to E3. Other useful relations are those describing orderings of the integers, where the integers are arranged so as to give integer analogs of our ideas of "preceding," "before," "after," or "greater than." We usually think of the integers as being "ordered" in the following way:

$$\ldots, -4, -3, -2, -1, 0, 1, 2, 3, 4, 5, \ldots$$

where all positive integers "follow" zero and zero "follows" or is "greater than"

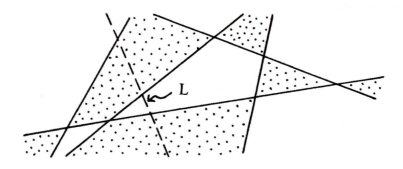

Figure 4: Painted with m lines.

any negative integer. Since trichotomy allows a determination of whether an integer is positive, we can define the relation $>$ contained in $Z \times Z$ as $(a, b) \in >$ or $a > b$ if and only if $a - b$ is positive. Because of trichotomy, $a - b$ is either positive, zero, or negative so that we can collect two of the possibilities into one category and say that $a \geq b$ if and only if $a - b$ is positive or $a - b$ is zero. This situation is equivalent to requiring that either $a > b$ or $a = b$. We have justified the following definition.

Definition 1.7 *For integers a and b, $a > b$ if and only if $a - b$ is positive; and $a \geq b$ if and only if $a > b$ or $a = b$. Also, $b < a$ if $a > b$ and $b \leq a$ if $a \geq b$.*

Obviously, $a > 0$ is equivalent to stating that a is positive because $a > 0$ if and only if $a - 0 = a$ is positive. Similarly, $a < 0$ is equivalent to stating that a is negative. All of the above are relations on Z that are defined using the axioms. Note that the definition above does not introduce new axioms since every mention of "$a > b$" can be replaced by "$a - b$ is positive."

Theorem 1.8 *For integers a and b:*

(a) *If $a \geq b$ and $b \geq a$, then $a = b$.*

(b) *If $a > b$ and $b > c$, then $a > c$.*

Proof. (a) If $a > b$ and $b > a$, then $a - b$ is positive and $b - a = -(a - b)$ is positive, contradicting N3, the Trichotomy Axiom. Thus, $a = b$.

(b) If $a > b$ and $b > c$, then $a - b$ and $b - c$ are positive. Hence $(a - b) + (b - c) = a - c$ is positive and $a > c$. ∎

Theorem 1.9 *Let $a, b, c,$ and d be integers.*

(a) *If $a > b$ and $c > d$, then $a + c > b + d$.*

(b) *If a is positive and $c > d$, then $ac > ad$.*

(c) If $a > b > 0$ and $c > d > 0$, then $ac > bd$.

(d) If $a \geq b \geq 0$ and $c \geq d \geq 0$, then $ac \geq bd$.

Proof. **(a)** If $a > b$ and $c > d$, then $a - b$ and $c - d$ are positive. Hence $(a - b) + (c - d)$ is positive. But $(a - b) + (c - d) = (a + b) - (c + d)$, and so $a + b > c + d$.

   (b) If $c > d$, then $c - d$ is positive. Since a is positive, $a(c - d) = ac - ad$ is positive and $ac > ad$.

   (c) If $a > b > 0$ and $c > d > 0$, then $ac > bc$ and $bc > bd$. By Theorem 1.8, $ac > bd$.

   (d) The proof is left to reader. ∎

   Similar theorems can be obtained by changing every $>$ and $\geq$ to $<$ and $\leq$, respectively.

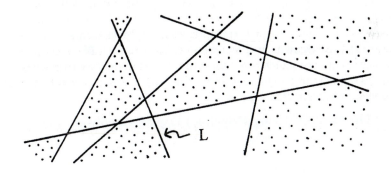

Figure 5: Painted with $m + 1$ lines.

   It is convenient to have the Trichotomy Axiom restated in terms of inequalities. The proof of the next theorem is left to the reader.

Theorem 1.10 (Trichotomy) *For integers a and b, one and only one of the following is true:*

(a) $a > b$.

(b) $a = b$.

(c) $b > a$.

   Consider the set $A = \{3, 7, 8\}$. The integer 3 has a unique property with respect to A; namely, 3 is in A, and 3 is less than or equal to every member of the set A. We say that 3 is the "least" integer in the set A. Not every set of integers has a least integer. The set $\varnothing$ has no least integer because it is empty and contains no integer at all. The set

$$\begin{aligned} C \;&=\; \{n : n \in Z \text{ and } n = 5k \text{ for some integer } k\} \\ &=\; \{\ldots, -15, -10, -5, 0, 5, 10, 15, 20, \ldots\} \end{aligned}$$

has no least integer; for if m is a least integer in C, then $m = 5 \cdot k$ for some integer k. But the integer $s = 5(k-1)$ is also in C and, clearly, $s = 5(k-1) < 5 \cdot k = m$ so that $s < m$, which contradicts the assumption that m is the least integer in C.

Definition 1.11 *The integer a is the* least integer *of a set A if and only if*

(a) *a is in A.*

(b) *If b is in A, then $a \leq b$.*

Theorem 1.12 *The least integer of a set is unique.*

Proof. Assume that a and b are both least elements of a set A; then by definition of least element, $a \leq b$ and $b \leq a$. Hence $a = b$. ∎

At this point, we include two alternatives to the Induction Principle. Both are equivalent to the Principle of Induction and, therefore, could have been used instead of Axiom N4. The proofs will be presented later.

(N5) (The Well-Ordering Principle) Every nonempty set of positive integers contains a least element.

(N6) (Second Principle of Induction) If S is a subset of N with the following properties:

(a) 1 is in S.

(b) Whenever k is in S for all $1 \leq k < n$, then n is in S.

Then $S = N$.

An example of the use of the Second Principle of Induction N6 will be given in Chapter 2. The following theorem is an example of a statement which seems too obvious to bother to prove; but since our "integers" are only defined by a set of rules, the statement may not be obvious or even true for our "integers." We can only find out by proving the theorem using the rules which we have set up. The proof is also a good example of an application of the Well-Ordering Principle, N5.

Theorem 1.13 *There is no positive integer a such that $0 < a < 1$.*

Proof. If there is such an integer, let S be the set of all positive integers between 0 and 1. Since S is nonempty, it has, by Axiom N5, a least positive integer, say a. Multiplying the inequality $0 < a < 1$ by a, we have $0 < a^2 < a$.

Also since $a < 1$, $a^2 < 1$. But this contradicts the fact that a is the least positive integer less than 1. Hence S is empty, and there is no positive integer between 0 and 1. ∎

Corollary 1.14 *If b is an integer, then there is no integer a having the property $b < a < b + 1$.*

In this section we gave axioms for the integers. In Euclid's Elements we find the first known logical arrangement of mathematics in terms of axioms, postulates, and propositions. It has been said that his works contain all known theorems of number theory in his time. Little is known of Euclid of Alexandria. He lived about 300 B.C. He may have been a Greek or an Egyptian who came to Greece to study. His work is a compilation of others as well as his own and it is difficult to determine how much of his work is original. He is credited with the Euclidean Algorithm for finding the greatest common divisor of two integers and hence must have been familiar with the Division Algorithm. He is also credited with the proof that there are infinitely many primes. Primes will be discussed in Chapter 2. He also knew that an integer could be factored uniquely into a product of primes (Fundamental Theorem of Arithmetic discussed in Chapter 2).

An important step toward an axiomatic treatment of the positive integers was made by Gottlob Frege (1848-1925). Frege, who has been called the founder of modern mathematical logic, used the work of George Boole (1815-1864) and others to develop mathematics based on set theory and logic. His main contribution to number theory was the axiomatic development of the cardinals.

Frege was a liberal Lutheran who loved mathematics and the monarchy, but hated socialism, democracy, the French, Catholics, Jews, and many of his fellow mathematicians. His work in mathematics was extremely difficult to read and he was criticized by others who had not or could not read his works. His strongest supporters were Guiseppe Peano and Bertrand Russell. Peano followed Frege's work on the cardinals with his axiomatic development of the integers. Russell's book, *Principia Mathematica*, followed Frege's tradition of writing mathematics which was difficult to read.

Guiseppe Peano (1858-1932) studied at the University of Turin and spent his entire career there, where he became associate professor of infinitesimal calculus and finally, full professor. He was originally known for his clear and logically rigorous lectures. As he began using more of the symbols of set theory, his lectures became much less popular and he stopped lecturing to concentrate on research.

Peano is best known for his five axioms which describe the natural numbers. These axioms are appropriately known as Peano's Axioms or Peano's Postulates (see Appendix B). It should be noted that one of the axioms is the Principle of Mathematical Induction. Peano also made other contributions to mathematics, including the discovery of the first known space-filling curve, the first explicit formulation of the Axiom of Choice, and the introduction of our current notation for set theory. The Axiom of Choice states that given any collection of nonempty sets, there exists a set containing one element from each set in the collection. Another statement of the Axiom of Choice is that given a collection of nonempty sets, one element can be selected from each set. It is easy to use this axiom without realizing it.

More formal developments of the positive integers have been made since Peano's Postulates. The first of these was in *Principia Mathematica* by Bertrand Russell and Alfred North Whitehead.

A more readable development was by Ernst Zermelo (1871-1953) in his paper, *Investigations into the Foundations of Set Theory*, which was amended by Abraham Fraenkel (1891-1965) and is known as the Zermelo-Fraenkel axioms for set theory. In this paper, Zermelo also proved the Well-Ordering Theorem for the positive integers.

A third set of axioms for the development of set theory is credited to Kurt Gödel (1906-1978), John von Neumann (1903-1957), and Paul Bernays (1888-1977). The set theory is appropriately known as Gödel-von Neumann-Bernays set theory. It is ironic that Bernays, a student of David Hilbert, and von Neumann worked with David Hilbert to try to prove that all of mathematics could be developed by formal set theory, while Gödel showed that this was impossible.

It is also interesting that Zermelo was criticized for using the Axiom of Choice to prove the Well-Ordering Theorem; and although Gödel used Gödel-von Neumann-Bernays set theory to show the consistency of the Axiom of Choice, Paul Cohen used Zermelo-Fraenkel set theory to show that denial of the Axiom of Choice was also consistent and hence the Axiom of Choice was a true axiom.

Exercises

1. Prove that for each integer a, $-(-a) = a$ (Theorem 1.1(c)).

2. Prove that for integers a and b, $(-a)(-b) = ab$ (Theorem 1.1(e)).

3. Prove that for integers a, b, c, and d, if $a \geq b \geq 0$ and $c \geq d \geq 0$, then $ac \geq bd$ (Theorem 1.9(d)).

4. Prove that 0 is the only integer equal to its additive inverse.

5. For each positive integer n show that there is a positive integer k such that $4^n - 1 = 3k$.

6. Prove that if a is any integer, then $a^{n+m} = a^n a^m$ for any positive integers n and m.

7. Prove that for each positive integer n:

   (a) $a^n - 1 = (a-1)(a^{n-1} + a^{n-2} + \cdots + a + 1)$

(b) $1 \cdot 2 + 2 \cdot 3 + \cdots + n(n+1) = \dfrac{n(n+1)(n+2)}{3}$

(c) $1^2 + 2^2 + 3^2 + \cdots + n^2 = \dfrac{n(n+1)(2n+1)}{6}$

(d) $1^3 + 2^3 + 3^3 + \cdots + n^3 = \left[\dfrac{n(n+1)}{2}\right]^2$

(e) $1 + 3 + 5 + \cdots + (2n-1) = n^2$

(f) $2^n > n$

8. Given that $1+a$ is positive, prove that $1+na \le (1+a)^n$ for each positive integer n.

9. Prove that there exists an integer n such that $n! \le 1000^{1000}$ and $(n+1)! \ge 1000^{1000}$. Here $n!$ means $n(n-1)(n-2)\cdots(3)(2)(1)$, the product of the first n positive integers. Note that $(n+1)! = (n+1) \cdot n!$ and, as a special case, $0! = 1$.

10. Let x and y be integers. Use mathematical induction to prove that $(x-y)^n$ has $x-y$ as a factor for every positive integer n.

11. Use mathematical induction to prove:

    (a) The formula for the geometric sum when $r \ne 1$:
    $$1 + r + r^2 + r^3 + \cdots + r^n = \frac{r^{n+1} - 1}{r - 1} \text{ for } n \ge 0$$

    (b) For $n \ge 1$
    $$2 \cdot 2^1 + 3 \cdot 2^2 + 4 \cdot 2^3 + \cdots + (n+1) \cdot 2^n = n \cdot 2^{n+1}$$

    (c) $\cos(n\pi) = (-1)^n$ for $n \ge 1$

12. Prove that a set of integers B has at most one greatest element. (An integer b_g is a greatest element of a set B of integers if and only if b_g is an element of B and $b \le b_g$ for all b in B.)

13. Prove that every nonempty set of negative integers has a greatest element.

14. Using the Well-Ordering Axiom prove that if a and b are positive integers, then there exists a positive integer n such that $na \ge b$ (Archimedean property).

15. For integers a and b, prove $ab = 0$ if and only if $a = 0$ or $b = 0$.

16. For integers a and b, prove that one and only one of these three relations holds: $a > b$, $a = b$, or $b > a$ (Theorem 1.10).

17. Prove that if $a \geq 0$, $b \geq 0$, and $a + b = 0$, then $a = b = 0$.

1.3   Principle of Induction

All of the axioms for the integers have been stated and several definitions and results derivable from the axioms were obtained. Before proceeding further, we will discuss induction in greater depth. Induction is the "workhorse" of number theory and is necessary to prove many of the results we need.

The inductive axioms encountered so far have dealt only with the set of positive integers N, which is a subset of Z. Axioms N4 and N6 are similar and address the integers starting with 1 and specifying conditions on integers greater than 1. The conclusion in both cases is that the set S, initially given as a subset of N, ends up being the set of all positive integers $\{n : n \in Z$ and $n \geq 1\}$. Axiom N5 appears to be somewhat unrelated to induction; nonetheless, it is equivalent to both N4 and N6. We will state two other variations upon the theme addressed by Axioms N4 and N6. Both N4 and N6 apply to integers greater than or equal to 1. Similar induction theorems apply to the integers greater than or equal to an integer j that is not necessarily the integer 1. We will refer to these as providing induction upon the integers instead of just upon the positive integers. Both follow from Axiom N4.

Theorem 1.15 (Induction Principle for Integers) *For a given integer j, let $T = \{t : t$ is an integer and $t \geq j\}$ be the set of all integers greater than or equal to j and R be a subset of T with the following properties:*

(a) $j \in R$.

(b) *For any $k \geq j$, if $k \in R$, then $k + 1 \in R$.*

Then $R = T$.

Proof. The proof follows from Axiom N4 by letting $n = t - j + 1$. ∎

Theorem 1.16 (Second Induction Principle) *For a given integer j, let $T = \{t : t$ is an integer and $t \geq j\}$ and R be a subset of T with the following properties:*

(a) $j \in R$.

(b) *If k is in R for all k such that $j \leq k \leq m$, then $m + 1$ is in R.*

Then $R = T$.

Example 1.17 *As an example of the use of the Induction Principle for Integers in Theorem 1.15, we consider how to show that a quantity can be defined or exists for every nonnegative integer n. Evidently, we cannot use the Principle of Induction, N4, directly because 0 is nonnegative and is not a positive* *integer. We define the quantity for $n = 0$; and then, using the expression for the quantity when $n = k$, define the expression for $n = k + 1$. The quantity we want to consider here is the factorial function $n!$, which is the product of the first n positive integers for $n > 1$. To define $n!$ we proceed as follows:*

$$0! = 1$$

$$(n + 1)! = (n + 1)(n!)$$

For example,

$$1! = 1 \cdot 0! = 1$$

$$2! = 2 \cdot 1! = 2$$

$$3! = 3 \cdot 2! = 6$$

For $j = 0$ and $R = \{n : n \in T$ and $n!$ is defined\}, we want to show that $R = T$. By definition, $0 \in R$. If $k \in R$, then $k + 1 \in R$ because $(k + 1)! = (k + 1)k!$. Therefore, by Theorem 1.16, $R = T$.

This type of definition is called an *inductive definition* or a *recursive definition*. In an inductive definition we are often interested in defining a sequence of integers b_j, b_{j+1}, b_{j+2}, b_{j+3}, ... or $\{b_n : \text{for } n \geq j\}$. The definition frequently proceeds by specifying the "first" member of the sequence as some known integer, as in $b_j = a$. Then, assuming that b_n can be calculated, b_{n+1} is specified in terms of b_n and n. By mathematical induction it follows that b_n is defined (i.e., can be calculated) for every $n \geq j$.

When $b_n = n!$, to obtain b_{n+1} only knowledge of b_n, the preceding term of the sequence, was required. This characteristic means that only the Principle of Induction, N4, or the Induction Principle for Integers in Theorem 1.15 is needed. More generally, the definition of a recursive sequence $b_j, b_{j+1}, b_{j+2}, \ldots$ may depend upon more than just the preceding term in the definition.

Example 1.18 *We define a sequence $c_1, c_2, c_3, \ldots$ by specifying the following:*

$$
\begin{aligned}
c_1 &= 5 \\
c_2 &= 8 \\
\text{if } c_n \text{ and } c_{n-1} \text{ exist, then } c_{n+1} &= 3 \cdot c_n + c_{n-1} \text{ for } n \geq 2
\end{aligned}
$$

Using the Second Induction Principle, we prove that c_n exists for every integer $n \geq 1$.

Let $R = \{n : n \in Z$ and c_n exists\}. Since c_1 is given, $1 \in R$. Assume that $m \geq 1$ and that $k \in R$ for all $1 \leq k \leq m$. We need to show that $m + 1$ is in R or that c_{m+1} exists. If $m = 1$, then $m + 1 = 2$ and c_2 is given. If $m \geq 2$, then $m > m - 1 \geq 1$. Since $1 \leq m - 1 < m$, the induction hypothesis states that both c_{m-1} and $c_m \in R$. Consequently, the definition gives $c_{m+1} = 3 \cdot c_m + c_{m-1}$ and $m + 1 \in R$. By the Second Induction Principle for Integers, $R = N$ and c_n exists for every integer $n \geq 1$.

Example 1.19 *Prove that* $n! > 2^n$ *for every integer* $n \geq 4$ *using the Induction Principle for Integers.*

If we want to use induction on n, we cannot start with $n = 1$ since the statement is not true for $n = 1$. Our starting place must be $n = 4$. In the notation of the Induction Principle for Integers, $j = 4$, $T = \{n : n \in Z \text{ and } n \geq 4\}$, and $R = \{n : n \in T \text{ and } n! > 2^n\}$.

First prove the statement for $n = 4$. When $n = 4$ we have $4! = 24$ and $2^4 = 16$, so $4! > 2^4$. Assume that the statement is true for $n = k$ where $k \geq 4$; i.e., assume that $k! > 2^k$. Next prove that $(k + 1)! > 2^{k+1}$. Since $k \geq 4$, we know that $k > 2$, and that by the induction hypothesis, $k! > 2^k$. Hence, $k \cdot k! > 2 \cdot 2^k$ and $(k + 1)! > 2^{k+1}$. Thus, $n! > 2^n$ for every $n \geq 4$.

A further application of an inductive or recursive definition is to define exponentiation. The laws of exponents are derived for nonnegative exponents.

Definition 1.20 *For integer a and positive integer n, define a^n as follows:*

$$\begin{aligned} a^0 &= 1 \text{ if } a \neq 0 \\ a^1 &= a \\ a^{n+1} &= a \cdot a^n \text{ if } n \geq 1 \end{aligned}$$

Theorem 1.21 *Exponentiation has the following properties.*

(a) *For all integers a and b and positive integer n, $(ab)^n = a^n b^n$.*

(b) *For any integer a and all positive integers m and n, $a^{mn} = (a^m)^n$.*

Proof. (a) For $n = 1$, $(ab)^1 = ab = a^1 b^1$ and the statement is true for $n = 1$. Assume that the equality is true for $n = k$, i.e., assume that

$$(ab)^k = a^k b^k$$

Now prove that the expression is true for $n = k + 1$; that is, show that

$$(ab)^{k+1} = a^{k+1} b^{k+1}$$

But

$$\begin{aligned} (ab)^{k+1} &= (ab)^k ab & \text{by definition of exponent} \\ &= a^k b^k ab & \text{by induction hypothesis} \\ &= a^k ab^k b & \text{by commutativity of multiplication} \\ &= a^{k+1} b^{k+1} & \text{by definition of exponent} \end{aligned}$$

(b) The proof is lengthy and is left to the interested reader. ∎

Exercises

1. Prove the following using mathematical induction, where n and k are integers:

(a) $n^2 > 2n + 1$ for $n \geq 3$.

(b) Find the largest set of positive integers for which it is true that $2^n > n^2$.

(c) Find the largest set of positive integers for which it is true that $2^n > n!$.

2. Prove that every integer $n \geq 8$ can be written as $n = 3k + 5m$ for some nonnegative integers k and m.

3. Prove that for each positive integer $n \geq 2$,

$$\frac{4^n}{n+1} < \frac{(2n)!}{(n!)^2}$$

4. Let $a_0 = 2, a_1 = 3$, and, for each positive integer k, $a_{k+1} = 3 \cdot a_k - 2 \cdot a_{k-1}$. Prove that:

(a) a_k is defined for every nonnegative integer k.

(b) $a_n = 2^n + 1$ for each integer $n \geq 0$.

5. Prove that for any integer a and positive integers m and n, $a^{mn} = (a^m)^n$ (Theorem 1.21(b)).

6. Using mathematical induction, prove that $n! > n^3$ for all $n \geq 6$.

7. Use mathematical induction to prove that:

(a)

$$\left(1 - \frac{1}{4}\right)\left(1 - \frac{1}{9}\right)\left(1 - \frac{1}{16}\right) \cdots \left(1 - \frac{1}{n^2}\right) = \frac{n+1}{2n} \text{ for } n \geq 2$$

(b)

$$\cos(\alpha) \cdot \cos(2\alpha) \cdot \cos(4\alpha) \cdots \cos(2^n\alpha) = \frac{\sin(2^{n+1}\alpha)}{2^{n+1}\sin(\alpha)} \text{ for } n \geq 0$$

8. Prove that if W is a nonempty subset of Z which has a least element, then every nonempty subset of W has a least element.

9. Prove that if W is a nonempty subset of Z and there is an integer x such that $x \leq w$ for all w in W, then W has a least element.

10. Prove that if W is a nonempty subset of Z and there is an integer y such that $w \leq y$ for all w in W, then W has a greatest element (see the Exercises of Section 1.2).

11. The method of "infinite descent" is sometimes used to prove that no positive integer can have a particular property. The method works as

follows. Suppose that there is a positive integer, say a_1, having the property. Then generate a second positive integer, say, a_2, also having the property but such that $a_2 < a_1$. Similarly, generate a positive integer a_3 having the property and such that $a_3 < a_2$, etc. Thus we have an "infinitely descending" sequence $a_1 > a_2 > a_3 > \cdots$ of positive integers. Evidently, no such sequence can exist. This sequence was generated by assuming that there was an integer having the given property; therefore, the assumption is false. Prove the following statement, which justifies the argument above: Assume that:

(1) T is a subset of N.

(2) If $n \in T$, then there is an $m \in N$ such that $m < n$ and $m \in T$.

Then $T = \varnothing$.

12. Use the method of infinite descent to prove that there are no positive integers p and q such that $p^2 = 3q^2$ by finding a positive integer $p' < p$ such that $(p')^2 = 3(q')^2$ for some positive integer q'. You may assume it is known that if a and b are positive integers such that $b^2 = 3 \cdot a$, then there is a positive integer k with $b = 3 \cdot k$. Although we are discussing only integers at this point, it is perhaps interesting to note that the lack of positive integers p and q such that $p^2 = 3q^2$ implies that 3 is irrational. The reals and rationals will be discussed later.

13. Prove the following statement related to the method of infinite descent: Let T be a nonempty set of nonnegative integers having the property that if $n \in T$ and $n > 0$, then there is a nonnegative integer $m \in T$ such that $0 \le m < n$. Then $0 \in T$.

14. For integers a and b, prove that

$$a^n - b^n = (a - b)(a^{n-1} + a^{n-2}b + a^{n-3}b^2 + \cdots + ab^{n-2} + b^{n-1})$$

for every integer $n \ge 1$ where the second factor on the right of the equality is defined to be 1 if $n = 1$.

15. Assume that $a_1, a_2, a_3, \ldots$ is a sequence of real numbers. Define sums and products recursively as follows:

$$\sum_{k=1}^{1} a_k = a_1$$

$$\sum_{k=1}^{n+1} a_k = \sum_{k=1}^{n} a_k + a_{n+1} \text{ for } n \ge 1$$

and

$$\prod_{k=1}^{1} a_k = a_1$$

$$\prod_{k=1}^{n+1} a_k = (\prod_{k=1}^{n} a_k) \cdot a_{n+1} \text{ for } n \geq 1$$

Use mathematical induction to prove that finite sums and products are defined for any number of terms or factors.

16. A sequence $a_1, a_2, a_3, \ldots$ is defined recursively as follows:

$$a_1 = 3$$
$$a_k = 5a_{k-1} + 2 \text{ if } k > 1$$

(a) Calculate a_2, a_3, a_4, and a_5.

(b) Based upon your experimental results in part (a), determine a formula for a_n for $n \geq 1$ in terms of n and constants. Use mathematical induction to prove that your formula is correct for every positive integer n.

17. Prove that for integers a and b and each positive integer n,

$$(a+b)^n = \sum_{k=0}^{n} \frac{n!}{k!(n-k)!} a^{n-k} b^k$$

$$= a^n + na^{n-1}b + \frac{n(n-1)}{2} a^{n-2}b^2 + \cdots + nab^{n-1} + b^n$$

This identity is known as the Binomial Theorem since $(a+b)$ is a "binomial," an expression with two terms. The formula $\frac{n!}{k!(n-k)!}$ is called the binomial coefficient and is often denoted by $\binom{n}{k}$.

18. Prove by mathematical induction that

$$\prod_{i=1}^{k}(1-b_i) = 1 - \sum_{j_1} b_{j_1} + \sum_{j_1 < j_2} b_{j_1}b_{j_2} - \cdots + (-1)^k b_1 b_2 \cdots b_k$$

where the summations are such that the sum with r subscripts $j_1, j_2, \ldots, j_r$ is over all such terms with $1 \leq j_s \leq k$ and $j_1 < j_2 < \cdots < j_r$.

1.4   Division

Many integers can be expressed as the product of smaller integers, and important characteristics of and relationships among integers can often be obtained by examining the decomposition. We will investigate the structure of integers in terms of their factors and also what remains after multiples of other integers are subtracted from these integers.

Definition 1.22 *An integer a is a* multiple *of the integer b if $a = bm$ for some integer m. An integer b* divides *an integer a, denoted by $b \mid a$, if a is a multiple of b. An integer b that divides an integer a is also said to be a* factor *of a or to be a* divisor *of a. Further, if $b \neq 0$ and $b \mid a$, then the integer m is denoted by $\dfrac{a}{b}$.*

From the definition, $9 \mid 27$ because $27 = 9 \cdot 3$ but 5 does not divide 12 since there is no integer m such that $12 = 5 \cdot m$. The integers $1, 2, 3, 4, 6$, and 12 are all divisors of 12 and are evidently the only positive divisors of 12. The integers $-1, -2, -3, -4t, -6$, and -12 are also divisors of 12.

Clearly, every integer a is divisible by itself and 1 since $a = 1 \cdot a$. Every integer divides 0, and 0 divides no integer except itself. For some problems in number theory one needs to know whether an integer is a divisor of another, whether a particular integer has any divisors besides itself and 1, and how many distinct positive divisors a given integer has.

Theorem 1.23 *Let a, b, and c be integers. Then:*

(a) *For all $a, a \mid a$.*

(b) *For all a, b, and c, if $a \mid b$ and $b \mid c$, then $a \mid c$.*

(c) *For all a, b, and c, $b \mid a$ and $b \mid c$ if and only if $b \mid (m \cdot a + n \cdot c)$ for all integers m and n.*

Proof. (a) Shown above.

(b) If $a \mid b$ and $b \mid c$, then $b = a \cdot m$ and $c = b \cdot n$. Hence $c = b \cdot n = (a \cdot m) \cdot n = a \cdot (m \cdot n)$ and $a \mid c$.

(c) Assume that $b \mid a$ and $b \mid c$; then $a = b \cdot p$ and $c = b \cdot q$ for some integers p and q. Hence $m \cdot a + n \cdot c = m \cdot (b \cdot p) + n \cdot (b \cdot q) = b \cdot (m \cdot p + n \cdot q)$ and $b \mid (m \cdot a + n \cdot c)$.

Conversely, assume that $b \mid (m \cdot a + n \cdot c)$ for all integers m and n. Then if $m = 1$ and $n = 0$, we have $b \mid a$; and if $m = 0$ and $n = 1$, we have $b \mid c$. ∎

Theorem 1.24 *If a and b are positive integers and $a \mid b$, then $a \leq b$.*

Proof. The proof follows from 1.9(d). ∎

Corollary 1.25 *If, for positive integers a and b, $a \mid b$ and $b \mid a$ then $a = b$.*

Theorem 1.26 *For integers a and b, if $a \mid b$ and $b \mid a$, then $a = b$ or $a = -b$.*

Proof. If $a > 0$ and $b > 0$, Corollary 1.25 gives $a = b$. If $a < 0$ and $b > 0$, then $-a > 0$ and $(-a) \mid b$ and $b \mid (-a)$ so that $b = -a$ or $a = -b$. The case $a > 0$ and $b < 0$ follows by symmetry. The case $a < 0$ and $b < 0$ is left to the reader. ∎

Theorems 1.23,1.24, and 1.26 imply that for positive integers, "divides" is reflexive, antisymmetric, and transitive (see Chapter 0) and hence is a partial ordering of the positive integers. Notice that the integers are not partially ordered by division because divides is not antisymmetric in the integers.

In the following we justify a process which occurs in elementary arithmetic. We can divide a positive integer a into a positive integer b and get a remainder r which is less than a. For example, if we divide 9 into 31, we get a quotient of 3 and a remainder of 4. Hence $31 = 3 \cdot 9 + 4$ where $4 < 9$.

Theorem 1.27 (Division Algorithm) *For positive integers a and b there exist unique nonnegative integers q and r with $0 \le r < b$ such that $a = bq + r$. The integers r and q are, respectively, called the remainder and quotient when a is divided by b.*

Proof. Let a and b be positive integers. Consider the set S of all nonnegative integers of the form $a - bq'$ for q' a nonnegative integer; that is,

$$S = \{x : x = a - bq', x \ge 0, and\, q' \ge 0\}$$

S is nonempty because for $q' = 0$ we have $a - bq' = a - b \cdot 0 = a > 0$ so that $a \in S$. The set S has a least element, say r', since if $0 \in S$ it is the least element, and if not, then S contains only positive integers and hence has a least element by the Well-Ordering Principle, N5. Since r' is in S, let q' be the nonnegative integer such that $r' = a - bq'$.

If $r' \ge b$, then $r'' = r' - b \ge 0$ and

$$
\begin{aligned}
a - bq' &= r' \\
a - bq' - b &= r' - b \\
a - b(q' + 1) &= r''
\end{aligned}
$$

so that r'' is in S. Also, $r'' < r'$ because

$$
\begin{aligned}
b &> 0 \\
-b &< 0 \\
r' - b &< r' + 0 = r'
\end{aligned}
$$

But $r'' < r'$ and $r'' \in S$ contradicts the assumption that r' is the least element of S. Thus $0 \le r' \le b$.

To show uniqueness, assume that $a = bq + r = bp + s$ where $0 \le r \le s < b$. Since $r \ge 0$, $-r \le 0$. Hence, $s - r < b + 0 = b$. But $s - r = b(q - p)$, so by Theorem 1.24, $s - r \ge b$ unless $q - p = 0$. Hence $q - p = 0$, $q = p$, and $r = s$. ∎

If $a < b$, then q will have to be 0 in order for $a = bq + r$ with $0 \le r < b$ and $q \ge 0$. For example, for $a = 4$ and $b = 7$, the Division Algorithm gives $q = 0$ and $r = a = 4$ so $4 = 7 \cdot 0 + 4$.

By the uniqueness of q and r, if we can obtain q and r by any means whatsoever with $a = bq + r$, $0 \le r < b$ and $q \ge 0$, then this q and r must be those guaranteed by the theorem.

A positive integer always has at least two positive divisors — itself and 1, although in the case of the integer 1, both divisors are the same. Since any positive divisor of a positive integer can be no greater than the integer, we can always find every positive divisor of an integer n simply by testing every integer $k, 1 \leq k \leq n$, to see whether it divides n. Thus the positive divisors of 12 were found to be $1, 2, 3, 4, 6,$ and 12. Similarly, the positive divisors of 90 may be shown to be $1, 2, 3, 5, 6, 9, 10, 15, 18, 30, 45,$ and 90.

By inspection, $1, 2, 3,$ and 6 are divisors of both 12 and 90. $1, 2, 3,$ and 6 are called common divisors of 12 and 90. Further, 6 is the largest or greatest of these common divisors. Notice also that all of the common divisors, $1, 2, 3,$ and 6, happen to divide the greatest common divisor 6. In the context of greatest common divisors we consider only positive divisors.

Definition 1.28 *A positive integer d is the* greatest common divisor *of integers a and b provided that (i) $d \mid a$ and $d \mid b$ and (ii) if $c \mid a$ and $c \mid b$, then $c \mid d$. The greatest common divisor of a and b will be denoted by $\gcd(a, b)$.*

Theorem 1.29 *If d and c are greatest common divisors of integers a and b, then $c = d$; that is, there is at most one greatest common divisor.*

Proof. By definition of greatest common divisor, $d \mid c$ and $c \mid d$. Hence $c = d$ or $c = -d$. Since c and d are both positive, $c = d$. ∎

Theorem 1.30 *The greatest common divisor of positive integers a and b exists. Further, the greatest common divisor can be written in the form*

$$u \cdot a + v \cdot b$$

for some integers u and v.

Proof. Let S be the set of all positive integers of the form $na + mb$. Let $d = ua + vb$ be the least element of S. Then $d \leq a$ since a is in S. Therefore, $a = qd + r$ for some q and r where $0 \leq r < d$. So $a = q(ua + vb) + r$. Solving for r, we have $r = (1 - qu)a + (-v)b$, so r is in S or $r = 0$. But r is less than d, which is the least element of S so that $r = 0$. Therefore, $d \mid a$. Similarly, $d \mid b$. If c is any divisor of both a and b, then since $d = ua + vb$, by Theorem 1.23(c), $c \mid d$. Hence d is the greatest common divisor of a and b. ∎

The proof of the following theorem is left to the reader.

Theorem 1.31 *For integers a and b which are not both zero, the greatest common divisor exists and can be written in the form $ua + vb$ for some integers u and v. In particular, if $\gcd(a, b) = 1$, then there are integers u and v such that $ua + vb = 1$.*

Although we have defined the greatest common divisor of two integers and have proved that it exists, we have not shown how to find it by other than exhaustive means. If we know all of the factors of the two integers, we can compare them and select the largest common factor. Fortunately, there is an easier method for finding the greatest common divisor of large numbers. This process is called the Euclidean Algorithm. An algorithm is simply a step-by-step method for performing a process. Before we can prove the Euclidean Algorithm and show how it works, we need the following lemma.

Lemma 1.32 *If $a = bq + c$, then $\gcd(a, b) = \gcd(b, c)$; or every divisor of a and b is a divisor of b and c, and conversely.*

Proof. Let $e = \gcd(a, b)$ and $f = \gcd(b, c)$. Since $e \mid a$ and $e \mid b$, then $e \mid c$ by Theorem 1.23 because $c = a - bq$. By definition of greatest common divisor, $e \mid f$.

Conversely, if $f \mid b$ and $f \mid c$, then $f \mid a$ by Theorem 1.23. By definition of greatest common divisor, $f \mid e$. Hence $e = f$. ∎

Theorem 1.33 (Euclidean Algorithm) *Assume that a and b are positive integers and the Division Algorithm is applied repeatedly, giving the following sequence:*

$$
\begin{array}{llll}
a & = & bq_0 + r_0 & \quad 0 \leq r_0 < b \\
b & = & r_0 q_1 + r_1 & \quad 0 \leq r_1 < r_0 \\
r_0 & = & r_1 q_2 + r_2 & \quad 0 \leq r_2 < r_1 \\
r_1 & = & r_2 q_3 + r_3 & \quad 0 \leq r_3 < r_2 \\
r_2 & = & r_3 q_4 + r_4 & \quad 0 \leq r_4 < r_3 \\
& \vdots & & \qquad \vdots \\
r_k & = & r_{k+1} q_{k+2} + r_{k+2} & \quad 0 \leq r_{k+2} < r_{k+1} \\
& \vdots & & \qquad \vdots
\end{array}
$$

There exists an $r_k = 0$. Let s be the first integer such that $r_s = 0$. Then $r_{s-1} = \gcd(a, b)$ if $s > 0$, and $b = \gcd(a, b)$ if $s = 0$.

Proof. First we prove that there exists an $r_k = 0$. Either $r_j > 0$ for all j or there is a k such $r_k = 0$. Assume that $r_j > 0$ for all j. Let $S = \{r_0, r_1, \ldots\}$. By the Well-Ordering Principle, S contains a least positive integer, say r_i. But $r_{i+1} < r_i$, which contradicts the assumption that r_i is the least element of S. Thus, there exists an integer k with $r_k = 0$. If $r_0 = 0$, let $s = 0$; otherwise, let s be the least positive k with $r_k = 0$ (see the Exercises of Section 1.3).

If $s > 0$, then, by Lemma 1.32,

$$\gcd(a, b) = \gcd(b, r_0) = \gcd(r_0, r_1) = \cdots = \gcd(r_{s-1}, r_s)$$

But since $r_s = 0$, $\gcd(r_{s-1}, r_s) = r_{s-1}$ so that $\gcd(a, b) = r_{s-1}$. If $s = 0$, then

$$a = bq_0 + r_0 = bq_0 + 0 = bq_0$$

showing that b is a factor of a and $\gcd(a, b) = b$. ∎

Example 1.34 *Using the Euclidean Algorithm to find* $\gcd(203, 91)$, *we proceed as follows: First divide* 91 *into* 203 *to obtain*

$$
\begin{aligned}
203 &= 91 \cdot 2 + 21 \\
a &= b \cdot q_0 + r_0
\end{aligned}
$$

We now take the remainder 21 *and divide it into the quotient* 91.

$$
\begin{aligned}
91 &= 21 \cdot 4 + 7 \\
b &= r_0 \cdot q_1 + r_1
\end{aligned}
$$

Now dividing 7 *into* 21, *we have*

$$
\begin{aligned}
21 &= 7 \cdot 3 + 0 \\
r_0 &= r_1 \cdot q_2 + r_2
\end{aligned}
$$

So $s = 2$ *and* $\gcd(203, 91) = \gcd(91, 21) = \gcd(21, 7) = 7 = r_{s-1} = r_1$.

Example 1.35 *Find* $\gcd(99, 205)$. *Proceeding as before, we divide* 99 *into* 205, *giving*

$$205 = 99 \cdot 2 + 7$$

We now divide the remainder 7 *into* 99, *giving*

$$99 = 7 \cdot 14 + 1$$

Next divide the remainder 1 *into* 7 *to get*

$$7 = 1 \cdot 7 + 0$$

Thus $\gcd(99, 205) = 1$.

Definition 1.36 *If the greatest common divisor of two integers* a *and* b *is* 1, *that is,* $\gcd(a, b) = 1$, *then* a *and* b *are said to be* relatively prime *or to be* coprime.

Example 1.35 has established that 99 and 205 are relatively prime since $\gcd(99, 205) = 1$. We know that the greatest common divisor of integers a and b can be expressed in the form $au + bv$ for integers u and v. Using the Euclidean Algorithm we can find the values of u and v. Using the notation of Theorem 1.33, $\gcd(a, b) = r_t$, where $r_{t-2} = r_{t-1} \cdot q_t + r_t$. Hence $r_t = r_{t-2} - r_{t-1} \cdot q_t$. Similarly, $r_{t-1} = r_{t-3} - r_{t-2} \cdot q_{t-1}$. Substituting in the previous equation, we have

$$r_t = r_{t-2} - (r_{t-3} - r_{t-2} \cdot q_{t-1}) \cdot q_t$$

Similarly, we can solve for r_{t-2} in terms of r_{t-3} and r_{t-2}, and substitute it in the preceding equation to eliminate r_{t-2}. Continuing in this manner, we eventually eliminate the r_i until we get back to a and b, and we have $\gcd(a, b)$ in the form $ua + vb$.

Example 1.37 *Express* $\gcd(85, 34)$ *in the form* $85u + 34v$. *First using the Euclidean Algorithm, we divide 34 into 85:*

$$85 \;=\; 34 \cdot 2 + 17$$

Dividing 17 into 34, we get

$$34 \;=\; 17 \cdot 1 + 0$$

Thus, $\gcd(85, 34) = 17$ *and* $\gcd(85, 34) = 17 = (85)(1) + (34)(-2)$.

Example 1.38 *Express* $\gcd(252, 580)$ *in the form* $252u + 580v$. *Divide 252 into 580, obtaining*

$$\begin{aligned} 580 &\;=\; 252 \cdot 2 + 76 \\ a &\;=\; b \cdot q_0 + r_0 \end{aligned}$$

Now divide 76 into 252:

$$\begin{aligned} 252 &\;=\; 76 \cdot 3 + 24 \\ b &\;=\; r_0 \cdot q_1 + r_1 \end{aligned}$$

Continuing, we get

$$\begin{aligned} 76 &\;=\; 24 \cdot 3 + 4 \\ r_0 &\;=\; r_1 \cdot q_2 + r_2 \end{aligned}$$

and

$$\begin{aligned} 24 &\;=\; 4 \cdot 6 + 0 \\ r_1 &\;=\; r_2 \cdot q_3 + r_3 \end{aligned}$$

Back substituting, we find that

$$\begin{aligned} 4 &\;=\; \underline{76} - \underline{24} \cdot 3 \\ &\;=\; \underline{76} - [\underline{252} - \underline{76} \cdot 3] \cdot 3 \\ &\;=\; \underline{76} \cdot 10 + \underline{252} \cdot (-3) \\ &\;=\; [\underline{580} - \underline{252} \cdot 2] \cdot 10 + \underline{252} \cdot (-3) \\ &\;=\; \underline{580} \cdot 10 + \underline{252} \cdot (-23) \end{aligned}$$

where the r_i *are emphasized with underlining.*

Example 1.39 *Express* $\gcd(252, 576)$ *in the form* $252u + 576v$. *First we divide 252 into 576.*

$$576 = 252 \cdot 2 + 72$$

Then we divide 72 into 252:

$$252 = 72 \cdot 3 + 36$$

Then, back substituting, we have

$$\begin{aligned} \underline{36} &= \underline{252} - \underline{72} \cdot 3 \\ &= \underline{252} - [\underline{576} - \underline{252} \cdot 2] \cdot 3 \\ &= (7)(\underline{252}) + (-3)(\underline{576}) \end{aligned}$$

Fractions may be added by adding their numerators provided that they have the same denominators. To add two fractions such as

$$\frac{7}{60} + \frac{1}{450}$$

the procedure is to multiply both numerator and denominator of $\frac{7}{60}$ by an integer r and both numerator and denominator of $\frac{1}{450}$ by an integer s so that $60r = 450s$, that is, so that the resulting denominators are the same. The easiest way is to use the least common denominator. But the least common denominator of two fractions is the least common multiple of the denominators, which we define next.

Definition 1.40 *A positive integer m is the* least common multiple *of integers a and b provided that (i) $a \mid m$ and $b \mid m$ and (ii) if n is any common multiple of a and b, then $m \mid n$. The least common multiple of a and b will be denoted by* $\mathrm{lcm}(a, b)$.

We need an algorithm to show us how to find the least common multiple of two integers. This goal is accomplished by finding the greatest common divisor of the integers (which we already have an algorithm for) and then using the following theorem.

Theorem 1.41 *For positive integers a and b, $\gcd(a, b) \cdot \mathrm{lcm}(a, b) = ab$.*

Proof. Let $d = \gcd(a, b)$, $e = \mathrm{lcm}(a, b)$, and consider $\dfrac{ab}{\gcd(a, b)} = \dfrac{ab}{d}$. It is divisible by a since $(ab/d)/a = \dfrac{ab}{ad} = \dfrac{b}{d}$ and by definition, $d \mid b$. Similarly, b divides $\dfrac{ab}{d}$. Hence by definition of least common multiple, $\mathrm{lcm}(a, b) \left| \dfrac{ab}{d} \right.$, $\mathrm{lcm}(a, b) \le \dfrac{ab}{d}$, and $\mathrm{lcm}(a, b) \cdot \gcd(a, b) \le ab$. Now consider $\dfrac{ab}{\mathrm{lcm}(a, b)} = \dfrac{ab}{e}$. This integer divides b since $b/(ab/e) = \dfrac{be}{ab} = \dfrac{e}{a}$ and, by definition of least common multiple, we know that $a \mid e$. Similarly, $\dfrac{ab}{e}$ divides a. Hence $\dfrac{ab}{e}$ divides $\gcd(a, b)$, $\dfrac{ab}{\mathrm{lcm}(a, b)} \le \gcd(a, b)$, and $ab \le \mathrm{lcm}(a, b) \cdot \gcd(a, b)$. Therefore, $ab = \mathrm{lcm}(a, b) \cdot \gcd(a, b)$. ■

To find $\mathrm{lcm}(91, 203)$ first find $\gcd(91, 203)$ using the Euclidean Algorithm as was done above and then divide it into the product of 91 and 203. Since $\gcd(91, 203) = 7$, we calculate

$$\mathrm{lcm}(91, 203) = \frac{(91)(203)}{7} = 2639$$

If we divide 91 and 203 by their greatest common divisor 7, obtaining $\dfrac{91}{7} = 13$ and $\dfrac{203}{7} = 29$, it is easy to see that the resulting integers 13 and 29 are relatively prime; that is, $\gcd(13, 29) = 1$. So it appears that if two integers are divided by their greatest common divisor, all common factors except 1 are removed. The proof of the following theorem is left to the reader.

Theorem 1.42 *Given nonzero integers a and b, then $\dfrac{a}{\gcd(a,b)}$ and $\dfrac{b}{\gcd(a,b)}$ are relatively prime; that is,* $\gcd\left(\dfrac{a}{\gcd(a,b)}, \dfrac{b}{\gcd(a,b)}\right) = 1.$

The concepts of greatest common divisor and least common multiple of two integers can easily be extended to apply to more than two integers. The greatest common divisor of integers $a_1, a_2, \ldots, a_n$, denoted by $\gcd(a_1, a_2, \ldots, a_n)$, is a positive integer d such that $d \mid a_i$ for $1 \leq i \leq n$ and if $c \mid a_i$ for $1 \leq i \leq n$, then $c \mid d$. The least common multiple of $a_1, a_2, \ldots, a_n$, denoted by $\text{lcm}(a_1, a_2, \ldots, a_n)$, is a positive integer m such that $a_i \mid m$ for $1 \leq i \leq n$ and if $a_i \mid s$ for $1 \leq i \leq n$, then $m \mid s$. If at least one $a_i \neq 0$, then $\gcd(a_1, a_2, \ldots, a_n)$ will exist; and if $a_i \neq 0$ for all $1 \leq i \leq n$, then $\text{lcm}(a_1, a_2, \ldots, a_n)$ will exist. The following theorem is a generalization of the case of $n = 2$ proved earlier. The proof will be left to the reader.

Theorem 1.43 *If $a_1, a_2, \ldots, a_n$ are integers with $a_j \neq 0$ for some j, then there exist integers $v_1, v_2, \ldots, v_n$ such that*

$$\gcd(a_1, a_2, \ldots, a_n) = a_1 v_1 + a_2 v_2 + \cdots + a_n v_n$$

and $\gcd(a_1, a_2, \ldots, a_n)$ is the least positive integer of this form.

There is a direct relationship between the greatest common divisor as originally defined for two integers and that just defined for n integers. The relationship is given by the next theorem. The proof is left to the reader.

Theorem 1.44 *If $a_1, a_2, \ldots, a_n$ are nonzero integers, then*

$$\gcd(a_1, a_2, \ldots, a_n) = \gcd(\gcd(a_1, a_2, \ldots, a_{n-1}), a_n)$$

When discussing the existence of the greatest common divisor of integers a and b, we found that $\gcd(a, b)$ was a linear combination of a and b so that there exist integers u and v such that $\gcd(a, b) = au + bv$. A more general, but similar, question is: Given integers a, b, and c, can we find integers x and y such that $ax + by = c$? Not all equations of the form $ax + by = c$ have integral solutions.

Theorem 1.45 *The equation $ax + by = c$, where a, b, and c are integers, has an integral solution (i.e., there exists integers x and y such that $ax + by = c$) if and only if c is divisible by $\gcd(a, b)$.*

When c is divisible by $\gcd(a, b)$, a solution of $ax + by = c$ is

$$x_0 = \frac{u \cdot c}{\gcd(a,b)} \qquad y_0 = \frac{v \cdot c}{\gcd(a,b)}$$

where u and v comprise any solution of $\gcd(a,b) = au + bv$.

Proof. We know that there exist integers u and v so that $au + bv = \gcd(a,b)$. If c is divisible by $\gcd(a,b)$, then $c = e \cdot \gcd(a,b)$ for some integer e. Hence $aue + bve = e \cdot \gcd(a,b) = c$ so that $x = u \cdot e$ and $y = v \cdot e$ are solutions to the equation. Conversely, if there exists x and y so that $ax + by = c$, then since $\gcd(a,b)$ divides both a and b, it divides $ax + by$ and hence divides c. ∎

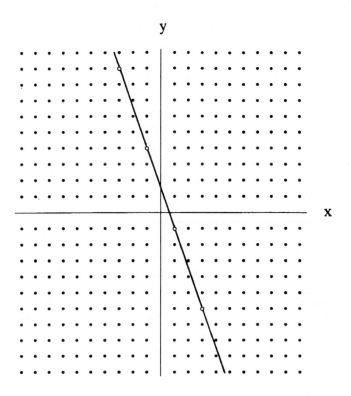

Figure 6: $85x + 34y = 51$.

Example 1.46 *Find a solution for* $85x + 34y = 51$. *In Example 1.37 we showed that* $\gcd(85, 34) = 17$ *and that* $au + bv = (85)(1) + (34)(-2) = 17$. *A solution is given by*

$$x_0 = \frac{u \cdot c}{\gcd(a,b)} = \frac{1 \cdot 51}{17} = 3$$

and

$$y_0 = \frac{v \cdot c}{\gcd(a,b)} = \frac{(-2) \cdot 51}{17} = -6$$

As a check, we calculate

$$ax_0 + by_0 = 85 \cdot 3 + 34 \cdot (-6) = 255 + (-204) = 51$$

Another way to generate the solution is to use the equation $au + bv = \gcd(a, b)$ directly. Since

$$au + bv = \gcd(a, b)$$

or

$$a \cdot (1) + b \cdot (-2) = 17$$

multiplying by 3 we obtain

$$a \cdot (3) + b \cdot (-6) = 51$$

We see there may be more than one solution by observing that

$$85 \cdot 5 + 34 \cdot (-11) = 425 + (-374) = 51$$

when $x = 5$ and $y = -11$.

Example 1.47 *Solve $252x + 580y = 20$. We showed that $\gcd(252, 580) = 4$ in Example 1.38 and also that $(252)(-23) + (580)(10) = 4$. Multiplying each term by 5, we have $(252)(-115) + (580)(50) = 20$. Hence $x = -115$ and $y = 50$ is a solution.*

We are now able to determine whether a solution exists and to find a specific solution for the equation $ax + by = c$ if a solution exists. We would like to be able to find all solutions for the equation. Before we can do so, we need to understand a certain characteristic of the decomposition of integers into products.

Consider the integers 28 and 30. 28 does not divide 30, but for some choices of c, $28 \mid (30c)$ without 28 dividing c. For example, for $c = 14$, $30c = 420$ and $28 \mid 420$ but $28 \nmid 14$. On the other hand, for integers 28 and 75, any choices of c such that $28 \mid (75c)$ implies that 28 must divide c.

Theorem 1.48 *If a, b, and c are integers, $\gcd(a, b) = 1$, and $a \mid bc$, then $a \mid c$.*

Proof. Since $\gcd(a, b) = 1$, there exist integers u and v such that $au + bv = 1$. Multiply each term by c giving $cau + cbv = c$. Then $a \mid cau$ and $a \mid cbv$ since $a \mid bc$. Hence $a \mid c$. ∎

Theorem 1.49 *If a and b are nonzero integers and (x_0, y_0) is a solution of the equation $ax + by = c$, then any other solution (x, y) has the form*

$$x = x_0 + \frac{b}{d}u$$

$$y = y_0 - \frac{a}{d}u$$

where u is an arbitrary integer and $d = \gcd(a, b)$.

Proof. If (x, y) and (x_0, y_0) are both solutions of $ax + by = c$, then $ax + by = ax_0 + by_0$. Therefore, $ax - ax_0 = by_0 - by$ and $a(x - x_0) = b(y_0 - y)$. Dividing both sides by d, we have $\frac{a}{d}(x - x_0) = \frac{b}{d}(y_0 - y)$. Since, by Theorem 1.42, $\gcd\left(\frac{a}{d}, \frac{b}{d}\right) = 1$, one obtains $\frac{a}{d} \mid (y_0 - y)$ and $\frac{b}{d} \mid (x - x_0)$, say $x - x_0 = u\left(\frac{b}{d}\right)$ and $(y_0 - y) = v\left(\frac{a}{d}\right)$. Then $\left(\frac{a}{d}\right) u \left(\frac{b}{d}\right) = \left(\frac{b}{d}\right) v \left(\frac{a}{d}\right)$ and hence $u = v$. Therefore, $x = x_0 + \left(\frac{b}{d}\right) u$ and $y = y_0 - \left(\frac{a}{d}\right) u$. We still need to show that the pair $\left(x_0 + \left(\frac{b}{d}\right) u, \ y_0 - \left(\frac{a}{d}\right) u\right)$ is a solution of $ax + by = c$. But

$$
\begin{aligned}
a\left(x_0 + \left(\frac{b}{d}\right) u\right) + b\left(y_0 - \left(\frac{a}{d}\right) u\right) &= ax_0 + a\left(\frac{b}{d}\right) u + by_0 - b\left(\frac{a}{d}\right) u \\
&= ax_0 + by_0 \\
&= c \qquad \blacksquare
\end{aligned}
$$

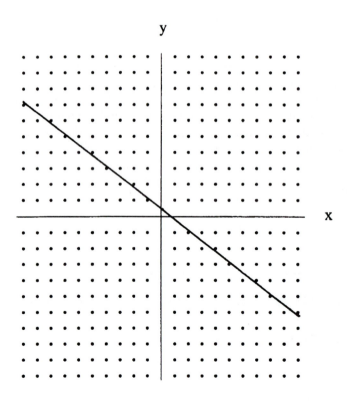

Figure 7: $4x + 6y = 3$.

Returning to Example 1.46 we see that the general solution to the equation $85x + 34y = 51$ is $x = 3 + 2u$ and $y = -6 - 5u$. In Example 1.47 the general solution to the equation $252x+580y = 20$ is $x = -115+145u$ and $y = 50-63u$. Below is a table of some of the solutions of the equation in Example 1.46.

All solutions comprise the set $\{(x,y) : (x,y) \in Z \times Z \text{ and } 85x+34y = 51\}$, which is the collection of all ordered pairs of integer solutions to the equation $85x+34y = 51$. If we sketch a Cartesian coordinate system, then the ordered pairs of $Z \times Z$ are the lattice points depicted in Figure 6. Thus, these lattice points are possible solutions in integers to the equation. To relate this result to the equation $85x+34y = 51$, where x and y are allowed to be real numbers (instead of just integers), we have drawn this line on the graph. We should note that this line intersects a lattice point when we obtain an integral solution to $85x + 34y = 51$. These solutions are circled in Figure 6.

u	$x = 3 + 2u$	$y = -6 - 5u$
-4	-5	14
-3	-3	9
-2	-1	4
-1	1	-1
0	3	-6
1	5	-11
2	7	-16
3	9	-21

Consider the equation $4x + 6y = 3$, $\gcd(4,6) = 2$ and 2 does not divide 3; therefore, $4x + 6y = 3$ has no integral solutions x and y. This result means that if we were to plot the line $4x + 6y = 3$ on a Cartesian coordinate system, as in Figure 7, allowing x and y to be real numbers, then the line would "miss" every lattice point.

Exercises

1. Prove that the relation R on the positive integers defined by aRb if $a \mid b$ is a partial ordering.

2. Use the Division Algorithm to show that 1 divides every integer.

3. Given the Division Algorithm $a = bq + r$, find q and r for the following values of a and b:

    (a) $a = 75, b = 8$

    (b) $a = 102, b = 5$

    (c) $a = 81, b = 9$

    (d) $a = 16, b = 25$

4. Find the greatest common divisor of the following pairs of numbers:

(a) 75, 25

(b) 27, 18

(c) 621, 437

(d) 289, 377

(e) 822, 436

5. Find the least common multiple of each of the pairs of numbers in Exercise 4.

6. Prove that if $a < b$ for $a, b \in Z$, then $a + 1 \le b$.

7. Find the remainder when $2^{38} - 1$ is divided by 31. (Use the formula for the geometric sum $(r - 1)(1 + r + r^2 + r^3 + \cdots + r^n) = r^{n+1} - 1$ in the Exercises of Section 1.2.)

8. Find the remainder when $2^{1101} - 1$ is divided by 31.

9. An integer n is odd if $n = 2 \cdot k + 1$ for some integer k. An integer n is even if $n = 2 \cdot k$ for some integer k. Prove that if n is an integer, then either n is even or n is odd but not both.

10. Prove that the sum of two even integers is even, the sum of two odd integers is even, and the sum of an even and an odd integer is odd.

11. Prove that the product of two even integers is even, the product of two odd integer is odd, and the product of an even integer and an odd integer is even.

12. Prove that if n is an integer, then one and only one of these three statements holds:

(a) $n = 3k$ for some integer k.

(b) $n = 3k + 1$ for some integer k.

(c) $n = 3k + 2$ for some integer k.

13. For a given integer n, prove that n^2 is even if and only n is even.

14. For a given integer n, prove that n^2 is odd if and only n is odd.

15. For a given integer n, prove that $3 \mid n^2$ if and only if $3 \mid n$.

16. Prove that if $a \mid c$, $b \mid c$, and $\gcd(a, b) = 1$, then $ab \mid c$.

17. For integers a and b, prove that $\gcd(a, ab) = a$.

18. For integers a and b and positive integer n, prove that if $\gcd(a, n) = 1$ and $\gcd(b, n) = 1$, then $\gcd(ab, n) = 1$.

19. Prove that if integers a and b are odd, then $2 \mid (a^2 + b^2)$ but $4 \nmid (a^2 + b^2)$.

20. It was shown by mathematical induction in Example 1.5 that $m^3 - m$ is divisible by 3 for every positive integer m. Use Exercise 16 to prove that $m^3 - m$ is divisible by 6 for every positive integer m.

21. For the following pairs of numbers a and b, find u, v, and d so that $au + bv = d$, where d is the greatest common divisor of a and b:

   (a) $83, 17$

   (b) $361, 418$

   (c) $25, 15$

   (d) $81, 9$

   (e) $216, 324$

22. For each of the following equations, find a solution:

   (a) $24x + 81y = 6$

   (b) $803x + 154y = 33$

   (c) $73x + 151y = 3$

   (d) $165x + 418y = 121$

   (e) $27x + 78y = 12$

23. Obtain the general solution for each of the equations in Exercise 22.

24. Prove that if a and b are integers which are not both zero, then the greatest common divisor of a and b exists and can be written in the form $ua + vb$ for appropriate integers u and v (Theorem 1.31).

25. Prove that if a and b are nonzero integers, then $\dfrac{a}{\gcd(a, b)}$ and $\dfrac{b}{\gcd(a, b)}$ are relatively prime (Theorem 1.42).

26. Prove that if $a_1, a_2, \ldots, a_n$ are integers with $a_j \neq 0$ for some j, then there exist integers $v_1, v_2, \ldots, v_n$ such that

$$\gcd(a_1, a_2, \ldots, a_n) = a_1 v_1 + a_2 v_2 + \cdots + a_n v_n$$

and $\gcd(a_1, a_2, \ldots, a_n)$ is the least positive integer of this form (Theorem 1.43).

27. Prove that if $a_1, a_2, \ldots, a_n$ are nonzero integers, then

$$\gcd(a_1, a_2, \ldots, a_n) = \gcd(\gcd(a_1, a_2, \ldots, a_{n-1}), a_n)$$

(Theorem 1.44).

28. In the Exercises of Section 1.3 it was shown by mathematical induction that for any integer $n \geq 8$ there are nonnegative integers k and m such that $n = 3k + 5m$.

(a) Using Theorem 1.49, show that n can be so represented if and only if there is an integer w such that

$$5w \geq 3n \quad \text{and} \quad 2n \geq 3w$$

(b) Use these inequalities to verify that there exist nonnegative integers k and m such that $n = 3k + 5m$ for $n = 3, 5, 6, 8, 9, 10,$ and 11.

(c) Plot the equations $5w = 3n$ and $2n = 3w$ on real axes using w as the vertical axis. Shade the region satisfying both $5w \geq 3n$ and $2n \geq 3w$. Determine what points correspond to solutions for $n = 3k + 5m$ in nonnegative integers. Verify that the values of n in part (b) correspond to points that graphically satisfy both inequalities in part (a). In this context, why can we be certain that there is a solution of $n = 3k + 5m$ for each $n \geq 15$?

29. Given integer a and positive integer b, prove that there are integers whose product is a and whose greatest common divisor is b if and only if $b^2 \mid a$.

30. Assume that
$$F(k) = 2^{2^k} + 1 \text{ for each } k \geq 0$$

(a) Prove that $F(m+1) = F(0)F(1)F(3) \cdots F(m) + 2$ for each $m \geq 0$.

(b) Prove that $F(m)$ and $F(n)$ are relatively prime if $m \neq n$.

31. If $\gcd(a, b)$ is expressed as $ax + by$, prove that x and y are relatively prime.

32. If $\gcd(a, b)$ is expressed as $ax + by$, show that x and y are not unique.

33. If $\gcd(a, b) = 1$, prove that $\gcd(a + nb, b) = 1$.

34. Prove that $n \cdot \gcd(a, b) = \gcd(na, nb)$.

35. Let S be the set all integers of the form $ax + by$. Show that $\gcd(a, b)$ divides all elements of S.

36. Prove this extension of the Division Algorithm of Theorem 1.27. For integers a and b with $b > 0$, there exist unique integers q and r with $0 \leq r < b$ such that $a = bq + r$.

37. Prove this extension of the Division Algorithm of Theorem 1.27. For integers a and b with $b \neq 0$, there exist unique integers q and r with $0 \leq r < |b|$ such that $a = bq + r$. Note that $|b|$ denotes the absolute value of b; that is, $|b| = b$ when $b \geq 0$ and $|b| = -b$ when $b < 0$.

38. Prove this extension of the Euclidean Algorithm of Theorem 1.33. Assume that a and b are integers and $b \neq 0$. If the Extended Division Algorithm (see Exercise 37) is applied repeatedly giving the following sequence:

$$
\begin{aligned}
a &= b \cdot q_0 + r_0 & 0 &\leq r_0 < |b| \\
b &= r_0 \cdot q_1 + r_1 & 0 &\leq r_1 < r_0 \\
r_0 &= r_1 \cdot q_2 + r_2 & 0 &\leq r_2 < r_1 \\
r_1 &= r_2 \cdot q_3 + r_3 & 0 &\leq r_3 < r_2 \\
r_2 &= r_3 \cdot q_4 + r_4 & 0 &\leq r_4 < r_3 \\
&\;\;\vdots & &\;\;\vdots \\
r_k &= r_{k+1} \cdot q_{k+2} + r_{k+2} & 0 &\leq r_{k+2} < r_{k+1} \\
&\;\;\vdots
\end{aligned}
$$

there exists an $r_k = 0$. Let s be the first integer such that $r_s = 0$. Then $r_{s-1} = \gcd(a, b)$ if $s > 0$, and $b = \gcd(a, b)$ if $s = 0$.

39. Suppose that a and b are positive integers. A recursive procedure generating a solution for

$$\gcd(a, b) = au + bv$$

is given as follows. Using the notation of the Euclidean Algorithm 1.33, let s be the least positive nonnegative integer such that $r_s = 0$. If $s = 0$, then $\gcd(a, b) = b$; so we can let $u = 0$ and $v = 1$. If $s \geq 1$, then let

$$Q_k = -q_{s-k} \quad \text{for } k = 1, 2, \ldots, s$$

and define W_k for $k = -1, 0, 1, 2, \ldots, s$ as follows:

$$
\begin{aligned}
W_{-1} &= 0 \\
W_0 &= 1 \\
W_k &= Q_k W_{k-1} + W_{k-2} \quad \text{for } k = 1, 2, \ldots, s
\end{aligned}
$$

Then

$$\gcd(a, b) = aW_{s-1} + bW_s$$

or

$$\gcd(a, b) = au + bv$$

where $u = W_{s-1}$ and $v = W_s$. Prove that this procedure produces $\gcd(a, b) = au + bv$ as described.

1.5   Representation

To make it easier to recognize the integers we use a representation for them that in effect gives names to the integers. This representation "scales" the integers and easily lets us compare them. Although we normally express integers in base 10 form, for example 4352 denotes $4 \cdot 10^3 + 3 \cdot 10^2 + 5 \cdot 10^1 + 2 \cdot 10^0$, we can actually denote an integer using any base as shown by the following theorem.

Theorem 1.50 *Given integers $b > 1$ and $c > 0$, there is a nonnegative integer m such that*

$$c = a_m b^m + a_{m-1} b^{m-1} + \cdots + a_1 b^1 + a_0 b^0$$

with $a_m > 0$ and $0 \leq a_i < b$ for each i. The coefficients $a_0, a_1, \ldots, a_m$ are the remainders obtained by successively dividing c and the resulting quotients by b.

Proof. Using induction on m, if $m = 1$, the Division Algorithm gives $c = q_1 b + r_0$ where $0 \leq r_0 < b$ and we let $q_0 = r_0$. Assume that $m = k$ and for each j, $1 \leq j \leq k$ there are nonnegative integers $q_0, q_1, \ldots, q_j$ and $r_0, r_1, \ldots, r_j$ such that

$$c = q_j b^j + r_{j-1} b^{j-1} + r_{j-2} b^{j-2} + \cdots + r_1 b^1 + r_0 b^0$$

where $q_j < q_{j-1}$ if $q_{j-1} \neq 0$ and $q_j = q_{j-1}$ if $q_{j-1} = 0$ and where $0 \leq r_j < b$. Thus,

$$c = q_k b^k + r_{k-1} b^{k-1} + r_{k-2} b^{k-2} + \cdots + r_1 b^1 + r_0 b^0$$

There exists a q_{k+1} so that $q_k = q_{k+1} b + r_k$ where $0 \leq r_k < b$ and $q_{k+1} < q_k$ if $q_k \neq 0$ and $q_{k+1} = q_k$ if $q_k = 0$. Substituting for q_k gives the desired result for $m + 1 = k + 1$. Hence, for all n, $c = q_n b^n + r_{n-1} b^{n-1} + r_{n-2} b^{n-2} + \cdots + r_1 b^1 + r_0 b^0$.

Let $S = \{q_0, q_1, q_2, \ldots\}$. S contains at least one positive q_i. By the Well-Ordering Principle, N5, there is a least positive integer in S, say q_n. Then $q_{n+1} = 0$ and

$$c = q_n b^n + r_{n-1} b^{n-1} + \cdots + r_1 b^1 + r_0 b^0$$

has the desired properties. ∎

This proof is very useful since it provides an algorithm for calculating the numbers a_i. Given the positive integers c and b with $b > 1$, just divide c by b, obtaining the quotient s_0 and remainder r_0 ($c = s_0 b + r_0$). Next divide this quotient s_0 by b, obtaining the quotient s_1 and remainder r_1 ($s_0 = s_1 b + r_1$). Continue the process of dividing the next quotient by b, obtaining a new quotient and remainder. The process stops when a quotient of zero is obtained. The representation of c in terms of powers of b is just the remainders obtained written from right to left:

$$c = r_n b^n + \cdots + r_2 b^2 + r_1 b^1 + r_0 b^0$$

The number b is called the *base* or *radix* of the representation.

Just as we would write the number $4 \cdot 10^3 + 5 \cdot 10^2 + 6 \cdot 10^1 + 3 \cdot 10^0$ as 4563 in base 10, we write the number $5 \cdot 6^3 + 3 \cdot 6^2 + 4 \cdot 6^1 + 3 \cdot 6^0$ as 5343 in base 6. This base six representation is often denoted 5343_6. Hence Theorem 1.50 assures us that we can write an integer in any base greater than 1. The proof also tells us how to convert an integer c to an arbitrary base b, since the a_i are simply the remainders obtained from dividing the quotients q_{i+1} by the base b where $q_{-1} = c$.

To convert 126 to base 4 we use the Division Algorithm to divide each quotient by 4 until a quotient of zero is obtained:

$$
\begin{aligned}
126 &= 31 \cdot 4 + 2 \\
31 &= 7 \cdot 4 + 3 \\
7 &= 1 \cdot 4 + 3 \\
1 &= 0 \cdot 4 + 1
\end{aligned}
$$

The remainders are

$$
\begin{aligned}
r_0 &= 2 \\
r_1 &= 3 \\
r_2 &= 3 \\
r_3 &= 1
\end{aligned}
$$

so that

$$
\begin{aligned}
126_{ten} &= r_3 4^3 + r_2 4^2 + r_1 4^1 + r_0 \\
&= 1332_4
\end{aligned}
$$

The representation of 126_{ten} in base 4 is thus 1332_4. Repeated substitution can also produce the same result:

$$
\begin{aligned}
126_{ten} &= 31 \cdot 4 + 2 \\
&= (7 \cdot 4 + 3) \cdot 4 + 2 \\
&= 7 \cdot 4^2 + 3 \cdot 4 + 2 \\
&= (1 \cdot 4 + 3) \cdot 4^2 + 3 \cdot 4 + 2 \\
&= 1 \cdot 4^3 + 3 \cdot 4^2 + 3 \cdot 4^1 + 2 \cdot 4^0 \\
&= 1332_4
\end{aligned}
$$

Converting from an arbitrary base to base 10 is also simple because all of the arithmetic can be done in base ten. For example, the number 1332 in base 4 is actually $1 \cdot 4^3 + 3 \cdot 4^2 + 3 \cdot 4 + 2 = 1 \cdot 64 + 3 \cdot 16 + 3 \cdot 4 + 2 = 64 + 48 + 12 + 2 = 126$, where all the numbers are expressed in base ten and all arithmetic is done in base ten.

When converting a number n from base b_1 to base b_2 if the arithmetic is being done in base b_3, convert the number first from base b_1 to base b_3 and then convert the base b_3 number to base b_2. The base b_3 is usually ten since we can do arithmetic computations in base ten easily. Sometimes b_1 and b_2 will be especially related so that the conversion can be effected without converting to base b_3.

For example, using b_1, b_2, and b_3, we convert $n = 4062_7$ to base 5. First we convert n in base 7 to n in base ten using base ten arithmetic.

$$
\begin{aligned}
n = 4062_7 &= [4 \cdot 7^3 + 0 \cdot 7^2 + 6 \cdot 7^1 + 2 \cdot 7^0]_{ten} \\
&= [4 \cdot 343 + 0 \cdot 49 + 6 \cdot 7 + 2 \cdot 1]_{ten} \\
&= [1372 + 0 + 42 + 2]_{ten} \\
&= 1416_{ten}
\end{aligned}
$$

where the brackets and subscript ten indicate that base ten is used for all numbers. Next convert $n = 1416_{ten}$ to base 5 using base ten arithmetic and base ten integer representations.

$$
\begin{array}{rcllcr}
1416 &=& 283 \cdot 5 + 1 & \text{giving } n &=& ----1_5 \\
283 &=& 56 \cdot 5 + 3 & \text{giving } n &=& ---31_5 \\
56 &=& 11 \cdot 5 + 1 & \text{giving } n &=& --131_5 \\
11 &=& 2 \cdot 5 + 1 & \text{giving } n &=& -1131_5 \\
2 &=& 0 \cdot 5 + 2 & \text{giving } n &=& 21131_5
\end{array}
$$

If we had available a base 5 multiplication and addition table, we could convert 4062_7 directly to base 5 starting with the following:

$$
\begin{aligned}
4062_7 &= [4 \cdot seven^3 + 0 \cdot seven^2 + 6 \cdot seven^1 + 2 \cdot seven^0]_7 \\
&= [4(12_5)^3 + 0 \cdot (12_5)^2 + 11_5 \cdot (12_5)^1 + 2 \cdot (12_5)^0]_5
\end{aligned}
$$

where

$$
seven = 12_5 = 1 \cdot (10_5)^1 + 2 \cdot (10_5)^0
$$

$$
six = 11_5 = 1 \cdot (10_5)^1 + 1 \cdot (10_5)^0
$$

The representation of integers in terms of powers of a base b as described by Theorem 1.50 is in fact unique. The next two theorems establish this uniqueness.

Theorem 1.51 *For integers $b > 1$ and $n \geq 0$, and for any set of integers $\{r_0, r_1, \ldots, r_n\}$ with $-b < r_i < b$ for each i, $1 \leq i \leq n$,*

$$
r_n b^n + r_{n-1} b^{n-1} + \cdots + r_1 b^1 + r_0 b^0 = 0
$$

if and only if

$$
r_i = 0 \text{ for each } i, 0 \leq i \leq n
$$

Proof. Using induction on m, if $m = 0$, then $r_0 b^0 = 0$ implies that $r_0 = 0$ and conversely. Assume that $m \geq 0$ and the theorem holds for $0 \leq n \leq m$. Then

$$
r_{m+1} b^{m+1} + r_m b^m + \cdots + r_1 b^1 + r_0 b^0 = 0
$$

implies that

$$
b(r_{m+1} b^m + r_m b^{m-1} + \cdots + r_1) = -r_0 b^0
$$

If $r_{m+1} b^m + r_m b^{m-1} + \cdots + r_1 \neq 0$, then $b \mid r_0$, implying that $r_0 = 0$. This result gives $b(r_{m+1} b^m + r_m b^{m-1} + \cdots + r_1) = 0$, which implies that $r_{m+1} b^m + r_m b^{m-1} + \cdots + r_1 = 0$; therefore, by the induction hypothesis,

$r_i = 0$ for $1 \leq i \leq m+1$. Clearly, $r_{m+1}b^{m+1} + r_m b^m + \cdots + r_1 b^1 + r_0 b^0 = 0$ when $r_i = 0$ for $0 \leq i \leq m+1$. Therefore, by induction, the theorem holds for all n. ∎

Theorem 1.52 *Let $b > 1$ and c be positive integers. The integer m and the sequence $a_m, a_{m-1}, \ldots, a_1, a_0$ given by Theorem 1.50 such that*

$$c = a_m b^m + a_{m-1}b^{m-1} + \cdots + a_1 b^1 + a_0 b^0$$

with $a_m > 0$ and $0 \leq a_i < b$ for all i is unique; that is, there is no other integer n and finite sequence $e_n, e_{n-1}, \ldots, e_1, e_0$ such that $e_n > 0$, $0 \leq e_i < b$ for all i, $0 \leq i \leq n$, and

$$c = e_n b^n + e_{n-1}b^{n-1} + \cdots + e_1 b^1 + e_0 b^0$$

Proof. Let $m, a_m, a_{m-1}, \ldots, a_1, a_0$ and $n, e_n, e_{n-1}, \ldots, e_1, e_0$ be two sets of integers as described in the theorem so that

$$c = a_m b^m + a_{m-1}b^{m-1} + \cdots + a_1 b^1 + a_0 b^0$$

and

$$c = e_n b^n + e_{n-1}b^{n-1} + \cdots + e_1 b^1 + e_0 b^0$$

We need to prove that $n = m$ and $a_i = e_i$ for all i.

If $m \geq n$, then the two equations above give

$$a_m b^m + a_{m-1}b^{m-1} + \cdots + a_{n+1}b^{n+1} + (a_n - e_n)b^n + \cdots + (a_0 - e_0)b^0 = 0$$

But $0 \leq a_i < b$ and $0 \leq e_i < b$ imply that $-b < a_i - e_i < b$. So letting $r_i = a_i - e_i$ for $0 \leq i \leq n$ and $r_i = a_i$ for $n+1 \leq i \leq m$ gives $r_m = 0$ using Theorem 1.51, so $n = m$ and $a_i = e_i$ for all i. ∎

If $c > 0$, then we have shown that it has a unique representation in terms of powers of $b > 1$ and nonnegative coefficients. If $d < 0$, then $c = -d > 0$ and c is uniquely given by

$$c = r_n b^n + r_{n-1}b^{n-1} + \cdots + r_1 b^1 + r_0 b^0$$

with $b > r_i \geq 0$, $0 \leq i \leq n$, and $r_n > 0$. So

$$d = -c = -\left[r_n b^n + r_{n-1}b^{n-1} + \cdots + r_1 b^1 + r_0 b^0 \right]$$

which is called the signed base b representation of d. Also,

$$
\begin{aligned}
d = -c &= (-r_n)b^n + (-r_{n-1})b^{n-1} + \cdots + (-r_1)b^1 + (-r_0)b^0] \\
&= s_n b^n + s_{n-1}b^{n-1} + \cdots + s_0 b^0
\end{aligned}
$$

where $-b < s_i \leq 0$, $0 \leq i \leq n$, and $s_n < 0$. This representation is unique for $d < 0$.

Combining the above, we have the following theorem:

Theorem 1.53 *For integers $b > 1$ and $c \neq 0$, if when $c < 0$, then $t_i \leq 0$ for all i and when $c > 0$, then $t_i \geq 0$ for all i, then there are unique integers n and t_i, $0 \leq i \leq n$, such that*

$$c = t_n b^n + t_{n-1} b^{n-1} + \cdots + t_0 b^0$$

Theorem 1.53 does not preclude different representations for a given integer in terms of powers of b without the restriction imposed by the theorem, as the next example shows.

$$\begin{aligned} 83 &= 3 \cdot 5^2 + 1 \cdot 5^1 + 3 \cdot 5^0 \\ &= 4 \cdot 5^2 - 4 \cdot 5^1 + 3 \cdot 5^0 \end{aligned}$$

This redundancy is not a violation of the theorem because, in one representation, some of the coefficients are positive and some are negative.

Since the unique representation of a negative integer d is just the representation of the positive integer $-d$ with the coefficients multiplied by -1, we usually just use the form $-(-d)$. For example,

$$\begin{aligned} -437_{ten} &= -[4 \cdot 10^2 + 3 \cdot 10^1 + 7 \cdot 10^0]_{ten} \\ &= [(-4)10^2 + (-3)10^1 + (-7)10^0]_{ten} \end{aligned}$$

However, for most applications we generally prefer to use the notation -437 and do not bother to distribute the (-1) factor among the terms of 437.

As a curiosity of representation, it can easily be shown in base ten that

$$12345679 \cdot 9 = 111111111$$

and that

$$12345679 \cdot 18 = 222222222$$

and so on. This pattern generalizes to other bases. For example, in base 5 we have $124 \cdot 4 = 1111$; and, in base 7 we have $12346 \cdot 6 = 111111$. For details of this curiosity, see the Exercises and "A Generalization of a Mathematical Curiosity" by James A. Anderson and Edwin F. Wilde [3].

Although no one knows when the concept of numbers and how to name them began, obviously there was a need to "count" things; that is, a set of items were put into one-to-one correspondence with the numbers (or fingers). Hence, man was using cardinals, a concept probably not clearly understood at that time. As pointed out by Bertrand Russell, "it must have required many years to discover that a brace of pheasants and a couple of days were both instances of the number two." During the period when man was a hunter, he probably did not have to count very high unless, of course, he was quite an exceptional hunter. Even when mankind became more advanced it is said that a man could detect that a sheep from the flock was missing without knowing the number in the flock. South African bushmen appear to have had only the numbers 1, 2, and many (which may say something about the hunting in that area). Early Australian aborigines could discern numbers up to 7.

It is impossible to say exactly when the first written "numbers" occurred. There is evidence of markings representing numbers in the stone age. In a much later period, the Chinese, Egyptians, Greeks, and Romans all had written numerals.

With regard to integer representations, most nations use the base 10 or decimal system, probably because humans have 10 fingers. Others have used the base 5 or quintary system, apparently using one hand to count the fingers on the other hand. A few peoples, including the Malayans, Aztecs, ancient Celtics, and Danes, apparently using both fingers and toes, used the base 20 or vigesimal system. There are also traces of its use in the English system, as evidenced by the word "score" for 20. The Babylonians used the base 60 or sexagesimal system. The sexagesimal system has an advantage over the decimal system in that 60 items may be exactly divided into 2, 3, 4, 5, 6, 10, 12, 15, 20, 30, and 60 equal parts. Portions of Babylonian sexagesimal multiplication tables for all products up to 20×20, the pairwise products of 30, 40, and 50, and the products of 30, 40, and 50 with 1 through 20, have been found. Sexagesimal representation and arithmetic were used extensively for astronomical calculations. At the other extreme, some primitive aborigine tribes in Australia developed a binary system. With the advent of computers, base 2, 8, and 16 representations have become important.

Instead of relying upon a written number, finger counting was a common way of communicating numbers. This was achieved by placing the fingers in certain positions as shown in Figure 8, which is rendered from a plate from a fifteenth century work, *Summa de arithmetica geometria*, of Luca di Pacioli. Finger counting also depended on the hand used. For example, the numbers 3 and 300 are identical except that opposite hands are used. The Arabs reversed the hands. It was apparently possible to perform operations such as multiplication and addition with finger counting.

Finger numbers were used by Greeks, Romans, Hindus, and Arabs during the classical period and Middle Ages. It was apparently considered to be quite an intellectual accomplishment to be able to finger count. The Italian mathematician Leonardo Fibonacci (Leonardo of Pisa) (c. 1170-1230) was a strong advocate of being able to finger count even though he is credited with popularizing the use of Arabic numerals through his book *Liber abaci*. Much of the information available on early finger counting was due to the works of the Venerable Bede (673-735).

The primary breakthrough for the written number was the use of the positional notation, which means that the value of a number symbol depended on its position. For example, today in base ten the 1 in the number 123 represents 100 because of its position. The first people to use positional numerical systems were the Babylonians with a sexagesimal representation written in cuneiform. The Mayans also developed a positional numerical system. It is believed that the Babylonians had a place holder to keep numbers in position but not a zero as such. The Mayans used a complete positional numerical system and apparently had a zero. The Hindus had a zero in the late seventh century, but the true origin of the zero is unknown. The ancient Egyptians did not use a positional representation of integers. Instead, the Egyptians used a repetitive principle which entailed the repetition of a symbol as many

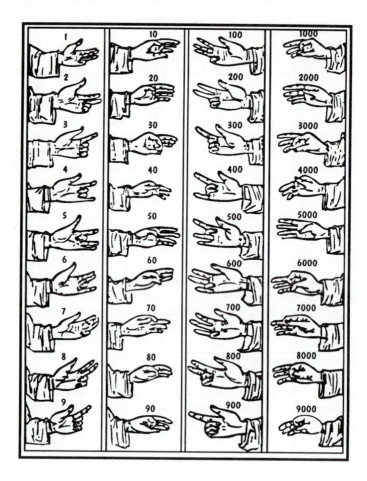

Figure 8: Finger counting.

times as it would take to represent the number of desired units. An example of the repetitive principle occurs with Roman numerals where, for example, XXX means three tens, or thirty. The lack of a positional representation for integers may have hindered the development of mathematics in ancient Egypt. Our system today is the Hindu-Arabic system, which was created by the Hindus and transmitted to us by the Arabs.

Exercises

1. Convert 3475_{ten} to base five, two, three, and eight.

2. Convert 6105_{eight} to base two, ten, and five.

3. Make a table of the first 50 positive integers in bases ten, three, five, and

eight. In other words, count to fifty in bases ten, three, five, and eight.

4. Construct multiplication and addition tables for base five arithmetic. Use base five representations throughout. Compute these numbers using only base five arithmetic:

$$24103_{five} + 24321_{five}$$

$$342_{five} \cdot 103_{five}$$

$$\frac{342_{five}}{13_{five}}$$

5. The usual digits in base 10 are $0, 1, 2, \ldots, 9$. If the base is greater than ten, then additional symbols need to be used. Often, letters are used as "digits" for some bases greater than ten. Thus the symbols

$$0, 1, 2, 3, 4, 5, 6, 7, 8, 9, A, B, C, D, E, F, G, H, \ldots, Z$$

will allow us to express integers in bases up to 36. Base sixteen (hexadecimal) uses the symbols $0, 1, 2, 3, 4, 5, 6, 7, 8, 9, A, B, C, D, E, F$, where A is ten, B is eleven, $\ldots$, and F is fifteen, so that, for example,

$$
\begin{aligned}
A3C_{hex} &= A(10_{hex})^2 + 3(10_{hex})^1 + C(10_{hex})^0 \\
&= 10_{ten}(16_{ten})^2 + 3(16_{ten})^1 + C \\
&= 2560_{ten} + 48_{ten} + 12_{ten} \\
&= 2620_{ten}
\end{aligned}
$$

(a) Convert the following base ten representations to hexadecimal representations:
$$256_{ten}, \ 32767_{ten}, \text{ and } 7203_{ten}$$

(b) Convert the following hexadecimal representations to decimal representations:
$$7A6F_{hex}, \ FF_{hex}, \text{ and } ABCD_{hex}$$

6. In Section 1.5 it is mentioned that in base ten $(12345679)(9) = 111111111$. Explain why we know then that $(12345679)(18) = 222222222$, $(12345679)(27) = 333333333$, and so on.

7. In base 5, we have $(124)(4) = 1111$ and in base 7 we have $(12346)(6) = 111111$. Make a statement which generalizes this process for base n.

1.6   Congruence

Sets of integers can be selected that are infinite in number but all have the same property. For instance, the set $A = \{\ldots, -6, -3, 0, 3, 6, 9, 12, \ldots\}$ is the set of all integers that are exactly divisible by 3. On the other hand, the difference of any two members of A is also divisible by 3:

$$
\begin{aligned}
(-3) - 6 &= -9 &&= 3 \cdot k &&\text{for } k = -3 \\
12 - (-3) &= 15 &&= 3 \cdot k' &&\text{for } k' = 5
\end{aligned}
$$

It is this second sense that we are interested in at the moment.

If we consider the set $B = \{\ldots, -5, -2, 1, 4, 7, 10, \ldots\}$, we see that none of the members of B is divisible by 3; however, their differences are.

$$
\begin{aligned}
(-5) - 7 &= -12 &&= 3k &&\text{for } k = -4 \\
10 - (-5) &= 15 &&= 3k' &&\text{for } k' = 5
\end{aligned}
$$

The same property of differences holds for members of the set

$$
C = \{\ldots, -7, -4, -1, 2, 5, 8, 11, \ldots\}
$$

The sets A, B, and C partition the integers Z since, evidently, A, B, and C are disjoint and $A \cup B \cup C = Z$. The equivalence relation generated by this partition (see Chapter 0) is called "congruence modulo 3." In fact, we can form such partitions for any positive integer n as we did for $n = 3$.

The concept of congruence in this example was introduced by giving the partition first and describing the congruence that results. We will now approach this concept from the opposite direction; namely, given a positive integer n, we define a congruence $\equiv$ first as a relation where $\equiv \in Z \times Z$ and then show that it is an equivalence relation. A different congruence, and, therefore, a different partition of the integers, is associated with each positive integer n.

Definition 1.54 *Let n be a positive integer. The integer a is congruent to the integer b modulo n, denoted $a \equiv b \pmod{n}$, if $n \mid (a - b)$.*

Theorem 1.55 *The relation $\equiv$ for fixed n is an equivalence relation on the set of integers; that is:*

(a) $a \equiv a \pmod{n}$ *for every integer a.*

(b) *If $a \equiv b \pmod{n}$, then $b \equiv a \pmod{n}$ for integers a and b.*

(c) *If $a \equiv b \pmod{n}$ and $b \equiv c \pmod{n}$, then $a \equiv c \pmod{n}$.*

Proof. (a) Since $a - a = 0 \cdot n$, $a \equiv a \pmod{n}$.

(b) If $a \equiv b \pmod{n}$, then $a - b = mn$ for some integer m. Hence $b - a = (-m)n$ and $b \equiv a \pmod{n}$.

(c) If $a \equiv b \pmod{n}$ and $b \equiv c \pmod{n}$, then $a - b = un$ for some integer u and $b - c = vn$ for some integer v. Hence $a - c = (a - b) + (b - c) = un + vn = (u + v)n$ and $a \equiv c \pmod{n}$. ■

The equivalence classes of congruence modulo n partition the integers into disjoint sets. If $n = 3$,

$$
\begin{aligned}
[0] &= \{a : a \equiv 0 \pmod 3\} \\
&= \{a : a - 0 = 3k \text{ for some integer } k\} \\
&= \{a : a = 3k \text{ for some integer } k\} \\
&= \{\ldots, -9, -6, -3, 0, 3, 6, 9, \ldots\}
\end{aligned}
$$

$$
\begin{aligned}
[1] &= \{a : a \equiv 1 \pmod 3\} \\
&= \{a : a - 1 = 3k \text{ for some integer } k\} \\
&= \{a : a = 3k + 1 \text{ for some integer } k\} \\
&= \{\ldots, -5, -2, 1, 4, 7, \ldots\}
\end{aligned}
$$

$$
\begin{aligned}
[2] &= \{a : a \equiv 2 \pmod 3\} \\
&= \{a : a - 2 = 3k \text{ for some integer } k\} \\
&= \{a : a = 3k + 2 \text{ for some integer } k\} \\
&= \{\ldots, -7, -4, -1, 2, 5, 8, 11, \ldots\}
\end{aligned}
$$

These equivalence classes are the sets A, B, and C, which we discussed above and which formed a partition of Z. Even though each equivalence class may be represented in infinitely many ways, as in $[0] = [3] = [-3] = [6] = \ldots$, we will soon see that each equivalence class modulo n can be represented uniquely by an r with $0 \leq r < n$.

Definition 1.56 *Let n be a positive integer. The set of all equivalence classes modulo n is denoted by Z_n and is called the* set of integers modulo n.

For example, for $n = 3$ there are three equivalence classes for congruence modulo 3 so that the set

$$Z_3 = \{[0], [1], [2]\}$$

has three members. The integers modulo n are new objects. We will eventually want to define operations between integers modulo n.

We will find that the congruence relation $\equiv$, as its notational similarity may suggest, has many properties that the equality relation $=$ has. With congruence modulo n there will often be more restrictions than apply with the analogous properties of equality. The next theorem generalizes some of the results of Examples 0.20 and 0.24.

Theorem 1.57 *The following properties of congruence hold.*

(a) *If $a \equiv b \pmod n$ and $c \equiv d \pmod n$, then $a + c \equiv b + d \pmod n$ and $ac = bd \pmod n$.*

(b) *If $ac \equiv bc \pmod n$ and $\gcd(c, n) = 1$, then $a \equiv b \pmod n$.*

(c) *If $a \equiv b \pmod n$, then $a^m \equiv b^m \pmod n$ for all positive integers m.*

(d) *For $c \neq 0$, $ac \equiv bc \pmod{n}$ if and only if $a \equiv b \left(\bmod \dfrac{n}{\gcd(c,n)} \right)$.*

(e) *If $a \equiv b \pmod{mn}$, then $a \equiv b \pmod{m}$ and $a \equiv b \pmod{n}$.*

(f) *If $a \equiv b \pmod{m}$, $a \equiv b \pmod{n}$, and $\gcd(m,n) = 1$, then $a \equiv b \pmod{mn}$.*

Proof. **(a)** By definition,

$$(a+c) - (b+d) = (a-b) + (c-d) = u \cdot n + v \cdot n = (u+v) \cdot n$$

for some integers u and v and $a + c \equiv b + d \pmod{n}$. Also, if $a \equiv b \pmod{n}$ and $c \equiv d \pmod{n}$, then $a - b = un$ and $c - d = vn$ for some integers u and v. Hence $a = b + un$ and $c = d + vn$, so

$$\begin{aligned} ac &= (b + (un))(d + (vn)) \\ &= bd + b(vn) + d(un) + uvn^2 \\ &= bd + (bv + ud + uvn)n \end{aligned}$$

Thus $ac - bd = (bv + ud + vun)n$ and $ac \equiv bd \pmod{n}$.

(c) Using induction on m, the theorem is certainly true when $m = 1$. Assume that the congruence is true when $m = k$, i.e. $a^k \equiv b^k \pmod{n}$. Then, since $a \equiv b$, $a \cdot a^k \equiv b \cdot b^k$ and $a^{k+1} \equiv b^{k+1}$ modulo n.

The proofs of parts (b), (d), (e), and (f) are left to the reader. ∎

Part (a) is analogous to Axiom I2 for equality. Part (b) is similar to Axiom I8, the multiplication cancellation property; however, for equality we can "cancel" any nonzero multiplicative factor common to both sides of an equality, but for congruences, we can "cancel" only multiplicative factors that are relatively prime to the modulus, n, of the congruence. Part (c) says that one can take powers of both sides of a congruence and still maintain a congruence. Part (f) is a partial converse of part (e). The requirement that m and n be relatively prime is necessary in part (f); for example, $14 \equiv 2 \pmod{6}$ and $14 \equiv 2 \pmod{4}$, but $14 \not\equiv 2 \pmod{24}$.

Example 1.58 *Let $n = 3$.*

$$\begin{aligned} 1 &\equiv 7 \pmod{3} \\ 2 &\equiv -4 \pmod{3} \end{aligned}$$

Theorem 1.57(a) implies that

$$1 + 2 \equiv 7 + (-4) \pmod{3}$$

Also it implies that

$$(1)(2) \equiv (7)(-4) \pmod{3}$$

which is true since $2 - (-28) = 30 = 3 \cdot 10 = 3 \cdot k$.

To illustrate 1.57(b), we see that

$$18 \equiv 30 \pmod{6}$$

because $18 - 30 = 6(-2)$, *so*

$$(6)(3) \equiv (6)(5) \ (\text{mod } 6)$$

But

$$3 \not\equiv 5 \ (\text{mod } 6)$$

because $3 - 5 = -2$ *is not divisible by* 6.
   On the other hand,

$$35 \equiv 65 \ (\text{mod } 6)$$

because $35 - 65 = -30$ *is divisible by* 6 *so*

$$7 \cdot 5 \equiv 13 \cdot 5 \ (\text{mod } 6)$$

Since $\gcd(5,6) = 1$,

$$7 \cdot 5 \equiv 13 \cdot 5 \ (\text{mod } 6) \ \textit{implies that} \ 7 \equiv 13 \ (\text{mod } 6)$$

which holds because $7 - 13 = -6 = 6(-1)$.

Theorem 1.59 *These congruence properties hold.*

(a) *If* $r \equiv r' \ (\text{mod } n)$ *and* $0 \leq r, r' < n$, *then* $r = r'$.

(b) *If* a *is any integer and* n *is a positive integer, then there is an* $r, \ 0 \leq r < n$, *such that* $a \equiv r \ (\text{mod } n)$. *The integer* r *is the remainder when* a *is divided by* n *($a = nq + r$).*

Proof. (a) Assume that $0 \leq r' \leq r < n$ so that $r - r' \geq 0$ and suppose that $r \equiv r' \ (\text{mod } n)$. By definition, $r - r' = nm$ for some integer m and $n \mid (r - r')$. But since $r - r' > 0$, $n \mid (r - r')$ implies that $n \leq r - r'$. Since $r < n$ and $-r' \leq 0$, we have $r - r' = r + (-r') < n$, which contradicts the assumption that $n \leq r - r'$. Hence $r = r'$.

   (b) The proof is left to the reader. ∎
   Part (a) guarantees that the equivalence classes modulo n have the property that those equivalence classes generated by the nonnegative integers less than n are distinct. Thus, $[0], [1], [2], \ldots, [n-2], [n-1]$ are different sets relative to the congruence modulo n.
   Part (b) shows that every integer a is congruent modulo n to one of the integers $0, 1, 2, \ldots, n-1$. So any integer a is in one of the n distinct equivalence classes $[0], [1], \ldots, [n-1]$. This discussion justifies the next theorem.

Theorem 1.60 *For a positive integer* n, Z_n, *the set of integers modulo* n, *consists of exactly the* n *distinct equivalence classes* $[0], [1], [2], \ldots, [n-1]$ *represented by the distinct remainders that can be obtained with division by* n. *Further, for* $0 \leq r < n$, *the equivalence class* $[r]$ *consists of exactly those integers* a *such that* $a \equiv r \ (\text{mod } n)$.

   The next theorem gives another property of integers that are equivalent modulo n.

Theorem 1.61 *If $a = nq + r$ and $b = nq' + r'$ for $0 \leq r, r' < n$, then $r = r'$ if and only if $a \equiv b \pmod{n}$.*

Proof. If $r = r'$, then $a - b = nq + r - (nq' + r') = nq - nq' + r - r' = n(q - q')$ and $a \equiv b \pmod{n}$. Conversely, if $a \equiv b \pmod{n}$, then $a - b = pn$ for some integer p. Hence $pn = n(q - q') + r - r'$ and $r - r' = (p - q + q')n$. Thus, we have $r \equiv r' \pmod{n}$. Since $0 \leq r, r' < n$, by Theorem 1.59, $r = r'$. ■

Example 1.62 *Since there are only five different nonnegative integers less than 5, then the equivalence classes modulo 5 form the set*

$$Z_5 \;\; = \;\; \{[0], [1], [2], [3], [4]\}$$

$$= \;\; \{[r] : 0 \leq r < 5\}$$

Specifically, $[2]$ is the set of all integers that have a remainder of 2 when divided by 5. So

$$[2] = \{\ldots, -13, -8, -3, 2, 7, 12, \ldots\}$$

For the set Z_n we can define the operations of addition and multiplication. If $[a]$ is the equivalence class containing a in the congruence modulo n and $[b]$ is the modulo n equivalence class contain b, we define addition and multiplication by

$$[a] \oplus [b] = [a + b]$$

$$[a] \odot [b] = [a \cdot b]$$

where the addition and multiplication on the right-hand side are between integers and the addition and multiplication on the left-hand side are between equivalence classes.

Using the preceding theorems we shall show that addition and multiplication of equivalence classes are well-defined; that is, the definitions are independent of the representatives of the equivalence classes. (See Chapter 0 for a detailed discussion.) For example, if $[a] = [c]$, then the result of addition must be the same whether we use $[a]$ or $[c]$. That is, if $[a] = [c]$ and $[b] = [d]$, then $[a] \oplus [b] = [c] \oplus [d]$ and $[a] \odot [b] = [c] \odot [d]$.

By definition $[a] \oplus [b] = [a + b]$ and $[c] \oplus [d] = [c + d]$.

$$[a] = [c] \quad \text{implies that} \quad a \equiv c \pmod{n}$$

$$[b] = [d] \quad \text{implies that} \quad b \equiv d \pmod{n}$$

Adding, we get

$$a + b \equiv c + d \pmod{n}$$

which implies that

$$[a + b] = [c + d]$$

and hence

$$[a] \oplus [b] = [c] \oplus [d]$$

Similarly, one can show that

$$[a] \odot [b] = [c] \odot [d]$$

Example 1.63 *For* $n = 5$, *and* $Z_5 = \{[0], [1], [2], [3], [4]\}$ *we see that*

$$[2] \oplus [4] = [2 + 4] = [6] = [1] \quad since \quad 6 \equiv 1 \pmod 5$$

$$[2] \odot [4] = [2 \cdot 4] = [8] = [3] \quad since \quad 8 \equiv 3 \pmod 5$$

By calculating all possible sums and products, we can produce "addition" and "multiplication" tables for the integers modulo 5.

$[a] \oplus [b]$	$[0]$	$[1]$	$[2]$	$[3]$	$[4]$
$[0]$	$[0]$	$[1]$	$[2]$	$[3]$	$[4]$
$[1]$	$[1]$	$[2]$	$[3]$	$[4]$	$[0]$
$[2]$	$[2]$	$[3]$	$[4]$	$[0]$	$[1]$
$[3]$	$[3]$	$[4]$	$[0]$	$[1]$	$[2]$
$[4]$	$[4]$	$[0]$	$[1]$	$[2]$	$[3]$

$[a] \odot [b]$	$[0]$	$[1]$	$[2]$	$[3]$	$[4]$
$[0]$	$[0]$	$[0]$	$[0]$	$[0]$	$[0]$
$[1]$	$[0]$	$[1]$	$[2]$	$[3]$	$[4]$
$[2]$	$[0]$	$[2]$	$[4]$	$[1]$	$[3]$
$[3]$	$[0]$	$[3]$	$[1]$	$[4]$	$[2]$
$[4]$	$[0]$	$[4]$	$[3]$	$[2]$	$[1]$

Theorem 1.60 shows that the integers may be partitioned into n disjoint sets or equivalence classes for any positive integer n. We will often focus on these sets as the objects of our attention, but sometimes we will be interested in representative integers taken from the n sets.

Definition 1.64 *For a positive integer* n, *if* $b \equiv r \pmod n$, *then we say that* r *is a* residue *of* b *modulo* n. *A* complete residue system *modulo* n *is a set* $S = \{r_1, r_2, \ldots, r_n\}$ *where the intersection of* S *with each equivalence class modulo* n *contains exactly one integer; that is,* S *contains one and only one representative from each such equivalence class. The complete residue system* $\{0, 1, 2, \ldots, (n-1)\}$ *is called the* primary residue system. *If* b *is an integer,* $b \equiv r \pmod n$, *and* $0 \leq r \leq n-1$, *then we denote this unique primary residue modulo* n *by* $r = [[b]]_n$. *A* reduced residue system *modulo* n *is the subset of a complete residue system consisting of only those integers that are relatively prime to* n; *that is,* $\{r : r \in S \text{ and } \gcd(r, n) = 1\}$.

A complete residue system is obtained by choosing one integer from each equivalence class $[0], [1], \ldots, [n-1]$ of Z_n. So for $n = 6$, $\{24, 7, -58, 40, 113\}$

is a complete residue system modulo 6 because

$$24 \equiv 0 \pmod 6 \qquad \text{giving} \qquad 24 \in [0]$$
$$7 \equiv 1 \pmod 6 \qquad \text{giving} \qquad 7 \in [1]$$
$$-58 \equiv 2 \pmod 6 \qquad \text{giving} \qquad -58 \in [2]$$
$$15 \equiv 3 \pmod 6 \qquad \text{giving} \qquad 15 \in [3]$$
$$40 \equiv 4 \pmod 6 \qquad \text{giving} \qquad 40 \in [4]$$
$$113 \equiv 5 \pmod 6 \qquad \text{giving} \qquad 113 \in [5]$$

Trivially, $\{0, 1, 2, 3, 4, 5\}$ is also a complete residue system modulo 6 and is the primary residue system. By definition $[[24]]_6 = 0$ and $[[-58]]_6 = 2$.

Since all members of an equivalence class modulo n are congruent to one another but to no member of another equivalence class, every integer is congruent to exactly one member of a complete residue system. This argument gives the following theorem.

Theorem 1.65 *If n is a positive integer, $\{r_1, r_2, \ldots, r_n\}$ is a complete residue system modulo n, and a is an integer, then $a \equiv r_k \pmod n$ for one and only one k, $1 \le k \le n$.*

The members of the complete residue system $\{24, 7, -58, 15, 40, 113\}$ and the primary (complete) residue system $\{0, 1, 2, 3, 4, 5\}$ modulo 6 that are relatively prime to $n = 6$ comprise the sets $\{7, 113\}$ and $\{1, 5\}$, respectively. Therefore, these are both reduced residue systems modulo 6. We say that $\{1, 5\}$ is the primary reduced residue system modulo 6.

The following theorem shows that a reduced residue system captures much of the essence of integers in Z that are relatively prime to n. The proof is left to the reader.

Theorem 1.66 *Let n be a positive integer and $\{r_1, r_2, \ldots, r_k\}$ be a reduced residue system modulo n. If a is an integer relatively prime to n, then $a \equiv r_j \pmod n$ for one and only one j, $1 \le j \le k$.*

Given one residue system, there are several ways to generate other residue systems. One way is to add the same integer to each residue in a complete residue system. Another is shown in the following theorem. The proof is left to the reader.

Theorem 1.67 *Let n be a positive integer and $\{r_1, r_2, \ldots, r_k\}$ be a complete [reduced] residue system modulo n. If a is an integer relatively prime to n, then $\{ar_1, ar_2, \ldots, ar_k\}$ is also a complete [reduced] residue system.*

Since $\gcd(6, 35) = 1$, given the complete residue system $\{0, 1, 2, 3, 4, 5\}$ and the reduced residue system $\{1, 5\}$ modulo 6, we have that $\{35 \cdot 0, 35 \cdot 1, 35 \cdot 2, 35 \cdot 3, 35 \cdot 4, 35 \cdot 5\}$ and $\{35 \cdot 1, 35 \cdot 5\}$ are complete and reduced residue systems, respectively. But

$$
\begin{aligned}
[[0]]_6 &= 0 \quad \text{and} \quad & [[35 \cdot 0]]_6 &= [[0]]_6 &= 0 \\
[[1]]_6 &= 1 & [[35 \cdot 1]]_6 &= [[35]]_6 &= 5 \\
[[2]]_6 &= 2 & [[35 \cdot 2]]_6 &= [[70]]_6 &= 4 \\
[[3]]_6 &= 3 & [[35 \cdot 3]]_6 &= [[107]]_6 &= 3 \\
[[4]]_6 &= 4 & [[35 \cdot 4]]_6 &= [[140]]_6 &= 2 \\
[[5]]_6 &= 5 & [[35 \cdot 5]]_6 &= [[175]]_6 &= 1
\end{aligned}
$$

Thus, the result of multiplying every member r_i of a complete or reduced residue system modulo n by an integer a relatively prime to n produces representatives of the same equivalence classes but possibly in a different order.

Earlier in this chapter we considered equations of the form $ax + by = c$, which have integers x and y for solutions. As a special case when $b = 0$, we found solutions for $ax = c$. We now seek solutions for the congruence $ax \equiv c \pmod{n}$, meaning that we want an integer x so that the integer ax is congruent to c modulo n. Stated in terms of equivalence classes, if $[a]$ and $[c]$ are equivalence classes modulo n, we want an equivalence class $[x]$ so that $[a] \odot [x] = [c]$, where the equality refers to equality for sets.

Given the multiplication tables for Z_5 above, we can solve the congruence

$$3x \equiv 1 \pmod 5$$

by considering (modulo 5)

$$[3] \odot [x] = [1]$$

By inspection of the multiplication table, we observe that $[x] = [2]$ will do since $[3] \odot [2] = [1]$. So we could let $x = 2$. Since $[2] = \{\ldots, -8, -3, 2, 7, 12, \ldots\}$, we note that, $x = -3$, $x = 7$, and $x = -8$ are all choices of x satisfying

$$3x \equiv 1 \pmod 5$$

The next two theorems give the conditions under which solutions to $ax \equiv c \pmod n$ exist and specify what the solutions are.

Theorem 1.68 *The congruence $ax \equiv c \pmod m$ has an integer x as a solution if and only if $\gcd(a, m) \mid c$. All integer solutions are given by*

$$x = x_0 + \frac{t \cdot m}{\gcd(a, m)}$$

where t is any integer and for x_0 there is a y_0, so (x_0, y_0) is a solution of $ax + my = c$.

Proof. By definition, $ax \equiv c \pmod m$ if and only if $ax - c$ is divisible by m if and only if there is an integer j so that $ax - c = jm$ if and only if

$$ax + my = c$$

has a solution. The integers a, m, and c are fixed and we want integers x and y so that $ax + my = c$. By Theorem 1.45, $ax + my = c$ has an integral solution if and only if $\gcd(a, m) \mid c$. Also, by this theorem, a solution is given by

$$x_0 = \frac{u \cdot c}{\gcd(a, m)}$$

$$y_0 = \frac{v \cdot c}{\gcd(a, m)}$$

where u and v are selected so that $au + mv = \gcd(a, m)$. By Theorem 1.49 all solutions are given by

$$x = x_0 + \frac{t \cdot m}{\gcd(a, m)}$$

$$y = y_0 - \frac{t \cdot m}{\gcd(a, m)}$$

for any integer t. For this application we need only the solutions for x. Thus, all integral solutions of $ax \equiv c \pmod{m}$ are of the form

$$x = x_0 + \frac{t \cdot m}{\gcd(a, m)}$$

where t may be any integer. ■

Theorem 1.68 implies that there are as many solutions to $ax \equiv c \pmod{m}$ as there are integers t. Many of these are equivalent to one another modulo m. The next theorem gives us the distinct solutions of $ax \equiv c \pmod{m}$. In fact, there is a finite set of integers that is the complete set of distinct solutions; and it is shown that every other solution is congruent modulo m to one in this set.

Theorem 1.69 *If $\gcd(a, m) \mid c$, then $ax \equiv c \pmod{m}$ has finitely many distinct solutions modulo m. These solutions are given by*

$$x_0 + \frac{t \cdot m}{\gcd(a, m)} \quad modulo\ m$$

for $t = 1, 2, 3, \ldots, \gcd(a, m)$, where for x_0 there is a y_0 so that (x_0, y_0) is a solution of $ax + my = c$.

Proof.   Let $d = \gcd(a, m)$. Assume that $d \mid c$ and (x_0, y_0) is a solution of $ax + my = c$. Recall that all integral solutions of $ax \equiv c \pmod{m}$ are of the form $x = x_0 + \dfrac{t \cdot m}{\gcd(a, m)}$. The proof will be in two parts:

(1) If $1 \leq t_1 < t_2 \leq d$, $x_1 = x_0 + \dfrac{t_1 \cdot m}{d}$, and $x_2 = x_0 + \dfrac{t_2 \cdot m}{d}$, then $x_1 \not\equiv x_2 \pmod{m}$.

(2) If t is not in the set $\{1, 2, \ldots, d\}$, then there exists an integer t', $1 \leq t' \leq d$ such that

$$x_0 + \frac{t \cdot m}{d} \equiv x_0 + \frac{t' \cdot m}{d} \pmod{m}$$

Proof of part (1). Assume that $1 \le t_1 < t_2 \le d$ and that x_1 and x_2 are as given above. If $x_1 \equiv x_2 \pmod{m}$, then $x_1 - x_2 = mk$ for some integer k and

$$x_1 - x_2 = \frac{t_1 \cdot m}{d} - \frac{t_2 \cdot m}{d}$$

$$= \frac{m(t_1 - t_2)}{d}$$

Thus,

$$mk = \frac{m(t_1 - t_2)}{d}$$

$$kd = t_1 - t_2$$

So $\gcd(a, m) = d$ divides $t_1 - t_2$, which contradicts $0 < t_1 - t_2 < d$. Thus $x_1 \not\equiv x_2 \pmod{m}$.

Proof of part (2). Let $x = x_0 + \dfrac{t \cdot m}{d}$ and $t \notin \{1, 2, \ldots, d\}$. By Theorem 1.59 there is an integer $t' \in \{1, 2, \ldots, d\}$ that is congruent to t modulo d, that is, $t \equiv t' \pmod{d}$ so that $t - t' = kd$ for some integer k. Note that we are using d instead of 0 in applying Theorem 1.60 since $d \equiv 0 \pmod{d}$.

Let

$$x' = x_0 + \frac{t' \cdot m}{d}$$

We need to show that $x \equiv x' \pmod{m}$.

$$x - x' = \frac{t \cdot m}{d} - \frac{t' \cdot m}{d}$$

$$= \frac{m \cdot (t - t')}{d}$$

$$= \frac{m \cdot k \cdot d}{d}$$

$$= m \cdot k$$

Thus, $x \equiv x' \pmod{m}$. ∎

Example 1.70 *For the congruence*

$$35 \cdot x \equiv 14 \pmod{84}$$

since $\gcd(35, 84) = 7$ *and* $7 \mid 14$, *the equation has exactly 7 distinct solutions modulo 84, which are of the form*

$$x_0 + \frac{84 \cdot t}{7} = x_0 + 12 \cdot t$$

where $t = 1, 2, 3, \ldots, 7$ and (x_0, y_0) is a solution of

$$35x + 84y = 14$$

which is equivalent to

$$5x + 12y = 2$$

By inspection a solution is given by $x_0 = -2$ and $y_0 = 1$. The seven distinct solutions modulo 84 are

t	$x_0 + 12t$
1	$-2 + 12 \cdot 1 = 10$
2	$-2 + 12 \cdot 2 = 22$
3	$-2 + 12 \cdot 3 = 34$
4	$-2 + 12 \cdot 4 = 46$
5	$-2 + 12 \cdot 5 = 58$
6	$-2 + 12 \cdot 6 = 70$
7	$-2 + 12 \cdot 7 = 82$

The distinct solutions could have been obtained by taking t from any complete residue system modulo $\gcd(a, m)$, say, $t \in \{0, 1, 2, 3, 4, 5, 6\}$ because $\gcd(a, m) = 7$.

When $\gcd(a, m) = 1$, there is a unique solution of $ax \equiv c \pmod{m}$. For example, consider

$$6x \equiv 7 \bmod 55)$$

$\gcd(6, 55) = 1$ and certainly 1 divides 7. There will be exactly one solution modulo 55 given by

$$x_0 + \frac{t \cdot m}{\gcd(6, 55)} = x_0 + \frac{1 \cdot 55}{1} = x_0 + 55 \equiv x_0 \pmod{55}$$

where (x_0, y_0) is a solution of $ax + my = c$ or $6x + 55y = 7$. To obtain x_0 and y_0 we begin by backtracking the Euclidean algorithm as shown in the examples following Theorem 1.33 to obtain $6(-9) + 55(1) = 1 = \gcd(6, 55)$. Multiplying each term by 7, we get $6(-63) + 55(7) = 7$, so $x_0 = -63$ and $x = -63 + 55 = -8$.

To check, calculate $6x = 6(-8) = -48$, which is congruent to 7 modulo 55. Note that any integer congruent to -8 modulo 55 can represent this solution, so we could use $-8 + 55 = 47$ instead. The only reason for the change is to represent a solution modulo m using a representative that lies between 0 and $m - 1$, inclusive.

Even though polynomials will be discussed in detail later, it is convenient to be able to use some of their properties now.

Definition 1.71 A polynomial in x of degree $n \geq 0$ with integral coefficients is a function of the form

$$f(x) = a_n x^n + a_{n-1} x^{n-1} + \cdots + a_1 x^1 + a_0$$

where $a_0, a_1, \ldots, a_{n-1}, a_n$ are integers and $a_n \neq 0$. The zero polynomial has no degree.

Examples illustrating the degree of a polynomial are:

$$f_1(x) = 5x^3 + (-3)x + 2 \qquad \text{has degree 3}$$
$$f_2(x) = 8 \qquad \text{has degree 0}$$
$$f_3(x) = 0 \qquad \text{has no degree}$$

Theorem 1.72 If $f(x)$ is a polynomial and $a \equiv b \pmod{m}$, then $f(a) \equiv f(b) \pmod{m}$.

Proof. Let $f(x)$ be a polynomial and $a \equiv b \pmod{m}$. The proof will be by induction on the degree of the polynomial. If $f(x)$ has degree 0, then $f(x) = a_0$, $f(a) = a_0$, and $f(b) = a_0$ so that trivially, $f(a) \equiv f(b) \pmod{m}$. Assume that $k \geq 0$ and for every polynomial $g(x)$ of degree k

$$g(a) \equiv g(b) \pmod{m}$$

Let $f(x)$ have degree $k+1$. $f(x) = x \cdot g(x) + a_0$ where $g(x)$ is a polynomial of degree k and a_0 is an integer. Then, by the induction hypothesis,

$$\begin{aligned} g(a) &\equiv g(b) \pmod{m} \\ ag(a) &\equiv bg(b) \pmod{m} \\ ag(a) + a_0 &\equiv bg(b) + a_0 \pmod{m} \\ f(a) &\equiv f(b) \pmod{m} \end{aligned}$$

Thus $f(a) \equiv f(b) \pmod{m}$ for a polynomial $f(x)$ of any degree $n \geq 0$. The case of the zero polynomial holds trivially. ∎

Example 1.73 *Solve the congruence*

$$623x \equiv -406 \pmod{84}$$

The integer 623 is greater than the modulus 84 and -406 is negative. Since we want solutions modulo 84, we select integers in the range $0, 1, 2, \ldots, 83$ because these are the possible remainders upon division by 84 and are the simplest representatives of the equivalence classes generated by congruence modulo 84. Using the Division Algorithm,

$$\begin{aligned} 623 &= 84 \cdot 7 + 35 \text{ so that } 623 \equiv 35 \pmod{84} \\ -406 &= 84(-5) + 14 \text{ so that } -406 \equiv 14 \pmod{84} \end{aligned}$$

Thus

$$35x \equiv 14 \pmod{84}$$

is a "reduced" version equivalent to the original $623x \equiv -406 \pmod{84}$. We solved the congruence $35x \equiv 14 \pmod{84}$ in a previous example.

The proof of the following theorem is left to the reader:

Theorem 1.74 *If* $a \equiv b \pmod{n_1}$, $a \equiv b \pmod{n_2}$, ..., $a \equiv b \pmod{n_k}$, *and* $n = \mathrm{lcm}(n_1, n_2, \ldots, n_k)$, *then* $a \equiv b \pmod{n}$; *and conversely.*

The next theorem relates solutions of one modulus to those of another modulus. The proof is left to the reader.

Theorem 1.75 *For positive integers n and m and $n = km$, integers $b + j \cdot m$ for $0 \leq j \leq k-1$ are incongruent modulo n; that is, $b + i \cdot m \not\equiv b + j \cdot m \pmod{n}$ for $i \neq j$. Further, if $x \equiv b + j \cdot m \pmod{n}$ for some integer j, $0 \leq j \leq k-1$, then $x \equiv b \pmod{m}$.*

For example, let $n = 12$ and $m = 3$ so that $k = \dfrac{n}{m} = 4$. If $x \equiv 7 \pmod{3}$, then, modulo 12, we have

$$x \equiv 7 + 0 \cdot 3 = 7$$
$$x \equiv 7 + 1 \cdot 3 = 10$$
$$x \equiv 7 + 2 \cdot 3 = 13 \equiv 1$$
$$x \equiv 7 + 3 \cdot 3 = 16 \equiv 4$$

Each of $7, 10, 1$, and 4 is congruent to 7 modulo 3. Thus, each residue congruent to 7 modulo 3 generates $k = 4$ incongruent residues modulo 12. Conversely, there are four incongruent residues modulo 12 that are congruent to the same integer modulo 3.

Exercises

1. For what values of n is $75 \equiv 35 \pmod{n}$?

2. Find all solutions of the following equations:

    (a) $4x \equiv 3 \pmod{7}$

    (b) $27x \equiv 12 \pmod{15}$

    (c) $28x \equiv 56 \pmod{49}$

    (d) $24x \equiv 6 \pmod{81}$

    (e) $91x \equiv 26 \pmod{169}$

3. Simplify the equation $234x \equiv 702 \pmod{169}$ by cancellation.

4. Prove that if a is an odd integer, then $a^2 \equiv 1 \pmod{8}$.

5. Find the least nonnegative residue modulo 17 of the following integers:
    (a) 86   **(b)** 153   **(c)** -98 **(d)** -12 **(e)** 5280

6. Find the least nonnegative residue of $1! + 2! + 3! + \cdots + 99!$ modulo the following integers:

(a) 6   (b) 12   (c) 24   (d) 48   (e) 88

7. List all of the positive integers that are both congruent to 5 modulo 3 and congruent to 5 modulo 2.

8. Prove that if a and n are integers and n is positive, then there is an r, $0 \leq r < n$ such that $a \equiv r \pmod{n}$.

9. Prove that if $a \equiv b \pmod{n_1}$, $a \equiv b \pmod{n_2}$, $\ldots$, $a \equiv b \pmod{n_k}$, and $n = \mathrm{lcm}(n_1, n_2, \ldots, n_k)$, then $a \equiv b \pmod{n}$ (Theorem 1.74).

10. For what value of n is $\{1, 15, 31, 26, 24, 72, 17, 38, 45, 10, 22, 21, 55\}$ a complete residue system modulo n? Explain why.

11. Prove Theorem 1.57(b).

12. Prove Theorem 1.57(d).

13. Prove Theorem 1.57(e).

14. Prove Theorem 1.57(f).

15. Prove that if $a \equiv b \pmod{n}$, then $\gcd(a, m) = \gcd(b, m)$.

16. Prove that if n is a positive integer and $\{r_1, r_2, \ldots, r_k\}$ is a reduced residue system modulo n, then any integer a is congruent to only one $r_j, 1 \leq j \leq k$ (Theorem 1.66).

17. Prove that if $\{r_1, r_2, \ldots, r_n\}$ is a complete residue system modulo n and a is an integer, then $\{r_1 + a, r_2 + a, \ldots, r_n + a\}$ is also a complete residue system modulo n. Give an example that shows that an analogous theorem does not hold for a reduced residue system.

18. Prove that if $\{r_1, r_2, \ldots, r_k\}$ is a complete (reduced) residue system modulo n, then $\{ar_1, ar_2, \ldots, ar_k\}$ is a complete, (reduced), residue system modulo n also (Theorem 1.67).

19. Prove Theorem 1.74.

20. Prove Theorem 1.75.

21. Find all primary residues modulo 56 that are congruent to 5 modulo 8.

22. Find all primary residues modulo 35 that are congruent to 3 modulo 7.

23. Prove that the $\gcd(a, m)$ distinct solutions of $ax \equiv c \pmod{m}$ obtained from $t \in \{1, 2, 3, \ldots, \gcd(a, m)\}$ in Theorem 1.69 are obtained if t is chosen from any complete residue system modulo $\gcd(a, m)$.

1.7   Application: Random Keys

A company has many salespersons who receive orders over the telephone. The sales staff have simultaneous access to the computer order entry system when performing their everyday activities of entering new orders and querying existing orders. Each new order receives a unique 10-digit order number which uses some of the digits to specify additional information such as the date. The order numbers are generated as consecutive integers within a year. Then, the order numbers and associated information such as name, address, items ordered, and so on, are stored in a commercially available database keyed by the order number.

The salespersons encounter a problem of being prevented from accessing certain orders because another person is actively modifying or viewing information from an order whose order number is numerically close to the one the first is attempting to activate. This clashing occurs because the database software stores the information for consecutive keys contiguously to one another and divides the whole database into groups of items, allowing only one item in a group to be accessed at a time. As more items are added to the database, the software automatically regroups the items to maintain size balance of the groups; however, within a group, all of the items in the database having keys between the largest and smallest keyed items of that group are "located" in that group. Thus, the items are grouped according to the key. Since salespersons tend to access recent orders more frequently, they often experience being locked out of the database and must wait until an active item in a group is released.

One solution to the problem would be to generate nonconsecutive order numbers and key the information according to this second number keeping the order number in the database as part of the information for this item. This solution is not viable because there is room for only one 10-digit number and because the order number contains other information, say, the date, about the order with respect to which we may want to query the database.

What is needed is a 10-digit number known to the outside world (the order number) and a 10-digit number known to the database according to which the items in the database are organized. This scheme would require that when we specify the order number, we can unambiguously obtain the corresponding database key (the index); and when we examine any item in the database, we can unambiguously obtain the proper order number from the key. Mathematically, what is needed is a function

$$f : S \to S$$

which is one to one and onto for the set $S = \{n : n \in Z \text{ and } 0 \leq n < 10^{10}\}$, the set of all 10-digit nonnegative integers. Additionally, we want the function f to "jumble" the integers in S so that for consecutive integers $n_1, n_2 =$

n_1+1, $n_3 = n_1+2$, $n_4 = n_1+3$, ..., the integers $f(n_1), f(n_2), f(n_3), f(n_4), \ldots$ are not consecutive.

A multiplicative congruential mapping can meet these criteria. Suppose that a and m are positive integers. Then define

$$f(n) = [[a \cdot n]]_m$$

for any n in S. In other words, if n is the order number and i is the corresponding index, then

$$f(n) \equiv i \pmod{m}$$

or

$$a \cdot n \equiv i \pmod{m}$$

where $0 \le i < m$ and $m = 10^{10}$ in this application. If an integer $b \in S$ can be found so that

$$b \cdot a \equiv 1 \pmod{m}$$

then the function f will be one to one and onto (see the Exercises). The integer b is called the *multiplicative inverse* of a modulo m and is usually denoted by $b = a^{-1}$. If such b exists, it is easy to show that f is one to one and onto. More directly, if n is given and i is calculated by

$$a \cdot n \equiv i \pmod{m}$$

then by multiplying both sides by b we can recover n:

$$\begin{aligned} b \cdot a \cdot n &\equiv b \cdot i \pmod{m} \\ 1 \cdot n &\equiv b \cdot i \pmod{m} \\ n &\equiv b \cdot i \pmod{m} \end{aligned}$$

Thus, just multiply i by $b = a^{-1}$ and reduce the product modulo m to get back the original order number n. In fact, once the pair a and $b = a^{-1}$ is obtained, either one could be used to encode the order number:

$$f(n) = [[a \cdot n]]_m$$

or

$$g(n) = [[a^{-1} \cdot n]]_m$$

We need to solve the congruence

$$a \cdot b \equiv 1 \pmod{10^{10}}$$

which, by Theorem 1.68, has a solution if and only if a and 10^{10} are relatively prime. Since $10^{10} = 2^{10}5^{10}$, evidently the only positive integers unequal to 1 that divide $m = 10^{10}$ must be divisible by 2 or 5. We will show in Chapter 2 that any number not divisible by 2 or 5 is relatively prime to 10^{10}.

For example, let $a = 8506017387$, which is divisible by neither 2 nor 5. In fact, we will show directly that $\gcd(8506017387, 10^{10}) = 1$ while solving the congruence

$$8506017387 \cdot x \equiv 1 \ (\text{mod } 10^{10})$$

First apply the Euclidean algorithm to solve

$$8506017387 \cdot x + 10^{10} \cdot y = 1$$

by finding $\gcd(8506017387, 10^{10})$:

$$
\begin{aligned}
10^{10} &= 8506017387 \cdot 1 + 1493982613 \\
8506017387 &= 1493982613 \cdot 5 + 1036104322 \\
1493982613 &= 1036104322 \cdot 1 + 457878291 \\
&\vdots \\
2784395 &= 1237509 \cdot 2 + 309377 \\
1237509 &= 309377 \cdot 4 + 1 \\
309377 &= 1 \cdot 309377 + 0
\end{aligned}
$$

Thus, $\gcd(8506017387, 10^{10}) = 1$ and 8506017387 and 10^{10} are relatively prime. To express the greatest common divisor, 1, in terms of a linear combination of 8506017387 and 10^{10} we "back substitute" as follows:

$$
\begin{aligned}
1 &= 1237509 - 309377 \cdot 4 \\
&= 1237509 - [2784395 - 1237509 \cdot 2] \cdot 4 \\
&= 1237509 \cdot 9 - 2784395 \cdot 4 \\
&\vdots \\
&= 1036104322 \cdot 1480 - 457878291 \cdot 3349 \\
&= 1036104322 \cdot 1480 - [1493982613 - 1036104322 \cdot 1] \cdot 3349 \\
&= 1036104322 \cdot 4829 - 1493982613 \cdot 3349 \\
&= [8506017387 - 14939826613 \cdot 5] \cdot 4829 - 1493982613 \cdot 3349 \\
&= 8506017387 \cdot 4829 - 14939826613 \cdot 27494 \\
&= 8506017387 \cdot 4829 - [10^{10} - 8506017387 \cdot 1] \cdot 27494 \\
&= 8506017387 \cdot 32323 - 10^{10} \cdot 27494
\end{aligned}
$$

Thus, the equation

$$ax + my = 1$$

has the solution

$$x = 32323$$

$$y = -27494$$

Since $\gcd(a, m) = 1$, there is only one solution modulo m for $a \cdot x \equiv 1 \ (\text{mod } m)$. So

$$a^{-1} = b = 32323$$

To check, note that

$$
\begin{aligned}
a \cdot a^{-1} &= 8506017387 \cdot 32323 \\
&= 274,940,000,000,001 \\
&\equiv 1 \ (\text{mod } 10^{10})
\end{aligned}
$$

since to compute modulo 10^{10} in base 10, just select the last 10 digits and discard the others.

We will convert order numbers n to indexed database keys using the one-to-one function $f : S \to S$ defined by

$$f(n) = [[a \cdot n]]_{10^{10}} = [[8506017387 \cdot n]]_{10^{10}} = i$$

The following table illustrates the use of this mapping:

n	$8506017387 \cdot n$	i
1983120423	16868456798552794701	8552794701
1983120424	16868456807058812088	7058812088
1983120425	16868456815564829475	5564829475
1983120426	16868456824070846862	4070846862
1983120427	16868456832576864249	2576864249
1983120428	16868456841082881636	1082881636
1983120429	16868456849588899023	9588899023
1983120430	16868456858094916410	8094916410
1983120431	16868456866600933797	6600933797

Obviously, n can be recovered by $g(i) = [[a^{-1} \cdot i]]_m$ for $i = 8552794701$ since

$$
\begin{aligned}
g(i) &= [[32323 \cdot 8552794701]]_{10^{10}} \\
&= [[276451983120423]]_{10^{10}} \\
&= 1983120423
\end{aligned}
$$

and $n = 1983120423$ for the first line of the table above.

Exercises

1. Let $S = \{n : n \in Z \text{ and } 0 \le n < m\}$, $a \in N$, and $f : S \to S$ be defined by $f(n) = [[a \cdot n]]_m$. Prove that if there is an integer b such that $b \cdot a \equiv 1 \pmod{m}$, then f is one to one and onto.

2. For $a = 8506017387$ discussed in this section, fill in the missing parts of the Euclidean Algorithm and back substitution in solving

$$850601787 \cdot x + 10^{10} \cdot y = 1$$

3. Define S as in Exercise 1, $m = 10^{10}$ and $a = 3^{10} = 59049$. Define $f : S \to S$ by $f(n) = [[59049 \cdot n]]_m$. Determine the inverse $f^{-1} : S \to S$ of f.

4. Show that for positive integers expressed in base ten notation, the modulo 10^s equivalent of an integer is obtained by keeping the rightmost s digits of the integer.

5. Let $a, b,$ and m be positive integers. Prove that if $\gcd(a, m) = 1$ and b is a solution of $a \cdot x \equiv 1 \pmod{m}$, then $\gcd(b, m) = 1$; that is, if a and

m are relatively prime and $a \cdot a^{-1} \equiv 1 \pmod{m}$, then m and a^{-1} are relatively prime.

1.8   Application: Random Number Generation I

Complex systems are difficult to describe exactly. Such systems often are non-deterministic and respond to components that change in a random or probabilistic manner. Exact mathematical descriptions and solutions of related equations are generally unobtainable or give only averaged results. Examples are phase changes in liquids and traffic flows in a system of highway interchanges. In such situations the method of simulation can often be used to obtain numerical estimations of the characteristics of the system. In a simulation, the system components are allowed to vary, often independently, according to rules reflecting physical constraints and component characteristics. The component changes over time follow some random process that controls the nature of the changes. Then the interaction of the components with one another in the system is "monitored" and information about the system performance is captured.

The advent of the computer has made simulations practical. Simulations depend upon the ability to generate "random" events or changes which occur in accordance to some probability distribution, say, the normal or "bell-shaped" distribution or the exponential distribution. A simulation often requires that many millions of "random" events be obtained; therefore, the computer must be able to produce these events in great profusion and very rapidly. A great deal of effort has been expended to make this process efficient.

The most common methods of generating events with a particular probability distribution rely upon the ability to produce real numbers, U, in the range $0 < U < 1$ so that the numbers obtained are uniformly distributed. Uniformly distributed on the real number interval $(0, 1)$ means that the probability that a number U is generated in a subinterval (a, b) of $(0, 1)$ is

$$PROB\{a < U < b\} = b - a$$

On the other hand, the real numbers U, $0 < U < 1$, are themselves obtained by producing integers k in the set $\{1, 2, \ldots, m - 1\}$ and then dividing by m to get U:

$$U = \frac{k}{m} \text{ where } 1 \leq k < m$$

For example, if $m = 10$, the distinct uniform random numbers that can be generated are

$$1/10 = 0.1, \ 2/10 = 0.2, \ 3/10 = 0.3, \ldots, 9/10 = 0.9$$

To mimic "randomness," the order in which the number U are chosen from the set $\{0.1, 0.2, 0.3, \ldots, 0.9\}$ must be "jumbled," which is tantamount to choosing integers k from the set $\{1, 2, 3, \ldots, 9\}$ in a jumbled or random way. This last problem, namely generating an integer from the set $\{1, 2, 3, \ldots, m-1\}$ in a random-like way, is the one we want to address here. We will assume that in a simulation, one would convert these integers to real uniformly distributed numbers U, $0 < U < 1$, as described above and then would use probability theory to obtain the events having the desired probability distribution. The details of how to use the uniformly distributed real numbers $U = \dfrac{k}{m}$ to obtain such events can be found in books on statistics and on simulation. (See *Operations Research* by Frederick S. Hillier and Gerald Lieberman [35].)

Random integers are most often generated using multiplicative congruential methods. Suppose that $m > 1$ and $S = \{1, 2, \ldots, m-1\}$ and let a be an integer in S. Define the multiplicative congruential mapping $L : S \to S$ by

$$L(x) = [[a \cdot x]]_m$$

for any x in S. Thus, $L(x) = [[a \cdot x]]_m$ is a primary residue modulo m. This function L can recursively generate a sequence $x_0, x_1, x_2, \ldots$ of integers in the set S as follows.

(a) Let x_0 be any integer in $S = \{1, 2, \ldots, m-1\}$.

(b) For each $i \geq 0$,

$$x_{i+1} = L(x_i)$$

or

$$x_{i+1} \equiv a \cdot x_i \pmod{m} \quad \text{and} \quad x_{i+1} = [[a \cdot x_i]]_m$$

The initial member of the sequence, x_0, is called the *seed* because, given a and m, x_0 completely determines the sequence $x_0, x_1, x_2, \ldots$. This result is evident since one can show by induction that

$$x_k \equiv a^k \cdot x_0 \pmod{m}$$

for each k.

For example, for $m = 10$, $a = 3$, and $x_0 = 2$, the corresponding multiplicative congruential generator is given by

$$x_0 = 2$$

$$x_{i+1} \equiv 3 \cdot x_i \pmod{10}$$

producing the table

i	x_i	$3x_i$	x_{i+1}	$U_i = x_i/10$
0	2	6	6	0.6
1	6	18	8	0.8
2	8	24	4	0.4
3	4	12	2	0.2
4	2	6	6	0.6
5	6	18	8	0.8

Examining the second column of the table, we see immediately that the sequence $x_0, x_1, x_2, \ldots$ begins repeating with $x_4 = x_0 = 2$. The length of the repeating subsequence is called the *period*. Of course, because S is finite, the sequence $x_0, x_1, x_2, \ldots$ must eventually repeat; however, ideally we would want all of the integers of S to appear in the sequence before a repeat occurs. In this example, if the seed x_0 had been chosen to be 5 instead of 2, then $5 = x_0 = x_1 = x_2 = \cdots$ which clashes with our idea of "random." Also, if $a = 2$ and $x_0 = 5$, then modulo 10 one gets $ax_0 = 10 \equiv 0 \notin S$. Thus, in order to make this method into a practical simulation tool, we would want the multiplicative congruential generator to have these properties:

(a) The modulus m should be as large as feasible so that there will be many different random numbers U and they can be thickly distributed in the real open interval $(0, 1)$.

(b) If possible, choose multiplier $a > 1$ and modulus m so that $ax_k \not\equiv 0 \pmod{m}$ for all k.

(c) The period of the sequence $x_0, x_1, x_2, \ldots$ should be as long as possible.

(d) The sequence $x_0, x_1, x_2, \ldots$ should produce real numbers $U_i = x_i/m$ that are uniformly distributed and, in addition, exhibit "randomness" in some sense.

To satisfy (b), we can ensure that for no k is $ax_k \equiv 0 \pmod{m}$ by requiring that a and m be relatively prime. If $\gcd(a, m) = 1$ and $x_k \equiv 0 \pmod{m}$ for some k, then

$$\begin{aligned} ax_k - 0 &= m \cdot j \quad \text{for some integer } j \\ a^k x_0 &= m \cdot j \quad \text{for some integer } j \end{aligned}$$

The last equality implies that $m \mid a^k x_0$. By induction one can show that m cannot divide a^k, therefore, m divides x_0 (see the Exercises). But this result contradicts the fact that $x_0 < m$ since the seed is in S.

To satisfy (c), it turns out that the period can be made to be the maximum possible provided that a and m are chosen properly. An adequate discussion of this aspect of multiplicative congruential generators will have to wait until later; however, if we choose m to be an odd prime and choose a to be such that the smallest k with $a^k \equiv 1 \pmod{m}$ is $k = m - 1$, then $x_0, x_1, x_2, \ldots$ will contain every member of S before repeating.

Ensuring in (d) that the sequence $x_0, x_1, x_2, \ldots$ (or, equivalently, the sequence $U_0, U_1, U_2, \ldots$) exhibits "randomness" is the most difficult to quantify. Indeed, the sequence is not random at all since every term of the sequence is completely determined once x_0, a, and m are specified. In fact, sequences of integers $x_0, x_1, x_2, \ldots$ and real numbers $U_0, U_1, U_2, \ldots$ are called *pseudo-random* sequences to reflect this characteristic, although we will loosely refer to them as random sequences in this discussion. Intuitively, we want the resulting random numbers $U_i = x_i/m$ to have the uniform probability distribution and to have no "patterns" or regularities at the scale we would be using them. We will not dwell on this aspect; however, the interested reader is referred to "A Statistical Evaluation of Multiplicative Congruential Random Number Generators with Modulus $2^{31} - 1$," by George S. Fishman and Louis R. Moore [27], and Donald Knuth, *The Art of Computer Programming* [43], for a discussion of the relevant issues.

Practically speaking, the modulus m is often chosen to be as large a prime integer as possible relative to the largest integer "native" to the computer used to implement the multiplicative congruential generator. A prime is defined to be an integer greater than one having only itself and one as factors. Since integers are commonly represented as 32-bit "words" on many computers, and since the integer arithmetic is optimized for speed for this type of integer, $m = 2^{31} - 1$ is the largest computer-representable positive signed integer in this context (see the discussion on two's complement). The integers in the range $-2^{31} = -2147483648$ to $2^{31} - 1 = 2147483647$ can be so represented. Fortunately, $m = 2^{31} - 1$ is prime so that every $a > 1$ in $S = \{1, 2, 3, \ldots, m-1\}$ is relatively prime to $2^{31} - 1$; and we have many choices of suitable multipliers.

The finite range of computer-representable integers provides an added complication; namely, it is possible to add or multiply two computer representable integers and the mathematical result be an integer out of this finite range. For example, if $a = 4$ and $x_0 = 2,000,000,000$, then $ax_0 = 8,000,000,000$, which is greater than $2^{31} - 1 = 2147483647$. Many computer implementations of multiplicative congruential generators rely upon "tricks" having to do with how such "overflows" are handled by the hardware arithmetic unit of the computer. This state of affairs is deceptively comforting because different computers may handle such arithmetic exceptions differently and, indeed, some computers have a larger than 32-bit integer word size. We want to implement a multiplicative congruential random number generator in such a way that the implementation can be used without modification across many kinds of computers. We need to be able to compute $x_{i+1} \equiv ax_i \pmod{m}$ with $m = 2^{31} - 1$ on any computer with integer word size at least 32 bits. Because we want to be able to generate random numbers very rapidly, we still want to use the "native" computer arithmetic but want the algorithm to be independent of the computer. Of course, for applications having the very highest speed requirements, it may still be necessary to tailor the method of generation to the hardware of the specific computer being used.

The Division Algorithm and "congruential operations" can be used to calculate $L(x) \equiv ax \pmod{m}$ with all intermediate results within the finite integer range of a computer provided that m is congruent to a small enough integer modulo a. (See, Stephen K. Park and Keith W. Miller, "Random Number Generators: Good Ones Are Hard to Find" [62].)

Theorem 1.76 *If*

(a) $a > 1$ *and* $\gcd(a, m) = 1$,

(b) $L(x) = [[a \cdot x]]_m$,

(c) *the Division Algorithm gives* $m = aq + r$ *with* $0 \le r < a$, *and*

(d) $r < q$,

then for $1 \le x \le m - 1$, $L(x)$ *can be calculated with all intermediate results being integers in the range* $-(m - 1)$ *to* $m - 1$. *The algorithm is as follows:*

Given $1 \le x \le m - 1$:

1. *Use the Division Algorithm to obtain integers* A *and* s *such that*

$$x = q \cdot A + s \text{ with } 0 \le s < q$$

2. *Calculate*

$$g = a \cdot s - r \cdot A$$

3. *If* $g > 0$, *then*

$$L(x) = g$$

else

$$L(x) = g + m$$

The arithmetic unit of the computer should efficiently calculate the integers A and s in step 1. It is a result of the proof of the theorem that the intermediate results $a \cdot s$ and $r \cdot A$ are in the range 0 to $m - 1$.

Example 1.77 *Let* $m = 2^{31} - 1 = 2147483647$. *A common multiplier is* $a = 16807$, *which must be relatively prime to* m *since* m *is prime. Applying the Division Algorithm to* a *and* m *to obtain* $m = aq + r$ *with* $0 \le r < a$, *we get*

$$
\begin{aligned}
m &= 2147483647 \\
a &= 16807 \\
q &= 127773 \\
r &= 2836
\end{aligned}
$$

The theorem applies since $r < q$. *For each* x,

1. *Use the Division Algorithm to obtain integers* A *and* s *such that* $x = q \cdot A + s = 127733 \cdot A + s$.

2. *Calculate*

$$
\begin{aligned}
g &= a \cdot s - r \cdot A \\
&= 16807 \cdot s - 2836 \cdot A
\end{aligned}
$$

3. *If* $g > 0$, *then*

$$L(x) = g$$

else

$$L(x) = g + 2147483647$$

The algorithm will be applied over and over again to obtain random integers x_0, x_1, x_2, *and* x_3. *Given the seed* $x_0 = 82310$, *we calculate*

$$x_{i+1} \equiv 16807 \cdot x_i \pmod{2147483647}$$

i	x_i	s	A	$a \cdot s$ $16807 \cdot s$	$r \cdot A$ $2836 \cdot A$	g
0	82310	82310	0	1383384170	0	$+$
1	1383384170	113672	10826	1910485304	30702536	$+$
2	1879782768	14711	114165	247247777	323771940	$-$
3	2070959484					

The outline of a proof of Theorem 1.76 is given below.

Proof. (Theorem 1.76) Assume that $\gcd(a, m) = 1$ and the Division Algorithm gives

$$m = a \cdot q + r \text{ with } 0 \leq r < a$$

and additionally, that $r < q$. Let $1 \leq x \leq m - 1$. We want to calculate $L(x) = [[a \cdot x]]_m$. The Division Algorithm gives

$$\begin{aligned} x &= q \cdot A + s \text{ with } 0 \leq s < q \\ a \cdot x &= m \cdot B + t \text{ with } 0 \leq t < m \end{aligned}$$

Note that $L(x) = t$ is the answer we seek since $ax \equiv t \pmod{m}$.

$$\begin{aligned} L(x) = t &= ax - mB \\ &= a(qA + s) - mB \\ &= aqA + as - mB \end{aligned}$$

Next, add and subtract mA and then substitute $m = aq + r$ to obtain

$$\begin{aligned} L(x) = t &= aqA + as - mB + mA - mA \\ &= aqA + as - mB + mA - (aq + r)A \\ &= (as - rA) + m(A - B) \\ &= g(x) + m \cdot d(x) \end{aligned}$$

where $g(x) = as - rA$ depends upon x through s and A and where $d(x) = A - B$ depends upon x through A and B.

Let $g(x) = as - rA$. We have $0 \leq as \leq m - 1$ because

$$0 \leq as < aq = m - r \leq m$$

Also, $0 \leq rA \leq m - 1$ because

$$rA < qA = x - s < m - s \leq m$$

Thus, both integers $a \cdot s$ and $r \cdot A$ can be computed without incurring a product exceeding $m - 1$. So $-(m - 1) \leq g(x) \leq m - 1$.

The other key result (see Park and Miller [62]) is that $d(x)$ has only one of two values:

$$d(x) = 0 \text{ or } 1$$

and

$$d(x) = 0 \quad \text{if and only if} \quad 1 \leq g(x) \leq m - 1$$
$$\text{if and only if} \quad g(x) > 0$$

$$d(x) = 1 \quad \text{if and only if} \quad -(m-1) \leq g(x) \leq -1$$
$$\text{if and only if} \quad g(x) < 0$$

so that the value of $d(x)$ can be obtained by observing whether $g(x)$ is positive or negative. ∎

Pseudo-random sequences generated by the multiplicative congruential method discussed here have a built-in shortcoming associated with the "randomness" of the resulting numbers. When n consecutive terms in the sequence $U_0, U_1, U_2, \ldots$, where $U_i = x_i/m$, are grouped as an n-tuple

$$(U_k, U_{k+1}, \ldots, U_{k+n-1})$$

then all such n-tuples are points that fall on a relatively few parallel hyperplanes in an n-dimensional cube. (See George Marsaglia's "Random Numbers Fall Mainly in the Planes" [49].) For many simulation tasks, this structure may not be of great concern; however, for precision simulations, one may want to modify the simple multiplicative congruential generator described here. See Pierre L'Ecuyer, "Efficient and Portable Combined Random Number Generators" [46], for a possible remedy to the deficiency.

Exercises

1. For positive integers a and m prove that if $\gcd(a, m) = 1$, then $\gcd(a^k, m) = 1$ for any positive integer k.

2. For positive integers a and m, assume that $m > 1$, $S = \{1, 2, \ldots, m-1\}$, $a \in S$, and $L(x) = [[a \cdot x]]_m$. If $x_0 \in S$ and $x_{i+1} = L(x_i)$, prove that the first term x_k in the sequence $x_0, x_1, x_2, \ldots$ that is repeated is when $x_k = x_0$.

3. Use the algorithm of Theorem 1.76 and Example 1.77 with $m = 2^{31} - 1$, $a = 16807$, and seed $x_0 = 38470531$ to generate the integers x_1, x_2, and x_3. Calculate also U_1, U_2, and U_3 where $U_i = x_i/m$. The U_i are three "pseudo"-random numbers in the real interval $(0, 1)$ from the uniform statistical distribution.

4. If $\gcd(a, m) = 1$ and $m > 1$, prove that $m \nmid a^k$ for every positive integer k.

1.9   Application: Two's Complement

In number theory, integers can be positive or negative and any size without limit. The axioms guarantee that if any two integers are multiplied, added, or subtracted, the result is another integer and is the correct product, sum, or difference. This universality is no longer applicable when computers do the arithmetic because any practical computer has finite limitations of memory, storage, and execution times. In fact, the "native" arithmetic performed by most computers is severely limited, namely, only finitely many integers may be represented, say only those between -32768 and 32767. This restriction means that if we were to add $32000 + 32000$, we would obtain 64000 in number theory but would obtain -1536 on some computers. This section discusses how integers are often represented in computers as two's complement integers and how the arithmetic is performed.

Integers are usually represented internally in computers as sequences of zeros and ones, such as the sequence

$$11101101$$

where each position is referred to as a bit and where the sequence is often variously called a nibble, byte, word, and so on, depending upon the computer and the length of the sequence. Sometimes the sequence is called a bit string. To be specific, we will give each bit a name

$$[11101101] = [a_7, a_6, a_5, a_4, a_3, a_2, a_1, a_0]$$

so that $a_7 = a_6 = a_5 = a_3 = a_2 = a_0 = 1$ and $a_4 = a_1 = 0$ for this 8-bit string. Because each bit is zero or one, the bit string is closely related to the base 2 representation of integers.

We saw in Theorem 1.52 that if $b > 1$, then any positive integer c can be uniquely represented as

$$c = a_m b^m + a_{m-1} b^{m-1} + \cdots + a_1 b^1 + a_0 b^0$$

with $a_m > 0$ and $0 \le a_i < b$ for all i. In particular, for $b = 2$, we have the binary (base or radix $= 2$) representation

$$c = a_m 2^m + a_{m-1} 2^{m-1} + \cdots + a_1 2^1 + a_0 2^0$$

with $a_m > 0$ and $0 \le a_i < 2$ for all i, $0 \le i \le m$. So for each i, $a_i = 0$ or $a_i = 1$ since these are the only nonnegative integers less than 2.

If there are N bits available to represent the number c and $N > m + 1$, then we could write

$$c = 0 \cdot 2^{N-1} + 0 \cdot 2^{N-2} + \cdots + 0 \cdot 2^{m+1} + a_m 2^m + a_{m-1} 2^{m-1} + \cdots + a_1 2^1 + a_0 2^0$$

which would be interpreted as the bit string

$$[0, 0, \ldots, 0, a_m, a_{m-1}, \ldots, a_1, a_0]$$

where there are possibly no leading zeros.

If we fix N, the number of bits we will allow in the bit string

$$[b_{N-1}, b_{N-2}, \ldots, b_1, b_0]$$

then there are exactly 2^N distinct such strings. This number is obtained by noticing that there are two choices of bit values for b_0, namely 0 or 1; and for whatever is chosen for b_0, there are two choices for b_1, and so on. Thus there are $2 \times 2 \times 2 \times \cdots \times 2 = 2^N$ ways to choose all of the bits. This result implies that there are also exactly 2^N integers of the form

$$c = a_{N-1}2^{N-1} + a_{N-2}2^{N-2} + \cdots + a_1 2^1 + a_0 2^0$$

with $a_i = 0$ or 1. In fact, $0 \le c \le 2^N - 1$. Thus, we can represent all the integers between 0 and $2^N - 1$ as bit strings of length N. We should distinguish between the radix 2 representation of the nonnegative number c, namely,

$$a_{N-1}2^{N-1} + a_{N-2}2^{N-2} + \cdots + a_1 2^1 + a_0 2^0$$

and the bit string corresponding to the number c by using the notation $B(c)$ for

$$[a_{N-1}, a_{N-2}, \ldots a_1, a_0]$$

The arithmetic processor in the central processing unit (CPU) of many computers calculates arithmetic operations assuming that the bits represent the nonnegative integers between 0 and $2^N - 1$. For example, if a is the nonnegative integer for the bit string $B(a)$ so that $B(a)$ is

$$[a_{N-1}, a_{N-2}, \ldots, a_1, a_0]$$

and b is the nonnegative integer for the bit string $B(b)$ so that $B(b)$ is

$$[b_{N-1}, b_{N-2}, \ldots, b_1, b_0]$$

then the bit string

$$[c_{N-1}, c_{N-2}, \ldots, c_1, c_0]$$

computed by computer addition

$$[a_{N-1}, a_{N-2}, \ldots, a_1, a_0] \oplus [b_{N-1}, b_{N-2}, \ldots, b_1, b_0]$$

is the bit string corresponding to the nonnegative integer

$$c \equiv a + b \pmod{2^N}$$

namely, $B(c)$, so that $0 \le c \le 2^N - 1$.

For example, suppose that $N = 4$, giving $2^N = 16$. If $a = 9$ so that $B(a) = [1001]$ and $b = 13$ so that $B(b) = [1101]$, then $B(a) \oplus B(b) = [0110]$ because $a + b = 22 \equiv 6 \pmod{2^4}$ and $B(6) = [0110]$.

We note that using a radix b to represent nonnegative integers makes it easy to compute integers modulo b^N. If

$$d = a_{M-1}b^{M-1} + a_{M-2}b^{M-2} + \cdots + a_N b^N + a_{N-1}b^{N-1} + \cdots + a_1 b^1 + a_0 b^0$$

then d modulo b^N is obtained by stripping off the leading terms, giving

$$[[d]]_{b^N} = a_{N-1}b^{N-1} + a_{N-2}b^{N-2} + \cdots + a_1 b^1 + a_0 b^0$$

and

$$B([[d]]_{b^N}) = [a_{N-1}, a_{N-2}, \ldots, a_1, a_0]$$

Now continuing the example above in which $N = 4$ and using binary place value notation (i.e., suppressing the powers of 2 just as in base or radix ten we normally suppress the powers of ten when we write the number twenty-seven as 27), we have

$$1001 + 1101 = 10110$$

so that

$$[[10110]]_{2^4} = 0110$$

The scheme above means that using the ordinary computer arithmetic processor, we cannot add two nonnegative integers a and b using the bit strings $B(a)$ and $B(b)$ and get the correct answer as the bit string $B(a + b)$ if $a + b \geq 2^N$ for N-bit strings. The answer we obtain is the modulo 2^N-bit string of the sum.

There is a more serious deficiency, namely, the method does not account for negative integers in the computer? Methods that have been used include using one of the bits as an indicator of whether the number is positive or negative. The "signed magnitude"-representation uses this method. There is perhaps a weakness in the signed-magnitude representation in that there are two representations of the integer 0. In this case, the arithmetic processor may have one set of circuits to add two positive integers and another to add a positive and a negative integer. The same complication occurs upon subtraction.

The computer arithmetic processor originally described above with regard to adding bit strings had an important property. This property was that it performed addition modulo 2^N "correctly" as long as the bit strings were considered to be nonnegative integers. It turns out that there is a way to map sets of both nonnegative and negative integers into bit strings of length N so that ordinary computer addition modulo 2^N gives the bit string corresponding to the integer sum whether the addends are negative or nonnegative. This representation is called the two's-complement representation of integers.

There are 2^N different bit strings $[b_{N-1}, b_{N-2}, \ldots, b_1, b_0]$ of length N so we can specify 2^N different integers using such a set of bit strings. Some of these bit strings will be selected to represent nonnegative integers and some to represent negative integers. About half of the bit strings will designate positive integers and the other half, negative integers. More specifically, if S is the set of bit strings of length N so that $S = \{[b_{N-1}, b_{N-2}, \ldots, b_1, b_0] : b_i = 0 \text{ or } 1\}$ and $R = \{i : -2^{N-1} \leq i \leq 2^{N-1} - 1\}$, then we define a function $T : S \to R$ as follows:

$$T([b_{N-1}, b_{N-2}, \ldots, b_1, b_0]) = -b_{N-1}2^{N-1} + b_{N-2}2^{N-2}$$

$$+ b_{N-3}2^{N-3} + \cdots + b_1 2^1 + b_0 2^0$$

That this rule defines a function is evident by substitution. The values of f are in R is obvious by noting that

$$-b_{N-1}2^{N-1} \leq -b_{N-1}2^{N-1} + b_{N-2}2^{N-2}$$

$$+ b_{N-3}2^{N-3} + \cdots + b_1 2^1 + b_0 2^0$$

$$\leq -b_{N-1}2^{N-1} + 2^{N-2} + 2^{N-3} + \cdots + 2^1 + 2^0$$

$$\leq -b_{N-1}2^{N-1} + 2^{N-1} - 1$$

$$\leq 2^{N-1} - 1$$

since $0 \leq b_i \leq 1$ and $2^{N-2} + 2^{N-3} + \cdots + 2^1 + 2^0 = 2^{N-1} - 1$.

There is an asymmetry in the range of positive and negative integers representable. For example, if $N = 16$, then $2^{N-1} - 1 = 2^{15} - 1 = 32767$ and $-2^{N-1} = -2^{15} = -32768$. All integers between -32768 and 32767 ($65536 = 2^{16}$ integers in all) may be represented in this manner.

As another example, suppose $N = 4$ so that $-2^{4-1} \leq i \leq 2^{4-1} - 1$ or $-8 \leq i \leq 7$.

Two's Complement $[b_3, b_2, b_1, b_0]$	$T([\cdots])$ Decimal	$T([\cdots])$ Signed Binary
0000	0	0000
0001	1	0001
0010	2	0010
0011	3	0011
0100	4	0100
0101	5	0101
0110	6	0110
0111	$7 = 2^3 - 1$	0111
1000	$-8 = -8 + 0 = -2^3$	-1000
1001	$-7 = -8 + 1$	-0111
1010	$-6 = -8 + 2$	-0110
1011	$-5 = -8 + 3$	-0101
1100	$-4 = -8 + 4$	-0100
1101	$-3 = -8 + 5$	-0011
1110	$-2 = -8 + 6$	-0010
1111	$-1 = -8 + 7$	-0001

The table illustrates that if $b_{N-1} = 0$, then

$$T([b_{N-1}, b_{N-2}, \ldots, b_1, b_0]) = b_{N-2}2^{N-2} + b_{N-3}2^{N-3} + \cdots + b_1 2^1 + b_0 2^0 \geq 0$$

but , if the leftmost bit of the bit string is one, that is, $b_{N-1} = 1$, then

$$\begin{aligned} T([b_{N-1}, b_{N-2}, \ldots, b_1, b_0]) &= -b_{N-1}2^{N-1} + b_{N-2}2^{N-2} \\ &\quad + b_{N-3}2^{N-3} + \cdots + b_0 2^0 \\ &\leq -2^{N-1} + 2^{N-2} + 2^{N-3} + \cdots + 2^1 + 2^0 \\ &= -2^{N-1} + 2^{N-1} - 1 = -1 \end{aligned}$$

so that $T([b_{N-1}, b_{N-2}, \ldots, b_1, b_0]) < 0$. Thus, if the leftmost bit of the bit string is 0, then the integer it represents is nonnegative; but if the leftmost bit of the bit string is 1, then the integer it represents is negative. It is also easy to see from the above string of inequalities that

$$\begin{aligned} T([1, b_{N-2}, \ldots, b_1, b_0]) &= -\big[(1 - b_{N-2})2^{N-2} + (1 - b_{N-3})2^{N-3} + \cdots \\ &\quad + (1 - b_0)2^0 + 1\big] \end{aligned}$$

Since $b_i = 0$ or 1, then $1 - b_i = 1$ or 0 so that the integer within brackets $[\cdots]$ is positive. This last formula can be used to generate the last column in the table above in case the leading bit of the bit string is 1. It also can be used to obtain the two's complement of negative integers.

Thus, to obtain the two's-complement bit string corresponding to a negative integer between -2^{N-1} and -1:

(a) Write the integer in binary.

(b) Change the zeros to ones and the ones to zeros.

(c) Add 1 to the result.

For example, suppose that we want the two's complement of the integer -5 assuming bit stings of length $N = 4$.

$$\begin{aligned} -5 &= -(1 \cdot 2^2 + 0 \cdot 2^1 + 1 \cdot 2^0) \\ &= -(0 \cdot 2^3 + 1 \cdot 2^2 + 0 \cdot 2^1 + 1 \cdot 2^0) \\ &= -(0101) \qquad \text{(signed binary)} \end{aligned}$$

Now, changing the bits (this transformation is known as generating the one's-complement form), we have

$$0101 \rightarrow 1010$$

Next, adding 1 we obtain

$$\begin{array}{r} 1010 \\ \underline{1} \\ 1011 \end{array}$$

So $[1011]$ is the two's-complement bit string corresponding to -5. We can write either $T([1011]) = -5$ or $T^{-1}(-5) = [1011]$.

The function $T : S \to R$ is one to one and onto R. Suppose that c is in R. If $c = 0$, then $T([0, 0, \ldots, 0]) = 0$. If $c > 0$, then by Theorem 1.50 there is an integer m and integer sequence $r_m, r_{m-1}, \ldots, r_0$ such that

$$c = r_m 2^m + r_{m-1} 2^{m-1} + \cdots + r_1 2^1 + r_0 2^0$$

with $0 \le r_i < 2$ and $r_m = 1$. But c in R implies that $c < 2^{N-1}$ so $m \le N - 2$. Thus, $T([0, \ldots, 0, r_m, r_{m-1}, \ldots, r_1, r_0]) = c$, where the leftmost bit is zero since $m < N - 1$. If $c < 0$, then $0 \le (-c) - 1 \le 2^{N-1} - 1$. According to the discussion above regarding positive or zero values of c, there is a bit string $[s_{N-1}, s_{N-2}, \ldots, s_1]$ in S such that $T([0, s_{N-2}, \ldots, s_1, s_0]) = -c - 1$. Then

$$-c - 1 \quad = \quad s_{N-2} 2^{N-2} + s_{N-3} 2^{N-3} + \cdots + s_1 2^1 + s_0 2^0$$

$$c \quad = \quad -1 - \left[s_{N-2} 2^{N-2} + s_{N-3} 2^{N-3} + \cdots + s_1 2^1 + s_0 2^0 \right]$$

But adding and subtracting $2^{N-1} = 2^{N-2} + 2^{N-3} + \cdots + 2^1 + 2^0 + 1$ and regrouping, we obtain

$$c = -2^{N-1} + (1 - s_{N-2}) 2^{N-2} + (1 - s_{N-3}) 2^{N-3} + \cdots + (1 - s_1) 2^1 + (1 - s_0) 2^0$$

Thus, $T([1, (1 - s_{N-2}), (1 - s_{N-3}), \ldots, (1 - s_0)]) = c$. An example from the table above is if $c = -5$. $-c - 1 = 4 = 0100_{binary}$. Then

$$-5 = c = T([1, (1 - 1), (1 - 0), (1 - 0)]) = T([1, 0, 1, 1,])$$

The function T is one to one. Suppose that each of $[b_{N-1}, b_{N-2}, \ldots, b_1, b_0]$ and $[c_{N-1}, c_{N-2}, \ldots, c_1, c_0]$ is in S and that

$$T([b_{N-1}, b_{N-2}, \ldots, b_1, b_0]) = T([c_{N-1}, c_{N-2}, \ldots, c_1, c_0])$$

We need to show that

$$[b_{N-1}, b_{N-2}, \ldots, b_1, b_0] = [c_{N-1}, c_{N-2}, \ldots, c_1, c_0]$$

Since

$$T([b_{N-1}, b_{N-2}, \ldots, b_1, b_0]) \quad = \quad -b_{N-1} 2^{N-1} + b_{N-2} 2^{N-2}$$

$$+ b_{N-3} 2^{N-3} + \cdots + b_1 2^1 + b_0 2^0$$

and

$$T([c_{N-1}, c_{N-2}, \ldots, c_1, c_0]) \quad = \quad -c_{N-1} 2^{N-1} + c_{N-2} 2^{N-2}$$

$$+ c_{N-3} 2^{N-3} + \cdots + c_1 2^1 + c_0 2^0$$

if we set the right-hand sides equal, we have

$$- (b_{N-1} - c_{N-1})2^{N-1} + (b_{N-2} - c_{N-2})2^{N-2} + \cdots + (b_0 - c_0)2^0 = 0$$

But because $0 \le b_i < 2$ and $0 \le c_i < 2$, we have $-2 < b_i - c_i < 2$ for each i, $0 \le i \le N-1$. For $i = N-1$ we have also that $-2 < -(b_{N-1} - c_{N-1}) < 2$. Thus, by Theorem 1.51, $b_i - c_i = 0$ or $b_i = c_i$ for all i; and consequently,

$$[b_{N-1}, b_{N-2}, \ldots, b_1, b_0] = [c_{N-1}, c_{N-2}, \ldots, c_1, c_0]$$

We now have established a one-to-one mapping, T, between the set S of bit strings of length N and the integers of the set $R = \{i : -2^{N-1} \le i \le 2^{N-1}-1\}$. The bit strings are the two's complement representations of the integers in R.

In considering the integers modulo b for $b > 0$ we saw that every integer a is equivalent to a unique integer r with $0 \le r < b$. The Division Algorithm gives $r \equiv a \pmod{b}$. These b integers were associated with the equivalence classes modulo b. The representatives most often used are $0, 1, 2, \ldots, b-1$. In fact, there are other sets of representatives that are useful. The following theorem describes such a set. The notation $\lfloor x \rfloor$ refers to the greatest integer less than or equal to x. For example, $\lfloor 3 \rfloor = 3$, $\lfloor 3.8 \rfloor = 3$, and $\lfloor -5.2 \rfloor = -6$.

Theorem 1.78 *If a and b are integers and $b > 0$, then there exist unique integers p and s such that*

$$a = pb + s$$

where $-\lfloor b/2 \rfloor \le s \le \lfloor b/2 \rfloor$ *if b is odd and* $-\lfloor b/2 \rfloor \le s < \lfloor b/2 \rfloor$ *if b is even. The integer s is denoted by $\langle\langle a \rangle\rangle_b$.*

Proof. Note that $\lfloor b/2 \rfloor = (b-1)/2$ if b is odd and $\lfloor b/2 \rfloor = b/2$ if b is even.

By the Division Algorithm, there are unique integers q and r such that $a = qb + r$ and $0 \le r < b$.

Assume that b is odd. If $r \le \lfloor b/2 \rfloor$, then let $s = r$ and $p = q$. If $r > \lfloor b/2 \rfloor$, then let $s = r - b$ and $p = q + 1$ so that

$$a = qb + r = (q+1)b + (r-b) = pb + s$$

In this second situation, $0 > s = r - b > \lfloor b/2 \rfloor - b = (b-1)/2 - b = -(b+1)/2$ or $0 > s > -(b+1)/2$. But if $d > c$, then $d \ge c + 1$ so that we have

$$0 > s \ge -\frac{b+1}{2} + 1 = -\left\lfloor \frac{b}{2} \right\rfloor.$$

Assume that b is even. If $r < \lfloor b/2 \rfloor$, then let $s = r$ and $p = q$. If $r \ge \lfloor b/2 \rfloor$, then let $s = r - b$ and $p = q + 1$ so that

$$a = qb + r = (q+1)b + (r-b) = pb + s$$

Again, in this second situation,

$$0 > s = r - b \ge \lfloor b/2 \rfloor - b = b/2 - b = -b/2 = -\lfloor b/2 \rfloor$$

It is left as an exercise to show that p and s are unique. ■

The proof of the following theorem is left to the reader.

Theorem 1.79 *Assume that $b > 0$, a and c are integers. Then $[[a]]_b = [[c]]_b$ if and only if $\langle\langle a \rangle\rangle_b = \langle\langle c \rangle\rangle_b$.*

Thus, the primary residue system modulo b, namely,

$$\{[[0]]_b, [[1]]_b, \ldots, [[(b-1)]]_b\} = \{0, 1, \ldots, (b-1)\}$$

contains one and only one representative of each equivalence class modulo b; and the set

$$\{\langle\langle 0 \rangle\rangle_b, \langle\langle 1 \rangle\rangle_b, \ldots, \langle\langle b-1 \rangle\rangle_b\}$$

is a complete residue system as well, but consists of the set of integers

$$\left\{ s : -\left\lfloor \frac{b}{2} \right\rfloor \leq s \leq \left\lfloor \frac{b}{2} \right\rfloor \right\}$$

if b is odd and the set

$$\left\{ s : -\left\lfloor \frac{b}{2} \right\rfloor \leq s < \left\lfloor \frac{b}{2} \right\rfloor \right\}$$

if b is even. The residues

$$\{\langle\langle 0 \rangle\rangle_b, \langle\langle 1 \rangle\rangle_b, \ldots, \langle\langle b-1 \rangle\rangle_b\}$$

are called the residues modulo b of least absolute value.

In order to make the two's-complement discussion easier, it is helpful to be able to convert between the two complete residue systems.

Corollary 1.80 *If $a \equiv r \pmod{2^N}$ so that $r = [[a]]_{2^N}$ and $s = \langle\langle a \rangle\rangle_{2^N}$, then*

$$s = \begin{cases} r & \text{if } r < 2^{N-1} \\ r - 2^N & \text{if } r \geq 2^{N-1} \end{cases}$$

or

$$\langle\langle a \rangle\rangle_{2^N} = \begin{cases} [[a]]_{2^N} & \text{if } [[a]]_{2^N} < 2^{N-1} \\ [[a]]_{2^N} - 2^N & \text{if } [[a]]_{2^N} \geq 2^{N-1} \end{cases}$$

The next theorem about the mapping T shows that computer modulo 2^N addition of bit strings (considered as nonnegative integers) in S faithfully reproduces integer addition for integers in the set $\{\langle\langle 0 \rangle\rangle_b, \langle\langle 1 \rangle\rangle_b, \ldots, \langle\langle b-1 \rangle\rangle_b\}$ with $b = 2^N$.

Theorem 1.81

$$T([a_{N-1}, a_{N-2}, \dots, a_1, a_0] \oplus [b_{N-1}, b_{N-2}, \dots, b_1, b_0])$$
$$= \langle\langle T([a_{N-1}, a_{N-2}, \dots, a_1, a_0]) + T([b_{N-1}, b_{N-2}, \dots, b_1, b_0]) \rangle\rangle_{2^N}$$

Before proving this theorem, two examples will illustrate how this theorem describes computer addition of integers using two's-complement representation of the integers and computer arithmetic modulo 2^N. Again , assume that there are $N = 4$ bits.

Assume that we want to calculate $3 - 5 = 3 + (-5)$ and note that $3_{ten} = 0011_{binary}$ and $5_{ten} = 0101_{binary}$. The corresponding two's-complement bit strings are $T^{-1}(0011) = [0011]$ and $T^{-1}(-(0101)) = [1011]$ and are the forms stored in the computer assuming that $N = 4$. The arithmetic unit performs the following addition using modulo 2^N arithmetic:

$$
\begin{aligned}
[0011] \oplus [1011] &\equiv 0011 + 1011 \ (\mathrm{mod}\ 2^N) \\
&= [1110]
\end{aligned}
$$

where [1110] is the bit string obtained as the result. It is the two's complement of the integer that is the sum $3 + (-5)$. Thus

$$
\begin{aligned}
T([1110]) &= -1 \cdot 2^3 + 1 \cdot 2^2 + 1 \cdot 2^1 + 0 \cdot 2^0 \\
&= -8 + 6 = -2
\end{aligned}
$$

which is the correct answer.

Next suppose that we want to calculate $5 + 6$ and note that $5_{ten} = 0101_{binary}$ and $6_{ten} = 0110_{binary}$. The corresponding two's-complement bit strings are $T^{-1}(0101) = [0101]$ and $T^{-1}(0110) = [0110]$ and are the forms stored in the computer assuming that $N = 4$. The arithmetic unit performs the following addition using modulo 2^N arithmetic:

$$
\begin{aligned}
[0101] \oplus [0110] &\equiv 0101 + 0110 \ (\mathrm{mod}\ 2^N) \\
&= [1011]
\end{aligned}
$$

where [1011] is the bit string obtained as the result. It is the two's complement of the integer that is given to be the sum $5 + 6$. Thus

$$
\begin{aligned}
T([1011]) &= -1 \cdot 2^3 + 0 \cdot 2^2 + 1 \cdot 2^1 + 1 \cdot 2^0 \\
&= -8 + 3 = -5
\end{aligned}
$$

The value reported for the sum is -5, which is not the correct sum of $5 + 6 = 11_{ten}$; however, $-5 \equiv 11 \ (\mathrm{mod}\ 2^4)$ using Corollary 1.80. In this case, the addition is said to have "overflowed." Thus, to obtain the correct sum using computer integer arithmetic, we need to insure that the integers we add are in the proper range not to "overflow."

Proof. (Theorem 1.81) Let

$$B(a) = [a_{N-1}, a_{N-2}, \dots, a_1, a_0] \text{ so that } a = a_{N-1} 2^{N-1} + \dots + a_1 2^1 + a_0 2^0$$

$$B(b) = [b_{N-1}, b_{N-2}, \dots, b_1, b_0] \text{ so that } b = b_{N-1} 2^{N-1} + \dots + b_1 2^1 + b_0 2^0$$

By definition of $\oplus$, if $c \equiv a + b \pmod{2^N}$, then $B(c) = B(a) \oplus B(b)$. Thus

$$
\begin{aligned}
a + b &= (a_{N-1}2^{N-1} + \cdots + a_1 2^1 + a_0 2^0) + (b_{N-1}2^{N-1} + \cdots + b_1 2^1 + b_0 2^0) \\
&= d_N 2^N + d_{N-1}2^{N-1} + \cdots + d_0 2^0 \\
&= d
\end{aligned}
$$

where $0 \le d < 2^{N+1}$ and $0 \le d_i \le 1$ for each i. Since $d \equiv c \pmod{2^N}$ with $0 \le c < 2^{N-1}$, we have

$$
\begin{aligned}
c &= d_{N-1}2^{N-1} + d_{N-2}2^{N-2} + \cdots + d_0 2^0 \\
&= c_{N-1}2^{N-1} + c_{N-2}2^{N-2} + \cdots + c_0 2^0
\end{aligned}
$$

where $c_i = d_i$ for $0 \le i \le N - 1$. Thus,

$$
\begin{aligned}
T([a_{N-1}, a_{N-2}, \ldots, a_0] \oplus [b_{N-1}, b_{N-2}, \ldots, b_0]) &= T(B(a) \oplus B(b)) \\
&= T(B(c))
\end{aligned}
$$

But

$$
\begin{aligned}
T(B(c)) &= T([d_{N-1}, d_{N-2}, \ldots, d_1, d_0]) \\
&= -d_{N-1}2^{N-1} + d_{N-2}2^{N-2} + \cdots + d_0 2^0
\end{aligned}
$$

which is in the set $R = \{i : -2^{N-1} \le i \le 2^{N-1} - 1\}$ by definition of the mapping T.

On the other hand,

$$
T([a_{N-1}, a_{N-2}, \ldots, a_1, a_0]) + T([b_{N-1}, b_{N-2}, \ldots, b_1, b_0])
$$

$$
= (-a_{N-1}2^{N-1} + a_{N-2}2^{N-2} + \cdots + a_0 2^0)
$$

$$
+ (-b_{N-1}2^{N-1} + b_{N-2}2^{N-2} + \cdots + b_0 2^0)
$$

$$
= -(a_{N-1} + b_{N-1})2^{N-1} + (a_{N-2} + b_{N-2})2^{N-2} + \cdots + (a_0 + b_0)2^0
$$

$$
= -(a_{N-1} + b_{N-1})2^N + d
$$

$$
\equiv c \pmod{2^N}
$$

But

$$
c = d_{N-1}2^{N-1} + d_{N-2}2^{N-2} + \cdots + d_0 2^0
$$

If $d_{N-1} = 0$, then $c = -d_{N-1}2^{N-1} + d_{N-2}2^{N-2} + \cdots + d_0 2^0$ and the conclusion of the theorem holds. If $d_{N-1} = 1$, then $c \ge 2^{N-1}$ so that, by the Corollary 1.80, we have

$$
\begin{aligned}
c &\equiv (d_{N-1}2^{N-1} + d_{N-2}2^{N-2} + \cdots + d_0 2^0) - 2^N \pmod{2^N} \\
&\equiv -d_{N-1}2^{N-1} + d_{N-2}2^{N-2} + \cdots + d_0 2^0 \pmod{2^N}
\end{aligned}
$$

and the conclusion of the theorem holds in this case also. ■

Exercises

1. If $b > 1$, $0 \le a_i < b$ for each i, $J > N - 1$, and

$$A = a_J b^J + a_{J-1} b^{J-1} + \cdots + a_N b^N + a_{N-1} b^{N-1} + \cdots + a_0 b^0$$

then prove that

$$A \equiv a_{N-1} b^{N-1} + a_{N-2} b^{N-2} \cdots + a_0 b^0 \pmod{b^N}$$

That is, if A has radix b representation, then the value of A modulo b^N is obtained by removing all terms with factor b^k for $k \ge N$; or, by keeping the rightmost N "digits" of the base b representation $a_J a_{J-1} \ldots a_N a_{N-1} \ldots a_0$.

2. Calculate:

    (a) $[[1011001101_{two}]]_{2^5}$

    (b) $[[1011001101_{three}]]_{3^5}$

    (c) $[[1011001101_{three}]]_{2^5}$

    (d) $[[4584357241_{ten}]]_{10^7}$ [$10 =$ ten here]

3. If the number of bits available for representing integers is 8, then:

    (a) How many distinct integers can be represented in two's-complement form?

    (b) What is the range of integers that can be represented in two's-complement form?

4. Repeat Exercise 3 if there are 32 bits available.

5. For 8-bit integers, let A be the least integer and B the greatest integer representable in two's complement. Find:

    (a) The decimal form

    (b) The signed binary form

    (c) The two's-complement bit string for these integers:

    $$A, A+1, A+2, -3, -2, -1, 0, 1, 2, 3, B-2, B-1, \text{ and } B$$

6. Complete the proof of Theorem 1.78 by showing that p and s are unique.

7. If, using the notation of this section, $c = a_{N-1}2^{N-1} + a_{N-2}2^{N-2} + \cdots + a_0 2^0$ with $a_i = 0$ or 1, prove that $0 \le c \le 2^N - 1$.

8. Using Theorem 1.78, find integers p and s with $a = pb + s$ or such that $\langle\langle a \rangle\rangle_b = s$ for the following situations:

    (a) $b = 8$ and $a = 0, 1, 2, 3, 4, 5, 6, 7, 8$

    (b) $b = 9$ and $a = 0, 1, 2, 3, 4, 5, 6, 7, 8, 9$

    (c) $b = 10$ and $a = 45, 54$

    (d) $b = 11$ and $a = 45, 54$

    (e) $b = 16$ and $a = 62, 63, 64, 65, 66$

    (f) $b = 256$ and $a = 5132, 7000$

9. Assume that a and b are integers with $0 \le a$, $b \le 2^N - 1$, $N = 8$ and we have bit strings

    $$B(a) = [10100110]$$

    $$B(b) = [00000101]$$

    where a is the nonnegative integer 10100110_{two} and b is the nonnegative integer $00000101_{two} = 101_{two}$. Calculate the bit string $B(c)$ that is the result generated by the computer arithmetic for adding bit strings $B(a)$ and $B(b)$, namely,

    $$B(c) = [10100110] \oplus [00000101]$$

    where $c \equiv a + b \pmod{2^N}$. Calculate the integers external to the computer that correspond to $B(a)$, $B(b)$, and $B(c)$, namely,

    $$T(B(a)), T(B(b)), \text{ and } T(B(c))$$

    Verify that $T(B(c)) \equiv T(B(a)) + T(B(b)) \pmod{2^N}$.

10. Repeat Exercise 9 for the following:

    (a) $B(a) = [00100110]$
        $B(b) = [00010111]$

    (b) $B(a) = [10101111]$
        $B(b) = [11011101]$

Chapter 2
PRIMES

2.1 Introduction

Some integers cannot be factored into a product of integers except in a trivial way. Such integers are called primes. Nonprime integers different from ± 1 can be factored into a product of primes which reveal important characteristics of and relationships among the integers. We can often find the characteristics of an integer in terms of the same characteristics of its prime factors. We reiterate the definition of this important class of integers that was given in Chapter 1.

Definition 2.1 *An integer greater than* 1 *is called* prime *if its only positive factors are itself and* 1. *A positive integer greater than* 1 *is* composite *if it is not a prime integer.*

Of the first 10 positive integers, only 2, 3, 5, and 7 are prime integers. On the other hand, the integers $4 = 2 \cdot 2$, $6 = 2 \cdot 3$, $8 = 2 \cdot 4$, $9 = 3 \cdot 3$, and $10 = 2 \cdot 5$ are composite. Thus, if $n = r \cdot s$ with $1 < r < n$ and $1 < s < n$, then n is composite. By definition, the integer 1 is neither prime nor composite. The integer 2 is the only even prime. It is easy to determine which of the small integers are prime by attempting a division by all smaller integers because the number of such possibilities is relatively small; however, deciding whether a large integer is prime can be a difficult task. We will discuss later some ways to reduce the amount of work involved for showing primeness.

2.2 Prime Factorization

The primes form a set of building blocks for the integers since any positive integer greater than 1 can be represented in terms of primes as follows:

Theorem 2.2 *Every positive integer is equal to* 1, *is a prime, or may be written as a product of primes.*

Proof. We prove this theorem using the second principle of induction. The theorem is certainly true for $n = 1$. Assume that it is true for all positive integers n less than k. If k is a prime, then the theorem is also true for k. If k is not a prime, then it is divisible by some integer p and $k = pq$ where neither p nor q is equal to either k or 1. Since, by Theorem 1.24, p and q are less than k, by the induction hypothesis they are primes or may be written as a product of primes. Hence $k = pq$ may be written as a product of primes. ∎

The integer 37 is prime. The integer $1554985071 = 3 \cdot 3 \cdot 4463 \cdot 38713$ is a product of four prime factors, two of which are the same prime.

In the next theorem we will see that if an integer n is divisible by a prime p, then it is not possible to factor n in such a way that p does not divide at least one of the factors of n.

Theorem 2.3 *If p is a prime and $p \mid ab$, where a and b are positive integers, then $p \mid a$ or $p \mid b$.*

Proof. If $p \mid a$, then the conclusion holds. On the other hand, assume that p does not divide a. Since p and a are relatively prime, $p \mid b$ by Theorem 1.49. ∎

Lemma 2.4 *If a prime number p divides a product of positive integers $q_1 q_2 \cdots q_n$, then p divides q_i for some i, $1 \le i \le n$.*

Proof. We prove this lemma using induction on n, the number of factors in the product. If $n = 1$, the lemma is obviously true. Assume that the lemma is true for $n = k$; that is, if p divides any product of k integers, then p divides one of the k factors. Assume that p divides a product of $k + 1$ integers, say, $p \mid q_1 q_2 \cdots q_k q_{k+1}$ so that $p \mid (q_1 q_2 \cdots q_k) q_{k+1}$. If p divides q_{k+1}, then we are done. If p does not divide q_{k+1}, then, by Theorem 2.3, $p \mid (q_1 \cdots q_k)$. But since $q_1 \cdots q_k$ is the product of k integers, by the induction hypothesis, $p \mid q_i$ for some $1 \le i \le k$. Hence $p \mid q_i$ for some i, $1 \le i \le k + 1$ and we are done. ∎

Lemma 2.5 *If a prime number p divides a product of primes q_1, q_2, ..., and q_n, then $p = q_i$ for some i, $1 \le i \le n$.*

Proof. By the Lemma 2.4, p divides q_i for some $1 \le i \le n$. Since p and q_i are both prime, $p = q_i$. ∎

The series of theorems above leads to the main result of this chapter which is called the Fundamental Theorem of Arithmetic or the Unique Prime Factorization Theorem.

Theorem 2.6 (Unique Prime Factorization) *Any positive integer m that is greater than 1 is a prime or can be written as a product of primes where this product is unique except for the arrangement of the primes.*

Proof. Since m can be written as a product of primes, assume that $q_1 q_2 \cdots q_n$ and $p_1 p_2 \cdots p_s$ are two ways of writing m as a product of primes. We shall prove the theorem using induction on n, the number of prime factors in the first product. If $n = 1$, the theorem is trivially true. Assume that the theorem is true when $q_1 q_2 \cdots q_k = p_1 p_2 \cdots p_s$; that is, if $m = q_1 q_2 \cdots q_k = p_1 p_2 \cdots p_s$, then $k = s$ and the product is unique up to the order of primes. Assume that $m = q_1 q_2 \cdots q_{k+1} = p_1 p_2 \cdots p_{s'}$. Since q_{k+1} divides $p_1 p_2 \cdots p_{s'}$, then $q_{k+1} = p_i$ for some $1 \le i \le s'$. Divide both products by q_{k+1} or use the cancellation property (Axiom I8). Then $q_1 q_2 \cdots q_k = p_1 p_2 \cdots p_{i-1} p_{i+1} \cdots p_{s'}$. But by induction, $k = s' - 1$ and the product is unique up to the order of primes. Hence $k + 1 = s'$ and the factorization of m is unique up to order of primes. ∎

For example,

$$n = 39616304 = 2 \cdot 13 \cdot 7 \cdot 2 \cdot 23 \cdot 13 \cdot 2 \cdot 13 \cdot 2 \cdot 7 = 2 \cdot 2 \cdot 2 \cdot 2 \cdot 7 \cdot 7 \cdot 13 \cdot 13 \cdot 13 \cdot 23$$

are two factorizations of n; however, the same primes are used the same number of times in both products. Only the order in which the factors are written differs. In fact, there are 12600 distinct factorizations of n using the 10 prime factors; but there are no factorizations without exactly four 2's, two 7's, three 13's, and one 23. Usually, the prime factors are grouped and combined using exponential notation as in

$$n = 2^4 7^2 13^3 23^1$$

Corollary 2.7 *Every positive integer m greater than 1 can be written uniquely, except for order, in the form $q_1^{k(1)} q_2^{k(2)} \cdots q_n^{k(n)}$, where $k(1)$, $k(2)$, ..., and $k(n)$ are positive integers.*

At this point we can see why 1 is not allowed to be a prime since we would not have the Unique Prime Factorization Theorem. When discussing the factorization of several integers using the representation given by the preceding corollary, it is often a notational convenience to allow a prime to have zero for an exponent. This practice normally causes no confusion since $q_i^0 = 1$ if $q_i \ge 1$. If the prime factorization of an integer is known, then the primes forming the factorization of any divisor of that integer are a subset of those of the dividend.

The proofs of the following two theorems are left to the reader.

Theorem 2.8 *If $a = p_1^{a(1)} \cdots p_k^{a(k)}$ and $b \mid a$, then $b = p_1^{b(1)} \cdots p_k^{b(k)}$ where $0 \le b(i) \le a(i)$ for all i; and, conversely.*

Theorem 2.9 *Let $a = p_1^{a(1)} p_2^{a(2)} p_3^{a(3)} \cdots p_k^{a(k)}$ and $b = p_1^{b(1)} p_2^{b(2)} p_3^{b(3)} \cdots p_k^{b(k)}$ where the p_i are primes contained in either a or b and some of the exponents may be 0. Let $m(i) = \min(a(i), b(i))$ and $M(i) = \max(a(i), b(i))$ for $1 \le i \le k$. Then*

$$\gcd(a, b) = p_1^{m(1)} p_2^{m(2)} p_3^{m(3)} \cdots p_k^{m(k)}$$

and

$$\operatorname{lcm}(a, b) = p_1^{M(1)} p_2^{M(2)} p_3^{M(3)} \cdots p_k^{M(k)}$$

As an application of Theorem 2.9, let $a = 195000$ and $b = 10435750$. The factorizations of a and b are

$$a = 2^3 3^1 5^4 13^1 \quad \text{and} \quad b = 2^1 5^3 13^3 19^1$$

Thus,

$$
\begin{aligned}
\gcd(195000, 10435750) &= 2^{\min(3,1)} 3^{\min(1,0)} 5^{\min(4,3)} 13^{\min(1,3)} 19^{\min(0,1)} \\
&= 2^1 3^0 5^3 13^1 19^0 = 2^1 5^3 13^1 = 3250
\end{aligned}
$$

and

$$
\begin{aligned}
\text{lcm}(195000, 10435750) &= 2^{\max(3,1)} 3^{\max(1,0)} 5^{\max(4,3)} 13^{\max(1,3)} 19^{\max(0,1)} \\
&= 2^3 3^1 5^4 13^3 19^1 = 626145000
\end{aligned}
$$

We could also have applied Theorem 1.33 using the Euclidean Algorithm as follows:

$$
\begin{aligned}
10435750 &= 195000 \cdot 53 + 100750 \\
195000 &= 10750 \cdot 1 + 94250 \\
100750 &= 94250 \cdot 1 + 6500 \\
94250 &= 6500 \cdot 14 + 3250 \\
6500 &= 3250 \cdot 2 + 0
\end{aligned}
$$

which gives $\gcd(195000, 10435750) = 3250$. Theorem 1.41 implies that

$$
\begin{aligned}
\text{lcm}(195000, 10435750) &= 195000 \cdot 10435750 / \gcd(195000, 10435750) \\
&= 195000 \cdot 10435750 / 3250 = 626145000
\end{aligned}
$$

In a numerical calculation of $\gcd(a, b)$ and $\text{lcm}(a, b)$, applying the Euclidean Algorithm is usually faster than obtaining the prime factorization of a and b, particularly for large integers. If one already has the prime factorizations available, then the method of Theorem 2.9 may be easier. It will be seen that the fact that there is a unique factorization in terms of a product of primes is a nearly indispensable tool for obtaining many important theoretical results in number theory.

Since primes seem to be so useful, if there were only a finite number of them, then number theory perhaps might be simpler; however, the following theorem shows that such finiteness is not the case.

Theorem 2.10 (Euclid) *There are infinitely many prime integers.*

Proof. Assume that there are only finitely many primes, say $p_1, p_2, \ldots, p_k$. Consider the integer $(p_1 p_2 \cdots p_k) + 1$. Let p_r be a prime and suppose that $p_r \mid ((p_1 p_2 \cdots p_k) + 1)$. But $p_r \mid (p_1 p_2 \cdots p_k)$, which implies that $p_r \mid 1$, a contradiction. Hence $(p_1 p_2 \cdots p_k) + 1$ is prime, also a contradiction since it is not one of the finitely many primes. Therefore, our assumption that there are only finitely many primes is false and there must be infinitely many prime integers. ∎

Because there are infinitely many primes and because the prime factorization of integers is important, it would be good to have quick and easy ways to decide whether a given positive integer is prime or composite. There are algorithms that offer significant improvements over brute force methods of testing integers for primeness. The next theorem shows that only some of the possible factors need to be considered to test an integer for primeness.

Theorem 2.11 *If the positive integer n is a composite integer, then n has a prime factor p such that $p^2 \leq n$.*

Proof. Let p be the smallest prime factor of n. If n factors into r and s, then $p \leq r$ and $p \leq s$. Hence $p^2 \leq rs = n$. ∎

For example, to determine whether $n = 521$ is prime, we only need to consider primes p that are less than or equal to 22 because $22^2 = 484$ and $23^2 = 529$. The primes less than or equal to 22 are 2, 3, 5, 7, 11, 13, 17, and 19. Trying each of these integers, we find that none of them divides 521. Therefore, 521 itself is prime by the preceding theorem.

An ancient method of determining primes is called the Sieve of Eratosthenes. We illustrate this method to determine the primes between 1 and 100. First list the integers between 1 and 100:

1, 2, 3, 4, 5, 6, 7, 8, 9, 10, 11, 12, 13, 14, 15, 16, 17, 18, 19, 20, 21, 22, 23, 24, 25, 26, 27, 28, 29, 30, 31, 32, 33, 34, 35, 36, 37, 38, 39, 40, 41, 42, 43, 44, 45, 46, 47, 48, 49, 50, 51, 52, 53, 54, 55, 56, 57, 58, 59, 60, 61, 62, 63, 64, 65, 66, 67, 68, 69, 70, 71, 72, 73, 74, 75, 76, 77, 78, 79, 80, 81, 82, 83, 84, 85, 86, 87, 88, 89, 90, 91, 92, 93, 94, 95, 96, 97, 98, 99, 100

Beginning with the first prime, 2, we mark in boldface all multiples of 2. This procedure is straightforward since every second integer larger than 2 is to be marked:

1, 2, 3, **4**, 5, **6**, 7, **8**, 9, **10**, 11, **12**, 13, **14**, 15, **16**, 17, **18**, 19, **20**, 21, **22**, 23, **24**, 25, **26**, 27, **28**, 29, **30**, 31, **32**, 33, **34**, 35, **36**, 37, **38**, 39, **40**, 41, **42**, 43, **44**, 45, **46**, 47, **48**, 49, **50**, 51, **52**, 53, **54**, 55, **56**, 57, **58**, 59, **60**, 61, **62**, 63, **64**, 65, **66**, 67, **68**, 69, **70**, 71, **72**, 73, **74**, 75, **76**, 77, **78**, 79, **80**, 81, **82**, 83, **84**, 85, **86**, 87, **88**, 89, **90**, 91, **92**, 93, **94**, 95, **96**, 97, **98**, 99, **100**

Continuing with the next prime, 3, we mark all multiples of 3. Again this procedure is straightforward since every third integer larger than 3 is to be marked:

1, 2, 3, **4**, 5, **6**, 7,**8**, **9**, **10**, 11, **12**, 13, **14**, **15**, **16**, 17, **18**, 19, **20**, **21**, **22**, 23,

24, 25, **26**, **27**, **28**, 29, **30**, 31, **32**, **33**, **34**, 35, **36**, 37, **38**, **39**, **40**, 41, **42**, 43,
44, **45**, **46**, 47, **48**, 49, **50**, **51**, **52**, 53, **54**, 55, **56**, **57**, **58**, 59, **60**, 61, **62**, **63**,
64, 65, **66**, 67, **68**, **69**, **70**, 71, **72**, 73, **74**, **75**, **76**, 77, **78**, 79, **80**, **81**, **82**, 83,
84, 85, **86**, **87**, **88**, 89, **90**, 91, **92**, **93**, **94**, 95, **96**, 97, **98**, **99**, **100**

We now mark all multiples of the next prime, 5:

1, 2, 3, **4**, 5, **6**, 7, **8**, **9**, **10**, 11, **12**, 13, **14**, **15**, **16**, 17, **18**, 19, **20**, **21**, **22**, 23,
24, 25, **26**, **27**, **28**, 29, **30**, 31, **32**, **33**, **34**, 35, **36**, 37, **38**, **39**, **40**, 41, **42**,
43, **44**, **45**, **46**, 47, **48**, 49, **50**, **51**, **52**, 53, **54**, **55**, **56**, **57**, **58**, 59, **60**, 61, **62**,
63, **64**, **65**, **66**, 67, **68**, **69**, **70**, 71, **72**, 73, **74**, **75**, **76**, 77, **78**, 79, **80**, **81**,
82, 83, **84**, **85**, **86**, **87**, **88**, 89, **90**, 91, **92**, **93**, **94**, **95**, **96**, 97, **98**, **99**, **100**

Next mark all multiples of the next prime, 7:

1, 2, 3, **4**, 5, **6**, 7, **8**, **9**, **10**, 11, **12**, 13, **14**, **15**, **16**, 17, **18**, 19, **20**, **21**, **22**, 23,
24, **25**, **26**, **27**, **28**, 29, **30**, 31, **32**, **33**, **34**, **35**, **36**, 37, **38**, **39**, **40**, 41, **42**, 43,
44, **45**, **46**, 47, **48**, **49**, **50**, **51**, **52**, 53, **54**, **55**, **56**, **57**, **58**, 59, **60**, 61, **62**,
63, **64**, **65**, **66**, 67, **68**, **69**, **70**, 71, **72**, 73, **74**, **75**, **76**, **77**, **78**, 79, **80**, **81**,
82, 83, **84**, **85**, **86**, **87**, **88**, 89, **90**, **91**, **92**, **93**, **94**, **95**, **96**, 97, **98**, **99**, **100**

Since 7 is the largest prime whose square is less than or equal to 100, we need continue no further. The numbers greater than 1 which are not marked are the primes less than 100.

The following theorem is helpful in attempting to factor large numbers by allowing divisibility by the primes less than or equal to eleven to be determined without actually performing the division.

Theorem 2.12 *Let n be an integer expressed in base ten.*

(a) *n is divisible by 2 if and only if its last digit is divisible by 2.*

(b) *n is divisible by 3 if and only if the sum of its digits is divisible by 3.*

(c) *n is divisible by 5 if and only its last digit is either 0 or 5.*

(d) *n, with final digit u, is divisible by 7 if and only if $\dfrac{n-u}{10} - 2u$ is divisible by 7.*

(e) *n is divisible by 9 if and only if the sum of its digits is divisible by 9.*

(f) *n with digits $n_1, n_2, n_3, \ldots, n_{k-1}$, and n_k is divisible by 11 if and only if $n_1 - n_2 + n_3 - \cdots + (-1)^k n_{k-1} + (-1)^{k+1} n_k$ is divisible by 11.*

Proof. **(a)** Let n have digits $d_k, d_{k-1}, \ldots, d_2, d_1$ so that $n = d_k \cdot 10^{k-1} + d_{k-1} \cdot 10^{k-2} + \cdots + d_2 \cdot 10 + d_1$. Since $10 \equiv 0 \pmod{2}$, $n \equiv d_1 \pmod{2}$. Hence n is even if and only if d_1 is.

(b) Let n have digits $d_k, d_{k-1}, \ldots, d_2, d_1$ so that $n = d_k \cdot 10^{k-1} + d_{k-1} \cdot 10^{k-2} + \cdots + d_2 \cdot 10 + d_1$. Since $10 \equiv 1 \pmod{3}$, $n \equiv d_k 1^{k-1} + d_{k-1} 1^{k-2} + \cdots + d_2 1 + d_1 \equiv d_k + d_{k-1} + \cdots + d_2 + d_1 \pmod{3}$. Hence n is divisible by 3 if and only if the sum of its digits is divisible by 3.

(c) Left to reader.

(d) Let $q = ((n-u)/10) - 2u$. Then $n - u = 10(q+2u)$ and $n = 10q + 21u$ so that $n \equiv 10q \pmod{7}$. Hence n is divisible by 7 if and only if $10q$ is divisible by 7. Since 7 and 10 are relatively prime, n is divisible by 7 if and only if q is divisible by 7.

(e) Left to reader.

(f) Left to reader. [*Hint:* $10 \equiv -1 \pmod{11}$.] ∎

To illustrate this theorem, suppose that we want to factor $n = 43821$. The prime $2 \nmid n$ because $2 \nmid 1$ and 1 is the last digit of n. Also, $3 \mid n$ because the sum of the digits of n is $4 + 3 + 8 + 2 + 1 = 18$ and $3 \mid 18$. Thus, $43821 = 3 \cdot 14607$. Again $3 \mid 14607$ because $1 + 4 + 6 + 0 + 7 = 18$ is divisible by 3 so that $n = 3^2 \cdot 4869$. Since $3 \mid (4 + 8 + 6 + 9)$, $3 \mid 4869$ and $n = 3^3 \cdot 1623$. Alternatively, we could have noticed that $9 \mid 14607$ because $9 \mid (1+4+6+0+7)$ in order to obtain $n = 3^3 \cdot 1623$. Since $3 \mid (1+6+2+3)$, $3 \mid 1623$ and $n = 3^4 \cdot 541$. $3 \nmid 541$ because $3 \nmid (5 + 4 + 1)$. $5 \nmid 541$ because the last digit of 541 is not 0 or 5. Consider $(541 - 1)/10 - 2 \cdot 1 = 52$. Since $7 \nmid 52$, $7 \nmid 541$. Since $7 \nmid 541$ and $n = 3^4 \cdot 541$, $7 \nmid n$. To check whether 11 is a factor of 541, we note that $5 - 4 + 1 = 2$ is not divisible by 11. These tests exhaust the divisibility criteria given in Theorem 2.12. However, since 23 is the largest prime with $23^2 \leq 541$, only the primes 13, 17, 19, and 23 need to be tested against 541. Since none of these divides 541, 541 is prime. Thus, the prime factorization of 43821 is $3^4 \cdot 541^1$.

The next theorem is the basis of another method of factoring primes called Fermat's Factorization Method.

Theorem 2.13 *An odd integer $n > 1$ is nonprime if and only if there are nonnegative integers p and q such that $n = p^2 - q^2$ with $p - q > 1$.*

Proof. Obviously, if n can be expressed as the difference of two squares of nonnegative integers, say $n = p^2 - q^2$, then n can be factored into $p - q$ and $p + q$. Since $p - q > 1$, then $p + q > 1$ also; and n is not prime.

Conversely, if $n = rs$ with $r \geq s > 1$, then n can be expressed as $((r + s)/2)^2 - ((r-s)/2)^2$ because, since n is odd, r and s are odd; and consequently, $r + s$ and $r - s$ are even. Letting $p = (r + s)/2$ and $q = (r - s)/2$ we see that p and q are nonnegative and $p - q = s > 1$. If $n = 1$, let $p = 1$ and $q = 0$. ∎

In using this method we try to find integers p and q such that $n = p^2 - q^2$ or, equivalently, such that $p^2 = n + q^2$ or $q^2 = p^2 - n$. If the first equation is used, we let $q = 1, 2, \ldots$ until $n + q^2$ is a perfect square. If we have not reached a perfect square before $q = (n-1)/2$, then we will when $q = (n-1)/2$, which gives $n + q^2 = ((n+1)/2)^2$ and factors n into $n \cdot 1$. Obviously, since q has the form $(r - s)/2$ where r and s are factors of n, q cannot exceed $(n - 1)/2$. Hence if we have not reached a perfect square before $q = (n - 1)/2$, n is a prime.

If the second equation is used, that is, $q^2 = p^2 - n$, then let m be the smallest integer such that $m^2 \geq n$, and let $p = m, m + 1, \ldots$ until $p^2 - n$ is a perfect square. As above, q cannot exceed $(n - 1)/2$; so if we have not reached a perfect square before $p = (n + 1)/2$, n is a prime. The advantage of using the second squares method is that we check smaller numbers to see if they are squares.

For example, consider using the form $p^2 = n + q^2$ to test $n = 527$ for being prime. We consider $q = 1, 2, \ldots, (n - 1)/2$.

q	$n + q^2$
1	$527 + 1 = 528$
2	$527 + 4 = 531$
3	$527 + 9 = 536$
4	$527 + 16 = 543$
5	$527 + 25 = 552$
6	$527 + 36 = 563$
7	$527 + 49 = 576 = (24)^2$

So $n = 527$ is composite and its factors may be calculated:

$$
\begin{aligned}
527 &= (24)^2 - 7^2 \\
&= (24 - 7)(24 + 7) \\
&= 17 \cdot 31
\end{aligned}
$$

Not every q from 1 to $(n - 1)/2$ needs to be checked unless n is prime.

After the Fundamental Theorem of Arithmetic, two important questions that remained in number theory with regard to primes were: (a) how to locate large primes and (b) find the distribution of the primes among the integers. Large tables were computed to determine primes and factorization of integers. Anton Felkel computed the factors of all numbers up to 408000 not divisible by 2, 3, or 5. It was published by the Imperial Treasury of Austria but was not a best seller. Hence most of the copies ended up as paper in the manufacture of cartridges for use against the Turks. J. P. Kulik (1773-1863), a professor of mathematics at the University of Prague, devoted twenty years of his life to preparing, without assistance, the factors of numbers up to 10^8. His work was not published. With the advent of computers, such published tables are unnecessary since primes in any publishable table on paper or magnetic computer storage media may be generated with the Sieve of Eratosthenes faster than the table may be read.

Exercises

1. Give the prime factorization of each of the following:

(a) 1080   (b) 539   (c) 955   (d) 583

(e) 349   (f) 9017   (g) 31752

2.   Determine which of the following are divisible by 3, 5, 7, 9, or 11 using methods described in the text:

(a) 1969   (b) 1421   (c) 116424

(d) 28350   (e) 17303   (f) 1089

3.   Show that the average of two consecutive primes is never a prime.

4.   Prove that any positive integer is congruent to its last digit modulo n when $n = 5$. For what other values of n is the statement true?

5.   Prove these theorems:

(a) A number n is a square if and only if all of the exponents in the prime factorization of n are even.

(b) State and prove a similar theorem characterizing n as a perfect cube.

6.   Determine all of the prime numbers less than 250.

7.   Prove that if $p \mid ab$ implies that $p \mid a$ or $p \mid b$ for all a and b, then p is a prime.

8.   Prove that an integer n is divisible by 5 if and only if its last digit is either 0 or 5 (Theorem 2.12(c)).

9.   Prove that an integer n is divisible by 9 if and only if the sum of its digits is divisible by 9 (Theorem 2.12(e)).

10.   Prove that the integer n with digits $n_1, n_2, n_3, \ldots, n_k$ is divisible by 11 if and only if $n_1 - n_2 + n_3 - \cdots + (-1)^k n_{k-1} + (-1)^{k+1} n_k$ is divisible by 11 (Theorem 2.12(f)).

11.   Prove Theorem 2.8. If $a = p_1^{a(1)} \cdots p_k^{a(k)}$ and $b \mid a$, then $b = p_1^{b(1)} \cdots p_k^{b(k)}$ where $0 \le b(i) \le a(i)$ for all i; and, conversely.

12.   Prove Theorem 2.9.

13.   Use the form $p^2 - n$ from the proof of Theorem 2.13 to determine whether $n = 1001$ is prime.

14.   Given any polynomial function $f(x)$ with integral coefficients, prove that there is an integer c so that $f(c)$ is not prime.

15.   Prove that the product of n consecutive positive integers is divisible by $n!$.

16.   If p and q are primes greater than or equal to 5, prove that either $p + q$ or $p - q$ is divisible by 3 and hence $p^2 - q^2$ is divisible by 24.

17.   Prove that if n is a nonprime integer greater than 4, then n divides $(n-1)!$

18.   Show there are no prime triplets, that is, three consecutive odd numbers each of which is prime, except 3, 5, and 7.

19.   Prove that the number of positive divisors of $a = p_1^{a(1)} \cdots p_k^{a(k)}$, where p_i is prime, is

$$\prod_{i=1}^{k} [a(i) + 1]$$

2.3   Distribution of the Primes

In Section 2.2 we saw that there are infinitely many primes. This fact means that given any integer m, there is a prime p that is greater than m; that is, $p > m$. Except for the pair of primes 2 and 3, the closest two primes can be is for there to be only one (even) integer n between them so that $n - 1$ and $n + 1$ are prime. Such pairs of primes are called *twin primes*. Of the positive integers less than or equal to 100, the Sieve of Eratosthenes revealed these twin primes.

n	Twins
4	3 and 5
6	5 and 7
12	11 and 13
18	17 and 19
30	29 and 31
42	41 and 43
60	59 and 61
72	71 and 73

The two large twin prime pairs

$$1706595 \cdot 2^{11235} \pm 1 \quad \text{and} \quad 571305 \cdot 2^{7701} \pm 1$$

were discovered by B. K. Parady, J. R. Smith, and S. Zarantonello [61]. There are exactly 224376048 twin primes up to $n = 10^{11}$. (See Richard P. Brent's "Irregularities in the Distribution of Primes and Twin Primes" [9]) It is conjectured that there are infinitely many twin primes; however, this has not been proved.

In spite of the seemingly frequent occurrence of twin primes, it is also true that there are arbitrarily long sequences of consecutive composite integers.

Theorem 2.14 *In the sequence of positive integers, there are sequences of consecutive composite integers of any finite length. Hence, there are gaps of arbitrary length between primes.*

Proof. For an arbitrary positive integer n greater than 1, the integers $n! + 2$, $n! + 3$, ... , and $n! + n$ are easily shown to be composite numbers. ■

The factorial function $n!$ increases rapidly with n so that to guarantee that there are 11 consecutive nonprime integers using the method of the proof of Theorem 2.14, we may have to go as far as

$$n! + 2 = 12! + 2 = 479001602$$

to find such a sequence. This method does not produce the first such sequence of 11 consecutive nonprime integers. In fact, there is a gap of 13 consecutive composite integers between 113 (prime) and 127 (prime). This gap is the first occurrence of at least 11 consecutive composites.

Theorem 2.14 gives information about how the primes occur among the positive integers. Related questions would be the following. Are the primes uniformly distributed among the positive integers? Do they tend to occur in clumps? Is the percent of the primes in the set $\{k : 1 \leq k \leq n\}$ approximately constant for all n or does the percent increase or decrease? The number of primes less than or equal to an integer x is germane to these questions. Although we could restrict the discussion to integers, it is convenient and customary to allow x to be a real number rather than be restricted to integers.

Definition 2.15 *For x a real number, $\pi(x)$ is the number of primes that are less than or equal to x; that is, $\pi(x)$ is the number of integers in the set $\{p : p$ is prime and $p \leq x\}$.*

Since the first few primes are $2, 3, 5, 7, 11, 13, \ldots$ we see that

$$
\begin{aligned}
\pi(1) &= 0 \\
\pi(2) &= 1 \\
\pi(2.7) &= 1 \\
\pi(3) &= 2 \\
\pi(9.2) &= 4 \\
\pi(-7) &= 0
\end{aligned}
$$

If n is an integer, then the fraction of the positive integers less than or equal to n is given by

$$\frac{\pi(n)}{n}$$

For $n = 10$, this fraction is $4/10 = 0.4$ since $\pi(10) = 4$ so that 40% of the positive integers less than or equal to 10 are prime. For $n = 100$, the fraction is $25/100 = 0.25$, giving 25% of the positive integers less than or equal to 100 are prime. A table for a range of powers of ten is given below. The result $\pi(10^8) = 5761455$ was determined by E. D. F. Meissel using a method based on the Sieve of Eratosthenes. Using this method, J. C. Lagarias, V. S. Miller, and A. M. Odlyzko [45] found that $\pi(4 \cdot 10^{16}) = 1075292778753150$.

n	$\pi(n)$	$\pi(n)/n$	$[\pi(n)/n] \log(n)$
10	4	0.500	0.921
10^2	25	0.250	1.151
10^3	168	0.168	1.161
10^4	1229	0.123	1.132
10^5	9592	0.096	1.104
10^6	78498	0.078	1.084
10^7	664579	0.066	1.071
10^8	5761455	0.058	1.061
10^9	50847534	0.051	1.054
10^{10}	455052511	0.046	1.048

It is known that for large x, the ratio of $\pi(x)$ and $x/\log(x)$ is close to one, where log is the logarithm to base e or the natural logarithm. This result is stated explicitly in the following theorem, whose proof is beyond the scope of this book. (See A. E. Ingham's *The Distribution of Prime Numbers* [36].)

Theorem 2.16 (Prime Number Theorem)

$$\lim_{x \to \infty} \frac{\pi(x)}{x/\log(x)} = 1$$

This limit implies that the fraction of integers less than or equal to n that are prime is asymptotic to $1/\log(n)$ so that the primes are more widely dispersed as the integers get larger.

Many mathematicians have contributed to the study of the behavior of $\pi(x)$ as x gets large. By looking at tables for $\pi(n)$, such as the one above, it was known early that $\pi(n)$ was related to $n/\log(n)$. Léonard Euler asserted that $x/\pi(x) \approx \log(x) - B$ where B has a slowly varying magnitude averaging about 1.08. Adrien-Marie Legendre stated that $x/(\log(x) - 1.08366)$ was a good approximation for $\pi(x)$ for $x < 1,000,000$. Peter Gustav Lejeune-Dirichlet suggested the formula $\sum_{n<x} \dfrac{1}{\log(n)}$. At age 14 Carl Friedrich Gauss asserted that $x/\log(x)$ was asymptotic to $\pi(x)$ or, more precisely, the formula given in Theorem 2.16. Later he asserted that $\pi(x)$ was asymptotic to $Li(x) = \int_2^\infty \dfrac{dt}{\log(t)}$, which is a better approximation to $\pi(x)$. Peter Lvovich Chebyshev (1821-1894) proved that $(0.92)(x/\log(x)) < \pi(x) < (1.11)(x/\log(x))$ and that $\lim_{x \to \infty} \dfrac{\pi(x)}{x/\log(x)} = 1$ if the limit exists.

Another giant step was taken by Léonard Euler when he proved that

$$\sum_{n=1}^{\infty} n^{-s} = \prod_p (1 - p^{-s})^{-1}$$

for real values $s > 1$ and prime p so that an equation is found with all integers on one side and only primes on the other.

Georg Friedrich Bernard Riemann was considered by many as the founder of analytic number theory. He extended the above Euler's series $\zeta(s) = 1 + 1/2^s + 1/3^s + \cdots$ to complex numbers s. This function $\zeta(s)$ is known as the Riemann zeta function. That the Riemann zeta function converges and is nonzero for all complex numbers whose real part is greater than or equal to 1 except at $s = 1$ can be used to prove the Prime Number Theorem. Riemann explored convergence and conjectured that all zeros for the zeta function for nonreal numbers occur when the real part is between 0 and 1. He believed they all occurred when the real part of s is equal to $1/2$. This is known as the Riemann Hypothesis and has never been proved. G. H. Hardy proved that there are infinitely many values of s with real part equal to $1/2$ such that the zeta function is zero. Riemann discovered six properties related to the zeta function. Jacques Hadamard proved three of them and H. von Mangoldt of Danzig proved two others. The remaining one is the Riemann hypothesis. Using these properties, Riemann sketched out a proof of the Prime Number Theorem, whose details were filled in by the efforts of several mathematicians. Finally, based on Riemann's work, Jacques Hadamard and de la Valleé-Poussin independently proved the Prime Number Theorem.

Georg Friedrich Bernard Riemann (1826-1866) was the last of the famous trilogy at Göttingen who worked on analytic number theory. The other two were Dirichlet and Gauss. Riemann received his doctorate at Göttingen under Gauss. His doctoral thesis was in complex variables, where he developed what is now known as the theory of Riemann surfaces. His thesis also contained the Cauchy-Riemann differential equations, which are used to test whether or not a complex function is analytic. To qualify for the position of Privatdozent (an official but unpaid lecturer) at Göttingen took him two years. In the work for this position, he developed Riemannian or elliptic non-Euclidean geometry, one of the two major forms of non-Euclidean geometry.

Jacques Hadamard (1865-1963) was born in Versailles, France. In addition to work in number theory, Hadamard made important contributions to complex analysis, functional analysis, and mathematical physics. Hadamard's most famous achievement was the proof of the Prime Number Theorem, which was made possible by his theorem on entire functions. Hadamard, an editor of a mathematics journal, apparently received some papers from an unknown mathematician. He was impressed with the papers and invited the author to dinner. The person wrote that he could not attend due to circumstances beyond his control, but invited Hadamard to dinner. Hadamard accepted and discovered the author in an asylum for the criminally insane.

Charles Jean Gustave Nicholas de la Valleé-Poussin (1866-1962) was a Belgian mathematician. After proving the Prime Number Theorem, he extended this work, established results about the distribution of primes in arithmetic sequences, an area made famous by Dirichlet's Theorem (Theorem 2.18 below), and developed error estimates related to the Prime Number Theorem. Poussin also wrote the well-known and important textbook, *Cours d'analyse*.

We also state without proof at this time (see Chapter 6) the following theorem about the occurrence of primes.

Theorem 2.17 (Bertrand's Postulate) *For each $n > 1$ there exists a prime p such that $n < p < 2n$.*

A practical import of Bertrand's Postulate is for a prime q the next largest prime, p, is less than twice q. For example, if $q = 3$, then $2q = 6$ and the next largest prime after q is $p = 5$. The number $2^{11213} - 1$ is prime. So there is a larger prime that is less than $2(2^{11213} - 1) = 2^{11214} - 2$.

We have been able to determine some information about how the primes are distributed among the positive integers. We may also ask whether they are expressible in certain forms; and, in particular, whether there are infinitely many primes having these forms. One result of this type is Dirichlet's Theorem. This theorem states that there are arithmetic progressions containing infinitely many primes. Note, however, that the theorem does not say that every integer in such a sequence is prime.

Theorem 2.18 (Dirichlet) *If a and b are relatively prime, then there are infinitely many primes of the form $a \cdot k + b$, where $k = 0, 1, 2, \ldots$.*

Two important special cases in the application of Dirichlet's Theorem are when $a = 4$ and $b = 1$ and when $a = 4$ and $b = 3$. These choices of a and b generate integers of the form $4k + 1$ and $4k + 3$ which are congruent to 1 and 3 modulo 4, respectively. So there are infinitely many primes congruent to 1 modulo 4 and infinitely many primes congruent to 3 modulo 4. The form $4k + 3$ could be respecified as $4k' - 1$ for $k' \geq 1$. These two cases include all odd integers because odd integers are congruent modulo 4 either to 1 or to 3. Obviously, every prime number greater than two, since it is odd, can be written either in the form $4n + 1$ or $4n + 3$. Although we will not prove Dirichlet's Theorem, we will prove a special case here.

Theorem 2.19 *There are infinitely many primes of the form $4k + 3$.*

Proof. Since any integer of the form $4k + 3$ is also of the form $4j - 1$, and conversely, through the relation $j = k+1$, it is sufficient to show that there are infinitely many primes of the form $4j - 1$ for $j \geq 1$. Assume that there are only finitely many primes of the form $4j - 1$ and that they are $p_1, p_2, p_3, \ldots, p_n$. Consider the positive integer $q = 4p_1p_2p_3 \cdots p_n - 1$. We claim that q has at least one prime divisor of the form $4j - 1$. If not, then every prime divisor must be of the form $4i + 1$ since q is not divisible by the prime 2. But the product of integers of the form $4i + 1$ must also be of the form $4i + 1$. Since q is of the form $4j - 1$, we have a contradiction. Thus, for some s, $p_s \mid q$; but this implies that $p_s \mid (-1)$, which cannot be the case. Therefore, q is prime, another contradiction since $q \neq p_i$ for all i. So there are infinitely many primes of the form $4j - 1$ and, equivalently, of the form $4k + 3$. ∎

Peter Gustav Lejeune-Dirichlet (1805-1859) was born into a French family near Cologne Germany. He studied at the University of Paris and then held posts at the University of Breslau and the University of Berlin before going to the University of Göttingen. He was a student of Carl Friedreich Gauss and truly idolized him. He was the first person to master Gauss's *Disquisitiones Arithmeticae*, and it is said that he took it everywhere he went. Dirichlet's book, *Vorlesungen über Zahlentheorie*, on number theory made Gauss's discoveries known to other mathematicians.

Dirichlet also did some very important work of his own. He is credited with expressing the Dirichlet Box Principle or Pigeonhole Principle (see the Exercises of Section 0.2). He proved Dirichlet's Theorem (Theorem 2.18). This proof was said to be the real beginning of modern analytic number theory (modern analytic number theory seems to have had several real beginnings). Dirichlet also worked on the class number of binary quadratics and proved a theorem on the existence of units in any algebraic field or domain. He and Adrien-Marie Legendre showed that Fermat's Last Theorem (Chapter 7) was true for $n = 5$. Ernst Kummer thought that he had proved Fermat's Last Theorem and several other mathematicians could find nothing wrong with it, but Dirichlet found the critical error. Kummer, however, did make significant contributions to the study of Fermat's Last Theorem. Dirichlet is said to have had a strong influence on the mathematician Leopold Kronecker, particularly in analytic number theory.

There is also a famous open problem regarding primes known as Goldbach's Conjecture, which may be stated as follows: If $n \geq 4$, then there are distinct primes p and q such that $p + q = 2n$.

Mathematicians have investigated primes since ancient times. Indeed, Euclid proved in his treatise *Elements* that there are infinitely many primes. Primes that have several specific forms have been given the names of some of the mathematicians — Mersenne, Fermat, Hilbert — who have worked extensively with them. The hope, in some cases, was that an integer having a certain form would always be prime. This hope is unfulfilled, but such work has lead to many important results in number theory. Aside from investigating primes for the sake of number theoretic study, large primes have recently become of great practical importance in cryptographic applications. These applications are discussed in Chapter 3.

As mentioned in Chapter 1, the study of numerology, which deals with the attempt to explain the meaning of the cosmos in terms of numbers, was an activity pursued by the ancient Greeks. A number is considered perfect in both sum and product provided that it is equal to the sum of all of its positive divisors other than itself. For example, 6 is perfect because the positive divisors of 6 are 6, 3, 2, and 1 and because $6 = 3 + 2 + 1$. 28 is perfect because the positive divisors of 28 are 28, 14, 7, 4, 2, and 1 and because $28 = 14 + 7 + 4 + 2 + 1$. The first few perfect numbers are 6, 28, 496, 8128, and 33550336.

Definition 2.20 *A positive integer is* perfect *if and only if it is the sum of all of its positive integer divisors other than itself.*

It is not known whether there are infinitely many perfect numbers or whether there is an odd perfect number. All known perfect numbers are even, and it can be shown that all even ones end in a 6 or an 8 in base ten. The first five perfect numbers are factored as

Perfect Number	Prime Factorization		
6	$2 \cdot 3$	$=$	$2^1 \cdot (2^2 - 1)$
28	$2^2 \cdot 7$	$=$	$2^2 \cdot (2^3 - 1)$
496	$2^4 \cdot 31$	$=$	$2^4 \cdot (2^5 - 1)$
8128	$2^6 \cdot 127$	$=$	$2^6 \cdot (2^7 - 1)$
33550336	$2^{12} \cdot 8191$	$=$	$2^{12} \cdot (2^{13} - 1)$

where 3, 7, 31, 127, and 8191 are prime. Three other similar prime factorizations — $2^3 \cdot (2^4 - 1) = 2^3 \cdot 15$, $2^5 \cdot (2^6 - 1) = 2^5 \cdot 63$, and $2^7 (2^8 - 1) = 2^7 \cdot 255$, where none of the three is perfect and the factor of the form $2^n - 1$ is not prime — suggest the following theorem.

Theorem 2.21 (Euclid) *If $2^n - 1$ is prime, then $2^{n-1} \cdot (2^n - 1)$ is perfect.*

Proof. If $p = 2^n - 1$ is prime, then $m = 2^{n-1}(2^n - 1) = 2^{n-1}p$. Using Theorem 2.6 (the Unique Prime Factorization Theorem), the divisors of $m = 2^{n-1}p$ are

$$1, 2, 2^2, 2^3, \ldots, 2^{n-1}$$

and

$$p, 2p, 2^2 p, 2^3 p, \ldots, 2^{n-1}p$$

The sum of these divisors of m (including $m = 2^{n-1}p$) is

$$
\begin{aligned}
\text{sum} &= 1 + 2 + 2^2 + \cdots + 2^{n-1} + p + 2p + \cdots + 2^{n-1}p \\
&= (2^n - 1) + (2^n - 1)p
\end{aligned}
$$

since $1 + 2 + 2^2 + \cdots + 2^k = 2^{k+1} - 1$. Thus,

$$
\begin{aligned}
\text{sum} &= (2^n - 1)(1 + p) = (2^n - 1)(1 + 2^n - 1) \\
&= 2 \cdot 2^{n-1} \cdot (2^n - 1) = 2m
\end{aligned}
$$

The sum of the divisors of m except for m is (sum $-m$) $= 2m - m = m$, which implies that m is perfect. ∎

From this theorem we see immediately that numbers of the form $2^n - 1$ appear to be closely related to perfect numbers. Every instance of a prime of the form $2^n - 1$ generates a perfect number.

Definition 2.22 *A Mersenne number is an integer of the form $M(n) = 2^n - 1$. If a Mersenne number is prime, it is called a Mersenne prime.*

The first few Mersenne numbers are

n	$M(n)$		
1	$2^1 - 1$	$=$	1
2	$2^2 - 1$	$=$	3
3	$2^3 - 1$	$=$	7
4	$2^4 - 1$	$=$	$15 = 3 \cdot 5$
5	$2^5 - 1$	$=$	31
6	$2^6 - 1$	$=$	$63 = 3^2 \cdot 7$
7	$2^7 - 1$	$=$	127
8	$2^8 - 1$	$=$	$255 = 3 \cdot 5 \cdot 17$
9	$2^9 - 1$	$=$	$511 = 7 \cdot 73$
10	$2^{10} - 1$	$=$	$1023 = 3 \cdot 11 \cdot 31$
11	$2^{11} - 1$	$=$	$2047 = 23 \cdot 89$

In this table, whenever $M(n)$ is prime, so is n. This observation suggests the possibility of

Theorem 2.23 *If $M(n)$ is prime, then n is prime.*

Proof. We first show that if $a^n - 1$ is prime for $n \geq 2$, then $a = 2$. Since $a^n - 1$ is divisible by $a - 1$ and since $a^n - 1$ is prime, then $a - 1 = 1$ and $a = 2$.

Now assume that $M(n)$ is prime and $n = pq$. Thus $M(n) = 2^{pq} - 1 = (2^p)^q - 1$ and since either $2^p = 2$ or $q = 1$, we have $p = 1$ or $q = 1$. ∎

We see that the statement

$$M(n) \text{ is prime if and only if } n \text{ is prime.}$$

is not true because $n = 11$ is prime but $M(11) = 23 \cdot 89$ is composite. So knowing that n is prime does not guarantee that $M(n)$ is prime, but only prime n need ever be considered for prime $M(n)$.

Mersenne numbers were named after Father Marin Mersenne (1588-1648), a Friar of the Order of Minim Brothers, who taught philosophy and theology at Nevers and Paris. He was an amateur mathematician who worked in several areas of mathematics but was best known for his work in the theory of numbers, specifically primes and perfect numbers. He was a mutual friend of Pierre de Fermat, Blaise Pascal, and other mathematicians and transmitted mathematics and science letters among them. He formed an informal group including Pascal, Fermat, René Descartes, and other well-known mathematicians whose regular weekly meetings grew into the French Academy. The reason Mersenne numbers were named after Mersenne is that he supposedly showed for primes $n \leq 257$ which Mersenne numbers $M(n)$ were prime and which were not. Unfortunately, some of Mersenne's primes could be factored. He made a total of five mistakes but it took 304 years to find all of them.

Mersenne numbers have been the primary source of large prime numbers. $M(19)$, shown by Cataldi to be prime in 1588, remained the largest verified Mersenne prime for 150 years. Euler verified that $M(31)$ was prime by checking all primes up to the square root of $M(31)$ as possible factors. In

1876 Edward Lucas proved that $M(67)$ was composite but did not find the factors. At the 1903 meeting of the American Mathematical Society, Frank Cole was scheduled to present a paper on the factorization of large numbers. Apparently, when called on to speak, Cole did not do so but silently computed $2^{67} - 1$ and then multiplied $193707721 \times 761838257287$. They were the same. Cole silently took his seat, receiving a standing ovation. R. E. Powers determined $M(89)$ to be prime, becoming the last person to determine an unknown Mersenne prime by hand. He also determined that $M(107)$ was prime. The largest Mersenne prime discovered without the use of computers was $M(127)$ by E. Lucas in 1876. $M(127)$ was the largest known prime for 75 years. In 1952 the Mersenne prime $M(521)$ was found by R. M. Robinson using a computer. In 1994 the Mersenne number $M(859433)$ was shown to be prime producing the gigantic perfect number $2^{859432}(2^{859433} - 1)$. It is not known if there are infinitely many Mersenne primes.

Referring again to the discussion of perfect numbers which led us to investigate Mersenne numbers, we saw that every Mersenne number $M(n)$ that is prime generates a perfect number $2^{n-1}M(n)$, where n will, of necessity, have to be prime. That these are the only kinds of even perfect numbers is given by the next theorem, which is the converse of Theorem 2.21.

Theorem 2.24 *Every even perfect number has the form*

$$2^{n-1} \cdot (2^n - 1) = 2^{n-1} \cdot M(n)$$

where $M(n)$ is a prime Mersenne number.

Proof. (Dickson) Let m be an even perfect number so that $m = 2^k d$, where $k \geq 1$ and d is odd. Certainly, $d > 1$ since 2^k cannot be perfect (see the Exercises). The Fundamental Theorem of Arithmetic implies that

$$m = 2^k d = 2^k p_1^{d(1)} p_2^{d(2)} \cdots p_t^{d(t)}$$

where $k, d(1), \ldots, d(t) \geq 1$. Using Theorem 2.8, the divisors of m are of two kinds: those involving positive exponents of only p_i's and those involving a factor 2^j with $j > 0$ along with, perhaps, some p_i's. The first kind are the odd divisors of m and the second kind are the even divisors. Let E be the sum of the even divisors of m and W be the sum of the odd divisors of m. Because of the Fundamental Theorem of Arithmetic, W is also the sum of all of the divisors of d, including the divisor d also. Notice that if v is any odd divisor of m, then $2v + 2^2 v + \cdots + 2^k v$ is the sum of all of the even divisors of m that include only the odd factor v. W is the sum of all of the v's. Since m is perfect, the sum of all of the divisors of m (including m itself) is

$$
\begin{aligned}
2m &= E + W \\
&= (2 + 2^2 + \cdots + 2^k)W + W \\
&= (1 + 2 + 2^2 + \cdots + 2^k)W \\
&= (2^{k+1} - 1)W
\end{aligned}
$$

But $2m = 2(2^k d) = 2^{k+1} d$ so that

$$
\begin{aligned}
2^{k+1} d &= (2^{k+1} - 1)W \\
(2^{k+1} - 1)d + d &= (2^{k+1} - 1)W
\end{aligned}
$$

which implies that $2^{k+1} - 1$ divides d and, therefore, that $d/(2^{k+1} - 1)$ is an integer and a divisor of d. Thus, dividing both sides of the above equation by $2^{k+1} - 1$, we get

$$W = d + \frac{d}{2^{k+1} - 1}$$

We recall that W is the sum of all of the divisors of d, including d itself, so d and $d/(2^{k+1} - 1)$ are the only divisors of d. But, $d > 1$ and 1 is a divisor of every integer so that $d/(2^{k+1} - 1) = 1$ or

$$d = 2^{k+1} - 1$$

Thus the only divisors of d are d and 1 so that $d = 2^{k+1} - 1$ is prime. ∎

Definition 2.25 *A* Fermat number *is an integer of the form* $F(n) = 2^{2^n} + 1$. *If a Fermat number is prime, it is called a* Fermat prime.

The first five Fermat numbers are $F(0) = 3$, $F(1) = 5$, $F(2) = 17$, $F(3) = 257$, and $F(4) = 65537$. All of the first five Fermat numbers are prime, but $F(5) = 4294967297$ is divisible by 641. In fact, it is known that $F(n)$ is composite for $5 \leq n \leq 22$. (See "The Twenty-Second Fermat Number Is Composite" [20].) Hence not all Fermat numbers are Fermat primes. It can be shown that the Fermat numbers are relatively prime to one another; that is, if $m \neq n$, then $F(m)$ is relatively prime to $F(n)$ (see the Exercises of Section 1.4).

Again, as in the case of Mersenne numbers, Fermat numbers were named after Pierre de Fermat because he had made an incorrect assumption. He believed that all Fermat numbers were prime after trying numbers 0, 1, 2, 3, and 4. He remained convinced of this until his death. Unfortunately, if he had calculated one more Fermat number he would have discovered his mistake. Léonard Euler showed that $F(5)$ could be factored. No further Fermat primes have been found. Hence the only known Fermat primes are 3, 5, 17, 257, and 65537. Therefore, unlike the Mersenne numbers, the Fermat numbers have not been a good source of primes. Fermat numbers have some applications; for example, Carl Friedrich Gauss proved that a regular polygon with n sides can be constructed by compass and straightedge if and only if n is of the form $2^k p_1 p_2 \cdots p_m$, where the p_i are all Fermat primes.

Pierre Fermat (1601-1665) was another of the famous "amateur" mathematicians. He was a member of the provincial parliament of Toulouse and spent his entire working life of 34 years working for the state, 17 years in Parliament. He became interested in the theory of numbers through Claude-Gaspar Bachet's translation of works of Diophantus. Instead of being published, his discoveries were given in letters to other mathematicians or noted on the pages of books he read. Fermat was famous in many areas of mathematics. He is said to have already conceived of the idea of analytic geometry in letters to Pascal before credit was given to Descartes, who discovered it independently later. He shared with Pascal the credit for the creation of the mathematical theory of probability. He first asserted that Pell's equation $x^2 - dy^2 = 1$ had an infinite number of integral solutions (see Chapter 5), but Euler accidentally gave credit to John Pell when he asserted that the solution of the equation $ax^2 + bx + c = y^2$ required the solution of Pell's equation. In optics Fermat showed the properties of the refraction of light through a medium.

In number theory Fermat proved the theorem known as Fermat's Little Theorem which states that if p is a prime then $a^{p-1} - 1$ is divisible by p for all a with $0 < a < p$ (see Chapter 3). He is also credited with using the "method of infinite descent" as a technique for proving theorems (see Chapter 1). His most famous theorem is known as Fermat's Last Theorem. It states that for $n > 2$, $a^n + b^n = c^n$ has no integral solutions. He claimed to have a proof but didn't write it down since there was no room in the margin of the book. Many people have tried in vain to prove Fermat's Last Theorem. The proof of Fermat's Last Theorem was completed in 1994 by Andrew Wiles and R. Taylor. Whether Fermat actually had a proof is not known but seems unlikely. Few of Fermat's proofs are preserved. Most of the knowledge we have about them comes from letters, marginal notes, and a manuscript by Huygens. We will discuss Fermat's equation $a^n + b^n = c^n$ in Chapter 7.

Definition 2.26 *Let H be the set of all positive integers of the form $4k + 1$. An integer p in H is called a* Hilbert prime *if it cannot be factored in H except as the product of itself and 1.*

The first Hilbert primes are 5, 9, 13, 17, 21, 29, 33, 37, 41, and 49. It follows from the definition that every member of H can be factored into primes (see the Exercises). The number $1617 = 4(404) + 1$ and hence belongs to H. However $1617 = 49 \cdot 33 = 77 \cdot 21$. We have already indicated that 49, 33, and 21 are Hilbert primes and it is easily seen that 77 is also a Hilbert prime.

This example illustrates that even though the members of H can be factored into Hilbert primes, there is no Unique (Hilbert) Prime Factorization Theorem for H. Thus, the uniqueness of the prime factorization of positive integers guaranteed by the Fundamental Theorem of Arithmetic is a very special property of the integers that may not be present in other "number" systems.

Exercises

1. Determine two numbers n and $n + 35$ which have no primes between them.

2. Prove that $\gcd(n, n+1) = 1$ for all $n > 1$.

3. Use the results of the preceding exercise to prove that if $n > 2$, then there exists a prime p such that $n < p < n!$. (*Hint:* Consider $n!$ and $n! - 1$.)

4. Prove that the product of two integers of the form $4n + 1$ is also of the form $4n + 1$.

5. Prove that every even perfect number ends in a 6 or an 8 in its base ten representation.

6. Classify the positive integers less than or equal to 100 of the set H (Definition 2.26) according to whether they are Hilbert primes. For each that is not a Hilbert prime, give a factorization into Hilbert primes. Find another example of an element of H besides 1617 in the discussion that has two factorizations into Hilbert primes.

7. Prove that for every positive integer k, 2^k is not perfect.

8. Prove that if p is prime, then for every positive integer k, p^k is not perfect.

9. Prove that every integer in the set H of Definition 2.26 can be factored into a product of Hilbert primes.

2.4   Elementary Algebraic Structures in Number Theory

The axioms for the integers were given in Chapter 1. These axioms dealt in part with properties of addition and multiplication and how addition and multiplication interact. Certain groupings of these properties have been studied extensively aside from their connection with the integers and form the subject matter of algebra — semigroups, groups, rings, and fields. Algebraic theory can be brought to bear upon problems in number theory provided that subsystems of the integers can be identified as having the structure of a known algebraic object. Some theorems about integers may be shown using only algebraic properties. Such analysis often clarifies what "causes" a theorem to be true. One proof of a certain theorem may rely upon integer properties and another may use algebraic characteristics to obtain the result. Several algebraic structures will be defined and some useful theorems about the structures and related terminology will be given.

A binary operation on a set takes two elements of the set and associates them with another element of the set. Examples of binary operations are addition and multiplication of integers. The least common multiple and greatest divisor are also examples of binary operations on the integers. This use of "binary" has no direct connection with the binary numbers.

Definition 2.27 *A* semigroup *is a nonempty set together with a binary operation (or product)* ∘ *on* $S \times S$ *such that*

$$a \circ (b \circ c) = (a \circ b) \circ c$$

for all a, b, and c in S (i.e., the operation on S is associative). The semigroup is denoted by $(S, \circ)$ *or simply S if the product is understood. If a semigroup S has the property that* $a \circ b = b \circ a$ *for all a,b in S, then S is called a* commutative *or* Abelian *semigroup.*

Notice that the set of $n \times n$ matrices is a semigroup with the binary operation multiplication but is not commutative.

Example 2.28 *Examples of semigroups include the following:*

1.   *The integers with the binary operation multiplication*

2.   *The positive integers with the binary operation multiplication*

3.   *The integers with the binary operation addition*

4.   *The positive integers with the binary operation addition*

5.   *The set H of all positive integers of the form* $4k + 1$ *with the binary operation multiplication*

6.   *The set of* $n \times n$ *matrices with the binary operation multiplication*

7.   *The set of* $n \times m$ *matrices with the binary operation addition*

8.   *The integers modulo a positive integer n,* Z_n, *with the binary operation addition*

9.   *The integers modulo a positive integer n with the binary operation multiplication*

In semigroup 5, $H = \{x : x = 4k + 1 \text{ for some integer } k \geq 0\}$ *and the operation* ∘ *of the definition of semigroup is integer multiplication. We first show that multiplication is a function from* $H \times H$ *into H; that is, multiplication in H is a binary operation on H. If* $h_1 \in H$ *and* $h_2 \in H$, *then, because* h_1 *and* h_2 *are integers and multiplication is a function from* $Z \times Z$ *into Z,* $h_1 \cdot h_2$ *is a unique integer. To be sure that multiplication is a function into H, we have only to show that* $h_1 \cdot h_2 \in H$ *(see the Exercises of Section 2.3). Thus* · *is a binary operation on H. Given* h_1, h_2, *and* h_3 *in H we have that* $(h_1 \cdot h_2) \cdot h_3 = h_1 \cdot (h_2 \cdot h_3)$ *simply because the* h_i *are integers and Axiom I4 applies to all integers, including those in H.*

Let a be an element of a semigroup S. Inductively define $a = a^1$ and $a^{n+1} = a^n \circ a$.

Theorem 2.29 *If S is a semigroup and $a \in S$, then $a^{m+n} = a^m \circ a^n$ for all positive integers m and n. Hence $a^m \circ a^n = a^n \circ a^m$ for all positive integers m and n.*

Proof. Using induction on n, we show that $a^{m+n} = a^m \circ a^n$ for all positive integers n. By definition $a^{m+1} = a^m \circ a = a^m \circ a^1$ and the hypothesis is true for $n = 1$. Assume that the theorem is true for $n = k$; i.e., assume that $a^{m+k} = a^m \circ a^k$. We show that $a^{m+(k+1)} = a^m \circ a^{k+1}$.

$$
\begin{aligned}
a^{m+(k+1)} &= a^{(m+k)+1} \\
&= a^{m+k} \circ a^1 && \text{by definition of exponent} \\
&= (a^m \circ a^k) \circ a^1 && \text{using the induction hypotheses} \\
&= a^m \circ (a^k \circ a^1) && \text{by associativity} \\
&= a^m \circ a^{k+1} && \text{by definition of exponent}
\end{aligned}
$$

and the theorem is proved. ∎

Theorem 2.30 $(a^m)^n = a^{mn}$ *for all positive integers m and n.*

Proof. This theorem may be proved using induction on n. ∎

Definition 2.31 *A group is a set G together with a binary operation (or product) $\circ$ on $G \times G$ which has the following properties:*

(a) $a \circ (b \circ c) = (a \circ b) \circ c$ *for all a, b, and c in G (i.e., the operation $\circ$ on S is associative).*

(b) *There exists an element 1 in G, called the identity, which has the property that $a \circ 1 = 1 \circ a = a$ for all a in G.*

(c) *For each element a in G, there exists an element a^{-1} in G such that $a \circ a^{-1} = a^{-1} \circ a = 1$.*

If a group G has the property that $a \circ b = b \circ a$ for all a,b in G, then G is called a commutative or Abelian group. If $G = \{a, a^2, \ldots, a^n\}$, then G is called a finite cyclic group and a is said to generate G.

Definition 2.32 *If G is a group with n elements, then n is called the order of the group G.*

Example 2.33 *Examples of groups include the following:*

1.   *The integers with the binary operation addition. Notice that multiplication does not work since, for example, given an integer, such as 5, there is no integer which can be multiplied times 5 to get the identity, 1.*

2.   *The set of $n \times m$ matrices with the binary operation addition.*

3.   *The integers modulo a positive integer n, Z_n, with the binary operation addition.*

   Note that every group is a semigroup. The converse, however, is not true. In fact, only the examples of semigroups given in Example 2.28 which are also given here are groups.

Theorem 2.34 *The identity of a group G is unique.*

Proof. Assume that 1 and e are both identities of G. Then $1 = 1 \circ e = e$ by definition of identity. ∎

Theorem 2.35 *In a group, the inverse of each element is unique.*

Proof. Let a be an element of a group G and assume that b and c are both inverses of a; then

$$b = b \circ 1 = b \circ (a \circ c) = (b \circ a) \circ c = 1 \circ c = c \qquad ∎$$

Theorem 2.36 *For each element a of a group G, $(a^{-1})^{-1} = a$.*

Proof. Since $a^{-1} \circ a = a \circ a^{-1} = 1$, a satisfies the definition of the inverse of a^{-1}. Since the inverse is unique, a is the inverse of a^{-1}. ∎

Theorem 2.37 *For elements a and b of a group G, $(a \circ b)^{-1} = b^{-1} \circ a^{-1}$.*

Proof. $(a \circ b) \circ (b^{-1} \circ a^{-1}) = a \circ (b \circ b^{-1}) \circ a^{-1} = a \circ 1 \circ a^{-1} = a \circ a^{-1} = 1$. Similarly, $(b^{-1} \circ a^{-1}) \circ (a \circ b) = 1$ and by definition and uniqueness of inverse, $(a \circ b)^{-1} = b^{-1} \circ a^{-1}$. ∎

   If a is an element of a group G, denote $(a^{-1})^k$ by a^{-k}. Let a^0 denote the identity 1.

   The proof of the following theorem is left to the reader.

Theorem 2.38 *Let G be a group and a be an element of G.*

(a) $a^n \circ a^{-n} = 1$ *for all positive integers n.*

(b) $a^{(m+n)} = a^m \circ a^n$ *for all integers n and m.*

(c) $(a^m)^n = a^{mn}$ *for all integers m and n.*

(d) $(a^{-n})^{-1} = a^n$ *for all integers n.*

Lemma 2.39 *If G is a finite group and a is an element of G, then $a^s = 1$ for some positive integer s.*

Proof. If G is a finite group and a is in G, then we know that $a^j = a^k$ for some positive integers j and k. Let t be the least positive integer such that $a^t = a^k$

for some k. Then $t = 1$; for if $t > 1$, then since $a^{t-1} \circ a = a^t = a^k = a^{k-1}a$, we have

$$
\begin{aligned}
(a^{t-1} \circ a) \circ a^{-1} &= (a^{k-1} \circ a) \circ a^{-1} \\
a^{t-1} \circ (a \circ a^{-1}) &= a^{k-1} \circ (a \circ a^{-1}) \quad \text{by associativity} \\
a^{t-1} \circ 1 &= a^{k-1} \circ 1 \quad\quad\quad \text{by definition of inverse}
\end{aligned}
$$

which implies that $a^{t-1} = a^{k-1}$. But this contradicts the definition of t. Hence $t = 1$. Let p be the least integer greater than 1 such that $a^{p+1} = a$. Then $a^p = a^{p+1} \circ a^{-1} = (a \circ a^{-1}) = 1$. ∎

Let p be the least integer greater than 1 such that $a^p = 1$. The elements $1, a, a^2, \ldots, a^{p-1}$ are distinct; for if $a^r = a^s$ for $0 \le r$, $s < p$, and $r > s$, then $a^{r-s+1} = a$, which contradicts the fact that p is the smallest positive integer such that $a^{p+1} = a$. Also if $n > p$, then $a^n = a^r$ for some $0 \le r < p$ since if $n = pq + r$ for some $0 \le r < p$, then

$$a^n = a^{pq+r} = a^{pq} \circ a^r = (a^p)^q \circ a^r = 1^q \circ a^r = 1 \circ a^r = a^r$$

Theorem 2.40 *Let G be a group and a be an element of G such that $a^s = 1$ for some s. If p is the least positive integer such that $a^p = 1$, then $p \mid s$. The integer p is called the* order *of a.*

Proof. Let p be the least positive integer such that $a^p = 1$. As in the discussion above, let $s = pq + r$ for some $0 \le r < p$; then $a^s = a^{pq+r} = a^{pq}a^r = (a^p)^q a^r = (1)^q a^r = a^r$. Hence $a^s = 1$ if and only if $r = 0$ because p is the least positive integer with $a^p = 1$. ∎

Definition 2.41 *A subset H of a group G is a* subgroup *of G if H with the same operation as G is also a group. For any a in G, $a \circ H = \{x : x = a \circ h$ for some h in $H\}$ is a* left coset *of H in G.*

Example 2.42 *We saw earlier that the set Z of integers with the binary operation of addition, $+$, is a group. Let H be the set of all multiples of 5 or, equivalently, the integers modulo 5:*

$$
\begin{aligned}
H &= \{n : n \in Z \text{ and } n = 5k \text{ for some integer } k\} \\
&= \{\ldots, -10, -5, 0, 5, 10, 15, 20, \ldots\}
\end{aligned}
$$

We show that H is a subgroup of Z. Clearly, $H \subseteq Z$. The addition on Z is also addition on H because if $s \in H$ and $t \in H$, then $s = 5i$ and $t = 5j$ for appropriate integers i and j. Then $s + t = 5i + 5j = 5(i + j) = 5k$ for $k = i + j$ so that $s + t$ is in H. The identity element of H is the identity element of G, namely 0. Every element of H has an inverse with respect to $+$ in H because if $t = 5k \in H$, then $t^{-1} = 5(-k)$. The inverse of $5k$ is $-5k = (-5)k$, which is also in the set.

We now consider the left cosets of H generated by various elements of Z. Let a ∘ H = a + H = {x : x = a + h for some h in H}. So

$$
\begin{aligned}
0 + H &= \{x : x = 0 + h \text{ for some } h \text{ in } H\} \\
&= \{x : x = h \text{ for some } h \text{ in } H\} = H \\
&= \{\ldots, -10, -5, 0, 5, 10, 15, 20, \ldots\}
\end{aligned}
$$

$$
\begin{aligned}
1 + H &= \{x : x = 1 + h \text{ for some } h \text{ in } H\} \\
&= \{\ldots, -9, -4, 1, 6, 11, 16, 21, \ldots\}
\end{aligned}
$$

$$
\begin{aligned}
2 + H &= \{x : x = 2 + h \text{ for some } h \text{ in } H\} \\
&= \{\ldots, -8, -3, 2, 7, 12, 17, 22, \ldots\}
\end{aligned}
$$

$$
\begin{aligned}
3 + H &= \{x : x = 3 + h \text{ for some } h \text{ in } H\} \\
&= \{\ldots, -7, -2, 3, 8, 13, 18, 23, \ldots\}
\end{aligned}
$$

$$
\begin{aligned}
4 + H &= \{x : x = 4 + h \text{ for some } h \text{ in } H\} \\
&= \{\ldots, -6, -1, 4, 9, 14, 19, 24, \ldots\}
\end{aligned}
$$

Thus we have again created Z_5 since $a + H = [a]$ and the five left cosets of the subgroup H form a partition of the original group Z.

Lemma 2.43 *For a fixed subgroup H of G, the left cosets of H in G are a partition of G.*

Proof. Each left coset is nonempty since for left coset $a \circ H$, $a = a \circ 1$ is in $a \circ H$. Assume that the intersection of $a \circ H$ and $b \circ H$ is nonempty; say, c is in the intersection. Hence $c = a \circ h = b \circ h'$ for some h and h' in H. Multiplying both sides of the equation by h^{-1}, we have $a \circ h \circ h^{-1} = b \circ h' \circ h^{-1}$; so by definition of inverse we have $a = b \circ (h' \circ h^{-1})$. Since $h' \circ h^{-1}$ is in H, a is in $b \circ H$. Hence $a \circ h$ is in $b \circ H$ for all h in H so that $a \circ H \subseteq b \circ H$. Similarly, $b \circ H \subseteq a \circ H$ and $b \circ H = a \circ H$. Hence the left cosets form a partition of G. ■

Lemma 2.44 *If G is a finite group and H a subgroup of G, then all cosets of H in G contain the same number of elements, namely, the number of elements that are in H.*

Proof. Let $a \circ H$ be a left coset of H in G. Define $f : H \to a \circ H$ by $f(h) = a \circ h$. It is left to the reader to show that f is one to one and onto. ■

Theorem 2.45 *If G is a finite group and H is a subgroup of G, then the order of H divides the order of G.*

Proof. If p is the order of H, q is the number of left cosets of H in G, and n is the order of G, then, because of the preceding two lemmas, $n = p + p + p + \cdots + p = pq$. ■

Theorem 2.46 *If g is in a group G, $g^n = 1$ for some n, and p is the least positive integer such that $g^p = 1$, then the set $\{g, g^2, \ldots, g^p\}$ is a subgroup of G.*

Proof. Left to the reader. ■

Theorem 2.47 *If G is a group of order n and g is in G, then $g^n = 1$.*

Proof. Let p be the least positive integer such that $g^p = 1$ and $H = \{g, g^2, \ldots, g^p\}$. Then H is a subgroup with p elements and hence p divides n, say $n = pq$. Thus, $g^n = g^{pq} = (g^p)^q = 1^q = 1$. ■

We now use group concepts to prove two theorems in number theory that will be proved again later using traditional number theoretic methods. In order to state the first theorem, we introduce a function $\phi : N \to N$, which will be discussed extensively in Chapter 3. We define $\phi(m)$ to be the number of positive integers less than m which are relatively prime to m. Thus, $\phi(m)$ is the number of elements in the primary reduced residue system modulo m.

Euler's Theorem states that if m is a positive integer and $\gcd(a, m) = 1$, then $a^{\phi(m)} \equiv 1 \pmod{m}$. To prove this theorem we first show that the subset $R_m = \{[a] : \gcd(a, m) = 1\}$ of Z_m forms a group under the usual multiplication in Z_m. Theorem 2.3 implies that the product of two elements of R_m is in R_m. The multiplication is associative since

$$
\begin{aligned}
([a] \odot [b]) \odot [c] &= [ab] \odot [c] \\
&= [(ab)c] \\
&= [a(bc)] \\
&= [a] \odot [bc] \\
&= [a] \odot ([b] \odot [c])
\end{aligned}
$$

The identity $[1]$ is in R_m. By Theorem 1.68 we know that a congruence $ax = 1 \pmod{m}$ has a unique solution if a and m are relatively prime; hence, each element $[a]$ of R_m has an inverse. Thus, R_m is a group of order $\phi(m)$. By Theorem 2.47, for each element $[a]$ of R_m, $[a]^{\phi(m)} = [1]$; therefore, $a^{\phi(m)} \equiv 1 \pmod{m}$.

The second theorem is known as Fermat's Little Theorem. It states that if p is a prime and a is an integer such that $0 < a < p$, then $a^{p-1} \equiv 1 \pmod{p}$. This theorem follows immediately from Euler's Theorem since $\phi(p) = p - 1$ if p is prime.

Often one group or semigroup has characteristics that are mirrored in a subgroup or subsemigroup of another group or semigroup. The technical device used to describe such similarities is the homomorphism.

Definition 2.48 *Let $(G, \bullet)$ and $(H, *)$ be semigroups, where $\bullet$ and $*$ are the operations on G and H, respectively. Let $f : G \to H$ be a function. The function f is a homomorphism if $f(g \bullet g') = f(g) * f(g')$ for all g and g' in G. The homomorphism f is said to be a monomorphism provided that f is one to one, an epimorphism provided that f is onto, and an isomorphism provided* *that f is both one to one and onto. Since every group is a semigroup, the definition applies when either of G or H is a group.*

Example 2.49 *In Example 2.42, where $H = \{n : n \in Z$ and $n = 5k$ for some integer $k\}$ is a subgroup of Z, the set of left cosets of H is*

$$W = \{0 + H, 1 + H, 2 + H, 3 + H, 4 + H\}$$

We can define addition of left cosets as follows:

$$(a + H) \oplus (b + H) = (a + b) + H$$

where the addition $(a+b)$ is addition of integers a and b. It is straightforward to show that the definition of $\oplus$ is independent of the representation of $a + H$ and $b + H$.

For the group $(Z_5, +)$ of equivalence classes modulo 5, it is easy to show that the function $f : Z_5 \rightarrow W$ defined by

$$f([k]) = k + H$$

is a homomorphism between the groups $(Z_5, +)$ and $(W, \oplus)$ and is also an isomorphism.

Consider the two groups $(Z, +)$ and $(Z_5, +)$ and the function $g : Z \rightarrow Z_5$ defined by

$$g(a) = [a]$$

g is a homomorphism since $g(a + b) = [a + b] = [a] + [b] = g(a) + g(b)$. g is an epimorphism since given any element $[c]$ of Z_5, clearly $g(d) = [d] = [c]$ for any $d \in [c]$. The function g is not an isomorphism because $g(1) = [1]$ and $g(6) = [6] = [1]$ so that g is not one to one.

Definition 2.50 *A ring is a nonempty set R together with binary operations called multiplication and addition, denoted by $\cdot$ and $+$, respectively, which satisfies the following conditions:*

1. *R is closed under addition; i.e., if $x \in R$ and $y \in R$, then $x + y \in R$.*

2. *Addition in R is associative; i.e., $x + (y + z) = (x + y) + z$ for all x, y, and z in R.*

3. *R has an additive identity 0 such that $x + 0 = 0 + x = x$ for all x in R.*

4. *R contains an additive inverse $-x$ for each element x in R such that $x + -x = -x + x = 0$.*

5. *Addition in R is commutative; i.e., $x + y = y + x$ for all x and y in R.*

6. *R is closed under multiplication; i.e., if $x \in R$ and $y \in R$, then $x \cdot y \in R$.*

7. *Multiplication in R is associative; i.e., $x \cdot (y \cdot z) = (x \cdot y) \cdot z$ for all x, y, and z in R.*

8. *The following distributive laws hold in R:*

$$x \cdot (y + z) = (x \cdot y) + (x \cdot z)$$

and

$$(y + z) \cdot x = (y \cdot x) + (z \cdot x)$$

for all x, y, and z in R.

If there is an element 1 in R such that $1 \cdot r = r \cdot 1 = r$ for all r in R, then R is called a ring with unity. If $r \cdot r' = r' \cdot r$ for all r and r' in R, then R is called a commutative ring.

For a commutative ring R, if $a, b \in R$, one says that a divides b, written $a \mid b$, provided there is a $c \in R$ such that $a \cdot c = b$. In this case, a is said to be a factor of b. If $a, b \in R$, then a greatest common divisor of a and b is a member d of R such that (i) $d \mid a$ and $d \mid b$ and (ii) if $c \in R$, $c \mid a$, and $c \mid b$, then $c \mid d$.

Note that a ring R is a group under addition and a semigroup under multiplication. In the ring of integers, Z, we defined $\gcd(a, b)$ to be positive so that it would be unique. For a general commutative ring R the definition above relaxes this requirement so that greatest common divisors of the same elements a and b of R differ only by what will be later defined as a unit factor (a divisor of 1). Therefore, applying this idea to Z, a greatest common divisor of 6 and 8 could be 2 or -2, which differ by a factor of -1, which is a unit in Z.

Definition 2.51 *A* field *is a commutative ring with unity where every nonzero element has a multiplicative inverse.*

Examples of rings are the integers, real numbers, rational numbers, and complex numbers under the usual addition and multiplication. Of these, only the integers do not form a field. Z_n with the usual multiplication and addition also forms a ring; if n is prime, the integers modulo n form a field. The set of $n \times n$ matrices under the usual matrix multiplication and addition form a ring.

Definition 2.52 *An* integral domain *is a commutative ring with unity such that $ab = 0$ implies that $a = 0$ or $b = 0$ and 1 is not equal to 0.*

The integers, the rational numbers, and the real numbers are examples of integral domains. The set of 2×2 matrices is not an integral domain since

$$\begin{bmatrix} 0 & 1 \\ 0 & 0 \end{bmatrix} \begin{bmatrix} 1 & 0 \\ 0 & 0 \end{bmatrix} = \begin{bmatrix} 0 & 0 \\ 0 & 0 \end{bmatrix}$$

so that the product of two nonzero matrices may be the zero matrix.

At this point let us examine Z_n with regard to being a ring, an integral domain, or a field. We can easily show that Z_n is always a ring. We have already shown that it satisfies all of the laws except the distributive laws, and that is easily shown. If we look at Z_6, however, we see that $[3] \odot [2] = [0]$ and hence Z_6 is not an integral domain. In fact, if n is not prime, say $n = pq$, then $[p] \odot [q] = [0]$ and Z_n is not an integral domain.

In n is a prime, then given a nonzero integer $a \not\equiv 0 \pmod{n}$ we know that $ax \equiv 1 \pmod{n}$ has a solution say a'. Hence, $[a] \odot [a'] = [1]$ and every element has an inverse. Hence Z_n is a field. Every field is an integral domain since if $ab = 0$ and $a \neq 0$, then $b = a^{-1}ab = a^{-1}0 = 0$. Hence if n is prime, Z_n is an integral domain.

If n is not prime, consider the subset $R = \{[x] : x$ is relatively prime to $n\}$. It is easily shown that this set forms a group under multiplication. The product of two integers relatively prime to n is relatively prime to n. The identity, 1, is relatively prime to n; and if b is relatively prime to n, then $bx \equiv 1 \pmod{n}$ has a unique solution so that $[b]$ has an inverse. R is not even a semigroup under addition, however, since the sum of two numbers relatively prime to n is not necessarily relatively prime to n.

Definition 2.53 *Let R and R' be rings and let $f : R \to R'$ be a function from R into R'. f is said to be a* ring homomorphism *if and only if*

$$\begin{aligned} f(a + b) &= f(a) + f(b) \\ f(a \cdot b) &= f(a) \cdot f(b) \end{aligned}$$

for all $a, b \in R$. The addition and multiplication are the ones in the respective rings. A ring homomorphism $f : R \to R'$ is said to be an epimorphism *if f is onto R', is said to be a* monomorphism *if f is one to one, and is said to be an* isomorphism *provided that f is both one to one and onto R'. Normally, when describing a homomorphism from a ring R with unity to a ring R' with unity, it is required that the multiplicative identity of R be mapped to the multiplicative identity of R'.*

For example, consider the ring of integers Z and the ring $Z_7 = \{[0], [1], [2], [3], [4], [5], [6]\}$ of integers modulo 7. Suppose that $f : Z \to Z_7$ is defined by $f(a) = [a]$; that is, $f(a)$ is the equivalence class of integers congruent to a modulo 7. If each of a and b is an integer, then $f(ab) = [ab] = [a][b] = f(a)f(b)$ and $f(a + b) = [a + b] = [a] + [b] = f(a) + f(b)$ using the definition of multiplication and addition of equivalences classes of Z_7. Thus, f is a homomorphism. f is clearly an epimorphism because $f(i) = [i]$ for $0 \leq i \leq 6$. f is not one to one because $7 \equiv 0 \pmod{7}$ and, consequently, $f(7) = [7] = [0] = f(0)$. Therefore, f is neither a monomorphism nor an isomorphism. The same argument may be applied, of course, to rings Z and Z_n for any positive integer $n > 1$.

Rings that are isomorphic have the same algebraic structure and differ only in the naming of their elements.

Theorem 2.54 *In a ring R, $a \cdot 0 = 0$ for all a in R.*

Proof. Left to reader. ∎

Definition 2.55 *A subset R' of a ring R is called a subring of R if R' is a ring with the same operations.*

Exercises

1. Write an addition and a multiplication table for Z_4.

2. Write an addition and a multiplication table for Z_7.

3. Evaluate $[1] \odot [2] \odot [3]$ in Z_4.

4. Evaluate $[1] \odot [2] \odot [3] \odot [4]$ in Z_7.

5. In Z_{10}, find the multiplicative inverse of $[7]$.

6. In Z_{12}, find all a such that $[a] \odot [6] = [0]$.

7. Let $Z_n = \{[0], [1], [2], \ldots, [n-1]\}$ for some integer n . Let $S = \{[k] : k$ is positive and relatively prime to $n\}$. Is S a group under multiplication?

8. Under what conditions is Z_n a group under addition?

9. Prove that the multiples of three form a subgroup of the integers under addition. Describe the cosets formed with respect to this subgroup.

10. Prove that the multiples of 3 form a semigroup under multiplication. Why are they not a group?

11. Prove that the integers modulo 6 form a group with respect to addition. Do they also form a group with regard to multiplication?

12. Prove that if $f : G \to H$ is a group homomorphism and 1 is the identity in G, then $f(1)$ is the identity in H.

13. Let $f : G \to H$ be a group homomorphism. Prove that:

    (a) If g' is the inverse of g in G, then $f(g')$ is the inverse of $f(g)$ in H.

    (b) If $g \in G$, then $f(g^{-1}) = (f(g))^{-1}$.

    (c) $f(G)$, the image of G in H, is a group.

14. Let $Z_5^R = \{[1], [2], [3], [4]\}$ be the set of nonzero elements of Z_5. These elements correspond to integers relatively prime to 5. With the multiplication, $\cdot$, of Z_5, $(Z_5^R, \cdot)$ is a group. Show that $(Z_5^R, \cdot)$ is isomorphic to $(Z_4, +)$ by exhibiting an explicit isomorphism $f : Z_5^R \to Z_4$.

15. Give necessary and sufficient conditions that Z_n be an integral domain.

16. Prove that a field is an integral domain.

17. Prove or disprove that every integral domain is a field.

18. Is the set of 2×2 matrices over the integers an integral domain?

19. Does the set of polynomials with integer coefficients form an integral domain?

20. An element a of a ring R is called an idempotent if $a^2 = a$. Prove that if R is an integral domain, then the only idempotents are 0 and 1.

2.5   Application: Pattern Matching

Pattern matching is a task that appears frequently in computer applications. A simple example of pattern matching is in word processing computer programs when one is writing or editing a document and wants to locate a certain word or phrase, say, the word "effect" in a document. If the document is more than a few paragraphs, then visually searching is tedious, time consuming, and error-prone. A more complex example is when one is presented with a two-dimensional rectangular image or picture represented within the computer application as a two-dimensional array of zero or one bits that indicate black or white. Such a digital image may be of size 1024×1024 bits and can be displayed on a computer display screen. In this case, the task is to search for a certain irregular shape or particular pattern of black or white, say a cross, ✠.

The "brute force" search method simply compares the pattern with every possibility. In the example of searching for the word "effect," we would compare the 6 characters in "effect" with every sequence of 6 characters in the document text. This action would require at worst six letters to be compared with every sequence of six contiguous characters in the document. For short patterns and short texts, this method may be quite efficient; however, usually the text is relatively long and the pattern may be long, making the brute force method impractical. By an efficient method we mean one that requires a relatively "short" time to complete or one that requires relatively "few" comparisons or computations. More specifically, we want a search algorithm that uses a minimum of storage, whose number of required comparisons/computations is linearly related to the lengths of the pattern and of the search text, and, perhaps for some applications, that executes in "real time" (accepts the search text one character at a time without having to know the entire text beforehand and requires the same computation time for each character of text). Aside from being efficient, it is desirable that an algorithm be extensible to many different types of pattern searching contexts. We present here a rather general pattern-matching algorithm due to Richard M. Karp and Michael O. Rabin [39] that relies upon choosing random primes. This algorithm has these desirable properties and is an example from a class of probabilistic methods.

All numbers and characters (digits and alphabetic letters) are represented in a computer as sequences of zeros and ones. For example, the representation of the letter "e" using ASCII 8 bit encoding is the bit sequence

$$[0\ 1\ 1\ 0\ 0\ 1\ 0\ 1]$$

which is treated as the base 2 integer (see Section 1.9)

$$0 \cdot 2^7 + 1 \cdot 2^6 + 1 \cdot 2^5 + 0 \cdot 2^4 + 0 \cdot 2^3 + 1 \cdot 2^2 + 0 \cdot 2^1 + 1 \cdot 2^0$$

The word "effect" may be represented as

$$[01100101][01100110][01100110][01100101][01100011][01110110]$$

which could be considered to be a single long bit string and could be treated as the base 2 integer

$$0 \cdot 2^{47} + 1 \cdot 2^{46} + 1 \cdot 2^{45} + \cdots + 1 \cdot 2^1 + 0 \cdot 2^0$$

The bit string $[01100101]$ for "e" could be represented in base sixteen or hexadecimal as 65_{hex} or in base eight or octal as 145_{octal}. In an actual implementation of the algorithm, a base besides 2 may be more natural; however, in the description of the Karp and Rabin algorithm, we will consider the "pattern" and "text" to be strings of bits.

Let X be the pattern and Y be the text so that

$$X = [x_1\ x_2\ x_3\ \cdots\ x_n] \quad \text{where } x_i = 0 \text{ or } 1$$
$$Y = [y_1\ y_2\ y_3\ \cdots\ y_m] \quad \text{where } y_i = 0 \text{ or } 1$$

The pattern has length n bits and the text has length m bits, where $m \geq n$. Let $Y(i)$ be a substring of n consecutive bits of Y starting at position i, $1 \leq i \leq m - n + 1$, so that

$$Y(i) = [y_i\ y_{i+1}\ y_{i+2}\ \cdots\ y_{i+n-1}]$$

Clearly, we have a match between the pattern and a part of the text for any i for which $X = Y(i)$. Note that X and $Y(i)$ may be considered to be elements of $\{0,1\}^n = \{0,1\} \times \{0,1\} \times \cdots \times \{0,1\}$, the cross product of the set $\{0,1\}$ taken n times or the set of all n-tuples of $\{0,1\}$. Let $N(X)$ and $N(Y(i))$ be the base 2 integers having these representations:

$$N(X) = x_1 2^{n-1} + x_2 2^{n-2} + \cdots + x_{n-1} 2^1 + x_n$$
$$N(Y(i)) = y_i 2^{n-1} + y_{i+1} 2^{n-2} + \cdots + y_{i+n-2} 2^1 + y_{i+n-1}$$

where we have a match if and only if $N(X) = N(Y(i))$.

Comparing $N(X)$ and $N(Y(i))$ requires just as much work as comparing X and $Y(i)$. We need to "contract" the information inherent in $N(X)$ and

$N(Y(i))$ to a more manageable amount so that the comparison requires a more manageable amount of work. One way of doing this contraction is to map the integers $N(X)$ and $N(Y(i))$ into smaller integers using modular arithmetic. Let p be a prime in the range $1 \leq p \leq M$ where M is some suitably large positive integer that will be specified later. Let N_p be the mapping from the set of bit strings of length n into the set of primary residues modulo p, $\{0, 1, 2, \ldots, (p-1)\}$, defined by

$$N_p(W) = [[N(W)]]_p$$

so that $N_p(X) = [[N(X)]]_p$ and $N_p(Y(i)) = [[N(Y(i))]]_p$. In a typical application, n may be around 200 and p may be between 2^{16} and 2^{64}. Many computer architectures can compare 16 to 64 bit integers efficiently. The function N_p produces a "proxy" number $N_p(W)$ for $N(W)$ that will be used in comparisons instead of W or of $N(W)$. Such a function is sometimes called a *fingerprint* function and $N_p(W)$ is said to be the fingerprint of W. Note that in this case, the fingerprint function N_p is not one to one so that it is possible to have two bit strings W and V of length n with the same fingerprint, that is, with $N_p(W) = N_p(V)$. If $N_p(W) = N_p(V)$ but $W \neq V$, we say that there is a false match. We will resolve this apparent difficulty presently.

If $N(W)$ is reduced modulo p to $N_p(W)$ starting from $N(W)$ for each new W, a great deal of time may be spent in this conversion. Fortunately, it will be possible to compute the sequence $N_p(Y(1))$, $N_p(Y(2))$, $N_p(Y(3))$, ... without such computational overhead. The reduction of $N(W)$ to $N_p(W)$ may be accomplished piecemeal. We note that

$$\begin{aligned} N(W) &= w_1 2^{n-1} + w_2 2^{n-2} + \cdots + w_{n-1} 2^1 + w_n \\ &= (\cdots((w_1 \cdot 2 + w_2) \cdot 2 + w_3) \cdot 2 + \cdots + w_{n-1}) \cdot 2 + w_n \end{aligned}$$

where in this last form the order of the operations of multiplication and addition is precisely specified. According to Theorem 1.72, if $a \equiv b \pmod{p}$, then b may be substituted for a in any expression involving multiplication, addition, and subtraction without changing the modulo p residue of the expression. We say that an operation $a \odot b$ for integers a and b is performed modulo p provided that after the operation is done, the result is reduced modulo p to its least nonnegative residue; that is, substitute the primary residue $[[a \odot b]]_p$ for $a \odot b$. For brevity, we will sometimes write $a \odot_p b$ instead of $[[a \odot b]]_p$ in order to emphasize that the reduction modulo p is to occur after the operation is performed. For example,

$$\text{substitute } w_1 \cdot_p 2 \text{ for } w_1 \cdot 2$$

and then

$$\text{substitute } (w_1 \cdot_p 2) +_p w_2 \text{ for } (w1 \cdot_p 2) + w_2$$

giving in the end

$$N_p(W) = (\cdots((w_1 \cdot_p 2 +_p w_2) \cdot_p 2 +_p w_3) \cdot_p 2 +_p \cdots +_p w_{n-1}) \cdot_p 2 +_p w_n$$

Thus, reduction of $N(W)$ to its primary residue modulo p, $N_p(W)$, may be done one operation at a time beginning with the innermost nested parentheses.

We still need to be able to compare X and $Y(i)$ for each i by comparing the proxy integers $N_p(X)$ and $N_p(Y(i))$ for each i; thus, unless $N_p(Y(i))$ can be calculated efficiently for each i, the overall search algorithm may not be practical. It is the case, however, that $N_p(Y(i+1))$ can be obtained from $N_p(Y(i))$ and y_{i+1} using only a few arithmetical operations without completely recalculating $N_p(Y(i+1))$ from scratch. Since

$$N(Y(i+1)) = (N(Y(i)) - 2^{n-1} \cdot y_i) \cdot 2 + y_{i+n}$$

we have, using only operations modulo p,

$$N_p(Y(i+1)) = (N_p(Y(i)) -_p [[2^{n-1}]]_p \cdot_p y_i) \cdot_p 2 +_p y_{i+n}$$

A formula such as this last one which allows the "next" value to be computed easily from the current value is called an *update formula*. Thus we say that $N_p(Y(i))$ admits an easy update.

The basic idea of the Karp and Rabin algorithm is to choose a prime p at random from a suitable set $1 \le p \le M$ and to compute the fingerprint $N_p(X)$ of the pattern bit string X. Next, compare $N_p(X)$ with the fingerprint $N_p(Y(i))$ of every sequence $Y(i)$ of n contiguous bits in the text bit string Y for $i = 1, 2, 3, \ldots, m - n + 1$. If it happens that $N_p(X) = N_p(Y(k))$ for some k, then we have evidence that $X = Y(k)$ and we stop. Ordinarily, if the algorithm is applied with only one prime p, we would check to see if $X = Y(k)$ because N_p is not one to one. If $X \ne Y(k)$, then we would continue with $i = k+1$; however, as we shall soon see, the use of several primes and several fingerprint functions exemplify the essence of the probabilistic methodology.

Pattern Search Algorithm (Karp and Rabin) *Suppose that we are given a pattern bit string X of length n, a text bit string Y of length m with $m \ge n$, and a positive integer M which will depend upon m and n. Let $S = \{p : p \text{ is prime and } 1 \le p \le M\}$. For a prime p in S, let $N_p : \{0,1\}^n \to \{0, 1, 2, \ldots, (p-1)\}$ be the fingerprint function. Choose k primes $p_1, p_2, \ldots, p_k$ at random from the set $\{1, 2, \ldots, M\}$ by choosing integers at random from $\{1, 2, 3, \ldots, M\}$, testing them for being prime, and stopping when k primes have been found. The search proceeds as follows.*

For each i, $1 \le i \le m - n + 1 = t$, compare $N_{p(j)}(X)$ with $N_{p(j)}(Y(i))$ for each prime $p_j = p(j)$ for $1 \le j \le k$ either until there is an i such that

$$N_{p(j)}(X) = N_{p(j)}(Y(i)) \text{ for all } j, 1 \le j \le k$$

at which time we declare that a "match" has occurred, or until $i = t$ with no "match" having occurred.

Note that a complete match requires that all k fingerprint functions must produce a match. The other important feature is that the primes $p_1, p_2, \ldots, p_k$ are chosen at random from a large enough set. It is still possible for a false match to occur, namely, for $N_{p(j)}(X) = N_{p(j)}(Y(i))$ for all j but with $X \ne Y(i)$; however, the nub of the matter is that the probability or chance of a false match can be made arbitrarily small. This last characteristic embodies what is meant by a probabilistic algorithm.

Example 2.56 *In order to fix the ideas of this algorithm, we will consider a smaller problem in this example than for which the method is intended to be applied. Suppose that*

$$X = [1001\ 1011]\ where\ n = 8$$

and

$$Y = [1011011001101100\cdots 0]\ where\ m = 300$$

Let $t = m - n + 1 = 293$ and $M = nt^2 = 686792$. Suppose that $k = 2$ and we have obtained these primes at random:

$$p_1 = 47\ and\ p_2 = 31$$

We first obtain the fingerprints of X by computing the primary residues

$$N_{p(1)}(X) = [[N(X)]]_{p(1)} = [[155_{ten}]]_{47} = 14$$
$$N_{p(2)}(X) = [[N(X)]]_{p(2)} = [[155_{ten}]]_{31} = 0$$

Starting with $i = 1$, we calculate the fingerprints of $Y(i)$ relative to the primes 47 and 31 and check for matches. Matches with either of the two fingerprints are indicated by an asterisk ($*$).

i	$Y(i)$	$N(Y(i))$	$N_{47}(Y(i))$		$N_{31}(Y(i))$		$Match?$
1	[1011 0110]	182	41		27		No
2	[0110 1100]	108	14	$*$	15		No
3	[1101 1001]	217	29		0	$*$	No
4	[1011 0011]	179	38		24		No
5	[0110 0110]	102	8		9		No
6	[1100 1101]	205	17		19		No
7	[1001 1011]	155	14	$*$	0	$*$	Yes

The first match occurs at position 7 in the bit string Y. We should note that because some of the numbers are small in this example, we performed direct calculations of the primary residues. In an actual computer implementation, we would need to utilize the modular operations and the updating formula described above.

Following Karp and Rabin, we now derive a bound on the probability of a false match when using the algorithm. First we need two concepts: probability and a number theoretic function.

Let T be a nonempty set of "possibilities," and any one of these possibilities may be chosen. A subset A of T contains the possibilities in which we are interested. Choose one element of T. The probability of choosing an element of A is defined to be

$$PROB\,(A) = \frac{\text{Number of elements in }A}{\text{Number of elements in }T}$$

Under these circumstances, we say that the element is chosen at random from the set T (any element of T is as likely to be chosen as any other element) and that the event A has occurred if the element chosen is in A. The probability of the event A is given by the formula above. For example, suppose that T is the set of faces of a cubical die so that $T = \{1, 2, 3, 4, 5, 6\}$ and we roll the die once (choose a side at random). Let A be the event (set of possible outcomes) of having a prime number of spots. $A = \{2, 3, 5\}$. Then $PROB(A) = 3/6 = 0.5$. Evidently, the probability of any event is between 0 and 1 inclusive.

Second, we recall from Definition 2.15 that $\pi(w)$ is the number of primes less than or equal to w. Thus $\pi(1) = 0$, $\pi(2) = 1$, $\pi(5) = 3$, and $\pi(20) = 8$.

We state without proof two results from J. B. Rosser and L. Schoenfeld [76].

Theorem 2.57 *If $w \geq 29$, then*

$$p_1 p \cdots p_{\pi(w)} > 2^w$$

where $p_1, p_2, p_3, \ldots, p_{\pi(w)}$ are the primes less than or equal to w.

Theorem 2.58 *If $w \geq 17$, then*

$$\frac{w}{\log(w)} \leq \pi(w) \leq \frac{1.25506w}{\log(w)}$$

Theorem 2.59 *If $w \geq 29$ and $b \leq 2^w$, then b has fewer than $\pi(w)$ prime divisors.*

Proof. Assume that $w \geq 29$ and $b \leq 2^w$ but that b has at least $\pi(w)$ prime divisors. Let $d_1 < d_2 < \cdots < d_r$ be these prime divisors of b so that $r \geq \pi(w)$. Let $p_1 < p_2 < \cdots < p_k$ be the first k primes for any k. Then

$$
\begin{array}{rcll}
2^w & \geq & b & \text{given} \\
b & \geq & d_1 d_2 d_3 \cdots d_r & \text{since } d_1 d_2 \cdots d_r \mid b \\
d_1 d_2 \cdots d_r & \geq & p_1 p_2 \cdots p_r & \text{since } d_i \geq p_i \text{ for each } i \\
p_1 p_2 \cdots p_r & \geq & p_1 p_2 \cdots p_{\pi(w)} & \text{since } r \geq \pi(w) \\
p_1 p_2 \cdots p_{\pi(w)} & > & 2^w & \text{by Theorem 2.57}
\end{array}
$$

But $2^w > 2^w$ is a contradiction. Thus, b must have fewer than $\pi(w)$ prime divisors. ∎

As mentioned before, this search algorithm has the possibility of terminating with a false match. The algorithm would not be of much use if a false match occurred often. The next theorem says that the chances of this algorithm producing a false match are no larger than a known value.

Theorem 2.60 *If the algorithm is applied to a search instance of X and Y with related characteristics n, m, t, M, k, and S, then the probability of a false match is less than or equal to*

$$t \left(\frac{\pi(n)}{\pi(M)} \right)^k \quad \text{when } n \geq 29$$

Proof. Assume that $n \geq 29$, $1 \leq r \leq t$, and $X \neq Y(r)$. Since $0 \leq N(X) < 2^n$ and $0 \leq N(Y(r)) < 2^n$, then $|N(X) - N(Y(r))| < 2^n$. By Theorem 2.59, the number of primes dividing $|N(X) - N(Y(r))|$ is less than $\pi(n)$; therefore, for any j, $1 \leq j \leq k$, since the prime p_j is chosen at random from S which has $\pi(M)$ elements,

$$PROB\,(p_j \text{ divides } |N(X) - N(Y(r))|) \leq \frac{\pi(n)}{\pi(M)}$$

Since the k primes $p_1, p_2, \ldots, p_k$ are randomly and independently chosen, it is a property of probability of the intersection of these k events that

$$PROB\,(p_j \text{ divides } |N(X) - N(Y(r))| \text{ for all } j) \leq \left(\frac{\pi(n)}{\pi(M)}\right)^k$$

Since there are t mutually exclusive instances of r, it is a property of probability that all the primes $p_1, p_2, \ldots, p_k$ divide $|N(X) - N(Y(r))|$ for some r is less than or equal to $t \cdot \left(\frac{\pi(n)}{\pi(M)}\right)^k$. ∎

Theorem 2.61 *Under the hypotheses of Theorem 2.60, if $M = nt^2$ with $n \geq 29$, then the probability of a false match is less than or equal to*

$$(1.25506)^k \, t^{-(2k-1)} \, (1 + 0.6 \log(t))^k$$

Proof. Theorem 2.60 implies that

$$PROB\,(\text{false match}) \leq t \cdot \left(\frac{\pi(n)}{\pi(M)}\right)^k$$

Theorem 2.58 implies that

$$\pi(n) \leq c\frac{n}{\log(n)} \quad \text{where } c = 1.25506$$

and that

$$\frac{nt^2}{\log(nt^2)} \leq \pi(nt^2)$$

Thus, we have

$$
\begin{aligned}
PROB\,(\text{false match}) \quad &\leq \quad t \cdot \left[\frac{c \cdot n}{\log(n)} \cdot \frac{\log(nt^2)}{nt^2}\right]^k \\
&= \quad t \cdot \left[\frac{c}{t^2} \cdot \left(1 + 2 \cdot \frac{\log(t)}{\log(n)}\right)\right]^k
\end{aligned}
$$

But for $n \geq 29$, $\log(n) \geq \log(29)$; therefore, the conclusion of the theorem holds since $\frac{2}{\log(n)} \leq \frac{2}{\log(29)} \leq 0.6$ by direct calculation. ∎

To illustrate Theorem 2.61, suppose that we have a string matching problem with the pattern X being of length $n = 200$, which corresponds to 25 eight-bit ASCII-encoded characters, and with the text string Y being of length 32000 corresponding to 4000 ASCII characters. Then $t = 31801$ and $M = nt^2 < 200(32000)^2 = 2^{11} \cdot 10^8 \approx 2^{11} \cdot 2^{26.6} < 2^{38}$ so that the base 2 computer representations of the primes p_i and corresponding residues will have fewer than 38 bits. If $k = 4$, then the probability of a false match will be less than about $2 \cdot 10^{-28}$.

The Karp and Rabin algorithm is a real-time algorithm requiring the same computation time for each text character. It also requires a constant amount of storage depending only upon n, k, and t. The algorithm can be proved to have a small probability of error. In addition, the method easily generalizes to apply to multidimensional problems. Other "fingerprinting" functions besides N_p may be appropriate.

We note that Karp and Rabin give other similar but different probabilistic pattern search algorithms in the reference cited earlier. For comparison purposes, information about some efficient nonprobabilistic searching algorithms may be found in (1) "Fast Pattern Matching in Strings," by D. E. Knuth, J. H. Morris, and V. R. Pratt [44], (2) "A Fast String Searching Algorithm", by R. S. Boyer and J. S. Moore [8], (3) Jack Purdum's "Pattern Matching Alternatives: Theory vs. Practice" [68], and (4) "A Very Fast Substring Search Algorithm," by Daniel M. Sunday [86].

Exercises

1. Prove that if W and V are bit strings of length n, then $|N(W) - N(V)| < 2^n$.

2. In Theorem 2.57, the product of all primes less than or equal to w was shown to be greater than 2^w as long as $w \geq 29$. Determine which positive integers $w < 29$ have the property of the conclusion of this theorem.

3. In Theorem 2.58, two inequalities were shown to hold for every positive integer $w \geq 17$. By direct computation, determine which positive integers $w < 17$ also satisfy these inequalities.

4. For a pattern X of length $n = 160$ and a text of length $m = 20000$, use Theorem 2.61 to determine an upper bound on the probability of a false match for $k = 1, 2, 3$, and 4.

5. Assume that the pattern length n bits is given, that the number of primes k has been given, that M is such that all fingerprinting primes and their residues have fewer than 32 bits, and that the updating described in the discussion will be used. Determine the number of 32-bit computer words of information storage are needed to implement the Karp and Rabin algorithm.

2.6   Application: Factoring by Pollard's ρ Method

For any positive integer $n > 1$, the Unique Prime Factorization Theorem guarantees that n may be written as a product of primes uniquely except for their arrangement; however, the theorem and its proof do not provide a way to find any prime divisors of n. The simple algorithm of dividing n by every integer $k \leq \sqrt{n}$ is certain to yield the factorization of n; however, this procedure of trial division is not practical for large n. In addition to the special cases of division by the small primes in Theorem 2.12, we discussed Fermat's method given in Theorem 2.13.

John M. Pollard in "A Monte Carlo Method for Factorization" [66] introduced the ρ (rho) method for factoring. It is aimed at finding the smallest prime factor of n, although it may produce other prime factors. It is referred to as a Monte Carlo method because it requires choosing integers in the range 0 to n at "random."

First, assume that n is composite with smallest prime factor p and that we have a method of choosing at random integers $x_0, x_1, \ldots, x_t$ so that $0 \leq x_i < n$. If n is large, p is small, and t is large relative to $\sqrt{p}$ but small relative to $\sqrt{n}$, then the probability is low that $x_i \equiv x_j \pmod{n}$ but high that $x_i \equiv x_j \pmod{p}$. The first probability statement holds because the number of distinct positive integers less than n is large relative to the number $t+1$, making it unlikely that we choose the same integer twice. On the other hand, $t+1$ is large relative to the number of integers less than p so that it is more likely that we choose x_i and x_j that have the same primary residue modulo p.

Suppose that through process described above we find x_r and x_s such that

$$
\begin{aligned}
&\text{(a)} \quad x_r \not\equiv x_s \pmod{n} \\
&\text{(b)} \quad x_r \equiv x_s \pmod{p}
\end{aligned}
$$

Then the (unknown) prime factor p of n has the property $p \mid (x_r - x_s)$. Consequently, $\gcd(x_r - x_s, n) > 1$. We want $x_r \not\equiv x_s \pmod{n}$ so that $x_r - x_s \neq 0$ in order that $\gcd(x_r - x_s, n)$ not equal n. Thus, the method is to generate a suitable number of distinct random integers $x_0, x_1, \ldots, x_t$ with $0 \leq x_i < n$ and then check all pairs x_i and x_j with $i \neq j$ to find x_r and x_s such that $\gcd(x_r - x_s, n) > 1$. Note that we cannot check whether $x_r \equiv x_s \pmod{p}$ since we do not know p; however, we choose t so that the probability that $x_i \equiv x_j \pmod{p}$ for some i and j will be high; therefore, the probability will be high that $\gcd(x_i - x_j, n) = d > 1$, giving a nontrivial factor of n. Note also that p may divide d.

Two difficulties arise. Given that the modulo n primary residues $x_0, x_1, \ldots, x_t$ have been generated, there are

$$
\binom{t+1}{2} = \frac{(t+1)t}{2} \approx t^2
$$

$\gcd(x_i - x_j, n)$ computations to be made. Since the Euclidean algorithm requires on the order of $(\log\ n)^3$ computer bit operations in this case, for $n \approx 10^{50}$, $\gcd(x_i - x_j, n)$ would require the order of $1.5 \cdot 10^6$ bit operations. Therefore, for large t and n, it would be impractical to test all pairs x_i and x_j.

In order that the probabilities be in favor of obtaining a nontrivial factor of n, we need to be able to generate random primary residues x_i of n. Satisfying this need will, in fact, help circumvent the problem of testing all pairs x_i and x_j.

Perfectly random selections are difficult to achieve or define. We saw in Section 1.8 that "random-like" integers (pseudo-random integers) may be produced using multiplicative congruential generators such as $f(x) = [[ax]]_n$ or linear ones such as $g(x) = [[ax + c]]_n$; however, we will see later these are not appropriate. In practice, generators of the form

$$f(x) = [[x^2 + c]]_n$$

where $c \neq 0$ or -2 appear to be sufficiently random.

We now combine the ideas above and use $f(x) = [[x^2 + 1]]_n$ for definiteness. Choose a "seed" x_0 and decide on t. Generate the sequence $x_0, x_1, \ldots, x_t$ as follows:

$$x_{i+1} = f(x_i)$$

Clearly, $0 \le x_i < n$. If it were to occur for some x_r and x_s pair with $r < s$ that $x_r \equiv x_s \pmod{p}$ (note that p is still unknown), then since $x^2 + 1$ is a polynomial, we get

$$x_{r+1} = f(x_r) \equiv f(x_s) = x_{s+1} \pmod{p}$$

Thus, the sequence $x_0, x_1, \ldots, x_t$ would be periodic modulo p with period $s - r$. Since $r < s$, there is a u with $r < u \le s$ such that $(s - r) \mid u$. Thus, if k is the smallest multiple of $s - r$ greater than r, then $2k$ is also a multiple of $s - r$. Consequently, $x_k \equiv x_{2k} \pmod{p}$. Therefore, instead of considering all pairs x_i and x_j with $0 \le i, j \le t$, we only need to compare x_i and x_{2i} for $0 \le i \le \sqrt{t}$.

Pollard's ρ Method consists of the following steps:

0. *Let n be a positive integer suspected or known to be composite.*

1. *Check by trial division whether n is divisible by known small primes, maybe up to 10^6.*

2. *Choose a seed, say $x_0 = 2$, a generating function, say, $f(x) = [[x^2 + 1]]_n$, and t that is not much bigger than $\sqrt{p}$, perhaps $t < 100\sqrt{p}$.*

3. *Generate the random integers $x_{i+1} = f(x_i) = [[x_i^2 + 1]]_n$ and simultaneously check x_k and x_{2k} by calculating $\gcd(x_k - x_{2k}, n)$.*

4. *If $\gcd(x_k - x_{2k}, n) = d$ with $1 < d < n$, then we have found a nontrivial factor of n. If $x_k \equiv x_{2k} \pmod{n}$ for some k or $k \ge \sqrt{t}$, then go to step*

2, *choosing a new seed* x_0 *or a new polynomial congruential generator, say* $f(x) = [[x^2 + 2]]_n$, *and repeat.*

Example 2.62 *Let* $n = 23083$. *We note that* n *is not divisible by 2, 3, 5, 7, or 11. Choose seed* $x_0 = 2$ *and generator* $x_{i+1} = f(x_i) = [[x_i^2 + 1]]_n$.

i	$x_i = \left[\left[x_{i-1}^2 + 1\right]\right]_n$		$\gcd(x_i - x_{2i}, n)$		
0	2		—		
1	5		$\gcd(5 - 26, n)$	$=$	1
2	26		$\gcd(26 - 19753, n)$	$=$	1
3	677		$\gcd(677 - 18574, n)$	$=$	1
4	19753	$= [[458330]]_n$	$\gcd(19753 - 20382, n)$	$=$	1
5	9061	$= [[390181010]]_n$	$\gcd(9061 - 16380, n)$	$=$	563
6	18574	$= [[82101722]]_n$			
7	18042	$= [[344993477]]_n$			
8	20382	$= [[325513765]]_n$			
9	1174	$= [[415425925]]_n$			
10	16380	$= [[1378277]]_n$			

Thus, $563 \mid n$. *Also, 563 is prime and* $n/563 = 41$ *is prime. The factorization of* n *is* $n = 23083 = 41 \cdot 563$. *Note that in this case, Pollard's Method did not yield the least prime factor of* n.

According to Pollard, a prime factor p of an integer n requires an order of $\sqrt{p}$ operations using this method. In general, the worst case would need the order of $n^{1/4}$ operations to factor n completely . The feature of using only x_k and x_{2k} in the algorithm is known as Floyd's cycle. The name ρ Method derives from the eventual periodicity of $x_0, x_1, \ldots$ modulo p as depicted in Figure 9 which has the appearance of the Greek letter ρ.

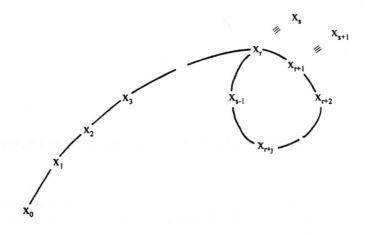

Figure 9: Pollard's ρ.

Sometimes generators such as $f(x) = [[x^m + c]]_n$ with $m > 2$ perform well. It is conjectured that if a factor p of n is such that $p \equiv 1 \pmod{m}$, then the use of $f(x) = [[x^m + c]]_n$ reduces the number of steps before success by a factor of $\sqrt{m-1}$. The use of a multiplicative or linear congruential generator may well be one to one modulo p, eliminating the chance for $x_i \equiv x_j \pmod{p}$ to occur.

The use of the generator $f(x) = [[x^2 + 1]]_n$ is not random among the integers $\{0, 1, \ldots, (n-1)\}$ because only about half of these integers are congruent to x^2 for any x. We will discuss this limitation in Section 3.9. Nevertheless, as evidenced by the successful application of Pollard's ρ Method, the integers generated by $f(x) = [[x^2 + 1]]_n$ appear to be random enough.

For a discussion of the computing complexity associated with factoring large integers, see Pollard's "Theorems on Factorization and Primality Testing" [65]. A more general introduction to factoring is given in Carl Pomerance and Stephen M. Gagola, Jr.'s *Lecture Notes on Primality Testing and Factoring* [67].

Richard P. Brent and J. M. Pollard report in "Factorization of the Eighth Fermat Number" [10] on the use of Brent's modification of Pollard's ρ Method to find the least prime factor of $F(n) = 2^{2^n} + 1$ when $n = 8$. The least prime factor was found to be

$$1238926361552897$$

which, according to Brent and Pollard, is rendered more easily rememberable via the mnemonic "I am now entirely persuaded to employ the method, a handy trick, on gigantic composite numbers."

Another important method of factoring is the number field sieve introduced by J. M. Pollard in 1988. A source of much of the relevant literature may be found in *The Development of the Number Field Sieve* by A. K. Lenstra and H. W. Lenstra, Jr. [47].

Exercises

1. Use Pollard's ρ Method to factor $n = 35183$

2. Use Pollard's ρ Method to factor $n = 87569$ (*Hint:* Use seed $x_0 = 3$.)

3. Use Pollard's ρ Method to factor $n = 14359$

Chapter 3
CONGRUENCES AND
THE FUNCTION ϕ

3.1   Introduction

The congruence relation $a \equiv b \pmod{n}$ has been discussed in several contexts in Chapters 0 to 2. We now summarize these results.

If n is a positive integer, then the Division Algorithm (Theorem 1.27) guarantees that for any integer a there are unique integers q and r such that $a = nq + r$ and $0 \le r < n$. We defined the congruence modulo n relation by $a \equiv b \pmod{n}$ if and only if $n \mid (a - b)$ and showed that $a \equiv b \pmod{n}$ if and only if a and b give the same remainder when divided by n. Congruence modulo n was shown to be an equivalence relation on the set of integers Z having exactly n equivalence classes: $[0], [1], [2], \ldots, [n-1]$. These equivalence classes comprise the set Z_n, called the integers modulo n. An equivalence class contains those integers that have the same remainder when divided by n. Operations of addition and multiplication of the equivalence classes in Z_n were defined by

$$[a] \oplus [b] \quad = \quad [a + b]$$
$$[a] \odot [b] \quad = \quad [a \cdot b]$$

where the operations $+$ and $\cdot$ are ordinary addition and multiplication among integers.

The set Z_n with the operation of addition $\oplus$ forms a group. Since a complete residue system contains exactly one integer from each equivalence class, this result is equivalent to saying that any complete residue system modulo n is a group under addition modulo n (meaning that the "group" sum of two residues in the complete residue system is the residue of that particular complete system that is congruent modulo n to the ordinary sum of the two residues). The complete residue system we often work with is the set of primary residues: $0, 1, 2, \ldots, (n - 1)$.

A reduced residue system modulo n is the subset of a complete residue system consisting of only the residues that are relatively prime to n. If $n > 1$, then the set of equivalence classes $\{[r_1], [r_2], \ldots, [r_k]\}$ corresponding to a reduced residue system $\{r_1, r_2, \ldots, r_k\}$ modulo n is a group under the multiplication operation $\odot$. We also say that any reduced residue system modulo n for $n > 1$ is a group under multiplication modulo n (meaning that the "group" product of two residues in the reduced residue system is the residue of that particular reduced residue system that is congruent modulo n to the ordinary product of the two residues).

If n is prime, then $\{1, 2, \ldots, (n-1)\}$ is a reduced residue system modulo n so that $\{[1], [2], \ldots, [n-1]\}$ with the operation $\odot$ forms a group. Thus, the set Z_n with operation $\odot$ and $[0]$ excluded is a group. If $n > 1$ is not prime, then Z_n is not a group whether or not $[0]$ is included. Also, $(Z_n, \oplus, \odot)$ is a commutative ring which is a field when n is prime.

In some discussions we will want to emphasize the equivalence classes, the operations $\oplus$ and $\odot$, and the group or semigroup properties. At other times we will want to emphasize complete and reduced residue systems and the operations $+$ and $\cdot$, where we understand that the result of addition or multiplication is the appropriate member of the given residue class that is congruent to the sum or product, respectively. In any case, we will feel free to call upon whatever interpretation that seems appropriate.

3.2   Chinese Remainder Theorem

In Chapter 1 we considered two types of equations with integer solutions. One was an equation of the form

$$ax + by = c$$

where we were given integers a, b, and c and wanted to find integers x and y satisfying the equation. The second was a congruence of the form

$$ax \equiv c \pmod{m}$$

where we were given integers a and c and positive integer m and wanted to find an integer x satisfying the congruence. In both cases we were able to construct explicit solutions using a, b, c, and m.

The equation of form $ax + by = c$ has an integral solution if and only if c is a multiple of the greatest common divisor of a and b. Furthermore, if (x_0, y_0) is a solution of $ax + by = c$, then the general solution has the form $x = x_0 + (b/d)u$ and $y = y_0 - (a/d)u$ where $d = \gcd(a, b)$ and $u \in Z$.

Since any solution of $ax \equiv c \pmod{m}$ is really a solution of $ax + mb = c$, $ax \equiv c \pmod{m}$ has a solution if and only if $\gcd(a, m) \mid c$ and each solution is given by $x(t) = x_0 + (tm)/\gcd(a, m)$ where t is any integer and for some y_0, (x_0, y_0) is a solution of $ax + by = c$. However, if $t = \gcd(a, m)$, then $x(t) \equiv x_0 \pmod{m}$. Hence there are $\gcd(a, m)$ distinct solutions for the congruence produced when $t = 1, 2, \ldots, \gcd(a, m)$. In particular, if a and m are relatively prime, then $ax \equiv u \pmod{m}$ has a unique solution; therefore, $ax \equiv 1 \pmod{m}$ has a unique solution and a has a "multiplicative inverse" modulo m.

Each of the situations above was a case of a single equation or congruence. We now wish to consider solutions of simultaneous congruences:

$$
\begin{aligned}
x &\equiv a_1 \pmod{m_1} \\
x &\equiv a_2 \pmod{m_2} \\
&\vdots \\
x &\equiv a_n \pmod{m_n}
\end{aligned}
$$

where the m_i are relatively prime in pairs. In other words, we want to find an integer x that leaves a remainder of a_1 when divided by m_1, a remainder of a_2 when divided by m_2, and so on, when the divisors $m_1, m_2, \ldots, m_n$ are relatively prime in pairs.

Solutions of simultaneous congruences were considered in ancient times. Often, whimsical word problems similar to the following were posed. Suppose that a group of monkeys contemplates a pile of coconuts. If the monkeys put the coconuts in piles of five each, there are four left over. Using piles of four, there are three left over. In piles of seven, there are two left. In piles of nine, there are six left over. What is the fewest number of coconuts possible?

If x is a possible number of coconuts in the pile, then having four left over from piles of five is expressed as

$$x \equiv 4 \pmod{5}$$

Similarly, the other conditions are

$$
\begin{aligned}
x &\equiv 3 \pmod{4} \\
x &\equiv 2 \pmod{7} \\
x &\equiv 6 \pmod{9}
\end{aligned}
$$

The smallest positive integer x satisfying all four congruences is the required solution. Solutions of such problems are given by the Chinese Remainder Theorem.

Theorem 3.1 (Chinese Remainder Theorem) *Let $m_1, m_2, \ldots, m_n$ be pairwise relatively prime, i.e., $\gcd(m_i, m_j) = 1$, for all i and j less than or equal to n where $i \neq j$. Then the system of congruences*

$$
\begin{aligned}
x &\equiv a_1 \pmod{m_1} \\
x &\equiv a_2 \pmod{m_2} \\
&\vdots \\
x &\equiv a_n \pmod{m_n}
\end{aligned}
$$

has a solution which is unique modulo the integer $m_1 m_2 \cdots m_n$. Further, if

$$M_j = \frac{\displaystyle\prod_{i=1}^{n} m_i}{m_j}$$

and z_j is a solution of $M_j z_j \equiv a_j \pmod{m_j}$ for each j, then the solution is given by

$$x = \left[\left[\sum_{j=1}^{n} M_j z_j \right] \right]_{m_1 m_2 \cdots m_n}$$

Proof. Let x be as defined in the theorem. Then for any k, $1 \leq k \leq n$,

$$
\begin{aligned}
x &= \left[\left[\sum_{j=1}^{n} M_j z_j \right] \right]_{m_1 m_2 \cdots m_n} &\equiv \sum_{j=1}^{n} M_j z_j \left(\operatorname{mod} \prod_{i=1}^{n} m_i \right) \\
&&\equiv \sum_{j=1}^{n} M_j z_j \pmod{m_k} \\
&&\equiv M_k z_k \pmod{m_k} \\
&&\equiv a_k \pmod{m_k}
\end{aligned}
$$

so that x satisfies the n congruences, $x \equiv a_k \pmod{m_k}$ for $1 \leq k \leq n$. Suppose that x' also satisfies these n congruences. Then

$$x - x' \equiv 0 \pmod{m_i} \quad \text{for } 1 \leq i \leq n$$

Since $\gcd(m_i, m_j) = 1$ for $i \neq j$, we obtain

$$x \equiv x' \left(\operatorname{mod} \prod_{i=1}^{n} m_i \right)$$

and the solution x is unique modulo $\displaystyle\prod_{i=1}^{n} m_i$. ∎

Example 3.2 *Solve the following set of congruences:*

$$
\begin{aligned}
x &\equiv 5 \pmod{4} \\
x &\equiv 7 \pmod{11}
\end{aligned}
$$

Since 4 and 11 are relatively prime, there exists an integer, namely 10, such that $(4)(10) \equiv 7 \pmod{11}$ and there exists an integer, namely 3, such that $(11)(3) \equiv 5 \pmod{4}$. Hence $(4)(10) + (11)(3) = 73$, which is congruent to 29 modulo 44, satisfies the congruences above.

Example 3.3 *We answer the monkey-coconut question by solving the following set of congruences:*

$$x \equiv 4 \pmod 5$$
$$x \equiv 3 \pmod 4$$
$$x \equiv 2 \pmod 7$$
$$x \equiv 6 \pmod 9$$

We have $M_1 = 4 \cdot 7 \cdot 9 = 252$, $M_2 = 5 \cdot 7 \cdot 9 = 315$, $M_3 = 180$, and $M_4 = 140$. Since 5 and 252 are relatively prime, there exists an integer z_1 such that $252z_1 \equiv 4 \pmod 5$ or equivalently, $2z_1 \equiv 4 \pmod 5$ or $z_1 \equiv 2 \pmod 5$. Hence z_1 can equal 7.

Since 4 and 315 are relatively prime, there exists an integer z_2 such that $315z_2 \equiv 3 \pmod 4$ or equivalently, $3z_2 \equiv 3 \pmod 4$. Hence z_2 can equal 1.

Since 7 and 180 are relatively prime, there exists an integer z_3 such that $180z_3 \equiv 2 \pmod 7$ or equivalently, $5z_3 \equiv 2 \pmod 7$. Hence z_3 can equal 6.

Since 9 and 140 are relatively prime, there exists an integer z_4 such that $140z_4 \equiv 6 \pmod 9$ or equivalently, $5z_4 \equiv 6 \pmod 9$. Hence z_4 can equal 3.

Hence $x = (7)(252) + (1)(315) + (6)(180) + (3)(140) \pmod{(5 \cdot 4 \cdot 7 \cdot 9)}$ or $x \equiv 3579 \pmod{1260}$ and $x = 1059$ is the least positive integer solution.

Exercises

1. Solve the following sets of simultaneous congruences:

   (a) $x \equiv 9 \pmod{12}$
        $x \equiv 6 \pmod{25}$

   (b) $x \equiv 3 \pmod 4$
        $x \equiv 5 \pmod 9$
        $x \equiv 10 \pmod{35}$

   (c) $x \equiv a \pmod{15}$
        $x \equiv b \pmod{16}$

   (d) $x \equiv 5 \pmod 7$
        $x \equiv 12 \pmod{15}$
        $x \equiv 18 \pmod{22}$

2. Let $p_1 < p_2 < \cdots < p_n$ be the least n primes so that $p_1 = 2$, $p_2 = 3$, and so on. Prove that there is an integer x with

$$x \equiv -1 \pmod{p_1^2}$$
$$x \equiv -2 \pmod{p_2^2}$$
$$\vdots$$
$$x \equiv -n \pmod{p_n^2}$$

Show that there are n consecutive integers each of which is divisible by a perfect square. (An integer y is a *perfect square* if $y = k^2$ for some integer k.)

3. Using the notation and assumptions of the Chinese Remainder Theorem and its proof, prove that if b_j is a solution of $M_j b_j \equiv 1 \pmod{m_j}$ for each j, $1 \le j \le n$, then

$$x = M_1 b_1 a_1 + M_2 b_2 a_2 + \cdots + M_n b_n a_n$$

is a solution of the system of n congruences of the Chinese Remainder Theorem system.

4. Assume that there are two gears, one (gear A) with 25 teeth and the other (gear B) with 54 teeth. Gear A has a "bad" tooth and gear B has a bad section between two teeth. The two gears are meshed together with gear A on the left. Gear A turns clockwise and gear B turns counterclockwise. At the beginning of rotation a tooth of gear A fits exactly with gear B, the bad tooth of gear A is three teeth before the mesh position; and the bad between-teeth section of gear B is 20 sections from reaching the mesh position. How many "teeth" must mesh before the bad tooth first meets the bad between-teeth section? How often after that do the defective parts meet?

5. If the marbles in a bag are lined up in rows of 15 each, there are 4 left in the bag. If the marbles are lined up in rows of 8 each, there are 3 left in the bag. If the marbles are lined up in rows of 23, there are 10 left in the bag. What is the least number of marbles that initially could have been in the bag; and how many rows of 15, 8, and 23 were there?

6. The Chinese Remainder Theorem may be generalized to moduli $m_1, m_2, \ldots, m_n$ that are not relatively prime as follows: The system of congruences

$$\begin{aligned} x &\equiv a_1 \pmod{m_1} \\ x &\equiv a_2 \pmod{m_2} \\ &\vdots \\ x &\equiv a_n \pmod{m_n} \end{aligned}$$

has a solution if and only if $\gcd(m_i, m_j)$ divides $a_i - a_j$ for all i and j with $1 \le i < j \le n$. When the solution exists, it is unique modulo $\operatorname{lcm}(m_1, m_2, \ldots, m_n)$.

(a) Prove this theorem for $n = 2$ by observing that if $x \equiv a_1 \pmod{m_1}$, then there is an integer k such that $x = a_1 + k m_1$. Next, substitute for x in the second congruence, and so on.

(b) Prove this theorem for all n by mathematical induction.

7. If possible, solve each of the following systems of congruences:

(a) $x \equiv 21 \pmod{36}$
     $x \equiv 5 \pmod{8}$

(b) $x \equiv 8 \pmod{12}$
     $x \equiv 5 \pmod{9}$
     $x \equiv 14 \pmod{15}$

(c) $x \equiv 19 \pmod{49}$
     $x \equiv 10 \pmod{14}$

3.3   Matrices and Simultaneous Equations

We give below a brief introduction of matrices because the treatment of some topics presented later can be given more clearly and succinctly using matrix representations.

Definition 3.4 *For positive integers n and m, an $n \times m$ matrix is a function $A : \{1, 2, \ldots, n\} \times \{1, 2, \ldots, m\} \rightarrow D$, where D is an integral domain, such as the real, complex, or rational numbers, or the integers. Therefore, for each i, $1 \leq i \leq n$, and each j, $1 \leq j \leq m$, there is a value $A(i, j)$, which, we think of as being in the i-th row and the j-th column of a rectangular array. The image $A(i, j)$ of the domain element (i, j), is usually shortened to A_{ij}. Thus the $n \times m$ matrix A is represented by a rectangular array where the images of $(i, j) \in \{1, 2, \ldots, n\} \times \{1, 2, \ldots, m\}$ under A are listed as follows:*

$$
A = \begin{bmatrix}
A_{11} & A_{12} & A_{13} & \cdots & A_{1m} \\
A_{21} & A_{22} & A_{23} & \cdots & A_{2m} \\
\vdots & & & & \vdots \\
A_{n1} & A_{n2} & A_{n3} & \cdots & A_{nm}
\end{bmatrix}
$$

where the first row of the representation consists of A_{1j} for $j = 1$ to m, the second row consists of A_{2j} for $j = 1$ to m, and so on. We say that A has n rows and m columns and is of dimension $n \times m$. Sometimes we write $A = [A_{ij}]$ for short or even $A = [a_{ij}]$, where sometimes a lowercase letter is used for easier visual discrimination. The value A_{ij} is called a component, entry, *or* element *of the matrix A. A matrix of dimension $1 \times m$ is called a* row matrix; *and a matrix of dimension $n \times 1$ is called a* column matrix. *If the number of rows of a matrix is the same as the number of columns, then the matrix is called a* square matrix. *If A is a row matrix, then the subscripts for the row are often suppressed and we write*

$$
A = \begin{bmatrix} A_{11} & A_{12} & \cdots & A_{1m} \end{bmatrix} = \begin{bmatrix} A_1 & A_2 & \cdots & A_m \end{bmatrix}
$$

Similarly, we write

$$A = \begin{bmatrix} A_{11} \\ A_{21} \\ \vdots \\ A_{n1} \end{bmatrix} = \begin{bmatrix} A_1 \\ A_2 \\ \vdots \\ A_n \end{bmatrix}$$

suppressing the column indices if A is a column matrix.

For example, $A = \begin{bmatrix} 2 & 1 & 7 \\ 4 & 0 & 6 \end{bmatrix}$ is a 2×3 matrix and $B = \begin{bmatrix} -2 & 5 & 7 \\ -3 & 9 & 0 \\ 25 & 2 & 9 \end{bmatrix}$

is a 3×3 square matrix. Thus, $A_{13} = 7$, $A_{21} = 4$, $B_{12} = 5$, and $B_{31} = 25$.

The matrix $C = \begin{bmatrix} 7 & 1 \\ 3 & 9 \end{bmatrix}$ is a 2×2 square matrix. In most of our uses of matrices the components are integers.

Definition 3.5 *Two $n \times m$ matrices $A = [A_{ij}]$ and $B = [B_{ij}]$ are equal if the corresponding elements are equal; that is, $A = B$ if and only if $A_{ij} = B_{ij}$ for all i with $1 \leq i \leq n$ and j with $1 \leq j \leq m$.*

For example, $\begin{bmatrix} A_{11} & A_{12} & A_{13} \\ A_{21} & A_{22} & A_{23} \\ A_{31} & A_{32} & A_{33} \end{bmatrix} = \begin{bmatrix} B_{11} & B_{12} & B_{13} \\ B_{21} & B_{22} & B_{23} \\ B_{31} & B_{32} & B_{33} \end{bmatrix}$ if and only if

$A_{ij} = B_{ij}$ for $1 \leq i, j \leq 3$.

We note that this definition of matrix equality is just a restatement of when two functions A and B are equal: $A((i,j)) = B((i,j))$ for all (i,j). That is, the functions have the same domain and the same values at each element of the domain.

Some definitions and theorems will be stated in full generality; however, in order to reduce the complexity of the presentation and for convenience and definiteness, others will be stated or proved only for 2×2 and 3×3 matrices. This practice will allow rather explicit proofs of the cases of those theorems that are most pertinent to the application here. In any case, any proof offered for a theorem involving determinants, introduced in the next definition, will be restricted in this book to 2×2 and 3×3 matrices. In nearly all cases, there are analogous definitions and theorems for matrices of larger sizes.

Definition 3.6 *If $A = \begin{bmatrix} a & b \\ c & d \end{bmatrix}$ is a 2×2 matrix, we define the determinant*

of A, denoted by $\begin{vmatrix} a & b \\ c & d \end{vmatrix}$, to be the number $ad - bc$. If

$$A = \begin{bmatrix} A_{11} & A_{12} & A_{13} \\ A_{21} & A_{22} & A_{23} \\ A_{31} & A_{32} & A_{33} \end{bmatrix}$$

then the determinant *of A, denoted by*

$$\begin{vmatrix} A_{11} & A_{12} & A_{13} \\ A_{21} & A_{22} & A_{23} \\ A_{31} & A_{32} & A_{33} \end{vmatrix}$$

is the number

$$A_{11} \cdot \begin{vmatrix} A_{22} & A_{23} \\ A_{32} & A_{33} \end{vmatrix} - A_{12} \cdot \begin{vmatrix} A_{21} & A_{23} \\ A_{31} & A_{33} \end{vmatrix} + A_{13} \cdot \begin{vmatrix} A_{21} & A_{22} \\ A_{31} & A_{32} \end{vmatrix}$$

The determinant *of A is denoted by* $\det(A)$ *for any* $n \times n$ *matrix A.*

It can be shown by direct computation that the determinant of the 3×3 matrix A of the definition is also equal to

$$A_{11} \cdot \begin{vmatrix} A_{22} & A_{23} \\ A_{32} & A_{33} \end{vmatrix} - A_{21} \cdot \begin{vmatrix} A_{12} & A_{13} \\ A_{32} & A_{33} \end{vmatrix} + A_{31} \cdot \begin{vmatrix} A_{12} & A_{13} \\ A_{22} & A_{23} \end{vmatrix}$$

Example 3.7 *Suppose that* $A = \begin{bmatrix} -1 & 3 \\ -2 & 7 \end{bmatrix}$ *and* $B = \begin{bmatrix} 2 & -3 & 1 \\ 5 & 0 & 4 \\ 7 & -8 & 6 \end{bmatrix}$. *Then*

$$\det(A) = \begin{vmatrix} -1 & 3 \\ -2 & 7 \end{vmatrix} = (-1)(7) - (3)(-2) = -1$$

$$\det(B) = \begin{vmatrix} 2 & -3 & 1 \\ 5 & 0 & 4 \\ 7 & -8 & 6 \end{vmatrix} = (2) \begin{vmatrix} 0 & 4 \\ -8 & 6 \end{vmatrix} - (-3) \begin{vmatrix} 5 & 4 \\ 7 & 6 \end{vmatrix} + (1) \begin{vmatrix} 5 & 0 \\ 7 & -8 \end{vmatrix}$$

$$= 64 + 6 + (-40) = 30$$

Definition 3.8

(a) *If V is either a row or column matrix with n entries and W is either a row or column matrix with n entries so that*

$$V = \begin{bmatrix} V_1 \\ V_2 \\ \vdots \\ V_n \end{bmatrix} \quad or \quad V = \begin{bmatrix} V_1 & V_2 & \cdots & V_n \end{bmatrix}$$

and

$$W = \begin{bmatrix} W_1 \\ W_2 \\ \vdots \\ W_n \end{bmatrix} \quad or \quad W = \begin{bmatrix} W_1 & W_2 & \cdots & W_n \end{bmatrix}$$

then the dot product or inner product of V and W, denoted by $V \bullet W$, is the number $V_1 W_1 + V_2 W_2 + \cdots + V_n W_n$.

(b) If

$$A = \begin{bmatrix} A_{11} & A_{12} & \cdots & A_{1p} \\ A_{21} & A_{22} & \cdots & A_{2p} \\ \vdots & & & \vdots \\ A_{n1} & A_{n2} & \cdots & A_{np} \end{bmatrix}$$

is an $n \times p$ matrix and

$$B = \begin{bmatrix} B_{11} & B_{12} & \cdots & B_{1m} \\ B_{21} & B_{22} & \cdots & B_{2m} \\ \vdots & & & \vdots \\ B_{p1} & B_{p2} & \cdots & B_{pm} \end{bmatrix}$$

is an $p \times m$ matrix, then the (matrix) product of A and B, denoted by AB, is the $n \times m$ matrix $C = [C_{ij}]$, where C_{ij} is the dot product of the i-th row of A and the j-th column of B; that is,

$$C_{ij} = \begin{bmatrix} A_{i1} & A_{i2} & A_{i3} & \cdots & A_{ip} \end{bmatrix} \bullet \begin{bmatrix} B_{1j} \\ B_{2j} \\ B_{3j} \\ \vdots \\ B_{pj} \end{bmatrix} = \sum_{k=1}^{p} A_{ik} B_{kj}$$

(c) If d is an integer and $A = [A_{ij}]$ is an $n \times m$ matrix, then dA is the $n \times m$ matrix $D = [D_{ij}]$, where $D_{ij} = dA_{ij}$, so that every component of D is obtained by multiplying the corresponding component of A by d. The product of a number d and a matrix A is called a scalar product and the number d is called a scalar.

Example 3.9 *Suppose that*

$$A = [A_{ij}] = \begin{bmatrix} 1 & -3 & 5 \\ 6 & 0 & -2 \end{bmatrix} \quad and \quad B = [B_{ij}] = \begin{bmatrix} -2 & 4 & 0 & 8 \\ 3 & -1 & -2 & 1 \\ 0 & 5 & 7 & 0 \end{bmatrix}$$

We want to calculate the matrix product $C = [C_{ij}] = AB$. The matrix A is 2×3 and B is 3×4 so that the product $AB = C$ is defined and will be a 2×4 matrix. C_{11} is the dot product of the first row of A and the first column of B and C_{23} is the dot product of the second row of A and the third column of B. Thus,

$$C_{11} = \begin{bmatrix} 1 & -3 & 5 \end{bmatrix} \bullet \begin{bmatrix} -2 \\ 3 \\ 0 \end{bmatrix} = (1)(-2) + (-3)(3) + (5)(0) = -11$$

$$C_{23} = \begin{bmatrix} 6 & 0 & -2 \end{bmatrix} \bullet \begin{bmatrix} 0 \\ -2 \\ 7 \end{bmatrix} = (6)(0) + (0)(-2) + (-2)(7) = -14$$

Continuing in this manner, we obtain

$$AB = \begin{bmatrix} 1 & -3 & 5 \\ 6 & 0 & -2 \end{bmatrix} \begin{bmatrix} -2 & 4 & 0 & 8 \\ 3 & -1 & -2 & 1 \\ 0 & 5 & 7 & 0 \end{bmatrix} = \begin{bmatrix} -11 & 32 & 41 & 5 \\ -12 & 14 & -14 & 48 \end{bmatrix}$$

To calculate the scalar product dA with $d = 7$, we have

$$7A = 7 \begin{bmatrix} 1 & -3 & 5 \\ 6 & 0 & -2 \end{bmatrix} = \begin{bmatrix} (7)(1) & (7)(-3) & (7)(5) \\ (7)(6) & (7)(0) & (7)(-2) \end{bmatrix}$$

$$= \begin{bmatrix} 7 & -21 & 35 \\ 42 & 0 & -14 \end{bmatrix}$$

Definition 3.10 *If A is an $n \times m$ matrix where the entry in the i-th row and j-th column is A_{ij}, then A^t, the* transpose *of A, is the $m \times n$ matrix with $A^t_{ij} = A_{ji}$. If A is an $n \times n$ matrix and $A_{ij} = A_{ji}$ for all $1 \le i, j \le n$, then A is* symmetric *or is said to be a symmetric matrix. An equivalent statement is that A is symmetric if and only if $A = A^t$.*

Example 3.11 *Let A be the 2×3 matrix of Example 3.9. Then A^t is the 3×2 matrix obtained from A by writing the rows of A as columns.*

$$A^t = \begin{bmatrix} 1 & -3 & 5 \\ 6 & 0 & -2 \end{bmatrix}^t = \begin{bmatrix} 1 & 6 \\ -3 & 0 \\ 5 & -2 \end{bmatrix}$$

Theorem 3.12 *If A and B are matrices whose product is defined, then $(AB)^t = B^t A^t$.*

Proof. The proof is left to reader. ∎

Theorem 3.13 *If A and B are both $n \times n$ matrices, then the determinant of AB is equal to the product of the determinant of A and the determinant of B; that is, $\det(AB) = \det(A) \cdot \det(B)$.*

Proof. The proofs for $n = 2$ and 3 may be shown by direct computation. ∎

Theorem 3.14 *If A is a square $n \times n$ matrix, then $\det(A) = \det(A^t)$.*

Proof. The proofs for $n = 2$ and 3 are left to the reader. ∎

Definition 3.15 *If $A = [A_{ij}]$ and $B = [B_{ij}]$ are $n \times m$ matrices, then $A+B$ is the $n \times m$ matrix $C = [C_{ij}]$ where $C_{ij} = A_{ij} + B_{ij}$, i.e., addition is component-wise. C is called the* matrix sum *of A and B. The notation $A - B$ means $A + (-1)B$.*

Example 3.16 *Let*

$$A = \begin{bmatrix} -1 & 3 \\ 2 & 7 \\ 4 & -5 \end{bmatrix} \quad and \quad B = \begin{bmatrix} 3 & 11 \\ -5 & 4 \\ 8 & 2 \end{bmatrix}$$

Then

$$A + B = \begin{bmatrix} -1 & 3 \\ 2 & 7 \\ 4 & -5 \end{bmatrix} + \begin{bmatrix} 3 & 11 \\ -5 & 4 \\ 8 & 2 \end{bmatrix} = \begin{bmatrix} (-1)+3 & 3+11 \\ 2+(-5) & 7+4 \\ 4+8 & (-5)+2 \end{bmatrix}$$

$$= \begin{bmatrix} 2 & 14 \\ -3 & 11 \\ 12 & -3 \end{bmatrix}$$

Theorem 3.17 *If A is a 2×2 matrix, B is a 3×3 matrix,*

$$I_2 = \begin{bmatrix} 1 & 0 \\ 0 & 1 \end{bmatrix}$$

and

$$I_3 = \begin{bmatrix} 1 & 0 & 0 \\ 0 & 1 & 0 \\ 0 & 0 & 1 \end{bmatrix}$$

then $AI_2 = I_2 A = A$ and $BI_3 = I_3 B = B$. I_2 and I_3 are called the 2×2 and 3×3 identity matrices, respectively. In general, the $n \times n$ identity matrix is denoted by $I_n = [\delta_{ij}]$, where $\delta_{ii} = 1$ and $\delta_{ij} = 0$ when $i \neq j$; therefore, if C is $n \times n$, then $I_n C = C I_n = C$.

Proof. Proofs for $n = 2$ and 3 may be shown by direct computation. In general, the definition of matrix product gives the result. ■

The proofs for $n = 2$ and 3 in the following theorem are left to the reader.

Theorem 3.18 *The set of $n \times n$ matrices forms a ring with unity, I_n, where the product is the matrix product and the sum is the matrix sum.*

Note that Theorem 3.18 says that the associative property holds for matrix addition and multiplication and that the commutative property of matrix addition holds; however, the ring of $n \times n$ matrices is not commutative for $n > 1$ since we do not have commutativity of multiplication.

We next consider simultaneous equations and look for integer solutions. Given the linear equations

$$\begin{aligned} 3x_1 + 2x_2 &= 7 \\ 2x_1 - 2x_2 &= -2 \end{aligned}$$

if we add the equations we have $5x_1 = 5$ so $x_1 = 1$; and substituting this value back into either equation we get $x_2 = 2$. This method of solving systems of equations is called the *addition/subtraction* method.

More generally, consider

$$ax_1 + bx_2 = e$$
$$cx_1 + dx_2 = f$$

If we want to solve for x_1, we may multiply the top equation by d and the bottom equation by b to get

$$adx_1 + bdx_2 = de$$
$$bcx_1 + bdx_2 = bf$$

Subtracting and solving for x_1, we get $x_1 = (de - bf)/(ad - bc)$ provided that $ad - bc \neq 0$. Similarly, we get $x_2 = (af - ce)/(ad - bc)$. Using determinants, we have

$$x_1 = \frac{\begin{vmatrix} e & b \\ f & d \end{vmatrix}}{\begin{vmatrix} a & b \\ c & d \end{vmatrix}} \quad \text{and} \quad x_2 = \frac{\begin{vmatrix} a & e \\ c & f \end{vmatrix}}{\begin{vmatrix} a & b \\ c & d \end{vmatrix}}$$

We can further say there is a unique solution if and only if $ad - bc = \begin{vmatrix} a & b \\ c & d \end{vmatrix}$ is not zero and we are guaranteed integral solutions if this determinant is ± 1. This method is called *Cramer's rule*.

Suppose that we have the following system of equations:

$$\begin{aligned} 3x_1 & + & 2x_2 & - & x_3 & = & 12 \\ x_1 & - & x_2 & + & x_3 & = & 2 \\ x_1 & - & 3x_2 & + & x_3 & = & -2 \end{aligned}$$

Again using the addition/subtraction method, if we add the first and second equations, we get $4x_1 + x_2 = 14$ and if we add the first and third equations, we get $4x_1 - x_2 = 10$. Note that in both cases it is essential to eliminate the same variable. Solving these two equations simultaneously as above, we have $x_1 = 3$ and $x_2 = 2$. Substituting these values into any of the original equations we find that $x_3 = 1$.

It may also be shown by a direct but rather tedious computation that if we have the equations

$$A_{11}x_1 + A_{12}x_2 + A_{13}x_3 = y_1$$
$$A_{21}x_1 + A_{22}x_2 + A_{23}x_3 = y_2$$
$$A_{31}x_1 + A_{32}x_2 + A_{33}x_3 = y_3$$

and solve for x_1, x_2, and x_3, we get, by Cramer's rule,

$$x_1 = \frac{\begin{vmatrix} y_1 & A_{12} & A_{13} \\ y_2 & A_{22} & A_{23} \\ y_3 & A_{32} & A_{33} \end{vmatrix}}{\begin{vmatrix} A_{11} & A_{12} & A_{13} \\ A_{21} & A_{22} & A_{23} \\ A_{31} & A_{32} & A_{33} \end{vmatrix}}, \quad x_2 = \frac{\begin{vmatrix} A_{11} & y_1 & A_{13} \\ A_{21} & y_2 & A_{23} \\ A_{31} & y_3 & A_{33} \end{vmatrix}}{\begin{vmatrix} A_{11} & A_{12} & A_{13} \\ A_{21} & A_{22} & A_{23} \\ A_{31} & A_{32} & A_{33} \end{vmatrix}}$$

and

$$x_3 = \frac{\begin{vmatrix} A_{11} & A_{12} & y_1 \\ A_{21} & A_{22} & y_2 \\ A_{31} & A_{32} & y_3 \end{vmatrix}}{\begin{vmatrix} A_{11} & A_{12} & A_{13} \\ A_{21} & A_{22} & A_{23} \\ A_{31} & A_{32} & A_{33} \end{vmatrix}}$$

Hence we have a unique solution if $\begin{vmatrix} A_{11} & A_{12} & A_{13} \\ A_{21} & A_{22} & A_{23} \\ A_{31} & A_{32} & A_{33} \end{vmatrix}$ is not 0 and we are

guaranteed a solution with integral coefficients if this determinant is ± 1.

Note in the numerator determinant when solving for x_i that the coefficients of x_i are replaced by the numbers y_i from the right side of the equal signs. This pattern holds both for two equations in two unknowns and three equations in three unknowns.

Theorem 3.19 *If B is a 2×2 matrix with nonzero determinant and A is a 3×3 matrix with nonzero determinant, then there exist a 2×2 matrix denoted by B^{-1} and a 3×3 matrix denoted by A^{-1} such that $BB^{-1} = B^{-1}B = I_2$ and $AA^{-1} = A^{-1}A = I_3$. A^{-1} and B^{-1} are called the* inverses *of A and B, respectively. (The forms of A^{-1} and B^{-1} are given in the proof.) In either case we are guaranteed that the inverse will have integral elements if the determinant is equal to ± 1. If the determinant of either A or B is zero, then its inverse does not exist.*

More generally, if C is an $n \times n$ matrix, then $\det(C) \neq 0$ if and only if there is an $n \times n$ matrix C^{-1}, called the inverse of C, such that $CC^{-1} = C^{-1}C = I_n$; and if $\det(C) = \pm 1$, the elements of C^{-1} will be integers.

Proof. In the proof we consider only the cases $n = 2$ and 3; and in these cases, the inverses are given explicitly. Let

$$B = \begin{bmatrix} a & b \\ c & d \end{bmatrix} \quad \text{and} \quad B^{-1} = \begin{bmatrix} \dfrac{d}{\det(B)} & \dfrac{-b}{\det(B)} \\[2ex] \dfrac{-c}{\det(B)} & \dfrac{a}{\det(B)} \end{bmatrix}$$

where $\det(B) = ad - bc$. It is straightforward to show by direct computation that $BB^{-1} = B^{-1}B = I_2$. Similarly, if

$$A = \begin{bmatrix} A_{11} & A_{12} & A_{13} \\ A_{21} & A_{22} & A_{23} \\ A_{31} & A_{32} & A_{33} \end{bmatrix}$$

let

$$\overline{A} = \frac{1}{\det(A)} \cdot \begin{bmatrix} \begin{vmatrix} A_{22} & A_{23} \\ A_{32} & A_{33} \end{vmatrix} & -\begin{vmatrix} A_{21} & A_{23} \\ A_{31} & A_{33} \end{vmatrix} & \begin{vmatrix} A_{21} & A_{22} \\ A_{31} & A_{32} \end{vmatrix} \\ -\begin{vmatrix} A_{12} & A_{13} \\ A_{32} & A_{33} \end{vmatrix} & \begin{vmatrix} A_{11} & A_{13} \\ A_{31} & A_{33} \end{vmatrix} & -\begin{vmatrix} A_{11} & A_{12} \\ A_{31} & A_{32} \end{vmatrix} \\ \begin{vmatrix} A_{12} & A_{13} \\ A_{22} & A_{23} \end{vmatrix} & -\begin{vmatrix} A_{11} & A_{13} \\ A_{21} & A_{23} \end{vmatrix} & \begin{vmatrix} A_{11} & A_{12} \\ A_{21} & A_{22} \end{vmatrix} \end{bmatrix}$$

and let $A^{-1} = \left(\overline{A}\right)^t$. It may be shown by direct computation that $AA^{-1} = A^{-1}A = I_3$.

Hence to find the inverse of A we define $\overline{A} = \left[\overline{A}_{ij}\right]$ by letting $\overline{A}_{ij}$ be formed by taking the determinant of the matrix obtained from $A = [A_{ij}]$ by removing the row and column containing A_{ij}, dividing this by the determinant of A, and multiplying by the correct sign, namely $(-1)^{i+j}$. Finally, we take the transpose of $\overline{A}$, to get A^{-1}; therefore, $A^{-1} = (\overline{A})^t$ and

$$A^{-1} = \frac{1}{\det(A)} \cdot \begin{bmatrix} \begin{vmatrix} A_{22} & A_{23} \\ A_{32} & A_{33} \end{vmatrix} & -\begin{vmatrix} A_{12} & A_{13} \\ A_{32} & A_{33} \end{vmatrix} & \begin{vmatrix} A_{12} & A_{13} \\ A_{22} & A_{23} \end{vmatrix} \\ -\begin{vmatrix} A_{21} & A_{23} \\ A_{31} & A_{33} \end{vmatrix} & \begin{vmatrix} A_{11} & A_{13} \\ A_{31} & A_{33} \end{vmatrix} & -\begin{vmatrix} A_{11} & A_{13} \\ A_{21} & A_{23} \end{vmatrix} \\ \begin{vmatrix} A_{21} & A_{22} \\ A_{31} & A_{32} \end{vmatrix} & -\begin{vmatrix} A_{11} & A_{12} \\ A_{31} & A_{32} \end{vmatrix} & \begin{vmatrix} A_{11} & A_{12} \\ A_{21} & A_{22} \end{vmatrix} \end{bmatrix}$$

Conversely, if the determinant of an $n \times n$ matrix M is 0 and $MM^{-1} = I_n$, then $0 = \det(M)\det(M^{-1}) = \det(I_n) = 1$, which is a contradiction. ∎

We also see from the proof above that if $\det(A) = \pm 1$ and A contains only integers, then $\det(A^{-1}) = \pm 1$ and A^{-1} contains only integers. Of course, if A contains only integer components and $\det(A)$ divides every 2×2 determinant in the expression for A^{-1} above, then A^{-1} will also contain only integral components.

A set of simultaneous equations may be expressed equivalently in terms of matrix operations in a matrix equation. Thus, if

$$\begin{aligned} ax + by &= e \\ cx + dy &= f \end{aligned}$$

then

$$\begin{bmatrix} ax + by \\ cx + dy \end{bmatrix} = \begin{bmatrix} a & b \\ c & d \end{bmatrix}\begin{bmatrix} x \\ y \end{bmatrix} = \begin{bmatrix} e \\ f \end{bmatrix}$$

If $ad - bc$ is not zero, then

$$\begin{bmatrix} a & b \\ c & d \end{bmatrix}^{-1}\begin{bmatrix} a & b \\ c & d \end{bmatrix}\begin{bmatrix} x \\ y \end{bmatrix} = \begin{bmatrix} a & b \\ c & d \end{bmatrix}^{-1}\begin{bmatrix} e \\ f \end{bmatrix}$$

and

$$\begin{bmatrix} 1 & 0 \\ 0 & 1 \end{bmatrix} \begin{bmatrix} x \\ y \end{bmatrix} = \begin{bmatrix} a & b \\ c & d \end{bmatrix}^{-1} \begin{bmatrix} e \\ f \end{bmatrix}$$

so that

$$\begin{bmatrix} x \\ y \end{bmatrix} = \begin{bmatrix} a & b \\ c & d \end{bmatrix}^{-1} \begin{bmatrix} e \\ f \end{bmatrix}$$

and we can solve for x and y. Obviously, this method works equally well for three equations in three variables; and indeed, it works for n equations in n variables. This matrix method works efficiently if the inverse of the matrix is already known or if there is a series of simultaneous equations to solve and only the e and f change.

We now consider simultaneous congruences modulo a prime. For example, suppose that we are given simultaneous congruences

$$3x + 2y \equiv 4 \;(\text{mod } 11)$$
$$x - 2y \equiv 2 \;(\text{mod } 11)$$

If we add, we have $4x \equiv 6 \;(\text{mod } 11)$. Since 11 is a prime, we are assured a solution. In this case, 3 is the inverse of 4 modulo 11, so if we multiply both sides by 3, we have $x \equiv 18 \equiv 7 \;(\text{mod } 11)$. Substituting in either of the original congruences, we get $y \equiv 8 \;(\text{mod } 11)$.

Next we will see that there are results for congruent matrices that are similar to those above for equality of matrices.

Definition 3.20 *If $A = [A_{ij}]$ and $B = [B_{ij}]$ are $n \times m$ matrices over the integers and s is a positive integer, then we write $A \equiv B \;(\text{mod } s)$ if $A_{ij} \equiv B_{ij} \;(\text{mod } s)$ for all i and j.*

We have defined the determinant only for square matrices of dimension $n = 2$ and $n = 3$; however, the next theorem is true for all positive values n.

Theorem 3.21 *Let $A = [A_{ij}]$ and $B = [B_{ij}]$ be $n \times n$ matrices of integers and assume that $A \equiv B \;(\text{mod } s)$. Then $\det(A) \equiv \det(B) \;(\text{mod } s)$.*

Proof. This follows for $n = 2$ and 3 from the fact that the determinant is calculated using only addition, subtraction, and multiplication, which all preserve congruences. ∎

When we use a/b in the following discussion we are assuming that b has an inverse, b^{-1}, modulo s and we write a/b for ab^{-1}.

As we saw just above, $4^{-1} \equiv 3 \;(\text{mod } 11)$ so that from $4x \equiv 6 \;(\text{mod } 11)$ we write

$$\begin{aligned} (1/4) \cdot 4x &\equiv (1/4) \cdot 6 \quad (\text{mod } 11) \\ 3 \cdot 4x &\equiv 3 \cdot 6 \quad\quad (\text{mod } 11) \\ x &\equiv 18 \equiv 7 \quad (\text{mod } 11) \end{aligned}$$

Using the same method as for solving simultaneous equations, we obtain the following theorem, whose proof is left to the reader.

Theorem 3.22 *If p is prime and* $\begin{vmatrix} a & b \\ c & d \end{vmatrix} \not\equiv 0 \;(\text{mod } p)$*, then the system of simultaneous congruences*

$$ax + by \equiv e \;(\text{mod } p)$$
$$cx + dy \equiv f \;(\text{mod } p)$$

has the solution

$$x \equiv \frac{\begin{vmatrix} e & b \\ f & d \end{vmatrix}}{\begin{vmatrix} a & b \\ c & d \end{vmatrix}} \;(\text{mod } p) \quad and \quad y \equiv \frac{\begin{vmatrix} a & e \\ c & f \end{vmatrix}}{\begin{vmatrix} a & b \\ c & d \end{vmatrix}} \;(\text{mod } p)$$

Using the same method as for equality of matrices of integers we have the following theorem whose proof is left to the reader.

Theorem 3.23 *If p is prime and $A \equiv \begin{bmatrix} a & b \\ c & d \end{bmatrix} \;(\text{mod } p)$, then $\det(A) \not\equiv 0 \;(\text{mod } p)$ if and only if the inverse A^{-1} modulo p of A exists. Further,*

$$A^{-1} \equiv \begin{bmatrix} \dfrac{d}{\det(A)} & \dfrac{-b}{\det(A)} \\[2ex] \dfrac{-c}{\det(A)} & \dfrac{a}{\det(A)} \end{bmatrix} \;(\text{mod } p)$$

Hence, if we have

$$ax + by \;\equiv\; e \;(\text{mod } p)$$
$$cx + dy \;\equiv\; f \;(\text{mod } p)$$

then

$$\begin{bmatrix} a & b \\ c & d \end{bmatrix} \begin{bmatrix} x \\ y \end{bmatrix} \equiv \begin{bmatrix} e \\ f \end{bmatrix} \;(\text{mod } p)$$

Thus,

$$\begin{bmatrix} a & b \\ c & d \end{bmatrix}^{-1} \begin{bmatrix} a & b \\ c & d \end{bmatrix} \begin{bmatrix} x \\ y \end{bmatrix} \equiv \begin{bmatrix} a & b \\ c & d \end{bmatrix}^{-1} \begin{bmatrix} e \\ f \end{bmatrix} \;(\text{mod } p)$$

and

$$\begin{bmatrix} x \\ y \end{bmatrix} = \begin{bmatrix} 1 & 0 \\ 0 & 1 \end{bmatrix} \begin{bmatrix} x \\ y \end{bmatrix} \equiv \begin{bmatrix} a & b \\ c & d \end{bmatrix}^{-1} \begin{bmatrix} e \\ f \end{bmatrix} \;(\text{mod } p)$$

We can then solve for x and y modulo p.

The next theorem deals with three congruences in three variables.

Theorem 3.24 *For a prime p and simultaneous congruences*

$$A_{11}x_1 + A_{12}x_2 + A_{13}x_3 \equiv y_1 \;(\text{mod } p)$$
$$A_{21}x_1 + A_{22}x_2 + A_{23}x_3 \equiv y_2 \;(\text{mod } p)$$
$$A_{31}x_1 + A_{32}x_2 + A_{33}x_3 \equiv y_3 \;(\text{mod } p)$$

if $\det(A) \not\equiv 0 \pmod{p}$, *then*

$$x_1 \equiv \frac{\begin{vmatrix} y_1 & A_{12} & A_{13} \\ y_2 & A_{22} & A_{23} \\ y_3 & A_{32} & A_{33} \end{vmatrix}}{\begin{vmatrix} A_{11} & A_{12} & A_{13} \\ A_{21} & A_{22} & A_{23} \\ A_{31} & A_{32} & A_{33} \end{vmatrix}} \pmod{p} \quad , \quad x_2 \equiv \frac{\begin{vmatrix} A_{11} & y_1 & A_{13} \\ A_{21} & y_2 & A_{23} \\ A_{31} & y_3 & A_{33} \end{vmatrix}}{\begin{vmatrix} A_{11} & A_{12} & A_{13} \\ A_{21} & A_{22} & A_{23} \\ A_{31} & A_{32} & A_{33} \end{vmatrix}} \pmod{p}$$

and

$$x_3 \equiv \frac{\begin{vmatrix} A_{11} & A_{12} & y_1 \\ A_{21} & A_{22} & y_2 \\ A_{31} & A_{32} & y_3 \end{vmatrix}}{\begin{vmatrix} A_{11} & A_{12} & A_{13} \\ A_{21} & A_{22} & A_{23} \\ A_{31} & A_{32} & A_{33} \end{vmatrix}} \pmod{p}$$

Proof. This result may be shown in the same manner as for simultaneous equations. ∎

We note that there are similar results for n simultaneous congruences in n variables with respect to a prime modulus p, but we will not state or prove them here.

Theorem 3.25 *If p is prime,*

$$A \equiv \begin{bmatrix} A_{11} & A_{12} & A_{13} \\ A_{21} & A_{22} & A_{23} \\ A_{31} & A_{32} & A_{33} \end{bmatrix} \pmod{p}$$

and $\det(A) \not\equiv 0 \pmod{p}$, *then*

$$A^{-1} \equiv \begin{bmatrix} \dfrac{\begin{vmatrix} A_{22} & A_{23} \\ A_{32} & A_{33} \end{vmatrix}}{\det(A)} & -\dfrac{\begin{vmatrix} A_{12} & A_{13} \\ A_{32} & A_{33} \end{vmatrix}}{\det(A)} & \dfrac{\begin{vmatrix} A_{12} & A_{13} \\ A_{22} & A_{23} \end{vmatrix}}{\det(A)} \\[4mm] -\dfrac{\begin{vmatrix} A_{21} & A_{23} \\ A_{31} & A_{33} \end{vmatrix}}{\det(A)} & \dfrac{\begin{vmatrix} A_{11} & A_{13} \\ A_{31} & A_{33} \end{vmatrix}}{\det(A)} & -\dfrac{\begin{vmatrix} A_{11} & A_{13} \\ A_{21} & A_{23} \end{vmatrix}}{\det(A)} \\[4mm] \dfrac{\begin{vmatrix} A_{21} & A_{22} \\ A_{31} & A_{32} \end{vmatrix}}{\det(A)} & -\dfrac{\begin{vmatrix} A_{11} & A_{12} \\ A_{31} & A_{32} \end{vmatrix}}{\det(A)} & \dfrac{\begin{vmatrix} A_{11} & A_{12} \\ A_{21} & A_{22} \end{vmatrix}}{\det(A)} \end{bmatrix} \pmod{p}$$

Proof. The argument is similar to that of Theorem 3.19. ∎

With the inverse we can solve three simultaneous congruences in three variables in the same way as for two simultaneous congruences in two variables. There is an analogous theorem for inverses modulo a prime p for $n \times n$ matrices over the integers, but we use only inverses of 2×2 and 3×3 matrices here, for which having explicit solutions is useful.

Theorem 3.24 deals with simultaneous congruences modulo a prime p. If we have n congruences in n unknowns modulo m, where m is not prime, the determination of solutions is not as simple as for a prime modulus. We show that we can create a chain of systems of congruences in which the solutions of one system are contained within those of the next.

Theorem 3.26 *Consider the following system of s simultaneous congruences in n variables modulo m:*

$$A_{11}x_1 + A_{12}x_2 + A_{13}x_3 + \cdots + A_{1n}x_n \equiv y_1 \ (\text{mod } m)$$
$$A_{21}x_1 + A_{22}x_2 + A_{23}x_3 + \cdots + A_{2n}x_n \equiv y_2 \ (\text{mod } m)$$
$$\vdots$$
$$A_{k1}x_1 + A_{k2}x_2 + A_{k3}x_3 + \cdots + A_{kn}x_n \equiv y_k \ (\text{mod } m)$$
$$\vdots$$
$$A_{s1}x_1 + A_{s2}x_2 + A_{s3}x_3 + \cdots + A_{sn}x_n \equiv y_s \ (\text{mod } m)$$

having the set of solutions $S_1 = \{(x_1, x_2, \ldots, x_n) \in Z^n : (x_1, x_2, \ldots, x_n)$ satisfies all congruences in the system above$\}$, where $Z^n = Z \times Z \times \cdots \times Z$, the product of Z taken n times. Construct a second system of s simultaneous congruences in n variables modulo m for which all congruences are the same as in the first system except that the k-th congruence is replaced by

$$(aA_{k1}+bA_{i1})x_1+(aA_{k2}+bA_{i2})x_2+\cdots+(aA_{kn}+bA_{in})x_n \equiv ay_k+by_i \ (\text{mod } m)$$

where $i \neq k$, where the new k-th congruence is formed by multiplying the k-th congruence of the first system by a and adding b times the i-th congruence, and where a and b are integers. Let the solution set of the second system be $S_2 \subseteq Z^n$. Then every solution of the first system is also a solution of the second system; that is, $S_1 \subseteq S_2$. If $\gcd(a, m) = 1$, then $S_1 = S_2$. If $\gcd(a, m) > 1$, then the second system may have solutions that are not solutions of the first; that is, $S_1 \subset S_2$.

Proof. If $(x_1, x_2, \ldots, x_n) \in S_1$, then $(x_1, x_2, \ldots, x_n)$ certainly satisfies every congruence of the second system except perhaps for the k-th congruence. But since $(x_1, x_2, \ldots, x_n)$ satisfies both the k-th and i-th congruences of the first system, Theorem 1.57(a) implies that

$$aA_{k1}x_1 + aA_{k2}x_2 + aA_{k3}x_3 + \cdots + aA_{kn}x_n \equiv ay_k \ (\text{mod } m)$$

and

$$bA_{i1}x_1 + bA_{i2}x_2 + bA_{i3}x_3 + \cdots + bA_{in}x_n \equiv by_i \ (\text{mod } m)$$

Thus, by Theorem 1.57(a) again,

$$(aA_{k1}+bA_{i1})x_1+(aA_{k2}+bA_{i2})x_2+\cdots+(aA_{kn}+bA_{in})x_n \equiv ay_k+by_i \ (\text{mod } m)$$

holds so that $(x_1, x_2, \ldots, x_n) \in S_2$. Therefore, $S_1 \subseteq S_2$.

Conversely, taking a solution $(x_1, x_2, \ldots, x_n) \in S_2$, we can reverse part of the process by replacing the k-th congruence of the second system with

$$(aA_{k1} + bA_{i1} - bA_{i1})x_1 + (aA_{k2} + bA_{i2} - bA_{i2})x_2 + \cdots$$
$$+ (aA_{kn} + bA_{in} - bA_{in})x_n \equiv ay_k + by_i - by_i \pmod{m}$$

giving a third system with solution set $S_3 \in Z^n$. Thus, the k-th congruence of the third system is equivalent to

$$aA_{k1}x_1 + aA_{k2}x_2 + aA_{k3}x_3 + \cdots + aA_{kn}x_n \equiv ay_k \pmod{m}$$

Further, $S_2 \subseteq S_3$ by the first part of this theorem just proved. According to Theorem 1.69, this last congruence will have more incongruent solutions modulo m than

$$A_{k1}x_1 + A_{k2}x_2 + A_{k3}x_3 + \cdots + A_{kn}x_n \equiv y_k \pmod{m}$$

if $\gcd(a, m) > 1$ and will have exactly the same solutions if $\gcd(a, m) = 1$ since in this case a has a multiplicative inverse modulo m. Thus, if $\gcd(a, m) = 1$, we have $S_3 \subseteq S_1$ and, therefore, $S_1 \subseteq S_2 \subseteq S_3 \subseteq S_1$, which implies that $S_1 = S_2$. If $\gcd(a, m) > 1$, then the extra solutions of $aA_{k1}x_1 + aA_{k2}x_2 + aA_{k3}x_3 + \cdots + aA_{kn}x_n \equiv ay_k \pmod{m}$ may also generate extra solutions for the second and third systems. ■

Any solutions in S_2 that are not in S_1 are called *extraneous* solutions of the first system. Thus, a system such as that given in Theorem 3.26 with $s = n$ may be converted from one system to another by the process given in the theorem, producing a final system of the form

$$a_1 x_1 \equiv b_1 \pmod{m} \quad \text{which has, say, } k_1 \text{ incongruent solutions}$$
$$a_2 x_2 \equiv b_2 \pmod{m} \quad \text{which has, say, } k_2 \text{ incongruent solutions}$$
$$\vdots \qquad\qquad\qquad\qquad \vdots$$
$$a_n x_n \equiv b_n \pmod{m} \quad \text{which has, say, } k_n \text{ incongruent solutions}$$

This last system will have exactly $k_1 k_2 \cdots k_n$ incongruent solutions modulo m in $Z_m \times Z_m \times \cdots \times Z_m = Z_m^n$. Because extraneous solutions may have been introduced by applying Theorem 3.26, each of these $k_1 k_2 \cdots k_n$ solutions must be checked in the original system of congruences to determine whether it is also a solution of the original system. By Theorem 3.26, no solution of the original system will have been lost; and all, if any, will be among the $k_1 k_2 \cdots k_n$ solutions just described. We note that if Theorem 3.26 can be applied with $\gcd(a, m) = 1$ at every stage, then no extraneous solutions will be generated.

Example 3.27 *Suppose that we want to solve the following system of congruences:*

$$3x_1 + 5x_2 \equiv 2 \pmod{12}$$
$$x_1 + 3x_2 \equiv 5 \pmod{12}$$

which has solution set S_1. Replacing the second congruence by the result of adding $b = 1$ times the first to $a = 9$ times the second gives

$$3x_1 + 5x_2 \equiv 2 \pmod{12}$$
$$(9 + 3)x_1 + (27 + 5)x_2 \equiv 45 + 2 \pmod{12}$$

or, equivalently, since $12 \equiv 0$, $35 \equiv 11$, and $47 \equiv 11 \pmod{12}$,

$$3x_1 + 5x_2 \equiv 2 \pmod{12}$$
$$0x_1 + 11x_2 \equiv 11 \pmod{12}$$

which has solution set S_2, say. Since 11 has a multiplicative inverse modulo 12, this last system is equivalent to

$$3x_1 + 5x_2 \equiv 2 \pmod{12}$$
$$0x_1 + 1x_2 \equiv 1 \pmod{12}$$

In this last system, if we replace the first congruence by the result of adding 7 times the second congruence to 1 times the first, we get

$$3x_1 + (5 + 7)x_2 \equiv 2 + 7 \pmod{12}$$
$$0x_1 + 1x_2 \equiv 1 \pmod{12}$$

or equivalently, since $12 \equiv 0 \pmod{12}$,

$$3x_1 \equiv 9 \pmod{12}$$
$$1x_2 \equiv 1 \pmod{12}$$

which has solution set S_3, say.

The congruence $3x_1 \equiv 9 \pmod{12}$ *is easily solved, giving $k_1 = 3$ incongruent solutions modulo 12, namely, $x_1 = 3$, 7, and 11. The second congruence has only $k_2 = 1$ incongruent solutions, namely $x_2 = 1$. Thus,*

$$S_3 = \{(3, 1), (7, 1), (11, 1)\}$$

Since Theorem 3.26 implies that $S_1 \subseteq S_2 \subseteq S_3$, we need to check every solution in S_3 in the original system of congruences. Doing so, we find that $S_1 = \varnothing$ so that all three solutions of the third system are extraneous for the first system.

Note that an exhaustive search for solutions would require testing $12 \times 12 = 144$ *pairs (x_1, x_2). With the application of Theorem 3.26 we had to test only three pairs.*

Exercises

1. Calculate:

(a) $\begin{vmatrix} -3 & 5 \\ 7 & 2 \end{vmatrix}$ **(b)** $\begin{vmatrix} 4 & 0 \\ -1 & 2 \end{vmatrix}$

(c) $\begin{vmatrix} 2 & -1 & 3 \\ 4 & 2 & 7 \\ -5 & 0 & 6 \end{vmatrix}$ **(d)** $\begin{vmatrix} 5 & 3 & 2 \\ -1 & 0 & 4 \\ 7 & 10 & 8 \end{vmatrix}$

2. Calculate:

(a) $\begin{bmatrix} 3 & -1 \\ 2 & 7 \\ -5 & 0 \end{bmatrix} \begin{bmatrix} 2 & -1 & 0 & 4 \\ 5 & 1 & 3 & 0 \end{bmatrix}$

(b) $\begin{bmatrix} 1 & -1 & 2 \\ 3 & 1 & 4 \\ 6 & -1 & 5 \end{bmatrix} \begin{bmatrix} 2 & -2 & 8 \\ 5 & 5 & 1 \\ 0 & 6 & 4 \end{bmatrix}$

(c) $\begin{bmatrix} 1 \\ 3 \\ 4 \\ 5 \end{bmatrix} \begin{bmatrix} -2 & 3 & -5 & 0 \end{bmatrix}$

(d) $\begin{bmatrix} -2 & 3 & -5 & 0 \end{bmatrix} \begin{bmatrix} 1 \\ 3 \\ 4 \\ 5 \end{bmatrix}$

(e) $(-2) \begin{bmatrix} -5 & 1 \\ 6 & -7 \end{bmatrix}$

(f) $\begin{bmatrix} -1 & 3 \\ 2 & 7 \end{bmatrix} - 8 \begin{bmatrix} 3 & 4 \\ -8 & 1 \end{bmatrix}$

3. Let $A = \begin{bmatrix} -2 & 4 & 7 \\ 6 & 1 & 5 \\ 2 & 3 & -4 \end{bmatrix}$. Calculate:

    (a) A^t **(b)** AA^t **(c)** A^tA

4. Let $A = \begin{bmatrix} -3 & 2 \\ 7 & 5 \end{bmatrix}$ and $B = \begin{bmatrix} 4 & 1 \\ -2 & 9 \end{bmatrix}$. Calculate:

    (a) AB **(b)** BA **(c)** AB^t **(d)** A^tB **(e)** $(AB)^t$

    (f) B^tA^t **(g)** $A - B$ **(h)** $B - A$ **(i)** $5A$ **(j)** $2B - 3A$

5. Let $A = \begin{bmatrix} 1 & 0 & 5 \\ 2 & -4 & 6 \end{bmatrix}$ and $B = \begin{bmatrix} -1 & 2 \\ 3 & 9 \\ 10 & 0 \end{bmatrix}$. Calculate the following:

   (a) AB (b) BA (c) A^t

   (d) B^t (e) $A^t B^t$ (f) $(BA)^t$

6. Show that when A is a 3×3 matrix, the two methods of calculating $\det(A)$ given in Definition 3.6 and in the discussion that followed are equivalent.

7. Prove Theorem 3.12.

8. Prove Theorem 3.13 for $n = 2$ and 3.

9. Prove Theorem 3.14 for $n = 2$ and 3.

10. Prove Theorem 3.17.

11. Prove Theorem 3.18 for $n = 2$ and 3.

12. Find A^{-1} if $A =$

   (a) $\begin{bmatrix} 2 & 3 \\ 5 & 7 \end{bmatrix}$ (b) $\begin{bmatrix} 3 & -5 \\ -4 & 7 \end{bmatrix}$

   (c) $\begin{bmatrix} 4 & -1 & 3 \\ -3 & 1 & -3 \\ 2 & 0 & 1 \end{bmatrix}$ d) $\begin{bmatrix} 7 & -7 & 3 \\ 4 & -3 & 2 \\ 2 & -2 & 1 \end{bmatrix}$

13. Find A^{-1} modulo 5 for these matrices A. (See Section 1.6 for operations tables for Z_5.)

   (a) $\begin{bmatrix} 4 & 2 \\ -3 & 2 \end{bmatrix}$ (b) $\begin{bmatrix} 2 & 3 & 2 \\ 4 & 0 & -3 \\ 2 & 4 & 1 \end{bmatrix}$

14. Solve these simultaneous congruences modulo 5:

   (a) Using both the methods of Theorems 3.22 and 3.23,

   $$4x + 2y \equiv 1 \pmod 5$$
   $$-3x + 2y \equiv 4 \pmod 5$$

   (b) Using both of the methods of Theorems 3.24 and 3.25,

   $$\begin{aligned} 2x_1 &+ 3x_2 &+ 2x_3 &\equiv 2 \pmod 5 \\ 4x_1 & & - 3x_3 &\equiv 1 \pmod 5 \\ 2x_1 &+ 4x_2 &+ x_3 &\equiv 3 \pmod 5 \end{aligned}$$

15. Solve:

(a) $4x + 2y \equiv 1 \pmod{15}$
$-3x + 2y \equiv 4 \pmod{15}$

(b) $2x_1 + 3x_2 + x_3 \equiv 2 \pmod{15}$
$4x_1 - 3x_3 \equiv 1 \pmod{15}$
$2x_1 + 4x_2 + x_3 \equiv 3 \pmod{15}$

(c) $2x_1 + 3x_2 + x_3 \equiv 2 \pmod{15}$
$4x_1 - 3x_3 \equiv 13 \pmod{15}$
$2x_1 + 4x_2 + x_3 \equiv 3 \pmod{15}$

(d) $3x_1 + 2x_2 \equiv 1 \pmod{36}$
$5x_1 + 10x_2 \equiv 3 \pmod{36}$

3.4   Polynomials and Solutions of Polynomial Congruences

This section develops procedures for solving polynomial congruences. More specifically, we want methods for finding all integers x such that, for example, $x^3 + 2x^2 + 3x - 22 \equiv 0 \pmod{56}$. Before attacking congruences, we will introduce some facts about polynomials. See Chapter 8 for a fuller discussion of polynomials.

Definition 3.28 *The expression* $a_n x^n + a_{n-1} x^{n-1} + \cdots + a_1 x + a_0$, *where the* a_i *are integers and* $a_n \neq 0$, *is called a* polynomial of degree n *over the integers. The numbers* $a_0, a_1, \ldots, a_n$ *are called the* coefficients *of the polynomial.*

A solution *or* root *of the polynomial equation* $f(x) = 0$ *is an integer* b *such that* $f(b) = 0$. *If* s *is a positive integer, then a* solution *or* root *of the polynomial congruence* $f(x) \equiv 0 \pmod{s}$ *is an integer* c *such that* $f(c) \equiv 0 \pmod{s}$. *If* $c \equiv d \pmod{s}$ *and* $f(c) \equiv 0 \pmod{s}$, *then* $f(d) \equiv 0 \pmod{s}$; *and we say that* c *and* d *are* congruent solutions *of* $f(x) \equiv 0 \pmod{s}$. *If* $c \not\equiv d \pmod{s}$ *and both* $f(c) \equiv 0 \pmod{s}$ *and* $f(d) \equiv 0 \pmod{s}$, *then we say that* c *and* d *are* incongruent solutions *of* $f(x) \equiv 0 \pmod{s}$.

Polynomials $f(x)$ *and* $g(x)$ *are* equal *provided that* $f(b) = g(b)$ *for every integer* b. *Polynomials* $f(x)$ *and* $g(x)$ *are* congruent modulo s, *written* $f(x) \equiv g(x) \pmod{s}$, *provided that* $f(b) \equiv g(b) \pmod{s}$ *for every integer* b. *If*

$$f(x) = a_n x^n + a_{n-1} x^{n-1} + \cdots + a_1 x + a_0$$

and

$$g(x) = b_n x^n + b_{n-1} x^{n-1} + \cdots + b_1 x + b_0$$

then

$$f(x) + g(x) = (a_n + b_n)x^n + (a_{n-1} + b_{n-1})_n x^{n-1} + \cdots + (a_1 + b_1)x + (a_0 + b_0)$$

and

$$f(x)g(x) = c_n x^n + c_{n-1} x^{n-1} + \cdots + c_1 x + c_0$$

where

$$c_k = \sum_{i+j=k} a_i b_j$$

It is clear that a polynomial over the integers is a function from Z into Z. The proof of the next theorem is left to the reader.

Theorem 3.29 *If $f(x)$ has degree n and $g(x)$ has degree m, then*

(a) $f(x) + g(x)$ *has degree* $\leq \max\{m, n\}$.

(b) $f(x)g(x)$ *has degree* $m + n$.

By induction it can be shown that for any k, if x and a are integers, then $x^k - a^k = (x - a) \cdot g(x)$ where $g(x)$ is a polynomial over the integers. In particular, $g(x) = 1$ if $k = 1$; and

$$g(x) = x^{k-1} + ax^{k-2} + a^2 x^{k-3} + \cdots + a^{k-2}x + a^{k-1}$$

is a polynomial of degree $k - 1$ when $k > 1$. Thus, $x^k - a^k$ is divisible by $x - a$ when $x \neq a$.

Theorem 3.30 *If $f(x)$ is a polynomial of degree n over the integers, $f(x) = a_n x^n + a_{n-1} x^{n-1} + \cdots + a_1 x + a_0$, and $f(a) = 0$, then $x - a$ is a factor of $f(x)$. Further, $f(x) = (x - a)g(x)$ where $g(x)$ has degree $n - 1$.*

Proof. $f(x) = f(x) - f(a) = a_n(x^n - a^n) + a_{n-1}(x^{n-1} - a^{n-1}) + \cdots + a_1(x - a)$. Since $x - a$ is a factor of each term of $f(x)$, it is a factor of $f(x)$. Further, $f(x) = (x - a)g(x)$, where $g(x) = \sum_{k=1}^{n} a_k g_k(x)$ has degree $n - 1$ and where

$$g_k(x) = \frac{x^k - a^k}{x - a}$$ has degree $k - 1$. ■

Theorem 3.31 *If $a_1, a_2, \ldots, a_n$ are integers, then the polynomial*

$$(x - a_1)(x - a_2) \cdots (x - a_n)$$

has degree n.

Proof. The proof follows easily using Theorem 3.29 and induction. ■

Theorem 3.32 *If $f(x)$ has degree n, then $f(x) = 0$ has at most n solutions.*

Proof. The proof follows directly from the preceding two theorems. ■

Theorem 3.33 *If $f(x)$ has degree $k > 0$ and $f(a) \equiv 0$ (mod s), then*

$$f(x) \equiv (x - a)g(x) \text{ (mod } s)$$

where $g(x)$ has degree $k - 1$.

Proof. $f(x) - f(a) \equiv f(x)$ (mod s) and, by Theorem 3.30, $f(x) \equiv (x - a)g(x)$ (mod s) for some polynomial $g(x)$ of degree $k-1$. Thus, $f(x) - f(a) \equiv (x-a)g(x)$ (mod s). Hence $f(x) - f(a) \equiv (x-a)g(x)$ (mod s) for every integer x. But $f(a) \equiv 0$ (mod s) and hence $f(x) \equiv (x-a)g(x)$ (mod s) for any integer x; that is, $f(m) \equiv (m - a)g(m)$ for all integers m. ∎

Theorem 3.34 (Lagrange) *If p is a prime and $f(x) = a_0 + a_1 x + a_2 x^2 + \cdots + a_n x^n$ is a polynomial of positive degree n over the integers with $a_n \not\equiv 0$ (mod p), then the congruence $f(x) \equiv 0$ (mod p) has at most n incongruent solutions modulo p.*

Proof. We use induction on n. If $n = 1$, we have $a_0 + a_1 x \equiv 0$ (mod p) or $a_1 x \equiv -a_0$ (mod p). From Theorem 1.69 we already know since $\gcd(a_1, p) = 1$ that there is only one solution modulo p. Assume that the theorem is true for any polynomial function of degree k and $f(x)$ has degree $k + 1$. Assume that $f(x)$ has at least $k + 2$ incongruent solutions $b_1, b_2, \ldots, b_{k+2}$. $f(x) \equiv (x - b_{k+2})g(x)$ where $g(x)$ has degree k. Since b_i and b_{k+2} for $i \neq k + 2$ are not congruent modulo p, $p \mid f(b_i)$ but p does not divide $(b_i - b_{k+2})$, we have $p \mid g(b_i)$ and therefore b_i is a solution of $g(x) \equiv 0$ (mod p). But then $g(x) \equiv 0$ (mod p) has $k + 1$ solutions, which contradicts the induction hypothesis. Hence the congruence $f(x) \equiv 0$ (mod p) of degree $k + 1$ has at most $k + 1$ solutions. ∎

The requirement in Lagrange's Theorem that the modulus p be prime is essential. For example, Theorem 1.69 says that if $d = \gcd(a, m)$ and $d \mid c$, then $ax \equiv c$ (mod m) has exactly d incongruent solutions modulo m. If $a \neq 0$ (mod m), then this congruence is equivalent to a first-degree polynomial congruence $f(x) \equiv 0$ (mod m), where $f(x) = ax - c$. Thus, if $d > 1$, then $f(x) \equiv 0$ (mod m) has more than one solution. For example, $8x \equiv 20$ (mod 12) has solutions $x = 1, 4, 7,$ and 10. Of course, there cannot be more than p incongruent solutions to $f(x) \equiv 0$ (mod p) for p prime, no matter what is the degree n of the polynomial $f(x)$.

Any polynomial congruence $f(x) \equiv 0$ (mod p) for prime p may be solved by exhaustion since only the integers $0, 1, 2, \ldots, p - 1$ need to be checked; however, this method is suitable only for "small" p.

Let $f(x) = a_n x^n + a_{n-1} x^{n-1} + \cdots + a_2 x^2 + a_1 x^1 + a_0$, where $a_n \not\equiv 0$ (mod m). We wish to find solutions of the congruence $f(x) \equiv 0$ (mod m). We know that, if $r_1 \equiv r_2$ (mod m), then $f(r_1) \equiv f(r_2)$ (mod m); and hence r_1 is a solution of $f(x) \equiv 0$ (mod m) if and only if r_2 is. We also know that if u and v are relatively prime, then $f(r) \equiv 0$ (mod uv) if and only if $f(r) \equiv 0$ (mod u) and $f(r) \equiv 0$ (mod v). Suppose that

$$x \equiv a_1 \text{ (mod } u), \quad \ldots, x \equiv a_j \text{ (mod } u)$$

are incongruent solutions to $f(x) \equiv 0 \pmod{u}$ and

$$x \equiv b_1 \pmod{v}, \ \ldots \ , x \equiv b_k \pmod{v}$$

are incongruent solutions to $f(x) \equiv 0 \pmod{v}$. Then using the Chinese Remainder Theorem on pairs of solutions $x \equiv a_i \pmod{u}$ and $x \equiv b_r \pmod{v}$, we can find solutions to $f(x) \equiv 0 \pmod{uv}$.

Thus, in looking for solutions to the polynomial congruence $f(x) \equiv 0 \pmod{m}$, we need only to factor $m = c_1 c_2 \cdots c_s$ with $\gcd(c_i, c_j) = 1$ for $i \neq j$. Then we solve

$$\begin{aligned} f(x) &\equiv 0 \pmod{c_1} \text{ giving, say, } k_1 \text{ incongruent solutions} \\ f(x) &\equiv 0 \pmod{c_2} \text{ giving, say, } k_2 \text{ incongruent solutions} \\ &\vdots \\ f(x) &\equiv 0 \pmod{c_s} \text{ giving, say, } k_s \text{ incongruent solutions} \end{aligned}$$

If

$$\begin{aligned} &r_1 \text{ is one of the } k_1 \text{ solutions above} \\ &r_2 \text{ is one of the } k_2 \text{ solutions above} \\ &\vdots \\ &r_s \text{ is one of the } k_s \text{ solutions above} \end{aligned}$$

then the Chinese Remainder Theorem will simultaneously solve

$$\begin{aligned} x &\equiv r_1 \pmod{c_1} \\ x &\equiv r_2 \pmod{c_2} \\ &\vdots \\ x &\equiv r_s \pmod{c_s} \end{aligned}$$

giving a solution of $f(x) \equiv 0 \pmod{c_1 c_2 \cdots c_s}$ or $f(x) \equiv 0 \pmod{m}$. Therefore, there will be $k_1 k_2 \cdots k_s$ incongruent solutions of $f(x) \equiv 0 \pmod{m}$.

Example 3.35 *Let*

$$f(x) = x^3 + 2x^2 + 3x - 22 \equiv 0 \pmod{56}$$

We first solve

$$x^3 + 2x^2 + 3x + 34 \equiv 0 \pmod{8}$$

and

$$x^3 + 2x^2 + 3x + 34 \equiv 0 \pmod{7}$$

By exhaustion, we find that $f(0) = -22$, $f(1) = -16$, $f(2) = 0$, $f(3) = 32$, $f(4) = 86$, $f(5) = 168$, $f(6) = 279$, and $f(7) = 440$. Using methods to determine if a number is divisible by 8 or 7, we find that $x = 1, 2, 3, 5$, and 7 are all solutions of $f(x) \equiv 0 \pmod{8}$ and $x = 2$ and 5 are solutions of $f(x) \equiv 0 \pmod{7}$. Hence solutions to $f(x) \equiv 0 \pmod{56}$ can be found using the Chinese Remainder Theorem on one congruence from the sequence $x \equiv 1 \pmod{8}$, $x \equiv 2 \pmod{8}$, $x \equiv 3 \pmod{8}$, $x \equiv 5 \pmod{8}$, $x \equiv 7 \pmod{8}$

and one congruence from the sequence $x \equiv 2 \pmod 7$, $x \equiv 5 \pmod 7$. There will be exactly $5 \times 2 = 10$ incongruent solutions modulo 56 of the original congruence.

Since $(1)(8) + (-1)(7) = 1$, the integer

$$(a)(1)(8) + (b)(-1)(7)$$

satisfies the congruences $x \equiv a \pmod 7$ and $x \equiv b \pmod 8$.

For $x \equiv 1 \pmod 8$ and $x \equiv 2 \pmod 7$, $(2)(1)(8) + (1)(-1)(7) = 9$ is a solution of $f(x) \equiv 0 \pmod{56}$.

For $x \equiv 1 \pmod 8$ and $x \equiv 5 \pmod 7$, $(5)(1)(8) + (1)(-1)(7) = 33$ is a solution of $f(x) \equiv 0 \pmod{56}$.

Continuing, we obtain:

Congruence Pairs	Solutions Modulo 56
$x \equiv 2 \pmod 8$ and $x \equiv 2 \pmod 7$	2
$x \equiv 2 \pmod 8$ and $x \equiv 5 \pmod 7$	26
$x \equiv 3 \pmod 8$ and $x \equiv 2 \pmod 7$	51
$x \equiv 3 \pmod 8$ and $x \equiv 5 \pmod 7$	19
$x \equiv 5 \pmod 8$ and $x \equiv 2 \pmod 7$	37
$x \equiv 5 \pmod 8$ and $x \equiv 5 \pmod 7$	5
$x \equiv 7 \pmod 8$ and $x \equiv 2 \pmod 7$	23
$x \equiv 7 \pmod 8$ and $x \equiv 5 \pmod 7$	47

Hence, the ten incongruent primary residue solutions to $f(x) \equiv 0 \pmod{56}$ are 2, 5, 9, 19, 23, 26, 33, 37, 47, and 51.

Using the techniques above and using the prime factorization of m, namely, $m = p_1^{\alpha_1} p_2^{\alpha_2} \cdots p_t^{\alpha_t}$, we can reduce the solution of the congruence $f(x) \equiv 0 \pmod m$ to solving the congruences

$$\begin{aligned} f(x) &\equiv 0 \pmod{p_1^{\alpha_1}} \\ f(x) &\equiv 0 \pmod{p_2^{\alpha_2}} \end{aligned}$$

$$\vdots$$

$$f(x) \equiv 0 \pmod{p_t^{\alpha_t}}$$

since $\gcd(p_i^{\alpha_i}, p_j^{\alpha_j}) = 1$ when $i \neq j$.

However, $p_i^{\alpha_i}$ can still be a very large number for checking solutions using the trial-and-error exhaustion method. It would be very helpful if one could use solutions to $f(x) \equiv 0 \pmod{p_i}$ rather than having to check all $p_i^{\alpha_i}$ possible solutions to $f(x) \equiv 0 \pmod{p_i^{\alpha_i}}$.

Definition 3.36 *The derivative of the polynomial*

$$f(x) = a_n x^n + a_{n-1} x^{n-1} + a_{n-2} x^{n-2} + \cdots + a_2 x^2 + a_1 x^1 + a_0$$

is the polynomial

$$f'(x) = n a_n x^{n-1} + (n-1) a_{n-1} x^{n-2} + (n-2) a_{n-2} x^{n-3} + \cdots + 2 a_2 x + a_1$$

Lemma 3.37 $(a+b)^n = a^n + na^{n-1}b + b^2m$ *for some integer m.*

Proof. The proof is evident from the Binomial Theorem (see the Exercises of Section 1.3). ∎

We will now show a way of finding solutions for $f(x) \equiv 0 \pmod{p_i^{k+1}}$ given solutions for $f(x) \equiv 0 \pmod{p_i^k}$. So, given solutions to $f(x) \equiv 0 \pmod{p}$, we use these to find solutions to $f(x) \equiv 0 \pmod{p^2}$. We then use these last solutions to find solutions to $f(x) \equiv 0 \pmod{p^3}$, etc. Note that we need only to consider incongruent primary residue solutions of polynomial congruences since all other integer solutions are simply related to the primary residue solutions by being congruent to them.

Lemma 3.38 *Let $r_1, r_2, \ldots, r_m$ be a complete set of incongruent primary residue solutions of the polynomial congruence $f(x) \equiv 0 \pmod{p^k}$ for prime p and $k \geq 1$. Let r be a primary residue solution of $f(x) \equiv 0 \pmod{p^{k+1}}$ so that $f(r) \equiv 0 \pmod{p^{k+1}}$. Then $r = r_i + q \cdot p^k$ for some i, $1 \leq i \leq m$, and some q, $0 \leq q < p$.*

Proof. Since r_i and r are primary residues of their respective moduli, $0 \leq r_i \leq p^k - 1$ and $0 \leq r \leq p^{k+1} - 1$. Because $f(r) \equiv 0 \pmod{p^{k+1}}$, it is also the case that $f(r) \equiv 0 \pmod{p^k}$. Therefore,

$$r \equiv r_i \pmod{p^k}$$

for some i, $1 \leq i \leq m$; and, there is some integer q with the property

$$r = r_i + q \cdot p^k$$

We need to show that $0 \leq q < p - 1$. Clearly,

$$qp^k \leq r_i + qp^k = r \leq p^{k+1} - 1 \leq p^{k+1}$$

so $q < p$. If $q < 0$, then $r = r_i + qp^k \leq p^k - 1 - p^k = -1$, which is also a contradiction. Therefore, $0 \leq q \leq p - 1$. ∎

By Lemma 3.37,

$$(r_i + qp^k)^j = r_i^j + jr_i^{j-1}qp^k + mp^{k+1} \equiv r_i^j + jr_i^{j-1}qp^k \pmod{p^{k+1}}$$

If $f(r) = \sum_{j=0}^{n} c_j r^j$, then, for each i,

$$
\begin{aligned}
f(r) &\equiv \sum_{j=0}^{n} c_j \left[r_i^j + jr_i^{j-1}qp^k \right] \\
&\equiv f(r_i) + qf'(r_i)p^k \pmod{p^{k+1}}
\end{aligned}
$$

Thus we have proved the following theorem.

Theorem 3.39 *If $f(r) \equiv 0 \pmod{p^{k+1}}$ and $r = r_i + qp^k$ where $f(x)$ is a polynomial, $0 \le r \le p^{k+1} - 1$, and $0 \le r_i \le p^k - 1$, then q with $0 \le q < p$ is a solution of*

$$f(r_i) + qf'(r_i) \cdot p^k \equiv 0 \pmod{p^{k+1}}$$

or, equivalently, q is a primary residue modulo p solution of

$$q \cdot f'(r_i) \equiv \frac{-f(r_i)}{p^k} \pmod{p}$$

In Theorem 3.39, note that $f(r_i)/p^k$ is an integer since $f(r_i) \equiv 0 \pmod{p^k}$. One need only solve the linear congruence of the theorem for q to find a possible solution for $f(x) \equiv 0 \pmod{p^{k+1}}$ if the solutions to $f(x) \equiv 0 \pmod{p^k}$ are known. Note also that the theorem gives only necessary conditions for r to be a solution of $f(x) \equiv 0 \pmod{p^{k+1}}$ so that any such solution r may be found by examining and appropriately modifying solutions of $f(x) \equiv 0 \pmod{p^k}$ according to Theorem 3.39.

Example 3.40 *We now reconsider part of our solution of the polynomial congruence in Example 3.35. In Example 3.35 we wanted to find a solution to*

$$f(x) = x^3 + 2x^2 + 3x - 22 \equiv 0 \pmod 8$$

Instead of using the exhaustion method, we could have first solved the congruence $x^3 + 2x^2 + 3x - 22 \equiv 0 \pmod 2$, which has solutions 0 and 1. Then solve

$$qf'(0) \equiv -f(0)/2 \pmod 2$$

and

$$qf'(1) \equiv -f(1)/2 \pmod 2$$

First consider solution 0. *Clearly, $f(0) \equiv -22$; and since $f'(x) = 3x^2 + 4x + 3$ gives $f'(0) = 3$ we have $q \cdot 3 \equiv 22/2 \pmod 2$ or $q \equiv 1 \pmod 2$ and $q = 1$ so that $0 + (1)(2) = 2$ is a possible solution to $f(x) \equiv 0 \pmod 4$. Indeed, $f(2) \equiv 0 \pmod 4$.*

Second consider solution 1. *We have $f(1) = -16$ and $f'(1) = 10$ and need to solve $q \cdot 10 \equiv -8 \pmod 2$. All possible choices of q satisfy this congruence so we have $q = 0$ and $q = 1$. For $q = 0$ we obtain $1 + (0)(2) = 1$ as a possible solution to $f(x) \equiv 0 \pmod 4$. For $q = 1$ we obtain $1 + 1(2) = 3$ as a possible solution to $f(x) \equiv 0 \pmod 4$. Hence the possible solutions for $f(x) \equiv 0 \pmod 4$ are 1, 2, and 3. Direct substitution shows that all three are indeed solutions.*

To find solutions for $f(x) = 0 \pmod 8$, *we need to solve*

$$qf'(1) \equiv -f(1)/4 \pmod 2 \text{ or } q \cdot 10 \equiv 16/4 \pmod 2$$

giving $q = 0$ and 1,

$$qf'(2) \equiv -f(2)/4 \pmod 2 \text{ or } q \cdot 23 \equiv 0 \pmod 2$$

giving $q = 0$, and

$$qf'(3) \equiv -f(3)/4 \,(\text{mod } 2) \; or \; q \cdot 42 \equiv 32 \,(\text{mod } 2)$$

giving $q = 0$ and 1. Hence the possible solutions of $f(x) \equiv 0 \,(\text{mod } 8)$ are

$$1 + 0 \cdot 4 = 1$$
$$1 + 1 \cdot 4 = 5$$
$$2 + 0 \cdot 4 = 2$$
$$3 + 0 \cdot 4 = 3$$
$$3 + 1 \cdot 4 = 7$$

Checking each one shows that they are, in fact, solutions as was shown in Example 3.40 by a more direct method. We examined all possibilities for solutions of $f(x) \equiv 0 \,(\text{mod } 8)$ via Lemma 3.38; therefore, there are no others.

Lagrange's contributions to polynomial congruences were important to the results of this section. Joseph Louis Lagrange (1736-1813) was born in the Italian city of Turin, although his father was of French descent. Apparently, his parents started out wealthy but did not stay that way; so during the first part of Lagrange's life, he struggled financially. He attended the University of Turin and at 18 was named professor of geometry at the Royal Artillery School in Turin, which apparently did not pay as well as the title might imply. At this point he was well known for his work in mechanics and was elected associate foreign member of the Berlin Academy of Science in 1756. In 1757 he helped found the Turin Academy of Science. When Euler left Berlin, Frederick the Great invited Lagrange to Berlin to replace him, claiming that the greatest king in Europe was to have at his court the greatest mathematician in Europe.

Twenty years later, when Frederick died, Lagrange accepted the invitation of Louis XVI to come to the French Academy in Paris. He was a favorite of Marie Antoinette. It is said he was so depressed in Paris that he was unable to work on his mathematics. He was brought out of his depression by the French Revolution. He managed to survive the Revolution and remained popular and respected, although he was bitter about friends and colleagues lost in the conflict. He helped found École Polytechnique in 1794, which became a successor to the now defunct French Academy. In 1795 he taught mathematics at École Normale with Pierre-Simon Laplace as his assistant. He was considered to be an excellent teacher and writer. His papers were considered very easy to read.

Lagrange was honored under Napoleon by being named to the Legion of Honor and later by being given the title Count of the Empire. Napoleon stated that he was the lofty pyramid of the mathematical sciences. He was best known for his contributions in building a sound foundation for calculus. As a side note, he created the notation f' and f'' for the first and second derivatives. He excelled in all branches of analysis and was said to be the first true analyst. It is also said that he brought the subject of the calculus of variations to maturity. He was well known for work in number theory, analytical mechanics, and celestial mechanics. In number theory, he investigated prime numbers, quadratic forms, and Pell's equation. He proved Wilson's Theorem and showed that every positive integer could be expressed as the sum of four squares. He also proved that if p is a prime and $f(x)$ is a polynomial of positive degree n over the integers, then the congruence $f(x) \equiv 0 \pmod{p}$ has at most p incongruent solutions modulo p, which theorem is named after him.

Lagrange once stated: "Astronomers are queer, they won't believe in a theory unless it agrees with their observations." Lagrange worked on the three-body problem and showed that if three bodies start from vertices of an equilateral triangle, they continue to move as though attached to the triangle while it rotates about the center of mass of the three bodies. In 1906 his theorem was found to apply to the sun, Jupiter, and an asteroid called Achilles.

Exercises

1. Prove Theorem 3.29.

2. Prove Theorem 3.31.

3. Prove Theorem 3.32.

4. Find all incongruent primary residue solutions for:

   (a) $2x^4 - 3x^3 + x^2 + 7x + 3 \equiv 0 \pmod{35}$

   (b) $x^3 - 5x^2 + x + 4 \equiv 0 \pmod{33}$

   (c) $x^5 - 4x^3 + x + 2 \equiv 0 \pmod{77}$

5. Find all incongruent primary residue solutions for:

   (a) $5x^4 + x^3 - 2x^2 + 1 \equiv 0 \pmod{5}$

   (b) $5x^4 + x^3 - 2x^2 + 1 \equiv 0 \pmod{25}$

   (c) $5x^4 + x^3 - 2x^2 + 1 \equiv 0 \pmod{125}$

   (d) $2x^3 + 4x^2 - 3x + 2 \equiv 0 \pmod{7}$

   (e) $2x^3 + 4x^2 - 3x + 2 \equiv 0 \pmod{49}$

   (f) $2x^3 + 4x^2 - 3x + 2 \equiv 0 \pmod{343}$

   (g) $4x^4 - x^3 + 3x^2 + 36 \equiv 0 \pmod{45}$

3.5   Properties of the Function ϕ

Let $n = p_1^{\alpha_1} p_2^{\alpha_1} \cdots p_k^{\alpha_k}$ be the prime factorization of n. Every positive divisor of n either is 1 or is divisible by p_i for some i, and every integer relatively prime to n has none of these primes as a factor. Some characteristics of n depend upon the number of integers s, $1 \le s \le n$, that do *not* contain any p_i as a factor. For example, if $n = 40 = 2^3 \cdot 5$, then the integers s, $1 \le s \le n$, and their factorizations are

$$
\begin{array}{llll}
\mathbf{1 = 1} & \mathbf{11 = 11} & \mathbf{21 = 3 \cdot 7} & \mathbf{31 = 31} \\
2 = 2 & 12 = 2^2 \cdot 3 & 22 = 2 \cdot 11 & 32 = 2^5 \\
\mathbf{3 = 3} & \mathbf{13 = 13} & \mathbf{23 = 23} & \mathbf{33 = 3 \cdot 11} \\
4 = 2^2 & 14 = 2 \cdot 7 & 24 = 2^3 \cdot 3 & 34 = 2 \cdot 17 \\
5 = 5 & 15 = 3 \cdot 5 & 25 = 5^2 & 35 = 5 \cdot 7 \\
6 = 2 \cdot 3 & 16 = 2^4 & 26 = 2 \cdot 13 & 36 = 2^2 \cdot 3^2 \\
\mathbf{7 = 7} & \mathbf{17 = 17} & \mathbf{27 = 3^3} & \mathbf{37 = 37} \\
8 = 2^3 & 18 = 2 \cdot 3^2 & 28 = 2^2 \cdot 7 & 38 = 2 \cdot 19 \\
\mathbf{9 = 3^2} & \mathbf{19 = 19} & \mathbf{29 = 29} & \mathbf{39 = 3 \cdot 13} \\
10 = 2 \cdot 5 & 20 = 2^2 \cdot 5 & 30 = 2 \cdot 3 \cdot 5 & 40 = 2^3 \cdot 5
\end{array}
$$

The integers highlighted in bold, none of which contains either 2 or 5 as a factor, are those relative prime to $n = 40$. The number of such s, $1 \le s \le n$, relatively prime to n is denoted by $\phi(40) = 16$.

Definition 3.41 *Let $\phi(n)$ be the number of positive integers less than n which are relatively prime to n; i.e., $\phi(n)$ is the number of reduced residues modulo n. ϕ is called* Euler's totient function *or the* Euler ϕ function.

The factorization table above also shows that

$$
\begin{array}{lll}
\phi(1) = 1 & \phi(5) = 4 & \phi(9) = 6 \\
\phi(2) = 1 & \phi(6) = 2 & \phi(10) = 4 \\
\phi(3) = 2 & \phi(7) = 6 & \phi(11) = 10 \\
\phi(4) = 2 & \phi(8) = 4 & \phi(12) = 4
\end{array}
$$

Every positive integer n can be written in terms of the number of relatively prime positive integers less than or equal to each divisor of n. For example, $6 = 2 \cdot 3$ has four divisors: 1, 2, 3, and 6. From the table above,

$$\phi(1) + \phi(2) + \phi(3) + \phi(6) = 1 + 1 + 2 + 2 = 6$$

This property is given by the following theorem.

Theorem 3.42 (Gauss) *If n be a positive integer, then*

$$\sum_{d \mid n} \phi(d) = n$$

where the divisors d are positive divisors of n.

Proof. Let d be a positive divisor of n. Let $C(d)$ equal the set of positive integers $1 \le m \le n$ where $\gcd(m, n) = d$. $C(d)$ and $C(d')$ are disjoint if $d \ne d'$ because an integer can have only one greatest common divisor with n. But by Theorem 1.42, $C(d)$ is also the set of positive integers m, $1 \le m \le n$ with $\gcd(m/d, n/d) = 1$. But this is the number of positive integers less than n/d and relatively prime to n/d; that is, this is $\phi(n/d)$. Since the union of all of these sets is the set of integers between 1 and n, $n = \sum_{d|n} \phi(n/d)$. But for every d that divides n, there is a corresponding n/d that divides n. Hence $\sum_{d|n} \phi(n/d) = \sum_{d|n} \phi(d) = n$, which proves the theorem. ∎

Example 3.43 *Let $n = 12$. The divisors of 12 are 1, 2, 3, 4, 6, and 12. From the table of Euler ϕ values above,*

$$\phi(1) + \phi(2) + \phi(3) + \phi(4) + \phi(6) + \phi(12) = 1 + 1 + 2 + 2 + 2 + 4 = 12$$

To illustrate the proof of Theorem 3.42, for $d = 1, 2, 3, 4, 6$, and 12, we see that the corresponding values of n/d are, respectively, $n/d = 12, 6, 4, 3, 2$, and 1 so that the two sums mentioned are equal.

We now proceed to determine how to evaluate $\phi(n)$ for any positive integer n. The next three theorems lead to this objective.

Theorem 3.44 *If m and n are relatively prime, then*

$$\phi(mn) = \phi(m)\phi(n)$$

Proof. Let m and n be relatively prime. An integer is relatively prime to mn if and only if it is relatively prime to both m and n. Let a be relatively prime to m and let $a < m$. Consider the sequence $a, a+m, a+2m, \ldots, a+(n-1)m$. No two of these numbers are congruent modulo n; for if $a+jm \equiv a+km \pmod{n}$, then $n \mid (jm - km)$. Hence $n \mid m(j - k)$. Since m and n are relatively prime, $n \mid (j - k)$, which is impossible. Therefore, this sequence is a complete residue system modulo n and each of the elements in the sequence is congruent modulo n to a positive integer less than n. Hence the number of these elements relatively prime to n is equal to $\phi(n)$. Since there are $\phi(m)$ of these sequences, there are $\phi(m)\phi(n)$ numbers that are relatively prime to both m and n, that are less than mn, and that are relative prime to mn. Therefore, $\phi(mn) = \phi(m)\phi(n)$. ∎

For example, let $m = 8$ and $n = 15$. Then $\phi(8) = 4$ since 1, 3, 5, and 7 are the only positive integers less than 8 and relative prime to 8. Also, $\phi(15) = 8$ since 1, 2, 4, 7, 8, 11, 13, and 14 are the only positive integers less than 15 and relatively prime to 15. Hence $\phi(120) = \phi(8)\phi(15) = 32$, which may be checked directly.

Because of Theorem 3.44, we say that ϕ is multiplicative for relatively prime factors. We will discuss other multiplicative functions in the next chapter. Next we see how to calculate $\phi(n)$ when n is a power of a single prime.

Theorem 3.45 *If p is a prime number, then $\phi(p^k) = p^k - p^{k-1}$.*

Proof. The numbers less than or equal to p^k which are not relatively prime to p^k are $p, 2p, 3p, \ldots, (p^{k-1})p$. Since there are p^{k-1} of these integers, then there are $p^k - p^{k-1}$ integers which are relatively prime to p^k. Hence $\phi(p^k) = p^k - p^{k-1}$. ∎

Corollary 3.46 *A positive integer p is prime if and only if $\phi(p) = p - 1$.*

Proof. If p is prime, it is immediate from Theorem 3.45 that $\phi(p) = p - 1$. On the other hand, if p is not prime, then p has a divisor d different from p and 1. Since by definition, $\phi(p) \leq p - 1$ and d is one of the $p - 1$ positive integers less than p, we have $\phi(p) \leq p - 2$, a contradiction. ∎

Corollary 3.47 $\phi(2^k) = 2^{k-1}$.

The multiplicative property $\phi(mn) = \phi(m)\phi(n)$ for m and n relatively prime and the result $\phi(p^k) = p^k - p^{k-1}$ may be combined to obtain an explicit formula for $\phi(n)$ for any positive integer n using the prime factorization of n. This formula is given in the next theorem whose proof is left as an exercise.

Theorem 3.48 *If n is a positive integer with prime factorization*

$$n = p_1^{\alpha_1} p_2^{\alpha_1} \cdots p_t^{\alpha_t}$$

then

$$\phi(n) = \prod_{i=1}^{t} [p_i^{\alpha_i - 1}(p_i - 1)] = n \prod_{i=1}^{t}\left(1 - \frac{1}{p_i}\right)$$

Example 3.49 *Since $n = 40 = 2^3 \cdot 5$, $\phi(40) = 40(1 - 1/2)(1 - 1/5) = 40(1/2)(4/5) = 16$ which agrees with an example at the beginning of this section. Also, in Chapter 2, we saw that $n = 39616304 = 2^4 \cdot 7^2 \cdot 13^3 \cdot 23^1$ so that*

$$\begin{aligned}
\phi(39616304) &= 2^3(2-1)7^1(7-1)13^2(13-1)23^0(23-1) \\
&= 8 \cdot 1 \cdot 7 \cdot 6 \cdot 169 \cdot 12 \cdot 1 \cdot 22 = 14990976
\end{aligned}$$

There are limitations on the number of integers s, $1 \leq s \leq n$ that are relatively prime to n. One such constraint is given by the next theorem.

Theorem 3.50 *If n is an integer greater than 2, $\phi(n)$ is even.*

Proof. If $n = 2^k m$ where m is an odd integer and $k > 1$, then $\phi(2^k m) = \phi(2^k)\phi(m) = 2^{k-1}\phi(m)$ and hence $\phi(n)$ is even. If $n = p^k m$, where p is an odd prime, and p^k and m are relatively prime, then $\phi(p^k m) = \phi(p^k)\phi(m) = (p^k - p^{k-1})\phi(m)$. But $p^k - p^{k-1} = p^{k-1}(p - 1)$ and $p - 1$ is an even number since p is odd. Hence $\phi(n)$ is even. ∎

Although the following result has been indicated previously, we state it here formally.

Theorem 3.51 *If n is an integer, then the nonzero reduced residue classes form a group under multiplication modulo n.*

Proof. Certainly, if a and b are relatively prime to n, then ab is also relatively prime to n and so we have closure. If a is relatively prime to n, then the congruence $ax \equiv 1 \pmod{n}$ has a unique solution and hence a has an inverse. ∎

Let p be a prime. Since $\{1, 2, \ldots, p-1\}$ is a set of reduced residues modulo p, $[1], [2], \ldots, [p-1]$ form a group under multiplication as discussed in Section 3.1. The next theorem shows that the product, $[1] \odot [2] \odot \cdots \odot [p-1]$, of all the nonzero residue classes is always the residue class $[p-1] = [-1]$. Stated in terms of congruences, we have equivalently that $1 \cdot 2 \cdots \cdots (p-1) \equiv -1 \pmod{p}$.

Theorem 3.52 (Wilson's Theorem) *The positive integer p is a prime if and only if $(p-1)! \equiv -1 \pmod{p}$.*

Proof. If p is prime, then $p \equiv 0 \pmod{p}$ and $p - 1 \equiv -1 \pmod{p}$. The nonzero residue classes modulo p form a group under multiplication when p is prime so that each residue class is paired with its inverse to yield the product $[1]$. Thus, if $1 \le u \le p - 1$, then there is a unique integer u^{-1}, $1 \le u^{-1} \le p - 1$, such that $u \cdot u^{-1} \equiv 1 \pmod{p}$. Either $u = u^{-1}$ or $u \ne u^{-1}$. Certainly, for $u = 1$, $u^{-1} = 1$ also so that $u^2 = uu^{-1} \equiv 1 \pmod{p}$. If there is an integer a, $1 < a \le p - 1$, such that $a^2 \equiv 1 \pmod{p}$ also, then $a^2 - 1 = (a-1)(a+1) \equiv 0 \pmod{p}$ and $p \mid (a-1)(a+1)$. Thus $p \mid (a-1)$ or $p \mid (a+1)$. Since $a - 1 \ne 0$ and $a < p$, we obtain $p \nmid (a-1)$. Thus, $p \mid (a+1)$ implies that $p \le a + 1$; and since $a \le p - 1$ implies that $a + 1 \le p$, we have $p = a + 1$ or $a = p - 1$. Thus, for $1 \le u \le p - 1$, only $u = 1$ and $u = p - 1$ have the property that $u^2 \equiv 1 \pmod{p}$. Hence, we obtain $(p-1)! = 1(u_1 u_1^{-1})(u_2 u_2^{-1}) \cdots (u_k u_k^{-1})(p-1) \equiv 1 \cdot 1 \cdot 1 \cdot \cdots \cdot 1 \cdot (p-1) = p - 1 \equiv -1 \pmod{p}$ and where u_j is one of the integers 2, 3, ..., or $(p-2)$, where $k = (p-3)/2$.

If p is not prime, then $p = r \cdot s$ with $1 < r, s < p$. Since $(p-1)!$ contains r as a factor, $(p-1)! \equiv 0 \pmod{r}$ so that $(p-1)! \not\equiv -1 \pmod{r}$. Thus p must be prime. ∎

For example, let $p = 5$. Then $(p-1)! = 4! = 24 \equiv -1 \pmod{5}$. Notice that the theorem says that the product $(p-1)!$ cannot be congruent to -1 unless p is prime. By the theorem we may check p for primeness by determining whether $(p-1)! \equiv -1 \pmod{p}$; however, this test is not used for large p because the calculation of $(p-1)!$ modulo p is not practical.

We now examine Wilson's Theorem from an algebraic point of view. We already know that $Z_p - \{[0]\}$ is a group under multiplication. Thus, every nonzero element of Z_p has a multiplicative inverse. The proof above of Wilson's Theorem shows that only $[1]$ and $[p-1]$ are their own inverses. Hence, in the product $[1][2][3] \cdots [p-1]$, each of the other elements is paired with its inverse so that $[1][2][3] \cdots [p-1] = [1][p-1] = [p-1]$ or, equivalently,

$$1 \cdot 2 \cdot 3 \cdots \cdots (p-1) \equiv p - 1 \pmod{p}$$
$$\equiv -1 \pmod{p}$$

In Section 3.7 we will show that $Z_p - \{[0]\}$ is a cyclic group. In a cyclic group of even order it is easy to show that there are only two elements that are their own inverses, which in this case are $[1]$ and $[p-1]$.

The ϕ function of this section is named for Léonard Euler, the most prolific writer in mathematics. Many of his theorems appear throughout this book. His works would fill more than 75 large volumes. He was active in virtually every branch of mathematics. In number theory, he did an immense amount of work, including proving several of Fermat's lesser theorems. He is credited with beginning the idea of topology as well as several branches of calculus. It is impossible here to describe the immense amount of important mathematics for which he was responsible. He won the prestigious biennial prize from the Académie des Sciences twelve times. Much of our mathematical notation today is due to Euler.

Léonard Euler (1707-1783) was the son of a Lutheran minister in Switzerland. His father was his first teacher and wanted him to enter the ministry. He had the fortune to become the student of Jean Bernoulli, one of Europe's best mathematicians. Since opportunities in Switzerland were limited, he, along with many other mathematicians from Europe, went to the newly organized Academy of St. Petersburg. While in Russia, he lost the sight in one eye. Shortly after he arrived, political repression began in Russia. After 14 years, Euler left Russia to head the mathematics division of the Berlin Academy. Apparently, when asked by the queen mother in Germany why he was so shy, he replied that he had just come from a country where he who speaks is hanged.

He was, however, still held in high esteem in Russia. When Russia attacked Germany in 1760, Euler's farm was destroyed by the Russians. When they learned of this, the loss was immediately made good; and an additional gift was added by the Empress. After a disagreement with Frederick II, Euler returned to Russia after 25 years in Germany to accept a generous offer by Catherine the Great. Four years after returning to Russia, he lost the sight in his other eye and was completely blind for the last seventeen years of his life; however, he did not cease to do mathematics. In 1771 a fire broke out and reached Euler's house. His Swiss servant, Peter Grimes, bravely dashed into the burning house and carried out the blind Euler. Catherine immediately built him a new house. It is said that on September 18, 1783, after spending the afternoon calculating the laws for the ascension of balloons and outlining the calculation for the orbit for the newly discovered Uranus, Euler was playing with his grandchild and smoking his pipe when he suffered a stroke. The pipe dropped from his mouth, he uttered the words "I die." and the career of Euler was ended.

There are several anecdotes about Euler. Reputedly, he could recite Virgil's *Aeneid* line by line, although he had not read it since he was a child. Thièbault relates that Diderot was invited to the Russian Court and, being an atheist, began spreading his ideas on atheism. A plot was contrived to silence the guest. Euler, a deeply religious Christian, walked up to Diderot and stated in French that "$(a + b^n)/n = x$, hence God exists: reply!". Diderot knew no mathematics and hence was silent. Apparently the crowd roared in laughter and Diderot was so embarrassed that he returned to France immediately.

Another story was that Euler, then blind and elderly, was invited by Princess Daschkoff to attend an address which the Princess would give to commence her directorship of the Imperial Academy of Sciences in Petersburg. Euler, accompanied by a son and a grandson, travelled with the Princess in her personal coach. After the address, in which Euler was highly praised, the Princess sat down, intending that Euler would occupy the seat of honor beside her. An arrogant professor, Schtelinn, grabbed the seat before Euler could be led to it. The Princess turned to Euler and told him to take any seat and that would be the seat of honor. This act pleased everyone except the arrogant professor.

Exercises

1. Find an example of positive integers m and n such that $\phi(mn) \neq \phi(m)\phi(n)$.

2. Prove that if p is prime and $p > 2$, then $(p-2)! \equiv 1 \pmod{p}$.

3. Prove that $1 \cdot 2 \cdot 3 \cdot \cdots \cdot 1007 \equiv 1 \pmod{1009}$.

4. Let
$$S(n) = \sin\left[\pi \cdot \frac{(n-1)! + 1}{n}\right]$$
   Show that n is prime if and only if $S(n) = 0$.

5. Prove Theorem 3.48.

6. Create a table of $\phi(n)$ for $1 \leq n \leq 50$.

7. Calculate $\phi(2025)$.

8. If n is composite and $n > 4$, prove that $(n-1)! \equiv 0 \pmod{n}$.

9. Prove that n is prime if and only if n divides $(n-1)! + 1$.

3.6   Order of an Integer

In this section we will determine integers j such that $a^j \equiv 1 \pmod{m}$. In particular, we will be interested in the least such positive integer j.

Theorem 3.53 (Euler) *If m is a positive integer and $\gcd(a, m) = 1$, then $a^{\phi(m)} \equiv 1 \pmod{m}$.*

Proof. Let m be a positive integer and a be relatively prime to m. If $\{x_1, x_2, \ldots, x_k\}$ is a reduced residue system modulo m, then since a and m

are relatively prime, $\{ax_1, ax_2, \ldots, ax_k\}$ is also a reduced residue system. Hence each x_i is congruent to only one ax_j modulo m. Therefore,

$$x_1 x_2 \cdots x_k \equiv ax_1 ax_2 \cdots ax_k \ (\text{mod } m)$$

or

$$a^{\phi(m)} x_1 x_2 \cdots x_k \equiv x_1 x_2 \cdots x_k \ (\text{mod } m)$$

and since m and $x_1 x_2 \cdots x_k$ are relatively prime, the x_i may be "cancelled" giving $a^{\phi(m)} \equiv 1 \ (\text{mod } m)$. ■

For example, let $a = 3$ and $m = 4$ so that $\phi(4) = 2$ and, hence, $3^2 = 9 \equiv 1 \ (\text{mod } 4)$.

If m is prime in Theorem 3.53, then every positive integer less than m is relatively prime to m so that $\phi(m) = m - 1$. The case for m prime was developed as a corollary to Theorem 3.45. Thus we then have the following theorem as a special case.

Theorem 3.54 (Fermat's Little Theorem) *If p is a prime, then for every integer a such that $0 < a < p$, $a^{p-1} \equiv 1 \ (\text{mod } p)$.*

For example, if $p = 7$, then $p - 1 = 6$. Then the sixth power of each positive integer less than $p = 7$ should be congruent to 1 modulo 7:

$$
\begin{aligned}
1^6 &= 1 \equiv 1 \ (\text{mod } 7) \\
2^6 &= 64 \equiv 1 \ (\text{mod } 7) \\
3^6 &= 729 \equiv 1 \ (\text{mod } 7) \\
4^6 &= 4096 \equiv 1 \ (\text{mod } 7) \\
5^6 &= 15625 \equiv 1 \ (\text{mod } 7) \\
6^6 &= 46656 \equiv 1 \ (\text{mod } 7) \\
7^6 &= 117649 \equiv 0 \ (\text{mod } 7)
\end{aligned}
$$

The converse of Fermat's Little Theorem is not true. For example, $3^{90} \equiv 1 \ (\text{mod } 91)$; however, $91 = 7 \cdot 13$ is composite. On the other hand, if p is a positive integer and $0 < a < p$ is such that $a^{p-1} \not\equiv 1 \ (\text{mod } p)$, then p cannot be prime. Thus, Fermat's Little Theorem comprises a partial primeness test since it can be used to show that a positive integer is not prime without finding a nontrivial divisor of p. Composite positive integers n such that $a^{n-1} \equiv 1 \ (\text{mod } n)$ for some a, $1 < a < n$, are somewhat prime-like; and for this reason, such a composite n is said to be a *pseudo-prime* to base a. Thus, $n = 91$ is a pseudo prime to base $a = 3$. However, if we had chosen $a = 2$ instead of $a = 3$, we would have found that $2^{90} \equiv 64 \not\equiv 1 \ (\text{mod } 91)$, showing that $n = 91$ is not prime. So 91 is a pseudo-prime to base 3 but not to base 2.

Definition 3.55 *Let n be a positive integer and a be an integer such that $\gcd(a, n) = 1$. The order of a modulo n, denoted by $\text{ord}_n(a)$, is the smallest positive integer k such that $a^k \equiv 1 \ (\text{mod } n)$.*

Theorem 3.56 *Let n be a positive integer, $\gcd(a, n) = 1$, and $k = \operatorname{ord}_n a$. Then:*

(a) *If $a^m \equiv 1 \pmod{n}$, where m is a positive integer, then $k \mid m$.*

(b) $k \mid \phi(n)$.

(c) *For integers r and s, $a^r \equiv a^s \pmod{n}$ if and only if $r \equiv s \pmod{k}$.*

(d) *No two of the integers $a, a^2, a^3, \ldots, a^k$ are congruent modulo k.*

(e) *If m is a positive integer, then the order of a^m modulo n is $\dfrac{k}{\gcd(k, m)}$.*

(f) *The order of a^m modulo n is k if and only if m and k are relatively prime.*

Proof. **(a)** If $a^m \equiv 1 \pmod{n}$ for a positive integer m, then by the Division Algorithm, $m = kq + r$ where $0 \le r < k$. Hence $a^m = a^{kq+r} = a^{kq} a^r$ so that $a^r \equiv 1 \pmod{n}$. But this contradicts the definition of the order of a unless $r = 0$. Hence $k \mid m$.

   (b) Since, by Theorem 3.53, $a^{\phi(n)} \equiv 1 \pmod{n}$, part (a) implies that $k \mid \phi(n)$.

   (c) Assume that $r > s$. Since a and n are relatively prime, $a^r \equiv a^s \pmod{n}$ if and only if $a^{r-s} \equiv 1 \pmod{n}$; and hence, by part (a), k divides $r - s$ and $r \equiv s \pmod{k}$.

   (d) This follows directly from part (c).

   (e) Let $d = \gcd(k, m)$ so that $k = ud$ and $m = vd$. $(a^m)^{k/\gcd(k,m)} = (a^m)^{ud/d} = a^{um} = a^{uvd} = a^{(ud)v} = a^{kv} \equiv 1 \pmod{n}$. Assume that t is such that $(a^m)^t \equiv 1 \pmod{n}$. Then $a^{mt} \equiv 1 \pmod{n}$ so that $k \mid mt$ because $\operatorname{ord}_n a = k$. Hence $ud \mid vdt$ and since u and v are relatively prime, $u \mid t$. Since $k = ud$, $(k/d) = k/\gcd(k, m)$ divides t and so, by definition of order, $k/\gcd(k, m)$ is the order of a^m.

   (f) This follows directly from part (e). ∎

Example 3.57 *To illustrate Theorem 3.56, suppose that $n = 14 = 2 \cdot 7$ so that $\phi(n) = (2-1)(7-1) = 6$. The primary reduced residue system for $n = 14$ is the set $\{1, 3, 5, 9, 11, 13\}$. Consider the following table of primary residues of powers of $a = 5$:*

m	$[[a^m]]_n$	m	$[[a^m]]_n$
1	5	8	11
2	11	9	13
3	13	10	9
4	9	11	3
5	3	12	1
6	1	13	5
7	5		

where we see that after $m = 6$, we merely repeat the same pattern. Thus, $k =$

$\text{ord}_{14}\, 5 = 6$. *For* $m = 12$, $a^m = 5^{12} \equiv 1 \pmod{14}$ *and* $k \mid m$, *in agreement with Theorem 3.56(a). Also,* $\text{ord}_{14}\, 5 \mid \phi(14)$ *since* $6 \mid 6$ *(Theorem 3.56(b)). We see that* $2 \equiv 8 \equiv 14 \pmod 6$ *and that* $5^2 \equiv 5^8 \equiv 5^{14} \equiv 11 \pmod{14}$ *(Theorem 3.56(c)). By inspection of the table, no two of the integers* 5^1, 5^2, 5^3, 5^4, 5^5, *and* 5^6 *are congruent modulo 14 (Theorem 3.56(d)). Since* $\text{ord}_n\, b \mid \phi(n)$ *for any integer* b *and since* $\phi(n) = 6$ *for* $n = 14$, *the order of every* b *in* $\{1, 3, 5, 9, 11, 13\}$ *can quickly be computed as we did for* $a = 5$.

b	$\text{ord}_n\, b$
1	1
3	6
5	6
9	3
11	3
13	2

If $m = 4$, $5^m \equiv 9 \pmod{14}$, *but* $\text{ord}_{14}\, 5/\gcd(\text{ord}_{14}\, 5, 4) = 6/\gcd(6, 4) = 6/2 = 3$. *According to the table of orders,* $\text{ord}_{14}\, 5^4 = 3$ *(Theorem 3.56(e)).*

Only $b = 3$ *and* $b = 5$ *have order 6 modulo 14. The exponents* m *in the table of powers above that produce an* a^m *that is congruent to either 3 or 5 are* $m = 1, 5, 7, 11$, *and 13. These are the only such* m's *that are relatively prime to* $n = 14$ *(Theorem 3.56(f)).*

Theorem 3.58 *If* $\gcd(a, n) = \gcd(b, n) = 1$ *and* $\text{ord}_n\, a$ *is relatively prime to* $\text{ord}_n\, b$, *then* $\text{ord}_n\, (ab) = (\text{ord}_n\, a) \cdot (\text{ord}_n\, b)$.

Proof. Let $\text{ord}_n\, a = R$ and $\text{ord}_n\, b = S$. Then $(ab)^{RS} = a^{RS}b^{RS} = (a^R)^S(b^S)^R = 1 \cdot 1 \equiv 1 \pmod n$. By Theorem 3.56, $\text{ord}_n\, (ab) \mid RS$. Since R and S are relatively prime, there are integers r and s with $\text{ord}_n\, (ab) = rs$, $r \cdot w = R$, and $s \cdot x = S$. We now show that $r = R$ and $s = S$. By definition of r and s,

$$(ab)^{rs} = a^{rs}b^{rs} \equiv 1 \pmod n$$
$$(a^{rs}b^{rs})^w \equiv 1^w = 1 \pmod n$$
$$(a^{rw})^s \cdot (b^{rw})^s \equiv 1 \pmod n$$

But since $a^{rw} \equiv 1 \pmod n$ and $rw = R$, we have

$$b^{Rs} \equiv 1 \pmod n$$

By Theorem 3.56(a), $\text{ord}_n\, b \mid Rs$ or, equivalently, $S \mid Rs$. Because $\gcd(R, S) = 1$, we have $S \mid s$; but $s \mid S$ also so that $S = s$. Similarly, $R = r$. Thus, $\text{ord}_n\, (ab) = (\text{ord}_n\, a) \cdot (\text{ord}_n\, b)$. ∎

Example 3.59 *If* $n = 11$, *then all the primary residues are relatively prime* *to n. The table of orders modulo 11 is*

Residue	Order	Residue	Order
1	1	6	10
2	10	7	10
3	5	8	10
4	5	9	5
5	5	10	2

If $a = 3$ and $b = 10$, then $ab = 30 \equiv 8 \pmod{11}$. Thus, $\operatorname{ord}_{11}(ab) = \operatorname{ord}_{11}(30) = \operatorname{ord}_{11} 8 = 10 = (\operatorname{ord}_{11} 3) \cdot (\operatorname{ord}_{11} 10)$. Of course, $\gcd(3, 11) = \gcd(10, 11) = 1$ and $\operatorname{ord}_{11} 3 = 5$ and $\operatorname{ord}_{11} 10 = 2$ are relatively prime. Note that if $a = 3$ and $c = 7$, then $\operatorname{ord}_{11} 3 = 5$ is not relatively prime to $\operatorname{ord}_{11} 7 = 10$. In this case, $\operatorname{ord}_{11}(ac) = \operatorname{ord}_{11} 21 = \operatorname{ord}_{11} 10 = 2 \neq (\operatorname{ord}_{11} 3) \cdot (\operatorname{ord}_{11} 7) = 50$.

Example 3.60 *The order of $a = 5$ modulo $n = 14$ was obtained in Example 3.57 by calculating a^m for $m = 1, 2, 3, \ldots, \phi(n)$ until finding that $a^m \equiv 1 \pmod{n}$. Theorem 3.56(b) implies that the order of a modulo n must divide $\phi(n)$; therefore, instead of testing each m, $1 \le m \le \phi(n)$, one at a time, test only the m's that divide $\phi(n)$. For $n = 14$, $\phi(n) = 6$, whose only positive divisors are 1, 2, 3, and 6. In this case the work to determine $\operatorname{ord}_{14} 5$ is reduced by only a small amount since in Example 3.57 we tested $m = 4$ and 5 also.*

However, for $n = 58$ and $a = 25$, we quickly obtain $\operatorname{ord}_{58} 25 = 7$ using Theorem 3.56(b). $\phi(58) = \phi(2 \cdot 29) = (2 - 1)(29 - 1) = 28 = 2^2 \cdot 7$. The only positive divisors of $2^2 \cdot 7$ are 1, 2, 4, 7, 14, and 28. The following table is easily generated:

m	$[[25^m]]_{58}$
1	25
2	45
4	53
7	1

Therefore, $\operatorname{ord}_{58} 25 = 7$ and we do not need to check $n = 14$ and 28.

The results obtained from Theorems 3.56 and 3.61 lead to a test for primeness called Lucas' Primality Test.

Theorem 3.61 (Lucas) *If n is a positive integer and there is an integer a such that*

$$a^{n-1} \equiv 1 \pmod{n}$$

and

$$a^{\frac{n-1}{p}} \not\equiv 1 \pmod{n}$$

for every prime p that divides $n - 1$, then n is prime.

Proof. $a^{n-1} \equiv 1 \pmod{n}$ implies that $\gcd(a, n) = 1$ and, by Theorem 3.56(a), that $\operatorname{ord}_n a \mid (n - 1)$. If p is a prime such that $p \mid (n - 1)$, then the congruence $a^{(n-1)/p} \not\equiv 1 \pmod{n}$ implies that $\mathrm{ord}_n\ a \nmid [(n-1)/p]$ because if $\mathrm{ord}_n\ a \mid [(n-1)/p]$, it would contradict $a^{\mathrm{ord}_n\ a} \equiv 1 \pmod{n}$. But $\mathrm{ord}_n\ a \mid (n-1)$ and $\mathrm{ord}_n\ a \nmid [(n-1)/p]$ for all p dividing $n-1$ imply that $\mathrm{ord}_n\ a = n-1$. By Theorem 3.56(b), $\phi(n) = n-1$. Therefore, by Corollary 3.46, n is prime. ∎

In order to use Lucas' test for testing n, we must be able to factor $n - 1$, which may itself be difficult. Additionally, an appropriate integer a must be found. We will find in Section 3.7 that the integer a of Theorem 3.61 is called a primitive root of n. One can show that the Mersenne number $n = 2^{31} - 1$ is prime using Lucas' test with $a = 7$ since $n - 1 = 2 \cdot 3^2 \cdot 7 \cdot 11 \cdot 31 \cdot 151 \cdot 331$. See the Exercises at the end of this section and Section 3.12.

If $\gcd(a, n) = 1$, Fermat's Theorem gives $a^{n-1} \equiv 1 \pmod{n}$ when n is prime. Its generalization, Euler's Theorem, gives $a^{\phi(n)} \equiv 1 \pmod{n}$ for any positive integer n. Aside from these and a few other special cases, the task of evaluating a^e modulo n or, more precisely, calculating $[[a^e]]_n$, the remainder when a^e is divided by n may appear formidable when e is large; for in such cases, just computing a^e and dividing by n is not practical.

In the small illustrative examples given so far, we have just written $e = e_1 + e_2 + \cdots + e_k$ for suitable e_i, calculated $[[a^{e_i}]]_n$, multiplied the results, and reduced the product modulo n. This method works because

$$[[st]]_n = [[\, [[s]]_n \cdot [[t]]_n \,]]_n$$

The e_i were chosen in some ad hoc fashion.

A much more efficient algorithm is similar but expresses the exponent e using a binary representation, that is, a base 2 representation. Thus,

$$\begin{aligned} e &= b_m 2^m + b_{m-1} 2^{m-1} + \cdots + b_1 2^1 + b_0 \\ &= [b_m b_{m-1} \cdots b_1 b_0]_{binary} \end{aligned}$$

where $b_i = 0$ or 1 and $b_m = 1$. Thus,

$$a^e = a^{b_m 2^m + b_{m-1} 2^{m-1} + \cdots + b_1 2^1 + b_0}$$

If the exponent is expressed with regrouping to reduce the number of multiplications, we obtain the Horner's Rule representation

$$e = (\cdots ((b_m \cdot 2 + b_{m-1}) \cdot 2 + b_{m-2}) \cdot 2 + \cdots + b_1) \cdot 2 + b_0$$

so that

$$a^e = (\cdots ((a^{b_m \cdot 2} \cdot a^{b_{m-1}})^2 \cdot a^{b_{m-2}})^2 \cdot \cdots \cdot a^{b_1})^2 \cdot a^{b_0}$$

We can calculate $[[a^e]]_n$ by evaluating this last expression "inside out" while reducing each product modulo n. Hence, for $e = [b_m b_{m-1} \cdots b_1 b_0]_{binary}$, start with

$$p_m = [[a]]_n$$

Then, for $k = m - 1, m - 2, \ldots, 2, 1$, and 0 calculate

$$p_k = \begin{cases} [[p_{k+1}^2]]_n & \text{if } b_k = 0 \\ \\ [[p_{k+1}^2 \cdot a]]_n & \text{if } b_k = 1 \end{cases}$$

The final result is $p_0 = [[a^e]]_n$. That is, beginning with $p_m = [[a]]_n$, obtain the next product p_k by squaring the previous product and reducing modulo n when $b_k = 0$ and by squaring the previous product, multiplying by a, and reducing modulo n when $b_k = 1$. The algorithm works because if a is squared k times, the result is a^{2^k}; and if $a^{2^k} b$ is squared j times, the result is $a^{2^{k+j}} b^{2^j}$.

Example 3.62 *Suppose that we want to evaluate* $[[3^{103}]]_{41}$. *Since*

$$103 = 2^6 + 2^5 + 2^2 + 2^1 + 1$$
$$= 1100111_{binary}$$

and $m = 6$, *we obtain*

k	b_k	$p_k = [[p_{k+1}^2 \cdot a^{b_k}]]_n$	
6	1	3	$\equiv 3$
5	1	$3^2 \cdot 3$	$= 27$
4	0	$(27)^2$	$= 729 \equiv 32$
3	0	$(32)^2$	$= 1024 \equiv 4$
2	1	$(40)^2 \cdot 3$	$= 4800 \equiv 3$
1	1	$(3)^2 \cdot 3$	$= 27$
0	1	$(27)^2 \cdot 3$	$= 2187 \equiv 14$

Therefore, $[[3^{103}]]_{41} = 14$. *By an ad hoc method using congruence modulo 41, we obtain*

$$3^{10} = 57049 \equiv 9$$
$$3^{50} = (3^{10})^5 \equiv 9^5 = 59049 \equiv 9$$
$$3^{103} = 3^{50} \cdot 3^{50} \cdot 3^3 \equiv 9 \cdot 9 \cdot 27 = 2187 \equiv 14$$

Exercises

1. Prove this variation of Fermat's Little Theorem: If p is prime and $a \not\equiv 0 \pmod{p}$, then $a^{p-1} \equiv 1 \pmod{p}$.

2. Prove that if $\gcd(a, m) = 1$, then the congruence $ax \equiv b \pmod{m}$ has the solution $x \equiv a^{\phi(m)-1}b \pmod{m}$.

3. Let p be prime, $p > 2$, and $J \equiv 0 \pmod{(p-1)}$. Prove that

$$1^J + 2^J + 3^J + \cdots + (p-1)^J \equiv -1 \pmod{p}$$

4. Let $\gcd(m, n) = 1$ and let the sets $\{r_1, r_2, \ldots, r_m\}$ and $\{s_1, s_2, \ldots, s_n\}$ be complete residue systems modulo m and n, respectively. Prove that $\{n \cdot r_i + m \cdot s_j : 1 \leq i \leq m \text{ and } 1 \leq j \leq n\}$ is a set of mn integers that is a complete residue system modulo mn.

5. Fill in the details of a proof of Fermat's Little Theorem using the Binomial Theorem and induction on a.

6. Use the method of Exercise 2 to solve these congruences:

    (a) $5x \equiv 8 \pmod{11}$

    (b) $7x \equiv 8 \pmod{25}$

    (c) $9x \equiv 13 \pmod{25}$

7. Prove that for every prime p,

$$(a + b)^p \equiv a^p + b^p \pmod{p}$$

8. Prove the converse of Theorem 3.56(a); that is, prove that if n is a positive integer, $\gcd(a, n) = 1$, $k = \operatorname{ord}_n a$, and $k \mid m$, then $a^m \equiv 1 \pmod{m}$.

9. Determine $\operatorname{ord}_n a$ for $1 \le a \le n - 1$ if:

    (a) $n = 9$ (b) $n = 20$ (c) $n = 27$

10. Show that:

    (a) $b^{10} - 1$ is divisible by 11 if b and 11 are relatively prime.

    (b) $b^{10k} - 1$ is divisible by 11 if $\gcd(b, 11) = 1$.

    (c) $b^7 - b$ is divisible by 42 for any integer b.

11. Show that:

    (a) $7^4 \equiv 1 \pmod{5}$

    (b) $7^4 \equiv 1 \pmod{2}$

    (c) $7^4 \equiv 1 \pmod{10}$

    (d) $7^{4k} \equiv 1 \pmod{10}$ for any positive integer k

    What is the last base ten digit of 7^{4000}?

12. Show that:

    (a) $7^{20} \equiv 1 \pmod{25}$

    (b) $7^2 \equiv 1 \pmod{4}$

    (c) $7^{20} \equiv 1 \pmod{4}$

    (d) $7^{20} \equiv 1 \pmod{100}$

    What are the last two base ten digits of 7^{500}?

13. Calculate the following residues:

(a) $[[3^{275}]]_{100}$

(b) $[[6^{5000}]]_{1000}$

(c) $[[11^{24681}]]_{83}$

(d) $[[3497^{100000}]]_{1234}$

(e) $\left[\left[7^{2\cdot3^2\cdot7\cdot11\cdot31\cdot331}\right]\right]_{2^{31}-1} = [[7^{14221746}]]_{2147483647}$ (The availability of extended integer precision computer software makes the solution of this part more tractable.)

14. Use Lucas' test to show that the following integers are prime:

   (a) 37

   (b) 199

15. Establish these statements:

   (a) Prove that if n is an odd positive integer and there is a positive integer a with the properties

   (i) $a^{\frac{n-1}{2}} \equiv -1 \pmod{n}$, and

   (ii) $a^{\frac{n-1}{p}} \not\equiv 1 \pmod{n}$ for every odd prime p that divides $(n-1)$,

   then n is prime.

   (b) Use the test of part (a) instead of Lucas' test to show that the integers in Exercise 14(a) and (b) are prime.

16. If $F(n) = 2^{2^n} + 1$ is a Fermat number and there is a positive integer a such that $a^{2^{2^n}} \equiv 1 \pmod{F(n)}$ and $a^{2^{(2^n-1)}} \not\equiv 1 \pmod{F(n)}$ prove that $F(n)$ is prime.

17. Use Exercise 16 to prove that the following integers are prime:

   (a) $F(3) = 257$

   (b) $F(4) = 65537$

3.7 Primitive Roots

If $\gcd(a, n) = 1$, recall that the order of an element a modulo n, namely $\text{ord}_n a$, is the smallest integer k such that $a^k \equiv 1 \pmod{n}$. Also we know from

Euler's Theorem that $a^{\phi(n)} \equiv 1 \pmod{n}$. These facts suggest the following definition.

Definition 3.63 *Let n be a positive integer and $\gcd(a,n) = 1$. If the order of an element a modulo n is $\phi(n)$, that is, if $\operatorname{ord}_n a = \phi(n)$, then we say that a is a* primitive root *of n.*

Theorem 3.64 *If a is a primitive root of n, then $\{a, a^2, a^3, \ldots, a^{\phi(n)}\}$ is a complete set of reduced residues modulo n. Hence the reduced residue set is a cyclic group.*

Proof. By definition of primitive root, a and n are relatively prime. Hence a^i and n are relatively prime for all $1 \le i \le \phi(n)$. Also, we know from Theorem 3.56 that the a^i are not congruent. Since there are only $\phi(n)$ positive integers less than n that are relatively prime to n, the set $\{a, a^2, a^3, \ldots, a^{\phi(n)}\}$ must be congruent to them. ∎

Theorem 3.65 *If a primitive root of a positive integer n exists, then there are exactly $\phi(\phi(n))$ incongruent primitive roots modulo n.*

Proof. Let a be a primitive root of n. Since $\{a, a^2, \ldots, a^{\phi(n)}\}$ is a reduced residue system modulo n, every primitive root of n must be congruent to a member of this set. However, by Theorem 3.56, a^m has the same order as a if and only if m and $\phi(n)$ are relatively prime. There are $\phi(\phi(n))$ of these numbers. ∎

Example 3.66 *For $n = 14$, the order of each reduced residue is given in the following table:*

Residue	Order	Residue	Order
1	1	8	*
2	*	9	3
3	6	10	*
4	*	11	3
5	6	12	*
6	*	13	2
7	*		

where the primary residues that are not relatively prime to $n = 14$ and that, therefore, have no order, are indicated by an asterisk. The set of $\phi(14) = 6$ reduced residues modulo 14 is $\{1, 3, 5, 9, 11, 13\}$. By Theorem 3.65, $\phi(\phi(14)) = \phi(6) = (2-1)(3-1) = 2$ of them are primitive roots of $n = 14$. The primitive roots of $n = 14$ are 3 and 5, each of which, according to the table, has order $\phi(14) = 6$.

Now consider the powers $\{a, a^2, \ldots, a^{\phi(n)}\}$ *for several primary residues modulo* 14.

k	$[[3^k]]_{14}$	$[[5^k]]_{14}$	$[[6^k]]_{14}$	$[[9^k]]_{14}$
1	3	5	6	9
2	9	11	8	11
3	13	13	6	1
4	11	9	8	9
5	5	3	6	11
$6 = \phi(n)$	1	1	8	1

For the two primitive roots 3 and 5, all the reduced residues are produced as powers; and 1 is produced by the exponent $\phi(14) = 6$. *The primary residue 6 is not relatively prime to* $n = 14$ *and has no order (no power of 6 is congruent to 1 modulo 14). 9 is relatively prime to 14 and has order 3; however, only three of the six reduced residues occur as powers of 9.*

Theorem 3.67 *Let p be a prime and let d be a positive integer that divides $p - 1$. Then $x^d - 1 \equiv 0 \pmod{p}$ has exactly d incongruent solutions.*

Proof. The congruence $x^{p-1} - 1 \equiv 0 \pmod{p}$ has exactly $p - 1$ solutions by Fermat's Theorem. Since $d \mid (p - 1)$, $p - 1 = dm$ for some m. Therefore, $x^{p-1} - 1 = x^{dm} - 1 = (x^d - 1)g(x)$ where $g(x)$ has degree $p - 1 - d$. If a is a root of $x^{p-1} - 1 \equiv 0 \pmod{p}$, then $p \mid (a^{p-1} - 1)$. Hence, either $p \mid (a^d - 1)$ or $p \mid g(a)$; therefore, any root of $x^{p-1} - 1 \equiv 0 \pmod{p}$ is a root of $x^d - 1 \equiv 0 \pmod{p}$ or of $g(x) \equiv 0 \pmod{p}$. Since $g(x) \equiv 0 \pmod{p}$ can have only $p - 1 - d$ solutions, $x^d - 1 \equiv 0 \pmod{p}$ must have at least $p - 1 - (p - 1 - d) = d$ solutions because $x^{p-1} - 1 \equiv 0 \pmod{p}$ has exactly $p - 1$ solutions. But, by Lagrange's Theorem, $x^d - 1 \equiv 0 \pmod{p}$ can have at most d solutions. Hence it has only d solutions. ■

Theorem 3.68 *Let p be a prime number and d be a positive integer that divides $p - 1$. Then there are exactly $\phi(d)$ incongruent integers having order d modulo p.*

Proof. Consider the set of integers a between 1 and $p - 1$, inclusive. Each number a of this set has an order d which divides $p-1$ by Theorem 3.56 since $a^{p-1} \equiv 1 \pmod{p}$. If we let C_d equal the number of elements which have order d, then

$$\sum_{d \mid (p-1)} C_d = p - 1$$

But by Gauss's Theorem (Theorem 3.42),

$$\sum_{d \mid (p-1)} \phi(d) = p - 1$$

If we can show that $C_d \leq \phi(d)$ for all d, then we can conclude that $C_d = \phi(d)$ for all d (see the Exercises). Assume that $C_d > 0$ since otherwise the statement is obvious, and let a have order d. Consider $a, a^2, a^3, \ldots, a^d$. By Theorem 3.56 these integers are not congruent modulo p. Since $(a^i)^d = (a^d)^i \equiv 1^i = 1 \pmod{p}$, these integers are all solutions of $x^d - 1 \equiv 0 \pmod{p}$. But by Lagrange's Theorem, these are the only solutions since there are at most d solutions modulo p. Further, by Theorem 3.56, a^i has order d if and only if i and d are relatively prime. Hence there are $\phi(d)$ elements of order d, and $C_d = \phi(d)$. ∎

Corollary 3.69 *Let p be a prime number. There are exactly $\phi(p-1)$ primitive roots of p.*

The prime $p = 11$ was considered in Example 3.59 and we have $p - 1 = 10 = 2 \cdot 5$. Since $\phi(p-1) = \phi(10) = (2-1)(5-1) = 4$, there should be exactly four primitive roots of $p = 11$, each of order $p - 1 = 10$. The four primitive roots of 11 by inspection of the table in Example 3.59, are 2, 6, 7, and 8. There is $\phi(2) = 1$ primary residue of order $d = 2$. There are $\phi(5) = 4$ of order $d = 5$, $\phi(10) = 4$ of order $d = 10$, and $\phi(1) = 1$ of order 1.

Once a primitive root of prime p has been found, the order of every integer not divisible by p, that is, of every integer not congruent to 0 modulo p, can easily be obtained. Let g be a primitive root of a prime p. Then $\{g, g^2, g^3, \ldots, g^{p-1} \equiv 1\}$ is a reduced residue system. Any integer $a \not\equiv 0 \pmod{p}$ is congruent to one of the reduced residues so that $a \equiv g^k \pmod{p}$ for some positive integer k, $1 \le k \le p - 1$. Then by Theorem 3.56, $\mathrm{ord}_p\, a = (p-1)/\gcd(p-1, k)$. There are $\phi(\mathrm{ord}_p\, a)$ residues of order $\mathrm{ord}_p\, a$.

Suppose we consider the primitive root $g = 7$ for $p = 11$ in Example 3.59. $2 \equiv 7^3 \pmod{11}$. Also, $(11-1)/\gcd(10, 3) = 10$, which is in agreement with the table in Example 3.59. So 2 is also a primitive root of $p = 11$. Note that $\mathrm{ord}_p\, a = p - 1$ when $\gcd(p-1, k) = 1$ so that each such k yields a primitive root g^k of p. Thus, $k = 1, 3, 7$, and 9 yield the primitive roots $7^1 \equiv 7$, $7^3 \equiv 2$, $7^7 \equiv 6$, and $7^9 \equiv 8$, respectively, of $p = 11$.

That not all positive integers have primitive roots is established by the following theorem.

Theorem 3.70 *If m and n are relatively prime positive integers greater than 2, then mn has no primitive roots.*

Proof. We first note that

$$\mathrm{lcm}(\phi(m), \phi(n)) = \frac{\phi(m)\phi(n)}{\gcd(\phi(m), \phi(n))}$$
$$= \frac{\phi(mn)}{\gcd(\phi(m), \phi(n))}$$

Since $\phi(m)$ and $\phi(n)$ are both even by Theorem 3.50, $\gcd(\phi(m), \phi(n)) \ge 2$. Hence $\mathrm{lcm}(\phi(m), \phi(n)) \le \phi(mn)/2$. But $\mathrm{lcm}(\phi(m), \phi(n)) = k\phi(m) = j\phi(n)$ for some integers k and j. Thus, for any integer a less than and relatively prime to mn, $a^{\mathrm{lcm}(\phi(m), \phi(n))} = (a^{\phi(m)})^k \equiv 1 \pmod{m}$ and $a^{\mathrm{lcm}(\phi(m), \phi(n))} =$

$(a^{\phi(n)})^j \equiv 1 \pmod{n}$. Hence $a^{\text{lcm}(\phi(m),\phi(n))} \equiv 1 \pmod{mn}$ since m and n are relatively prime. Finally, $\text{ord}_n\, a \le \phi(mn)/2$ so that no integer has order $\phi(mn)$ modulo mn and mn has no primitive roots. $\blacksquare$

Thus, we can say that $n = 15 = 3 \cdot 5$ has no primitive roots. In fact, the highest-order reduced residue modulo 15 is 4 instead of the necessary $\phi(15) = 8$. In the case of $n = 14 = 2 \cdot 7$ of Example 3.66, there is an integer of order $\phi(14) = 6$, namely the integer 3. This result means that 3 is a primitive root of 14 and shows that the hypothesis of Theorem 3.70 about both factors being greater than 2 is necessary. The case of even moduli divisible by 4 is characterized by the following corollary.

Corollary 3.71 *The integer* $2^k m$, *where* $k > 1$ *and* m *is an odd integer greater than 2, has no primitive roots.*

We will proceed in the next few theorems to establish exactly which positive integers have primitive roots.

Theorem 3.72 *There are primitive roots of* 2^n *if and only if* n *is a positive integer less than three.*

Proof. $\phi(2) = 1$ and 1 is a primitive root of 2. $\phi(4) = 2$ and 3 is a primitive root of 4 since $3 \not\equiv 1 \pmod{4}$ but $3^2 \equiv 1 \pmod{4}$.

We will show that if b is odd, then $b^{2^{n-2}} \equiv 1 \pmod{2^n}$ for $n \ge 3$, in which case b is not a primitive root of 2^n because $2^{n-2} < 2^{n-1}$.

Let b be odd so that $b = 2j+1$ for some integer j. Thus, $b^2 = 4j^2+4j+1 = 4(j^2 + j) + 1$. But $j^2 + j \equiv 0 \pmod{2}$ for any j so that $b^2 \equiv 1 \pmod{2^3}$. Assume that $k \ge 3$ and $b^{2^{k-2}} \equiv 1 \pmod{2^k}$. Then $b^{2^{k-2}} = 2^k s + 1$ for some integer s so that

$$\left(b^{2^{k-2}}\right)^2 = 2^{2k} s^2 + 2^{k+1} s + 1 = 2^{k+1}\left[2^{k-1} s^2 + s\right] + 1$$

Thus, $b^{2^{(k+1)-2}} \equiv 1 \pmod{2^{k+1}}$. Therefore, $b^{2^{n-2}} \equiv 1 \pmod{2^n}$ for $n \ge 3$. $\blacksquare$

The next step in classifying which moduli have primitive roots is to show that powers of odd primes have them. It is perhaps reasonable to search for primitive roots of p^k among the primitive roots of p. A primitive root g of p^k will have order $\phi(p^k) = p^k - p^{k-1} = p^{k-1}(p-1)$ according to Theorem 3.45. Thus, we investigate the integers of the form $g^{p^{k-1}(p-1)}$ in the following lemma.

Lemma 3.73 *For an odd prime* p, *there is a primitive root* g *of* p *such that for every positive integer* k,

$$g^{p^{k-1}(p-1)} = 1 + p^k a_k$$

where $\gcd(p, a_k) = 1$. *If* r *is a primitive root of* p *such that* $r^{p-1} = 1 + pa$ *with* $\gcd(p, a) = 1$, *then we may choose* $g = r$. *If* r *is a primitive root of* p *not of the foregoing form, then we may choose* $g = r + p$.

Proof. If p is an odd prime so that $p - 1 \geq 2$, by Corollary 3.69, p has a primitive root r. Since $\operatorname{ord}_p r = p - 1$, $r^{p-1} \equiv 1 \pmod{p}$ or, equivalently, $r^{p-1} = 1 + p \cdot a$ for some integer a. If $\gcd(a, p) = 1$, let $g = r$ and $a_1 = a$ so that $g^{p-1} = 1 + p \cdot a_1$. Trivially, g is a primitive root of p. On the other hand, if $\gcd(a, p) \neq 1$, let $g = r + p$. Since $g \equiv r \pmod{p}$, g is a primitive root of p in this case also. Using the Binomial Theorem (see the Exercises of Section 1.3) and recalling that $p - 1 \geq 2$,

$$
\begin{aligned}
g^{p-1} &= (r + p)^{p-1} \\
&= r^{p-1} + (p - 1)r^{p-2}p + [\text{terms with the factor } p^2] \\
&\equiv r^{p-1} + (p - 1)r^{p-2}p \pmod{p^2} \\
&\equiv 1 + (p - 1)r^{p-2}p \pmod{p^2}
\end{aligned}
$$

Thus, for some integer t,

$$
g^{p-1} = 1 + (p - 1)r^{p-2}p + tp^2 = 1 + p((p - 1)r^{p-2} + tp)
$$

But, $p \nmid r$ because r is a primitive root of p and $\gcd(r, p) = 1$. So $p \nmid r^{p-2}$ and, consequently, $p \nmid (p - 1)r^{p-2} + tp$. Let $a_1 = (p - 1)r^{p-2} + tp$. Thus, in either case, there is a primitive root g of p and an integer a_1 such that $g^{p-1} = 1 + p \cdot a_1$ and $\gcd(p, a_1) = 1$.

Having proved the lemma for $k = 1$, we proceed by induction. Assume that the equation of the lemma is true for $k = n$ so that there is an integer a_n such that $\gcd(p, a_n) = 1$ and $g^{p^{n-1}(p-1)} = 1 + p^n a_n$. Again using the Binomial Theorem for the case $n + 1$,

$$
\begin{aligned}
g^{p^n(p-1)} &= (g^{p^{n-1}(p-1)})^p \\
&= (1 + p^n a_n)^p \\
&= 1 + p^{n+1} a_n + p^{n+1} \cdot W(p) \\
&= 1 + p^{n+1} a_{n+1}
\end{aligned}
$$

where $W(p)$ is divisible by p, and where $a_{n+1} = a_n + W(p)$ has the property $\gcd(a_{n+1}, p) = 1$ since $\gcd(a_n, p) = 1$. Thus, by mathematical induction the equation claimed by the lemma is true for every positive integer k. ∎

Theorem 3.74 *Let p be an odd prime. Then p^k has a primitive root for every positive integer k. If r is a primitive root of p such that $r^{p-1} = 1 + pa$ with $\gcd(p, a) = 1$, then we may choose $g = r$ to be a primitive root of p^k. If r is a primitive root of p not of the form above, then we may choose $g = r + p$ to be a primitive root of p^k.*

Proof. Let k be a positive integer and g be a primitive root of p having the property of Lemma 3.73. We will show that g is a primitive root of p^k. For g to be a primitive root of p^k, it must be true that $\operatorname{ord}_{p^k} g = \phi(p^k) = p^{k-1}(p - 1)$.

Let $u = \operatorname{ord}_{p^k} g$. We will show that $u = p^{k-1}(p - 1)$. Because $u = \operatorname{ord}_{p^k} g$ and $g^{\phi(p^k)} \equiv 1 \pmod{p^k}$, Theorem 3.56 implies that $u \mid p^{k-1}(p - 1)$. Since p

is prime, $u = p^s h$ where $s \leq k-1$ and $h \mid (p-1)$. We will show that $h = p-1$ and $s = k - 1$.

$g^u \equiv 1 \pmod{p^k}$ since u is the order of g modulo p^k so that $g^u \equiv 1 \pmod{p}$ is evident. Since g is a primitive root of p, $(p-1) \mid u$, but $\gcd(p-1, p^s) = 1$ so $(p-1) \mid h$. Thus $h = p-1$.

$g^u \equiv 1 \pmod{p^k}$ implies that $1 \equiv g^{p^s h} \equiv g^{p^s(p-1)} \pmod{p^k}$ since $h = p-1$. Since, by Lemma 3.73, $g^{p^s(p-1)} = 1 + p^{s+1} a_{s+1}$ with $\gcd(p, a_{s+1}) = 1$, we have

$$1 + p^{s+1} \cdot a_{s+1} \equiv 1 \pmod{p^k}$$
$$p^{s+1} \cdot a_{s+1} \equiv 0 \pmod{p^k}$$
$$p^{s+1} \cdot a_{s+1} = t \cdot p^k \text{ for some integer } t$$

Since $p \nmid a_{s+1}$, $p^k \mid p^{s+1}$ so that $k \leq s+1$ or $k-1 \leq s$. Thus $s = k - 1$.

Therefore, $\text{ord}_{p^k} g = u = p^s h = p^{k-1}(p-1) = \phi(p^k)$ and g is a primitive root of p^k. ∎

Theorem 3.75 *There are primitive roots of integers of the form $2p^k$ where p is an odd prime. If r is a primitive root of p^k, then the choice of $g = r$ is a primitive root of $2p^k$ if r is odd and the choice of $g = r + p^k$ is a primitive root of $2p^k$ if r is even.*

Proof. Let g be a primitive root of p^k. We may assume g is odd since if it is even, $g + p^k$ is an odd primitive root of p^k because $g + p^k$ is congruent to g modulo p^k. Hence, $\gcd(g, 2p^k) = 1$ since g is relatively prime to both 2 and p^k. $\phi(2p^k) = \phi(2)\phi(p^k) = \phi(p^k)$ since $\phi(2) = 1$. If n is the least integer such that $g^n \equiv 1 \pmod{2p^k}$, then n divides $\phi(p^k)$. But since $g^n \equiv 1 \pmod{p^k}$, $\phi(p^k)$ divides n. Hence $n = \phi(p^k) = \phi(2p^k)$ and g is a primitive root of $2p^k$. ∎

The last few theorems completely classify the positive integers that have primitive roots. The results are summarized in the next theorem.

Theorem 3.76 *The only positive integers that have primitive roots are 1, 2, 4, and integers of the forms p^k and $2p^k$ with p an odd prime and k a positive integer.*

Example 3.77 *Suppose that we want to find primitive roots of $11^2 = 121$. Since $r = 7$ is a primitive root of 11 and since $r^{10} = 1 + 11 \cdot 25679568$ and $\gcd(25679568, 11) = 1$, Theorem 3.74 implies that $g = r = 7$ is a primitive root of 11^2. By Theorem 3.75, $g = 7$ is a primitive root of $2 \cdot 11^2 = 242$.*

In addition, $r = 2$ is another primitive root of 11. Since $r^{10} = 1 + 11 \cdot 93$ and $\gcd(93, 11) = 1$, $g = 2$ is a primitive root of 11^2. Since $g = 2$ is even, Theorem 3.75 implies that $g = 2 + 11^2 = 123$ is a primitive root of $2 \cdot 11^2$.

If $p = 29$, then $r = 14$ is a primitive root of p. The integer r^{p-1} is not of the form $1 + pa$ with $\gcd(p, a) = 1$. Thus, by Theorem 3.74, $g = r + p = 43$ is a primitive root of 29^k for k a positive integer. Note that $p = 29$ is the smallest prime having a primitive root r such that r^{p-1} is not of the form $1 + pa$ with $\gcd(p, a) = 1$. The next larger such prime p is $p = 37$ with primitive root $r = 18$.

There is no general procedure for discovering primitive roots besides trial and error; however, if we know one primitive root, a second one may easily be obtained. If g is a primitive root of n, then so is its inverse modulo n. Thus, $g^{\phi(n)} \equiv 1 \pmod{n}$ implies that $g \cdot g^{\phi(n)-1} \equiv 1 \pmod{n}$ so that $g^{-1} \equiv g^{\phi(n)-1} \pmod{n}$ since $\gcd(g, n) = 1$. Clearly, $(g^{-1})^{\phi(n)} \equiv 1 \pmod{n}$. If $(g^{-1})^t \equiv 1 \pmod{n}$ with $1 \leq t < \phi(n)$, then

$$1 \equiv (g^{-1})^t \cdot g^t \equiv 1 \cdot g^t \equiv g^t \pmod{n}$$

a contradiction of the assumption that $\operatorname{ord}_n g = \phi(n)$. Thus, $\operatorname{ord}_n g^{-1} = \phi(n)$ also.

Referring to Example 3.59 where $n = 11$, we saw that $g = 2$ was a primitive root of 11. So $2^{\phi(n)-1} = 2^{10-1} = 2^9 \equiv 6 \pmod{11}$ and $g \cdot g^{-1} = 2 \cdot 6 = 12 \equiv 1 \pmod{11}$. Also from the table in Example 3.59, $\operatorname{ord}_{11} 6 = 10$.

Consider $n = 14$ which was discussed in Examples 3.66 and 3.57. The primitive root $g = 5$ of order 6 has inverse $5^{\phi(n)-1} = 5^{6-1} = 5^5 \equiv 3 \pmod{14}$. From the example table, $\operatorname{ord}_{14} 3 = 6$.

A more general statement about generating primitive roots once one primitive root is known is given by the following theorem.

Theorem 3.78 *Let g be a primitive root of $n > 1$. Then g^s is a primitive root of n if and only if $\gcd(s, \phi(n)) = 1$.*

Proof. We need to show that $\operatorname{ord}_n g^s = \phi(n)$ if and only if $\gcd(s, \phi(n)) = 1$. Since g is a primitive root of n, $\operatorname{ord}_n g = \phi(n)$. Theorem 3.56(e) implies that

$$\operatorname{ord}_n g^s = \frac{\operatorname{ord}_n g}{\gcd(\operatorname{ord}_n g, s)} = \frac{\phi(n)}{\gcd(\phi(n), s)}$$

Thus, $\operatorname{ord}_n g^s = \phi(n)$ if and only if $\gcd(\phi(n), s) = 1$. ∎

We observed above that $g = 2$ is a primitive root of $n = 11$. Since $n = 11$ is prime, $\phi(11) = 11 - 1 = 10$. Thus, for $1 \leq s < \phi(11) = 10$, $1 = \gcd(s, \phi(n)) = \gcd(s, 10)$ if and only if $s = 1, 3, 7$, or 9. But $2^1 = 2$, $2^3 = 8$, $2^7 = 128 \equiv 7$, and $2^9 = 512 \equiv 6$ modulo 11. Thus, $\phi(11 - 1) = \phi(10) = 4$ incongruent primitive roots of $n = 11$ are $2, 6, 7$, and 8. Of course, one of these, 6, was found above and is the multiplicative inverse modulo 11 of $g = 2$.

Exercises

1. Let $0 \leq a_i \leq b_i$ for $i = 1, 2, \ldots, n$ and

$$\sum_{i=1}^{n} a_i = \sum_{i=1}^{n} b_i$$

Prove that $a_i = b_i$ for $1 \leq i \leq n$.

2. Find all primitive roots of:

    (a) 3   **(b)** 5   **(c)** 7

    (d) 10   **(e)** 19   **(f)** 23

3. Find two primitive roots of:

    (a) 7^2 and $2 \cdot 7^2$ **(b)** 19^k and $2 \cdot (19)^k$

    (c) 3^2, 3^3, and $2 \cdot 3^2$ **(d)** 5^4 and $2 \cdot 5^4$

3.8 Indices

Given an integer n, if n has a primitive root r, then $r, r^2, r^3, \ldots, r^{\phi(n)}$ form a complete set of reduced residues modulo n so that r is a generator of the cyclic group of reduced residues modulo n. Under these conditions we can form the equivalent of a logarithm with base r in the following sense.

Definition 3.79 *If b and n are relatively prime and n has a primitive root r, then $b \equiv r^i \pmod{n}$ for only one i, $1 \le i \le \phi(n)$. The exponent i is called the index of b base r modulo n and is denoted by $\text{ind}_r\, b$. Thus, $b \equiv r^{\text{ind}_r\, b} \pmod{n}$ by definition of index.*

The function ind_r is sometimes called the *discrete logarithm* and is, in this case, denoted by $\log_r$ so that we write $b \equiv r^{\log_r\, b} \pmod{n}$. That the function ind_r has many of the properties of real base r logarithm function $\log_r$ is shown by the following theorem.

Theorem 3.80 *If n has a primitive root r and each of a and b is relatively prime to n, then*

(a) $a \equiv b \pmod{n}$ *if and only if* $\text{ind}_r\, a \equiv \text{ind}_r\, b \pmod{\phi(n)}$.

(b) $\text{ind}_r\, 1 = \phi(n)$ *and hence* $\text{ind}_r\, 1 \equiv 0 \pmod{\phi(n)}$.

(c) $\text{ind}_r\, (ab) \equiv \text{ind}_r\, (a) + \text{ind}_r\, (b) \pmod{\phi(n)}$.

(d) $\text{ind}_r\, a^k \equiv k \cdot \text{ind}_r\, a \pmod{\phi(n)}$ *for any positive integer k.*

(e) $\text{ind}_r\, r = 1$ *and hence* $\text{ind}_r\, r \equiv 1 \pmod{\phi(n)}$.

Proof. The proof is left to the reader. ∎

Example 3.81 *Suppose that $n = 13$. Since n is prime, all positive integers less than n are relatively prime to n. Since $r = 6$ is a primitive root of $n = 13$, we will use $r = 6$ as the base modulo 13. The reduced residues of $n = 13$ are* congruent to $r^1, r^2, \ldots, r^{\phi(13)}$ where $\phi(13) = 12$. The indices of the reduced residues may be found simply by calculating r^m modulo n for $m = 1, 2, \ldots, 12$.

m	$[[r^m]]_n$	m	$[[r^m]]_n$
1	6	7	7
2	10	8	3
3	8	9	5
4	9	10	4
5	2	11	11
6	12	12	1

where the residues may be calculated recursively. For example, once $[[6^4]]_{13} = 9$ has been calculated,

$$[[6^5]]_{13} = [[6 \cdot [[6^4]]_{13}]]_{13} = [[6 \cdot 9]]_{13} = [[54]]_{13} = 2$$

By rearranging the table in order of increasing reduced residues, we obtain the following table of indices base 6 modulo 13:

b	$\text{ind}_6\, b$	b	$\text{ind}_6\, b$
1	12	7	7
2	5	8	3
3	8	9	4
4	10	10	2
5	9	11	11
6	1	12	6

$15 \equiv 2 \pmod{13}$ and $\text{ind}_6\, 2 = 5$ so that $\text{ind}_{13}\, 15 = 5$ also. $\text{ind}_6\, 1 = 12 \equiv 0 \pmod{12}$ since $\phi(13) = 12$, in agreement with Theorem 3.80(b). We see that $\text{ind}_{13}\, 3 = 8$, $\text{ind}_6\, 4 = 10$, $3 \cdot 4 = 12$, and $\text{ind}_6\, 12 = 6$. But $\text{ind}_6\, 3 + \text{ind}_6\, 4 = 8 + 10 = 18 \equiv 6 \pmod{12}$ (Theorem 3.80(c)). For $2^3 = 8$, we see that $\text{ind}_6\, 2 = 5$ and $\text{ind}_6\, 8 = 3$ so that $3 \cdot \text{ind}_6\, 2 = 3 \cdot 5 = 15 \equiv 3 \pmod{12}$ in agreement with Theorem 3.80(d). From the table, $\text{ind}_6\, 6 = 1$.

Of course, if we were to choose another base modulo $n = 13$, say the primitive root $r = 2$, then we would obtain a different set of indices than given in the table for base 6.

Example 3.82 *Once a table of indices is known, we can solve exponential congruences in much the same way as we would solve exponential equations when we know the logarithms to some base. Suppose that we want to find all integer solutions for the congruence*

$$5^{2x+1} \equiv 7^{5x+2} \pmod{13}$$

We will use the primitive root 6 as the index base along with the table of relevant indices modulo 13 given in Example 3.81. Note that $\phi(13) = 12$. Using Theorem 3.80(a), a solution x exists if and only if

$$\text{ind}_6\, 5^{2x+1} \equiv \text{ind}_6\, 7^{5x+2} \pmod{12}$$

By Theorem 3.80(d),

$$(2x + 1) \cdot \mathrm{ind}_6 \, 5 \equiv (5x + 2) \cdot \mathrm{ind}_6 \, 7 \pmod{12}$$

Using the table in Example 3.81,

$$
\begin{aligned}
(2x + 1) \cdot 9 &\equiv (5x + 2) \cdot 7 \pmod{12} \\
18x + 9 &\equiv 35x + 14 \pmod{12} \\
-17x &\equiv 5 \pmod{12} \\
7x &\equiv 5 \pmod{12}
\end{aligned}
$$

Since $\gcd(12, 7) = 1$, *which divides 5, there is a unique solution modulo 12, namely* $x \equiv 11 \pmod{12}$ (*Theorem 1.69*).

We can check as follows: $5^{2x+1} \equiv 5^{23} \equiv 6^{23\,\mathrm{ind}_6\,5} \pmod{13}$ *and* $7^{5x+2} \equiv 7^{57} = 6^{57\,\mathrm{ind}_6\,7} \pmod{13}$. *But* $23 \cdot \mathrm{ind}_6 \, 5 \equiv 23 \cdot 9 \equiv 207 \equiv 3 \pmod{12}$ *and* $57 \cdot \mathrm{ind}_6 \, 7 \equiv 57 \cdot 7 \equiv 399 \equiv 3 \pmod{12}$. *Therefore, by Theorem 3.80(a),* $5^{2x+1} \equiv 7^{5x+2} \pmod{13}$ *when* $x \equiv 11 \pmod{12}$.

Example 3.83 *Suppose that we want to solve the congruence*

$$6x \equiv 10 \pmod{14}$$

Since $2 = \gcd(2, 14)$, *Theorem 1.57 implies that the congruence above holds for any integer* x *satisfying*

$$3x \equiv 5 \pmod{7}$$

In contrast to the method of Theorem 1.68, we will solve this second congruence using indices, although one would not ordinarily solve $3x \equiv 5 \pmod{7}$ *using indices.*

In the Exercises of Section 3.7 it was shown that $r = 3$ *is a primitive root of 7. A table of indices for base* $r = 3$ *modulo 7 is easily constructed:*

a	$\mathrm{ind}_3 \, a$
1	6
2	2
3	1
4	4
5	5
6	3

Since $\phi(7) = 6$, *the congruence* $3x \equiv 5 \pmod{7}$ *is equivalent* (*Theorem 3.80*) *to*

$$
\begin{aligned}
\mathrm{ind}_3 \, (3x) &\equiv \mathrm{ind}_3 \, 5 \pmod{6} \\
\mathrm{ind}_3 \, 3 + \mathrm{ind}_3 \, x &\equiv \mathrm{ind}_3 \, 5 \pmod{6} \\
\mathrm{ind}_3 \, x &\equiv \mathrm{ind}_3 \, 5 - \mathrm{ind}_3 \, 3 \pmod{6} \\
&\equiv 5 - 1 \equiv 4 \pmod{6}
\end{aligned}
$$

Thus, from the table above, $x \equiv 4 \pmod{7}$. *So any integer congruent to 4 modulo 7 is a solution for* $3x \equiv 5 \pmod{7}$ *and also a solution to* $6x \equiv$

10 (mod 14), *but we want distinct solutions modulo* 14. *Theorem* 1.75 *implies that* $x \equiv 4$ (mod 7) *only if* $x \equiv 4 + j \cdot 7$ (mod 14) *for some* j, $j = 0, 1$. *Thus, the only solutions modulo* 14 *of* $6x \equiv 10$ (mod 14) *are* $x \equiv 4$ *and* $x \equiv 11$ (mod 14).

The function ind_r maps the set of positive integers relatively prime to n to the set $A = \{1, 2, \ldots, \phi(n)\}$. If the integers are considered together with the operation $\cdot$ and A is considered with the operation of addition modulo $\phi(n)$, then ind_r is a homomorphism from the positive integers to A.

Note that $x^k \equiv b$ (mod n) where $\gcd(b, n) = 1$ if and only if $k \cdot \mathrm{ind}_r x \equiv \mathrm{ind}_r b$ (mod $\phi(n)$). By Theorem 1.68, this congruence has a solution if and only if $d = \gcd(k, \phi(n))$ divides $\mathrm{ind}_r b$. There are d incongruent solutions to $k \cdot \mathrm{ind}_r x \equiv \mathrm{ind}_r b$ (mod $\phi(n)$) and hence d incongruent solutions to $x^k \equiv b$ (mod n). This argument proves the following theorem.

Theorem 3.84 *Let* r *be a primitive root of* n. *Then* $x^k \equiv b$ (mod n), *where* $\gcd(b, n) = 1$, *has a solution if and only if* $d = \gcd(k, \phi(n))$ *divides* $\mathrm{ind}_r b$ *and in this case there are* d *incongruent solutions.*

This theorem leads to the following definition.

Definition 3.85 *Given positive integers* n *and* k *and* $\gcd(n, b) = 1$, *the integer* b *is said to be a* k-th power residue *of* n *if* $x^k \equiv b$ (mod n) *has a solution.*

Theorem 3.86 *If* r *is a primitive root of* n, $\gcd(b, n) = 1$, *and* $d = \gcd(k, \phi(n))$, *then* $x^k \equiv b$ (mod n) *has a solution if and only if* $b^{\phi(n)/d} \equiv 1$ (mod n).

Proof. Let x be a solution of $x^k \equiv b$ (mod n). Since $\gcd(b, n) = 1$, $\gcd(x, n) = 1$ also. Then

$$b^{\phi(n)/d} \equiv \left(x^k\right)^{\phi(n)/d} \equiv \left(x^{\phi(n)}\right)^{k/d} \equiv 1^{k/d} \equiv 1 \pmod{n}$$

Note that n is not required to have a primitive root for this part.

Conversely, if $b^{\phi(n)/d} \equiv 1$ (mod n), then $r^{(\mathrm{ind}_r \, b)\phi(n)/d} \equiv 1$ (mod n). Since $\mathrm{ord}_n r = \phi(n)$, $\phi(n) \mid (\mathrm{ind}_r b)\phi(n)/d$ and, therefore, $d \mid \mathrm{ind}_r b$ because $(\mathrm{ind}_r b)/d$ must be an integer. By Theorem 3.84 $x^k \equiv b$ (mod n) has a solution. ∎

Exercises

1. Prove Theorem 3.80.

2. Generate indices modulo 13 for the base $r = 2$ and compare with those in Example 3.81 for the base $r = 6$.

3. Find all solutions of the following congruences:

(a) $3^{5x} \equiv 11^{3x+1} \pmod{13}$

(b) $3^{4x} \equiv 11^{x+3} \pmod{13}$

(c) $9^x \equiv 2 \pmod{13}$

4. Show that the solutions found in Exercise 3 are the same using indices with base 2 modulo 13 as for indices with base 6 modulo 13.

5. Generate a table of indices for n:

(a) $n = 19$

(b) $n = 23$

6. Let p be an odd prime and r a primitive root of p. Prove that the following three statements are equivalent:

(a) $x^n \equiv b \pmod{p}$ has a solution x.

(b) $n \cdot \mathrm{ind}_r \, x \equiv \mathrm{ind}_r \, b \pmod{(p-1)}$.

(c) $\gcd(n, p-1)$ divides $\mathrm{ind}_r \, b$.

Further, prove that in the case of $\gcd(n, p-1) \mid \mathrm{ind}_r \, b$, there are exactly $\gcd(n, p-1)$ solutions.

7. Use the table of indices obtained in Exercise 5 to solve these congruences:

(a) $3x \equiv 17 \pmod{19}$

(b) $15x \equiv 12 \pmod{57}$

(c) $7x^2 \equiv 8 \pmod{19}$

(d) $x^2 - 6x + 9 \equiv 5 \pmod{19}$

(e) $x^2 - 4x - 7 \equiv 0 \pmod{19}$

(f) $x^8 \equiv 5 \pmod{19}$

8. Let g and r be primitive roots of the prime p and let b be relatively prime to p. Prove that

$$\mathrm{ind}_r \, b \equiv \mathrm{ind}_g \, b \cdot \mathrm{ind}_r \, g \pmod{(p-1)}$$

3.9   Quadratic Residues and the Law of Reciprocity

Consider quadratic congruences of the form

$$ax^2 + bx + c \equiv 0 \ (\text{mod } p)$$

where p is a prime and $p \nmid a$. We require that $p \nmid a$ because if $p \mid a$, then $a \equiv 0 \ (\text{mod } p)$, making the "quadratic" congruence equivalent to the linear congruence $bx + c \equiv 0 \ (\text{mod } p)$, which was solved in Chapter 1. According to Lagrange's Theorem (Theorem 3.34), the quadratic congruence above has at most 2 incongruent solutions modulo p for p prime.

If $p = 2$, then any quadratic congruence $ax^2 + bx + c \equiv 0 \ (\text{mod } 2)$ is equivalent to one of the following four congruences: $x^2 \equiv 0$, $x^2 + 1 \equiv 0$, $x^2 + x \equiv 0$, and $x^2 + x + 1 \equiv 0 \ (\text{mod } 2)$ (see the Exercises). All solutions of these four congruences may be obtained by inspection; thus, the case of integer solutions to quadratic congruences modulo 2 is completely known. We next discuss quadratic congruences modulo p with p an odd prime.

Theorem 3.87 *Let p be an odd prime and $p \nmid a$. Then*

$$ax^2 + bx + c \equiv 0 \ (\text{mod } p)$$

has an integer solution x if and only if

$$y^2 \equiv d \ (\text{mod } p)$$

has an integer solution y, where $d = b^2 - 4ac$. For each solution y of $y^2 \equiv d \ (\text{mod } p)$, a solution x of

$$2ax + b \equiv y \ (\text{mod } p)$$

is a solution of $ax^2 + bx + c \equiv 0 \ (\text{mod } p)$.

Proof. Let x_0 be a solution of $ax^2 + bx + c \equiv 0 \ (\text{mod } p)$. Since p is odd and $p \nmid a$, $p \nmid 4a$ also. Thus, multiplying by $4a$, we obtain

$$4a^2x_0^2 + 4abx_0 + 4ac \equiv 0 \ (\text{mod } p)$$

Adding and subtracting b^2, rearranging, and factoring give

$$(2ax_0 + b)^2 \equiv b^2 - 4ac \ (\text{mod } p)$$

If $y_0 \equiv 2ax_0 + b \ (\text{mod } p)$ and $d = b^2 - 4ac$, then

$$y_0^2 \equiv d \ (\text{mod } p)$$

so that $y^2 \equiv d \ (\text{mod } p)$ has a solution. On the other hand, if $y^2 \equiv d \ (\text{mod } p)$ has a solution, say y_0, then $2ax + b \equiv y_0 \ (\text{mod } p)$ can be solved because

$\gcd(2a, p) = 1$, giving a solution, say, x_0. Thus $(2ax_0 + b)^2 \equiv b^2 - 4ac \pmod{p}$, which is equivalent to $ax_0^2 + bx_0 + c \equiv 0 \pmod{p}$ because $\gcd(4a, p) = 1$. ∎

The quantity $d = b^2 - 4ac$ is called the *discriminant* related to the quadratic form $ax^2 + bx + c$. In the context of complex number solutions of the equation $ax^2 + bx + c = 0$ with a, b, and c real numbers and $a \neq 0$, the discriminant "discriminates" among possible types of solutions. For quadratic congruences modulo and odd prime p, the discriminant also plays a role. If the discriminant $d = b^2 - 4ac$ is a second power residue of p (Definition 3.85) making $y^2 \equiv d \pmod{p}$ solvable, then the corresponding quadratic congruence is solvable. Because of the importance of second power residues, a special terminology is used for them.

Definition 3.88 *If n is a positive integer and $\gcd(n, b) = 1$, then the integer b is a* quadratic residue *of n if and only if $x^2 \equiv b \pmod{n}$ has a solution. An integer b is a* quadratic nonresidue *of n if and only if $x^2 \equiv b \pmod{n}$ has no solution.*

Notice that if $b \equiv 0 \pmod{n}$, then $x^2 \equiv b \pmod{n}$ has a solution; however, b is not a quadratic residue of n because $\gcd(b, n) \neq 1$. For $n = p$, a prime, $\gcd(p, b) = 1$ unless $p \mid b$ or $b \equiv 0 \pmod{p}$. Clearly, $x^2 \equiv 0 \pmod{p}$ has a solution; however, the definition classifies such an integer b as neither a quadratic residue nor a quadratic nonresidue.

Theorem 3.87 could be restated by saying that for p an odd prime with $p \nmid a$, $ax^2 + bx + c \equiv 0 \pmod{p}$ has a solution if and only if $d = 0$ or $d = b^2 - 4ac$ is a quadratic residue of p. Determining whether a given integer is a quadratic residue or quadratic nonresidue will comprise much of the remainder of this section.

Every prime p has a primitive root r. If $p \nmid b$, Theorem 3.84 gave necessary and sufficient conditions for b to be a second power residue modulo p: b is a quadratic residue modulo p if and only if $\gcd(2, p-1) = 2$ divides $\text{ind}_r b$ (since $\phi(p) = p - 1$) if and only if $\text{ind}_r b$ is even. Thus we have the following theorem.

Theorem 3.89 *Let p be prime with primitive root r and let $p \nmid b$. Then b is a quadratic residue of p if and only if the index of b to base r is even. Also, b is a quadratic nonresidue of p if and only if the index of b to base r is odd.*

Thus, if we have a primitive root r of p and a table of indices for base r modulo p, then we can determine whether any quadratic congruence $ax^2 + bx + c \equiv 0 \pmod{p}$ has a solution and can find all the solutions.

Example 3.90 *Suppose that $p = 13$ and we want to solve the congruence*

$$5x^2 - 3x + 7 \equiv 0 \pmod{13}$$

$d = b^2 - 4ac = (-3)^2 - 4(5)(7) = -131 \equiv 12 \pmod{13}$ *is the discriminant. Consider*

$$y^2 \equiv 12 \pmod{13}$$

$r = 6$ *is a primitive root of* $p = 13$ *and the indices base* 6 *modulo* 13 *are given in Example 3.81.* $\text{ind}_6\, 12 = 6$, *which is even; so* 12 *is a quadratic residue of* $p = 13$. *To solve for* y, *we first solve*

$$2 \cdot \text{ind}_6\, y \;\equiv\; \text{ind}_6\, 12 \;(\text{mod}\,(13 - 1))$$
$$2 \cdot \text{ind}_6\, y \;\equiv\; 6 \;(\text{mod}\, 12)$$

The last congruence is solved by considering any solution of

$$\text{ind}_6\, y \equiv 3 \;(\text{mod}\, 6)$$

or

$$w \equiv 3 \;(\text{mod}\, 6)$$

whose only solution modulo 6 *is* $\text{ind}_6\, y = w = 3$. *Thus, by Theorem 1.75, the only solutions of* $2w \equiv 6 \;(\text{mod}\, 12)$ *are*

$$\text{ind}_6\, y = 3 \qquad \text{and} \qquad \text{ind}_6\, y = 3 + 6 = 9$$

Using the table of indices base 6 *modulo* 13,

$$y_0 = 8 \qquad \text{and} \qquad y_1 = 5$$

are the only incongruent least positive residue solutions of $y^2 \equiv 12 \;(\text{mod}\, 13)$. *For illustrative purposes, we used a general method of solving* $ax \equiv b \;(\text{mod}\, n)$, *where* $\gcd(a, n) \neq 1$, *which would not ordinarily be used for such small integers. For example, there are only two cases to consider:* $\text{ind}_6\, y = 3$ *or* 9.

   According to Theorem 3.87, we want to solve $2ax + b \equiv y_i \;(\text{mod}\, 13)$, *that is,* $2(5)x + (-3) \equiv y_0 = 8 \;(\text{mod}\, 13)$ *and* $2(5)x + (-3) \equiv y_1 = 5 \;(\text{mod}\, 13)$. *These two congruences reduce to*

$$10x \equiv 11 \;(\text{mod}\, 13)$$

and

$$10x \equiv 8 \;(\text{mod}\, 13)$$

Since $10 \cdot 4 = 40 \equiv 1 \;(\text{mod}\, 13)$ *shows that* 4 *is the multiplicative inverse of* 10 *modulo* 13, *multiplying the two congruences by* 4 *gives*

$$x \equiv 44 \equiv 5 \;(\text{mod}\, 13) \qquad \text{and} \qquad x \equiv 32 \equiv 6 \;(\text{mod}\, 13)$$

Alternatively, we could have used Theorem 1.68 if the inverse of 10 *were not so obvious. Thus,* $x = 5$ *and* $x = 6$ *are the only incongruent solutions modulo* 13 *of the quadratic* $5x^2 - 3x + 7 \equiv 0 \;(\text{mod}\, 13)$. *These solutions may be checked directly.*

   The distinct indices for a base r of an odd prime p are the integers $1, 2, 3, 4, \ldots, (p - 1)$. Since $(p - 1)$ is even, exactly half of these indices are even and half are odd. Thus we have the following corollary to Theorem 3.89.

Corollary 3.91 *If p is an odd prime, then there are exactly $(p-1)/2$ incongruent quadratic residues and exactly $(p-1)/2$ incongruent quadratic nonresidues of p.*

For $p = 13$ and for primitive root $r = 6$ of p, the table in Example 3.90 gives the corresponding indices. Considering only the primary residues $\{0, 1, 2, \ldots, 12\}$ modulo 13 , the set of quadratic residues of $p = 13$ is $\{1, 3, 4, 9, 10, 12\}$, each element having an even index. The set of quadratic nonresidues of $p = 13$ is $\{2, 5, 6, 7, 8, 11\}$, each element having an odd index.

If p is an odd prime and b is a quadratic residue of p so that $x^2 \equiv b \pmod{p}$ has a solution s, then $p - s$ is also a solution since

$$(p - s)^2 \equiv (-s)^2 \equiv s^2 \equiv b \pmod{p}$$

Further, $s \equiv -s \pmod{p}$ implies that $2s \equiv 0 \pmod{p}$, which implies that $s \equiv 0 \pmod{p}$. Thus, $p \mid s$ and, therefore, $p \mid b$, a contradiction since $\gcd(p, b) = 1$. Thus, $x^2 \equiv b \pmod{p}$ has two incongruent solutions whenever it has one, that is, when b is a quadratic residue. For example, $x^2 \equiv 9 \pmod{13}$ has the immediate solution $x = 3$. That $x = 13 - 3 = 10$ is also a solution is verified by $(10)^2 = 100 \equiv 9 \pmod{13}$.

We saw earlier that having a table of indices of an odd prime p for a primitive root r of p would allow us both to determine whether an arbitrary quadratic congruence modulo p has a solution and to find such solutions. For large primes such tables are nonexistent and may be infeasible to construct; therefore, it would be useful to have a simple test for whether an integer is a quadratic residue for an odd prime p without knowing the index table for a primitive root r of p. The following theorem is an example of such a test.

Theorem 3.92 (Euler's Criterion) *Let p be an odd prime and $p \nmid b$. Then b is a quadratic residue of p if and only if*

$$b^{(p-1)/2} \equiv 1 \pmod{p}$$

and b is a quadratic nonresidue of p if and only if

$$b^{(p-1)/2} \equiv -1 \pmod{p}$$

Proof. If p is an odd prime and $p \nmid b$, then $\gcd(2, \phi(p)) = \gcd(2, p - 1) = 2$ because $p-1$ is even. Theorem 3.86 implies that $x^2 \equiv b \pmod{p}$ has a solution if and only if $b^{(p-1)/2} \equiv 1 \pmod{p}$. But Euler's Theorem 3.53, implies that for any b with $\gcd(p, b) = 1$,

$$b^{(p-1)} - 1 \equiv 0 \pmod{p}$$

Since $p - 1$ is even, we can factor the left side, giving

$$\left(b^{(p-1)/2} + 1\right) \left(b^{(p-1)/2} - 1\right) \equiv 0 \pmod{p}$$

Thus, only one of the following two situations occurs for any integer b with $\gcd(p, b) = 1$:

$$b^{(p-1)/2} \equiv 1 \ (\mathrm{mod}\ p) \qquad \text{or} \qquad b^{(p-1)/2} \equiv -1 \ (\mathrm{mod}\ p)$$

Since $b^{(p-1)/2} \equiv 1 \ (\mathrm{mod}\ p)$ occurs only in case b is a quadratic residue of p, $b^{(p-1)/2} \equiv -1 \ (\mathrm{mod}\ p)$ must occur only when b is a quadratic nonresidue of p. $\blacksquare$

Example 3.93 *Suppose that $p = 13$ and we apply Euler's Criterion to every reduced residue b of the odd prime $p = 13$. The results are given in the following table:*

b	$b^{(p-1)/2}$	$\left[\left[b^{(p-1)/2}\right]\right]_p$	b	$b^{(p-1)/2}$	$\left[\left[b^{(p-1)/2}\right]\right]_p$
1	1	1	7	117649	12
2	64	12	8	262144	12
3	729	1	9	531441	1
4	4096	1	10	1000000	1
5	15625	12	11	1771561	12
6	46625	12	12	2985984	1

The quadratic residues of $p = 13$ are 1, 3, 4, 9, 10, and 12; and the quadratic nonresidues of $p = 13$ are 2, 5, 6, 7, 8, and 11. Comparing these results with the table of indices in Example 3.90 we see that $b^{(p-1)/2} \equiv 1 \ (\mathrm{mod}\ p)$ for $p = 13$ only in case b has an even index and $b^{(p-1)/2} \equiv -1 \ (\mathrm{mod}\ p)$ only in case b has an odd index. Note, however, that just knowing via Euler's Criterion that $x^2 \equiv b \ (\mathrm{mod}\ 13)$ has a solution x or that b is a quadratic residue does not also provide a solution x; but knowing that b is a quadratic nonresidue of $p = 13$ tells us not to search further for such a solution.

Since the character of $b^{(p-1)/2}$ is related to whether b is a quadratic residue, we define an indicator of whether b is a quadratic residue and write it in terms of $b^{(p-1)/2}$.

Definition 3.94 *For an odd prime p, the* Legendre symbol, *$\left(\dfrac{b}{p}\right)$, is defined by*

$$\left(\frac{b}{p}\right) = \begin{cases} 1 & \text{if } b \text{ is a quadratic residue of } p \\ -1 & \text{if } b \text{ is a quadratic nonresidue of } p \\ 0 & \text{if } p \mid b \end{cases}$$

Alternate notations are (b/p) and $(b \mid p)$. In this context, the symbol (b/p) and the other varieties do not mean b divided by p. Euler's Criterion, Theorem 3.92, may be restated in terms of the Legendre symbol as follows.

Theorem 3.95 *If p is an odd prime, then*

$$\left(\frac{b}{p}\right) \equiv b^{(p-1)/2} \ (\mathrm{mod}\ p)$$

The next theorem gives several properties of the Legendre symbol that aid in calculations.

Theorem 3.96 *Let p be an odd prime and $\gcd(p, a) = \gcd(p, b) = 1$.*

(a) *If $a \equiv b \pmod{p}$, then $\left(\dfrac{a}{p}\right) = \left(\dfrac{b}{p}\right)$.*

(b) $\left(\dfrac{ab}{p}\right) = \left(\dfrac{a}{p}\right)\left(\dfrac{b}{p}\right)$.

(c) $\left(\dfrac{a^2}{p}\right) = 1$.

(d) $\left(\dfrac{1}{p}\right) = 1$.

(e) $\left(\dfrac{-1}{p}\right) = \begin{cases} 1 & \text{if } p \equiv 1 \pmod{4} \\ -1 & \text{if } p \equiv 3 \pmod{4}. \end{cases}$

Proof. The proofs of parts (a) to (d) are left to the reader. For part (e), Theorem 3.95 gives

$$\left(\frac{-1}{p}\right) = (-1)^{(p-1)/2}$$

Since p is odd, either $p \equiv 1$ or 3 modulo 4. Thus either $p = 4k + 1$ or $p = 4k + 3$ for some integer k. If $p = 4k + 1$, then $(p - 1)/2 = (4k)/2 = 2k$ so that $(-1)^{2k} = 1$ and -1 is a quadratic residue of p. If $p = 4k + 3$, then $(p - 1)/2 = (4k + 2)/2 = 2k + 1$ so that $(-1)^{(p-1)/2} = (-1)^{2k+1} = -1$, implying that -1 is a quadratic nonresidue of p. ■

Part (d) of Theorem 3.96 states that $x^2 \equiv 1 \pmod{p}$ always has a solution. Of course, $x = 1$ is a solution and so is $p - 1 \equiv -1 \pmod{p}$. Part (e) states that $x^2 \equiv -1 \pmod{p}$ has a solution when $p \equiv 1 \pmod{4}$ and does not have one when $p \equiv 3 \pmod{4}$. For $p = 13 = 4(3) + 1$, $x^2 \equiv -1 \equiv 12 \pmod{13}$ has a solution. From the table in Example 3.90, $\text{ind}_6\, 12 = 6$ and $x = 6^3 \equiv 8 \pmod{13}$ is a solution which is verified by $8^2 = 64 \equiv 12 \equiv -1 \pmod{13}$. Also, $x = p - 8 = 13 - 8 = 5$ is a solution which is verified by $5^2 = 25 \equiv 12 \equiv -1 \pmod{13}$. But since $p = 19 \equiv 3 \pmod{4}$, $x^2 \equiv -1 \pmod{19}$ has no solutions.

Theorem 3.96 provides for the reduction of the integers in the Legendre symbol to smaller positive integers for which the evaluation of the Legendre symbol may be easier. For example

$$\left(\frac{233}{29}\right) =_{3.96(a)} \left(\frac{1}{29}\right) =_{3.96(d)} 1$$

because $233 \equiv 1 \pmod{29}$. The subscripts 3.96(a) and 3.96(d) indicate that parts (a) and (d) of Theorem 3.96 were used to justify the respective equalities.

Thus $x^2 \equiv 233 \pmod{29}$ has a solution.

$$\left(\frac{147}{29}\right) =_{3.96(a)} \left(\frac{2}{29}\right) =_{3.95} 2^{(29-1)/2} = 2^{14} = 16384 \equiv 28 \equiv -1 \pmod{29}$$

so that $x^2 \equiv 147 \pmod{29}$ has no solution. Notice that the process is not unique since we could have written

$$\left(\frac{147}{29}\right) = \left(\frac{7^2 \cdot 3}{29}\right) =_{3.96(b)} \left(\frac{7^2}{29}\right)\left(\frac{3}{29}\right) =_{3.96(c)} \left(\frac{3}{29}\right) \equiv_{3.95} 3^{14}$$

But $3^{14} = 4782969 \equiv 28 \equiv -1 \pmod{29}$ and the result is the same as before. The Law of Quadratic Reciprocity soon to be stated will provide another method to evaluate the Legendre Symbol.

Example 3.97 *Theorem 1.67 proved that if $\{r_1, r_2, \ldots, r_{\phi(n)}\}$ is a reduced residue system modulo n with $1 < r_i < \phi(n)$, then $\{ar_1, ar_2, \ldots, ar_{\phi(n)}\}$ is also a reduced residue system provided that n and a are relatively prime. Thus, if $x_i = [[ar_i]]_n$ so that $0 < x_i < n$, then the integers $x_1, x_2, \ldots, x_{\phi(n)}$ are just a rearrangement of the integers $r_1, r_2, \ldots, r_{\phi(n)}$. A specialization of this property to n an odd prime p by examining the integers x_i yields a remarkable characterization of whether the integer a is a quadratic residue. If $n = p$ is prime, then $\phi(p) = p - 1$ and the least reduced residues, r_i, are the integers $1, 2, 3, \ldots, (p - 1)$, which are all relatively prime to p.*

To illustrate this property, we consider the odd prime $p = 13$ for which $\phi(p) = p - 1$. The following table contains the products $a \cdot i$ for $1 \leq i \leq (p-1)/2$ for each a, $1 \leq a \leq 12$. Each product $a \cdot i$ is reduced to the least positive residue congruent to it: $x_i = [[a \cdot i]]_p$ for $1 \leq i \leq (p-1)/2 = 6$. For example, if $a = 11$ and $i = 4$, then $x_i = [[a \cdot i]]_p = [[11 \cdot 4]]_{13} = [[44]]_{13} = 5$. For each particular a, the last column, γ, counts the number of x_i that are greater than or equal to $(p+1)/2 = 7$.

a	$[[a \cdot 1]]_p$	$[[a \cdot 2]]_p$	$[[a \cdot 3]]_p$	$[[a \cdot 4]]_p$	$[[a \cdot 5]]_p$	$[[a \cdot 6]]_p$	γ
1	1	2	3	4	5	6	0
2	2	4	6	8	10	12	3
3	3	6	9	12	2	5	2
4	4	8	12	3	7	11	4
5	5	10	2	7	12	4	3
6	6	12	5	11	4	10	3
7	7	1	8	2	9	3	3
8	8	3	11	6	1	9	3
9	9	5	1	10	6	2	2
10	10	7	4	1	11	8	4
11	11	9	7	5	3	1	3
12	12	11	10	9	8	7	6

Notice that every integer a that is a quadratic residue of $p = 13$ (see Example 3.90) corresponds to an even value of γ and every integer a that is a quadratic nonresidue of $p = 13$ corresponds to an odd value of γ. This result is the essence of the theorem known as Gauss's Lemma; namely, if p is an odd prime and $\gcd(p, a) = 1$, then $\left(\dfrac{a}{p}\right) = (-1)^\gamma$.

Before proceeding to the statement and proof of Gauss's Lemma, we need to illustrate another property of the residues $x_i = [[a \cdot i]]_p$ for $p = 13$. In this case $(p-1)/2 = 6$ and $(p+1)/2 = 7$. Suppose that $a = 10$ and

$$x_1 = 10,\ x_2 = 7,\ x_3 = 4,\ x_4 = 1,\ x_5 = 11,\ \text{and}\ x_6 = 8$$

As expected, the x_i are distinct least positive reduced residues of $p = 13$. There are 2 of the x_i that are less than $(p+1)/2$ and $\gamma = 4$ of the x_i that are greater than or equal to $(p+1)/2$. Rename those x_i that are less than $(p+1)/2 = 7$ using the notation s_k, and rename those x_i that are greater than or equal to $(p+1)/2 = 7$ using the notation t_k. Let

$$s_1 = x_3 = 4 \ \text{and}\ s_2 = x_4 = 1$$

Which s_k equals which x_i is immaterial except for requiring that $x_i < (p+1)/2$. Let

$$t_1 = x_1 = 10,\ t_2 = x_2 = 7,\ t_3 = x_5 = 11,\ \text{and}\ t_4 = x_6 = 8$$

Thus, we have residues $s_1, s_2, \ldots, s_u$ and $t_1, t_2, \ldots, t_v$ where $u = 2$, $v = 4$, and $u + v = (p-1)/2 = 6$.

We next consider integers of the form $p - t_k$. For $a = 10$, they are

$$p - t_1 = 3,\ p - t_2 = 6,\ p - t_3 = 2,\ \text{and}\ p - t_4 = 5$$

We note that $p - t_k < (p+1)/2 = 7$ for each k and that no integer of the form $p - t_k$ is any one of the s_j. Further, the set $\{s_1, s_2, p - t_1, p - t_2, p - t_3, p - t_4\} = \{4, 1, 3, 6, 2, 5\} = \{1, 2, 3, 4, 5, 6\}$, the set of positive integers less than or equal to $(p-1)/2 = 6$. These properties hold for every odd prime p and integer a with $\gcd(a, p) = 1$ and are summarized in the following lemma.

Lemma 3.98 Let p be an odd prime and let a be an integer with $\gcd(p, a) = 1$. For $1 \leq i \leq (p-1)/2$, let $a \cdot i = w_i \cdot p + x_i$ and $0 \leq x_i < p$. Define $X = \{x_1, x_2, \ldots, x_{(p-1)/2}\}$ to be the set of these remainders. Let $s_1, s_2, \ldots, s_u$ be those elements x_i of X such that $1 < x_i < (p+1)/2$, and let $t_1, t_2, \ldots, t_v$ be those elements x_i of X such that $(p+1)/2 \leq x_i < p$. Then

(a) X contains $(p-1)/2$ different integers and $1 \leq x_i < p$ for all i.

(b) $u + v = (p-1)/2$.

(c) $1 \leq p - t_k \leq (p-1)/2$ for all k.

(d) $p - t_k \neq s_j$ for any k and j.

(e) $\{s_1, s_2, \ldots, s_u, p - t_1, p - t_2, \ldots, p - t_v\} = \{1, 2, 3, \ldots, (p-1)/2\}$.

Proof. (a) For any reduced residue system $\{r_1, r_2, \ldots, r_{(p-1)}\}$ of prime p, the set $\{ar_1, ar_2, \ldots, ar_{(p-1)}\}$ for $\gcd(a, p) = 1$ is also a reduced residue system consisting of $p - 1$ distinct integers each of which is relatively prime to p (Theorem 1.67). Therefore, the set $\{a \cdot 1, a \cdot 2, \ldots, a \cdot (p - 1)\}$ is a complete residue system of $p - 1$ distinct and incongruent integers. Thus, the set

$$\{a \cdot 1, a \cdot 2, \ldots, a \cdot ((p - 1)/2)\}$$

contains $(p - 1)/2$ distinct and incongruent integers. Since $a \cdot i \equiv x_i \pmod{p}$ for $1 \le i \le (p - 1)/2$, $X = \{x_1, x_2, \ldots, x_{(p-1)/2}\}$ consists of $(p - 1)/2$ distinct incongruent integers. Also, each x_i is congruent to a reduced residue and is relatively prime to p, $1 \le x_i < p$ and part (a) is proved.

(b) Certainly, $x_i \ne x_j$ for $i \ne j$. Since an element x_i is either an s_j or a t_k but not both, $u + v = (p - 1)/2$.

(c) $t_k < p$ implies that $0 < p - t_k$ so that $1 \le p - t_k$. Also, $t_k \ge (p + 1)/2$ implies that $p - t_k \le p - (p + 1)/2 = (p - 1)/2$.

(d) Assume that there are integers k and j such that $p - t_k = s_j$. Since $s_j = x_c$ and $t_k = x_d$ for appropriate integers c and d, we have by the Division Algorithm and the definition in the statement of the lemma that

$$a \cdot c \equiv x_c = s_j \pmod{p}$$

and

$$a \cdot d \equiv x_d = t_k \pmod{p}$$

Adding these two congruences and rearranging give

$$a \cdot c + a \cdot d \equiv s_j + t_k \equiv p \pmod{p}$$

or

$$(c + d) \cdot a \equiv 0 \pmod{p}$$

so that $p \mid [(c + d) \cdot a]$. Since $0 < c, d \le (p - 1)/2$, we have $0 < c + d \le p - 1 < p$. If $p \mid (c + d)$, then $c + d = p$, which is a contradiction since $0 < c, d \le (p - 1)/2 < p$. Thus $p \mid a$, which is also a contradiction of the assumption $\gcd(p, a) = 1$ so that $p - t_k \ne s_j$ for all k and j.

(e) We note that the elements of $\{s_1, s_2, \ldots, s_u, p - t_1, p - t_2, \ldots, p - t_v\}$ are distinct, number exactly $(p - 1)/2$, and are all members of the set $\{1, 2, 3, \ldots, (p - 1)/2\}$ which contains exactly $(p - 1)/2$ elements; therefore, part (e) is proved. ∎

Theorem 3.99 (Gauss's Lemma) *If p is an odd prime, a is an integer such that $\gcd(p, a) = 1$, and γ is the number of least positive residues of the integers $a \cdot 1, a \cdot 2, \ldots, a \cdot ((p-1)/2)$ that are greater than or equal to $(p+1)/2$, then*

$$\left(\frac{a}{p}\right) = (-1)^{\gamma}$$

Proof. Let p and a be as given in the theorem and let s_j and t_k be as given in Lemma 3.98. Let v of the lemma be γ. Then, by Lemma 3.98(e),

$$s_1 \cdot s_2 \cdots \cdot s_u \cdot (p - t_1) \cdot (p - t_2) \cdots \cdot (p - t_\gamma) = 1 \cdot 2 \cdot 3 \cdots \cdot [(p-1)/2]$$

where the products on the two sides of the equation have the same factors in, perhaps, a different order. Since $p - t_k \equiv (-1)t_k \pmod{p}$ for each k, the equation above yields the congruence

$$(-1)^\gamma \cdot s_1 \cdot s_2 \cdots \cdot s_u \cdot t_1 \cdot t_2 \cdots \cdot t_\gamma \equiv [(p-1)/2]! \pmod{p}$$

But since $s_1 s \cdots s_u t_1 t_2 \cdots t_\gamma = x_1 x_2 \cdots x_{(p-1)/2}$ and $x_i \equiv a \cdot i \pmod{p}$ for each i,

$$(-1)^\gamma (a \cdot 1)(a \cdot 2) \cdots (a \cdot [(p-1)/2]) \equiv [(p-1)/2]! \pmod{p}$$

Because all the integers less than p are relatively prime to p, we can cancel the factors $1, 2, 3, \ldots,$ and $(p-1)/2$ giving

$$(-1)^\gamma a^{(p-1)/2} \equiv 1 \pmod{p}$$

By multiplying by $(-1)^\gamma$, we get

$$a^{(p-1)/2} \equiv (-1)^\gamma \pmod{p}$$

Since $\left(\dfrac{a}{p}\right) \equiv a^{(p-1)/2} \pmod{p}$ by Theorem 3.95, the proof is complete. ∎

Gauss's Lemma relates whether the integer a is a quadratic residue to a property of the remainders $[[a \cdot i]]_p$ for $1 \leq i \leq (p-1)/2$. In the Division Algorithm $a \cdot i = w_i \cdot p + x_i$, where the remainder x_i and quotient w_i are unique. In Gauss's Lemma we use the x_i, but in the following theorem we use the w_i.

Theorem 3.100 *Let p be an odd prime and a an odd integer such that $\gcd(a, p) = 1$; and let the quotient, w_i, when $a \cdot i$ is divided by p be given by $a \cdot i = w_i \cdot p + x_i$ with $1 \leq x_i < p$ for $1 \leq i \leq (p-1)/2$. Then*

(a) $\left(\dfrac{a}{p}\right) = (-1)^{\sum\limits_{i=1}^{(p-1)/2} w_i}$.

(b) $\left(\dfrac{2}{p}\right) = (-1)^{(p^2-1)/8}$.

Proof. Let $\gcd(p, a) = 1$ for an odd prime p. Let $x_i, s_j, t_k, u,$ and v be as given in Lemma 3.98. Because part (e) of this lemma we can write

$$\sum_{i=1}^{(p-1)/2} i = \sum_{k=1}^{v}(p - t_k) + \sum_{j=1}^{u} s_j = v \cdot p - \sum_{k=1}^{v} t_k + \sum_{j=1}^{u} s_j$$

and

$$\sum_{i=1}^{(p-1)/2} a \cdot i = \sum_{i=1}^{(p-1)/2} (w_i \cdot p + x_i) = \sum_{i=1}^{(p-1)/2} w_i \cdot p + \sum_{j=1}^{u} s_j + \sum_{k=1}^{v} t_k$$

Subtracting these two equations gives

$$(1 - a) \cdot \sum_{i=1}^{(p-1)/2} i = v \cdot p - \sum_{i=1}^{(p-1)/2} w_i \cdot p - 2 \cdot \sum_{k=1}^{v} t_k$$

It was shown previously that $\displaystyle\sum_{i=1}^{n} i = \frac{n(n+1)}{2}$ for any positive integer n so

that $\displaystyle\sum_{i=1}^{(p-1)/2} i = \frac{p^2 - 1}{8}$. Substitution gives

$$(1 - a)\frac{p^2 - 1}{8} = p \cdot \left(v - \sum_{i=1}^{(p-1)/2} w_i \right) - 2 \cdot \sum_{k=1}^{v} t_k$$

But $p \equiv 1 \pmod{2}$ and $2 \equiv 0 \pmod{2}$ so that

$$(1 - a)\frac{p^2 - 1}{8} \equiv v - \sum_{i=1}^{(p-1)/2} w_i \pmod{2}$$

This last congruence holds as long as $\gcd(p, a) = 1$. If a is odd, then $1 - a \equiv 0 \pmod{2}$ so that

$$v \equiv \sum_{i=1}^{(p-1)/2} w_i \pmod{2}$$

By Gauss's Lemma we obtain

$$\left(\frac{a}{p} \right) = (-1)^v = (-1)^{\sum\limits_{i=1}^{(p-1)/2} w_i}$$

which is the result of part (a). We point out that in the argument above, the seemingly abrupt introduction of modulo 2 congruences is appropriate because the value of $(-1)^v$ depends upon whether v is odd or even.

On the other hand, if $a = 2$, then $w_i = 0$ for each i since

$$p - 1 = 2 \cdot \left[\frac{p - 1}{2} \right] \geq 2 \cdot i = w_i p + x_i.$$

Thus

$$v \equiv (-1)\frac{p^2 - 1}{8} \equiv \frac{p^2 - 1}{8} \pmod{2}$$

because $(1 - a) = (-1) \equiv 1 \pmod{2}$. By Gauss's Lemma we obtain

$$\left(\frac{2}{p}\right) = (-1)^{(p^2 - 1)/8}$$

which is part (b). ∎

Theorem 3.101 (Law of Quadratic Reciprocity) *If p and q are differ-ent odd primes, then*

$$\left(\frac{p}{q}\right)\left(\frac{q}{p}\right) = (-1)^{[(p-1)/2][(q-1)/2]}$$

or equivalently,

$$\left(\frac{p}{q}\right) = \left(\frac{q}{p}\right)(-1)^{[(p-1)/2][(q-1)/2]}$$

Proof. From Theorem 3.100,

$$\left(\frac{p}{q}\right)\left(\frac{q}{p}\right) = (-1)^{\left(\sum\limits_{k=1}^{(q-1)/2} w'_k + \sum\limits_{i=1}^{(p-1)/2} w_i\right)}$$

where

$$q \cdot i = w_i p + x_i \quad \text{for } 1 \leq i \leq (p-1)/2$$
$$p \cdot k = w'_k q + x'_k \quad \text{for } 1 \leq k \leq (q-1)/2$$

w_i and w'_k are the quotients and x_i and x'_k are the remainders when $q \cdot i$ and $p \cdot k$ are divided by p and q, respectively. We will show that

$$\sum_{k=1}^{(q-1)/2} w'_k + \sum_{i=1}^{(p-1)/2} w_i = \frac{(p-1)}{2}\frac{(q-1)}{2}$$

There are $\dfrac{(p-1)}{2}\dfrac{(q-1)}{2}$ pairs of integers (i, k) that generate quotients w_i and w'_k.

First, we count the pairs (i, k) for which $q \cdot i < p \cdot k$. For each k, $1 \leq k \leq (q-1)/2$, the allowable values of i are such that $q \cdot i < p \cdot k = w'_k \cdot q + x'_k$. Since $x'_k < q$, we have $q \cdot i \leq w'_k \cdot q$ or, equivalently, $i \leq w'_k$. Since $1 \leq i$ also, there are exactly w'_k such i for this k. Thus, the number of pairs (i, k) with $q \cdot i < p \cdot k$ is $\sum\limits_{k=1}^{(q-1)/2} w'_k$.

Second, in a similar way, we count the pairs (i, k) for which $q \cdot i > p \cdot k$. Thus, for each i, $1 \leq i \leq (p-1)/2$, the allowable values of k are such that $p \cdot k < q \cdot i = w_i \cdot p + x_i$. Since $x_i < p$, we have $p \cdot k \leq w_i \cdot p$ or, equivalently, $k \leq w_i$. Since $1 \leq k$ also, there are exactly w_i such k for this i. Thus, the number of pairs (i, k) with $q \cdot i > p \cdot k$ is $\sum\limits_{i=1}^{(p-1)/2} w_i$.

The third case is the set of (i, k) with $q \cdot i = p \cdot k$. Let i and k be such that $q \cdot i = p \cdot k$. Since p and q are different primes, $q \nmid p$, so it is necessary that $q \mid k$. But $1 \leq k \leq (q-1)/2 < q$ so that $q \nmid k$, a contradiction. Thus, there are no pairs (i, k) with $q \cdot i = p \cdot k$. Hence, the total number of pairs (i, k) is

$$\frac{(p-1)}{2} \frac{(q-1)}{2} = \sum_{k=1}^{(q-1)/2} w'_k + \sum_{i=1}^{(p-1)/2} w_i$$

and the proof is complete. ■

The Law of Quadratic Reciprocity (LQR) is so named by virtue of the "reciprocal" nature of the two Legendre symbols $\left(\dfrac{p}{q}\right)$ and $\left(\dfrac{q}{p}\right)$. We showed $\left(\dfrac{147}{29}\right) = \left(\dfrac{3}{29}\right)$ in the discussion following Theorem 3.96. Continuing but using LQR, we obtain

$$\left(\frac{147}{29}\right) = \left(\frac{3}{29}\right) =_{LQR} \left(\frac{29}{3}\right)(-1)^{14\cdot2} =_{3.96(a)} \left(\frac{2}{3}\right) =_{3.100} (-1)$$

Therefore, 147 is a quadratic nonresidue of 29, which is in agreement with the result obtained earlier.

Euler's Criterion (Theorems 3.92 and 3.95), Theorems 3.96 and 3.100, and the Law of Quadratic Reciprocity (Theorem 3.101) may be used to evaluate $\left(\dfrac{b}{p}\right)$ for any odd prime p in order to settle the question of whether $x^2 \equiv b \pmod{p}$ has a solution.

Example 3.102 *Much of the earlier discussion dealt with determining what integers b are quadratic residues for a given odd prime p. We now reverse the question by asking for what odd primes p has a given b for a quadratic residue. For example, for $b = 19$, find all odd primes p such that $x^2 \equiv 19 \pmod{p}$ has a solution or such that 19 is a quadratic residue of p. We want an odd prime p such that $\left(\dfrac{19}{p}\right) = 1$. Thus,*

$$1 = \left(\frac{19}{p}\right) = \left(\frac{p}{19}\right)(-1)^{[(p-1)/2]\cdot9} = \left(\frac{t}{19}\right)(-1)^{9\cdot(p-1)/2} = \left(\frac{t}{19}\right)(-1)^{(p-1)/2}$$

where $[[p]]_{19} = t$. If $t = t_1$ is a quadratic residue of 19 so that $\left(\dfrac{t_1}{19}\right) = 1$, then $(-1)^{(p-1)/2}$ must be 1 also. Thus, $(p-1)/2 \equiv 0 \pmod{2}$ or $p \equiv 1 \pmod{4}$. If $t = t_2$ is a quadratic nonresidue of 19 so that $\left(\dfrac{t_2}{19}\right) = -1$, then $(-1)^{(p-1)/2}$ must be -1. Thus, $(p-1)/2$ is odd and $p \equiv 3 \pmod{4}$. Therefore, a prime p is suitable provided that either

(a) *$p \equiv t_1 \pmod{19}$ with t_1 a quadratic residue of 19 and $p \equiv 1 \pmod{4}$*

or

(b) $p \equiv t_2 \pmod{19}$ with t_2 a quadratic nonresidue of 19 and $p \equiv 3 \pmod 4$.

The two simultaneous congruences in (a) may be solved using the Chinese Remainder Theorem (Theorem 3.1) giving

$$p \equiv 57 + 20 \cdot t_1 \pmod{76}$$

for any quadratic residue t_1 of 19 with $1 \le t_1 < 19$. Similarly, the solution for the two simultaneous congruences in (b) is

$$p \equiv 19 + 20 \cdot t_2 \pmod{76}$$

for any quadratic nonresidue t_2 of 19 with $1 \le t_2 < 19$.

If t_1 *is a quadratic residue of 19, then $19 - t_1$ is a quadratic nonresidue of 19 (see the Exercises). Thus, as t_1 ($1 \le t_1 < p$) runs through the quadratic residues of 19, $t_2 = 19 - t_1$ runs through the quadratic nonresidues of 19 and $1 \le t_2 < p$. The solution of (b) can be written*

$$
\begin{aligned}
p &\equiv 19 + 20 t_2 &\equiv&\ 19 + 20(19 - t_1) \\
&\equiv 19 - 20 t_1 &\equiv&\ -57 - 20 t_1 \pmod{76}
\end{aligned}
$$

because $19 \equiv -57 \pmod{76}$. Any odd prime p that is congruent to

$$p \equiv \pm(57 + 20 t_1) \pmod{76}$$

for some quadratic residue t_1 of 19 with $1 \le t_1 < p$ has the property that $x^2 \equiv 19 \pmod p$ has a solution. Since the quadratic residues of 19 are 1, 4, 5, 6, 7, 9, 11, 16, and 17 (see the Exercises of Section 3.8), t_1 may be any one of these 9 integers.

For example, for $t_1 = 1$, if p is prime and $p \equiv 57 + 20 = 77 \equiv 1 \pmod{76}$, then $x^2 \equiv 19 \pmod p$ is solvable. We, therefore, examine primes p of the form $1 + k \cdot 76$. $p = 229$, 457, and 761, generated by $k = 3$, 6, and 10, are three such primes. Next, selecting the negative sign, any prime p with $p \equiv -57 - 20 = -77 \equiv 75 \pmod{76}$ also has $x^2 \equiv 19 \pmod p$ solvable. Checking $p = 457$ and $p = 151$, we obtain

$$\left(\frac{19}{457} \right) = \left(\frac{457}{19} \right) (-1)^{228 \cdot 9} = \left(\frac{1}{19} \right) = 1$$

and

$$\left(\frac{19}{151} \right) = \left(\frac{151}{19} \right)(-1)^{75 \cdot 9} = \left(\frac{18}{19} \right)(-1) = \left(\frac{-1}{19} \right)(-1) = (-1)(-1) = 1$$

Mersenne numbers, $M(n) = 2^n - 1$, were introduced in Section 2.3. We were interested in what n made $M(n)$ prime. Only prime n need be considered since if $M(n) = 2^n - 1$ is prime, then so is n. If $2^n - 1$ is composite, then it is divisible by some integer m. We will show that any divisor m of $M(n)$ must be of the form $m = \pm 1 \pmod 8$.

Theorem 3.103 *For an odd prime q,*

(a) $q \mid (2^{(q-1)/2} - 1)$ *if and only if* $2^{(q-1)/2} \equiv 1 \pmod{q}$ *if and only if* $q \equiv \pm 1 \pmod 8$.

(b) $q \mid (2^{(q-1)/2} + 1)$ *if and only if* $2^{(q-1)/2} \equiv -1 \pmod{q}$ *if and only if* $q \equiv \pm 3 \pmod 8$.

Proof. (a) Using the symbol "$\Leftrightarrow$" for "if and only if",

$$q \mid (2^{(q-1)/2} - 1) \quad \Leftrightarrow \quad 2^{(q-1)/2} \equiv 1 \pmod q$$

$$\Leftrightarrow \quad \left(\frac{2}{q}\right) = 1$$

$$\Leftrightarrow \quad (q^2 - 1)/8 \equiv 0 \pmod 2$$

$$\Leftrightarrow \quad q^2 - 1 \equiv 0 \pmod{16}$$

$$\Leftrightarrow \quad (q-1)(q+1) \equiv 0 \pmod{16}$$

$$\Leftrightarrow \quad q \equiv 1 \pmod{16} \text{ or } q \equiv -1 \pmod{16}$$

Since both $q - 1$ and $p - 1$ are even.

(b) The proof is similar and is left to the reader. ∎

Part (a) means that 2 is a quadratic residue for primes of the form $8k+1$, and part (b) means that 2 is a quadratic nonresidue for primes of the form $8k + 3$.

Theorem 3.104 *If p is an odd prime and* $d \mid M(p) = 2^p - 1$, *then* $d \equiv \pm 1 \pmod 8$.

Proof. Let p be an odd prime and $d \mid M(p) = 2^p - 1$. Since d is odd, let q be a prime factor of d. Then q must be odd also, so let $q = 2Q + 1$. Since $q \mid (2^p - 1)$ we have, equivalently, $2^p \equiv 1 \pmod q$ and $2^{p+1} \equiv 2 \pmod q$ when multiplied by 2. If $T = 2^{(p+1)/2}$, we obtain

$$T^2 \equiv 2 \pmod q$$

Thus, 2 is a quadratic residue of q so that $\left(\frac{2}{q}\right) = 1$, which implies that $q \equiv \pm 1 \pmod 8$. Therefore, d is the product of prime factors of the form $q \equiv \pm 1 \pmod 8$ so that $d \equiv \pm 1 \pmod 8$ also. ∎

Gauss made many contributions to number theory, including congruence and the first proof of the Law of Quadratic Reciprocity, which was first stated, but not proved, by Legendre. Carl Friedrich Gauss (1777-1855) was born in Brunswick, Germany. He was the son of a day laborer. Some sources say that Gauss's father was a bricklayer; others say that he was a canal foreman, gardener, and street butcher. His father was opposed to Gauss's receiving an

education and wished him to follow in his trade (whatever it was). It is said
that Gauss's father made him go to bed early in the winter to save on light
and fuel but that Gauss made a candle out of a hollowed-out turnip so that
he could study. Gauss was too brilliant to remain undiscovered. He taught
himself reading and arithmetic. It is said that he found an error in his father's
ledgers at age 3. In his first mathematics class the teacher, J. G. Büttner,
in order to keep his class occupied, asked students to sum the integers from
1 to 100. Gauss solved the problem in seconds having discovered a formula
for summing the first n integers. Büttner redeemed himself in history for
assigning such a drudging problem by ordering special mathematics books for
Gauss and having his assistant, John Martin Bartells, tutor him.

Gauss was brought to the attention of the Duke of Brunswick, who became
his patron until his death. This support allowed Gauss to attend Collegium
Carolinum, a college preparatory school, from 1792 to 1795. During this period he formulated the method of least squares. In 1795 he went to Göttingen
but had not decided between studying philosophy or mathematics. At nine-
teen, while at Göttingen, he discovered that a regular 17-sided polygon could
be constructed using only straight edge and compass. This discovery, which
had been unsolved for centuries, made Gauss decide to enter mathematics.
Apparently, the teaching of mathematics at Göttingen during this period was
rather poorly done; so Gauss worked alone and proved some of his greatest
theorems. After completing his course at Göttingen in 1798, Gauss returned
to Brunswick, where he was able to work supported by his patron. Gauss
was granted his doctorate in absentia in 1801. His dissertation was nomi-
nally written under Johann Frederick Pfaff, the best known mathematician
in Germany, whom he met at a library. His dissertation was the proof of
the Fundamental Theorem of Algebra (every polynomial of positive degree
n with real coefficients has exactly n solutions over the complex field). He
later produced several other proofs of this theorem, using complex numbers
as coefficients in the final proof. When the Duke of Brunswick died, Gauss
accepted a position at Göttingen. He decided to take the chair in astronomy
so he would have more time to work. Gauss did not like to teach but pre-
ferred to do mathematics; however, he had several famous students, including
Möbius, who will be discussed in Chapter 4. After accepting this chair, Gauss
only left Göttingen twice, once to attend a meeting.

When he was twenty four years old, he began work on *Disquisitiones Arithmeticae*, which established number theory as a branch of mathematics. In this
fantastic book he established congruence of numbers, including proving the
Law of Quadratic Reciprocity. Gauss gave eight proofs of the Quadratic Reciprocity Law over a period of seventeen years. He also showed that a regular
polygon of p sides can be constructed with straightedge and compass if p is of
the form $2^{2^n} - 1$ and proved a generalization of Wilson's Theorem. The book,
which contains too many important theorems to list here, was divided into
seven chapters and became known as the book of seven seals because it was so
difficult to read. Fortunately, Peter Gustav Lejeune-Dirichlet broke the seals
when he rewrote the contents of the book in readable form as *Vorlesungen*

über Zahlentheorie. It is said that due to the bankruptcy of a publisher, only a few copies of Gauss's book were available. Even some of his students didn't have a copy. Dirichlet slept with his copy under his pillow.

When Gauss, at the age of 19, discovered that a seventeen-sided regular polygon could be constructed using straightedge and compass, he began keeping a scientific diary with this theorem as his first entry. During the years he kept all of his theorems and proofs in it, but he published relatively few of them. One of the most famous unpublished entries in his journal was the discovery of non-Euclidean geometry and Gaussian curvature, which resulted in the theory of relativity. Gauss's notes were not published until 1901. Many theorems proved by mathematicians during the period before this had already been proven by Gauss.

Gauss, known as the Prince of Mathematics, was equally well known in physics, mechanics, and theoretical astronomy. He was able to calculate the orbits so accurately that one small planet was rediscovered using his calculations. He invented the heliotrope and, along with Weber, a physicist at Göttingen, invented and used the first telegraph at a distance of a mile.

Gauss's last two doctoral students were Riemann and Wilhelm Dedekind (1831-1916). Dedekind is best known for the concepts of Dedekind cuts, which are used to formally define irrational numbers. In algebra, he formally defined the ring, field, and ideal of a ring.

As mentioned above, Legendre first stated the Law of Quadratic Reciprocity. Discussions in this section related to the Reciprocity Law used the Legendre symbol. Adrien-Marie Legendre (1752-1833) was the son of wealthy parents who lived in the south of France but lived most of his life in Paris. He was educated at Collége Mazarin in Paris and was Professor of Mathematics at École Militaire for five years. He resigned to devote more time to his research. Two years later he won a prize at the Berlin Academy for an essay on the path of a projectile. He was appointed to the Académie des Science, where he remained until its closing ten years later. He refused to yield to the government when they tried to dictate to the Academy. He lost his pension and died in poverty.

Legendre was not particularly liked or respected although he was a great mathematician. He felt that he was not given the credit he deserved and he was probably right. He was particularly irritated with Gauss. Gauss felt that if he had demonstrated a proof of a theorem in his diary, then it was his. Legendre claimed that the person who published the proof should get the credit. He first stated the theorem that became known as Legendre's Law of Quadratic Reciprocity but could not prove it. Gauss proved it and ignored Legendre's contributions. Legendre published the first proof of the method of least squares, which has been said to be the first satisfactory proof. Again Gauss claimed prior credit.

Legendre is best known for his book on Euclidean geometry, *Èléments de géometrie*, which is considered to be one of the best textbooks ever written. Also, his book *Théorie des nombres* was considered a dominant work on the subject. He worked in many areas but was mainly known for his work in number theory, celestial mechanics, and theory of elliptic functions. His other work in number theory included proving Fermat's Last Theorem for $n = 5$ and estimating the distribution of the primes. He was one of three commissioners to supervise triangulations necessary for determining the standard meter.

Another anecdote which may indicate some of Legendre's unpopularity involved the mathematician Niels Henrik Abel. Abel had the misfortune of making great discoveries to which no one would pay any attention. One of his major papers was submitted to the French Academy which gave it to Augustin-Louis Cauchy and Legendre to examine. Cauchy laid it aside to do his own work and Legendre apparently forgot about it. After Abel's death, the mathematician Jacobi asked Legendre about the paper. Legendre claimed that the paper was badly written and illegible and that he had returned it to Abel to rewrite, which he never did.

Exercises

1. Show that any quadratic congruence of the form $ax^2 + bx + c \equiv 0 \pmod 2$ is equivalent to one of the following four congruences:

   (a) $x^2 \equiv 0 \pmod 2$

   (b) $x^2 + 1 \equiv 0 \pmod 2$

   (c) $x^2 + x \equiv 0 \pmod 2$

   (d) $x^2 + x + 1 \equiv 0 \pmod 2$

   Find all solutions modulo 2 for each of the congruences above.

2. Determine whether these congruences have any solutions. If so, find all the solutions.

   (a) $2x^2 + 8x + 5 \equiv 0 \pmod{13}$

   (b) $3x^2 - 11x - 5 \equiv 0 \pmod{13}$

   (c) $8x^2 + 2x + 5 \equiv 0 \pmod{13}$

   (d) $5x^2 + 7 \equiv 0 \pmod{13}$

   (e) $16x^2 + x + 10 \equiv 0 \pmod{19}$

   (f) $-x^2 + x + 1 \equiv 0 \pmod{19}$

   (g) $15x^2 - x + 9 \equiv 0 \pmod{19}$

3. Classify every integer a, $1 \le a < 19$ as to whether it is a quadratic

residue or nonresidue modulo 19 by examining a table of indices (see the Exercises of Section 3.8).

4. Use Euler's Criterion to determine whether there exist solutions to

(a) $x^2 \equiv 7 \pmod{11}$

(b) $x^2 - 18 \equiv 0 \pmod{41}$

5. Prove parts (a) to (d) of Theorem 3.96.

6. Evaluate each of the following Legendre symbols using Gauss's Lemma (Theorem 3.99):

(a) $\left(\dfrac{3}{19}\right)$ (b) $\left(\dfrac{9}{19}\right)$ (c) $\left(\dfrac{16}{23}\right)$

7. Evaluate each of the following Legendre symbols using the formulas and methods of Theorem 3.100:

(a) $\left(\dfrac{3}{19}\right)$ (b) $\left(\dfrac{9}{19}\right)$ (c) $\left(\dfrac{16}{23}\right)$ (d) $\left(\dfrac{2}{1777}\right)$

8. Evaluate:

(a) $\left(\dfrac{21}{23}\right)$ (b) $\left(\dfrac{30}{31}\right)$ (c) $\left(\dfrac{16}{31}\right)$

(d) $\left(\dfrac{165}{73}\right)$ (e) $\left(\dfrac{157}{41}\right)$

9. Prove that $\displaystyle\sum_{i=1}^{p-1} \left(\dfrac{i}{p}\right) = 0$ for p an odd prime.

10. Let p be an odd prime and b be a quadratic residue of p.

(a) Prove that $p - b$ is a quadratic residue if and only if $\left(\dfrac{-1}{p}\right) = 1$ if and only if $p \equiv 1 \pmod 4$.

(b) Prove that $p - b$ is a quadratic nonresidue if and only if $\left(\dfrac{-1}{p}\right) = -1$ if and only if $p \equiv 3 \pmod 4$.

(c) For $p = 13$ for which $p \equiv 1 \pmod 4$, show by direct calculation (see the Exercises of Section 3.8) that the pairs 1 and 12, 2 and 11, 3 and 10, etc. are alike with respect to the quadratic residue characteristic.

(d) For $p = 19$ for which $p \equiv 3 \pmod 4$, show by direct calculation (see the Exercises of Section 3.8) that the pairs 1 and 18, 2 and 17, 3

and 16, and so on, have one quadratic residue and one quadratic
nonresidue.

11. Prove Theorem 3.103(b).

12. Let $q > 3$ be an odd prime and let $q = 2Q + 1$. Prove that:

    (a) $q \mid (3^Q - 1)$ if and only if $q \equiv \pm 1 \pmod{12}$.

    (b) $q \mid (3^Q + 1)$ if and only if $q \equiv \pm 5 \pmod{12}$.

13. For a prime $p > 3$, prove that the sum of the quadratic residues of p is divisible by p. $\left[\textit{Hint: } \sum_{i=1}^{n} i^2 = n(n+1)(2n+1)/6. \right]$

14. Let p be an odd prime and $\gcd(p, a) = \gcd(p, b) = 1$. Prove that ab is a quadratic residue only in case both of a and b are quadratic residues or both are quadratic nonresidues.

15. Find an example of two quadratic nonresidues a and b of the nonprime 12 such that ab is not a quadratic residue of 12.

16. Determine the form of all odd primes for which 5 is a quadratic residue.

17. Determine the form of all odd primes for which 7 is a quadratic residue.

18. Given a prime p of the form $4k + 3$, prove that $(3p - 1)/4$ is a quadratic nonresidue of p.

3.10   Jacobi Symbol

In Section 3.9 we developed tools to determine whether an integer b is a quadratic residue modulo n. In particular we wanted to know whether $x^2 \equiv b \pmod{n}$ had a solution when n is a prime p and b and p are relatively prime. The Legendre symbol $\left(\dfrac{b}{p} \right)$, which equals 1 when b is a quadratic residue of p and equals -1 when b is a quadratic nonresidue of p, was introduced. Several of the tools were in the form of formulas that facilitated evaluation of $\left(\dfrac{b}{p} \right)$. In this section we introduce the Jacobi symbol, which generalizes the Legendre symbol $\left(\dfrac{b}{p} \right)$ to cases when $p > 1$ is odd but not prime. Many of the properties of the Legendre symbol are similar to those of the Jacobi symbol.

Definition 3.105 *For an odd integer Q, let $Q = \displaystyle\prod_{i=1}^{s} q_i$ be the factorization of Q into primes q_i where the q_i are not necessarily different primes; and let*

b be an integer such that $\gcd(b, Q) = 1$. The Jacobi symbol is defined by

$$\left(\frac{b}{Q}\right)_J = \prod_{i=1}^{s} \left(\frac{b}{q_i}\right)_L$$

where the subscripts J and L refer to the Jacobi and Legendre symbols, respectively.

Clearly, if Q is an odd prime, then

$$\left(\frac{b}{Q}\right)_J = \left(\frac{b}{q_1}\right)_L = \left(\frac{b}{Q}\right)_L$$

Thus we have proved the next theorem.

Theorem 3.106 *If p is an odd prime and $\gcd(b, p) = 1$, then*

$$\left(\frac{b}{p}\right)_J = \left(\frac{b}{p}\right)_L$$

Because of Theorem 3.106, we will usually drop the subscripts J and L and simply write

$$\left(\frac{b}{Q}\right) = \prod_{i=1}^{s} \left(\frac{b}{q_i}\right)$$

in Definition 3.105 applying either the Jacobi or Legendre definition according as Q is composite or prime.

The next theorem establishes some properties of the Jacobi symbol that are similar to those of the Legendre symbol.

Theorem 3.107 *Let $P > 1$ and $Q > 1$ be odd integers and let a and b be integers with $\gcd(ab, PQ) = 1$. Then*

(a) *If $a \equiv b \pmod{P}$, then $\left(\dfrac{a}{P}\right) = \left(\dfrac{b}{P}\right)$.*

(b) $\left(\dfrac{ab}{P}\right) = \left(\dfrac{a}{P}\right)\left(\dfrac{b}{P}\right)$.

(c) $\left(\dfrac{a}{PQ}\right) = \left(\dfrac{a}{P}\right)\left(\dfrac{a}{Q}\right)$.

(d) $\left(\dfrac{a^2}{P}\right) = \left(\dfrac{a}{P^2}\right) = 1$.

(e) $\left(\dfrac{1}{P}\right) = 1$.

(f) *If the prime factorization of P is $\prod_{i=1}^{t} p_i^{\alpha_i}$, where the p_i are distinct primes,*

then

$$\left(\frac{a}{P}\right) = \prod_{i=1}^{t} \left(\frac{a}{p_i}\right)^{\alpha_i}$$

Proof. If $\gcd(ab, PQ) = 1$, then each of a or b is relatively prime to each of P or Q.

(a) If $P = \prod_{i=1}^{s} q_i$ is the factorization of P into primes according to Definition 3.105, then

$$\left(\frac{a}{P}\right) = \prod_{i=1}^{s} \left(\frac{a}{q_i}\right) = \prod_{i=1}^{s} \left(\frac{b}{q_i}\right) = \left(\frac{b}{P}\right)$$

where we have applied Theorem 3.96(a) to obtain $\left(\frac{a}{q_i}\right) = \left(\frac{b}{q_i}\right)$ for all i.

The proofs of parts (b) to (f) are direct applications of Theorem 3.96 and are left to the reader. ∎

Even though the Jacobi symbol is a generalization of the Legendre symbol, the close association of the value of the Legendre symbol $\left(\frac{b}{q}\right)$ with whether b is a quadratic residue of prime q is not maintained with the Jacobi symbol $\left(\frac{b}{Q}\right)$. Assume that $Q = \prod_{i=1}^{s} q_i > 1$ is odd and $\gcd(b, Q) = 1$. We note that $\left(\frac{b}{Q}\right) = \pm 1$ because $\left(\frac{b}{Q}\right)$ is the product of Legendre symbols, each of which is ± 1. If $\left(\frac{b}{Q}\right) = -1$, then $\left(\frac{b}{q_i}\right) = -1$ for some i; and therefore, b is a quadratic nonresidue of q_i. Thus, in this case, $x^2 \equiv b \pmod{q_i}$ has no solution; therefore, $x^2 \equiv b \pmod{Q}$ has no solution. Consequently , b cannot be a quadratic residue of Q. Thus, $\left(\frac{b}{Q}\right) = -1$ implies that b is not a quadratic residue of Q.

On the other hand, if $\left(\frac{b}{Q}\right) = 1$ we cannot necessarily conclude that b is a quadratic residue of Q. For example,

$$\left(\frac{2}{33}\right) = \left(\frac{2}{3}\right)\left(\frac{2}{11}\right) = (-1)^{(9-1)/8}(-1)^{(121-1)/8} = (-1)^{16} = 1$$

but it is easy to show that $x^2 \equiv 2 \pmod{33}$ has no solution since $x^2 \equiv 2 \pmod 3$ has no solution. Thus, $\left(\frac{b}{Q}\right) = 1$ does not guarantee that b is a quadratic residue of Q. The Jacobi symbol, therefore, does not perfectly generalize this aspect of the Legendre symbol; however, the Jacobi symbol as defined does generalize the formulas related to the Legendre symbol, including the Reciprocity Law.

Since several of the remaining properties of the Jacobi symbol are related to powers of (-1), congruence modulo 2 is important. The following lemma will be useful.

Lemma 3.108 *If $P > 1$ is odd and $P = \prod_{i=1}^{t} p_i^{\alpha_i}$ is the prime factorization of P, then*

(a) $P \equiv 1 + \sum_{i=1}^{t} \alpha_i(p_i - 1) \pmod{4}$.

(b) $\dfrac{P-1}{2} \equiv \sum_{i=1}^{t} \alpha_i \left(\dfrac{p_i - 1}{2} \right) \pmod{2}$.

(c) $P^2 \equiv 1 + \sum_{i=1}^{t} \alpha_i(p_i^2 - 1) \pmod{64}$.

(d) $\dfrac{P^2-1}{2} \equiv \sum_{i=1}^{t} \alpha_i \left(\dfrac{p_i^2 - 1}{8} \right) \pmod{8}$.

Proof. (a) Let $t = 1$. Then using the Binomial Theorem we obtain

$$
\begin{aligned}
P &= p_1^{\alpha_1} = [1 + (p_1 - 1)]^{\alpha_1} \\
&= 1 + \alpha_1(p_1 - 1) + \binom{\alpha_1}{2}(p_1 - 1)^2 + \cdots \\
&\equiv 1 + \alpha_1(p_1 - 1) \pmod{4}
\end{aligned}
$$

because $(p_1 - 1)$ is even and so $(p_1 - 1)^k \equiv 0 \pmod{4}$ for $k \geq 2$.

Next, assume that part (a) holds for *any* prime factorization involving $t = r$ distinct primes. If the prime factorization of $P = \prod_{i=1}^{r+1} p_i^{\alpha_i}$ contains $r + 1$ distinct primes, then

$$
\begin{aligned}
P &= \prod_{i=1}^{r+1} p_i^{\alpha_i} = \left[\prod_{i=1}^{r} p_i^{\alpha_i} \right] [p_{r+1}^{\alpha_{r+1}}] \\
&\equiv \left[1 + \sum_{i=1}^{r} \alpha_i(p_i - 1) \right] [1 + \alpha_{r+1}(p_{r+1} - 1)] \pmod{4}
\end{aligned}
$$

by the induction hypothesis for $t = r$ and $t = 1$

$$
\equiv 1 + \sum_{i=1}^{r+1} \alpha_i(p_i - 1) \pmod{4}
$$

since $(p_i - 1)(p_{r+1} - 1) \equiv 0 \pmod 4$ for each i. Thus, part (a) holds by mathematical induction.

Part (b) follows directly from part (a). The proofs of parts (c) and (d) are left to the reader. ∎

Theorem 3.109 *If $P > 1$ and $Q > 1$ are odd and $\gcd(P, Q) = 1$, then*

(a) $\left(\dfrac{-1}{P}\right) = (-1)^{(P-1)/2}$.

(b) $\left(\dfrac{2}{P}\right) = (-1)^{(P^2-1)/8}$.

(c) $\left(\dfrac{P}{Q}\right)\left(\dfrac{Q}{P}\right) = (-1)^{[(P-1)/2][(Q-1)/2]}$ *or* $\left(\dfrac{P}{Q}\right) = \left(\dfrac{Q}{P}\right)(-1)^{[(P-1)/2][(Q-1)/2]}$.

Proof. (a) Let P have prime factorization $\displaystyle\prod_{i=1}^{t} p_i^{\alpha_i}$. Using Theorems 3.107(f) and 3.95 we obtain

$$
\begin{aligned}
\left(\frac{-1}{P}\right) &= \prod_{i=1}^{t}\left(\frac{-1}{p_i}\right)^{\alpha_i} = \prod_{i=1}^{t}(-1)^{[(p_i-1)/2]\alpha_i} \\
&= (-1)^{\sum_{i=1}^{t}\alpha_i(p_i-1)/2} = (-1)^{(P-1)/2}
\end{aligned}
$$

where we have applied Lemma 3.108(b).

(b) This proof is similar to that of part (a) except that Theorem 3.100(b) is used. The details are left to the reader.

(c) This proof is also similar to that of part (a) except that the Law of Quadratic Reciprocity (Theorem 3.101) is used. The details are also left to the reader. ∎

Example 3.110 *The evaluation of $\left(\dfrac{b}{p}\right)$ with p an odd prime and $\gcd(b, p) = 1$ can, of course, be done using the Legendre symbol and related theorems of Section 3.9; however, if $\left(\dfrac{b}{p}\right)$ is evaluated using the Jacobi symbol, the result is often obtained more quickly.*

First, using methods of Section 3.9 we have, since 1993 is prime,

$$
\left(\frac{231}{1993}\right) = \left(\frac{3}{1993}\right)\left(\frac{7}{1993}\right)\left(\frac{11}{1993}\right) = (1)(-1)(-1) = 1
$$

because

$$
\left(\frac{3}{1993}\right) =_{LQR} \left(\frac{1993}{3}\right) =_{3.96(a)} \left(\frac{1}{3}\right) = 1
$$

and because, similarly, $\left(\dfrac{7}{1993}\right) = -1$ *and* $\left(\dfrac{11}{1993}\right) = -1$.

Alternately, using the Jacobi symbol we obtain

$$\left(\frac{231}{1993}\right) = \left(\frac{1993}{231}\right) =_{3.107(a)} \left(\frac{4}{231}\right) = \left(\frac{2^2}{231}\right) =_{3.107(d)} 1$$

The application of Theorems 3.107 and 3.109 to evaluate $\left(\dfrac{b}{p}\right)$ can vary so that which theorem to apply must be decided upon at each point in the calculation. It would be desirable to have a more direct procedure — one that would not require choosing among so many alternatives at each stage. Such an procedure is provided by the next two theorems which depend upon a Euclidean-like algorithm (see Theorem 1.33).

For integers a and b such that $a > b > 1$ and $\gcd(a, b) = 1$, the Division Algorithm guarantees that there are positive integers q and r with $0 < r < b$ such that

$$a = bq + r$$

If we factor the highest power of 2 from r, we obtain

$$r = 2^\alpha s$$

where s is odd; and if r is even, then $\alpha \geq 1$, otherwise $\alpha = 0$. In addition, $\gcd(b, s) = 1$; for otherwise, we would have $\gcd(a, b) \neq 1$. Then, using only the "odd" part s of r and provided that $s > 1$, we can apply the Division Algorithm again and divide b by s giving

$$b = sq' + 2^{\alpha'} r'$$

where $q' > 0$, r' is odd, and $r' < s < r$. Also, $\gcd(s, r') = 1$. Thus, a straightforward induction gives the next theorem whose proof is left to the reader (compare the proof of Theorem 1.33).

Theorem 3.111 *Let a and b be positive integers with $a > b > 1$ and $\gcd(a, b) = 1$. If $s_0 = a$ and $s_1 = b$, then repeated application of the Division Algorithm produces the following equations:*

$$\begin{aligned}
s_0 &= s_1 q_1 + 2^{\alpha_1} s_2 \\
s_1 &= s_2 q_2 + 2^{\alpha_2} s_3 \\
&\vdots \\
s_{n-3} &= s_{n-2} q_{n-2} + 2^{\alpha_{n-2}} s_{n-1} \\
s_{n-2} &= s_{n-1} q_{n-1} + 2^{\alpha_{n-1}} s_n \\
&= s_{n-1} q_{n-1} + 2^{\alpha_{n-1}} \cdot 1
\end{aligned}$$

where

(a) $q_i > 0$ *is the quotient and* $2^{\alpha_i} s_{i+1}$ *the remainder when s_{i-1} is divided by s_i for $1 \leq i \leq n - 1$.*

(b) $\alpha_i \geq 0$ *for $1 \leq i \leq n - 1$.*

(c) $s_2, s_3, \ldots, s_{n-1}, s_n = 1$ *is a strictly decreasing sequence of odd integers.*

(d) $\gcd(s_i, s_{i-1}) = 1$ *for* $1 \leq i \leq n$.

Theorem 3.112 *Let a and b be positive integers such that $a > b > 1$, $\gcd(a, b) = 1$, and b is odd; and let $\alpha_1, \alpha_2, \ldots, \alpha_{n-1}$ and $s_0, s_1, \ldots, s_{n-1}, s_n = 1$ be the sequences given in Theorem 3.111. Then*

$$\left(\frac{a}{b}\right) = (-1)^{A+B}$$

where

$$A = \frac{1}{8} \sum_{i=1}^{n-1} \alpha_i(s_i^2 - 1)$$

$$B = \frac{1}{4} \sum_{i=1}^{n-2} (s_i - 1)(s_{i+1} - 1)$$

Proof. If $n = 2$, then

$$\left(\frac{a}{b}\right) = \left(\frac{s_0}{s_1}\right) = \left(\frac{2^{\alpha_1} s_2}{s_1}\right) \quad \text{since } s_0 \equiv 2^{\alpha_1} s_2 \pmod{s_1}$$

$$= \left(\frac{2^{\alpha_1}}{s_1}\right) \quad \text{since } s_2 = 1$$

$$= \left(\frac{2}{s_1}\right)^{\alpha_1} = (-1)^{[(s_1^2-1)/8]\alpha_1} = (-1)^{A+B}$$

where $A = \alpha_1(s_1^2 - 1)/8$ and $B = 0$.

If $n = 3$, then $s_3 = 1$ and

$$\left(\frac{a}{b}\right) = \left(\frac{2^{\alpha_1} s_2}{s_1}\right) = \left(\frac{2^{\alpha_1}}{s_1}\right)\left(\frac{s_2}{s_1}\right)$$

$$= \left(\frac{2}{s_1}\right)^{\alpha_1} \left(\frac{s_1}{s_2}\right)(-1)^{[(s_1-1)/2][(s_2-1)/2]}$$

$$= \left(\frac{2}{s_1}\right)^{\alpha_1} \left(\frac{2^{\alpha_2} s_3}{s_2}\right)(-1)^{[(s_1-1)/2][(s_2-1)/2]}$$

$$= \left(\frac{2}{s_1}\right)^{\alpha_1} \left(\frac{2}{s_2}\right)^{\alpha_2} (-1)^{[(s_1-1)/2][(s_2-1)/2]}$$

$$= (-1)^{[\alpha_1(s_1^2-1)/8]}(-1)^{[\alpha_2(s_2^2-1)/8]}(-1)^{[(s_1-1)/2][(s_2-1)/2]}$$

$$= (-1)^{A+B}$$

where

$$A = \frac{1}{8}\sum_{i=1}^{2}\alpha_i(s_i^2-1)$$

$$B = \frac{1}{4}\sum_{i=1}^{1}(s_i-1)(s_{i+1}-1)$$

and the formula of the theorem holds for $n=3$.

If $m>3$ and the formula holds for any appropriate a and b with $n=k$ and $3 \le k < m$, let Theorem 3.111 be applied to some positive integers a and b meeting the conditions of Theorem 3.112 and terminating in $s_m = 1$. Then

$$\left(\frac{a}{b}\right) = \left(\frac{2^{\alpha_1}s_2}{s_1}\right) = \left(\frac{2^{\alpha_1}}{s_1}\right)\left(\frac{s_2}{s_1}\right)$$

$$= \left(\frac{2}{s_1}\right)^{\alpha_1}\left(\frac{s_1}{s_2}\right)(-1)^{[(s_1-1)/2][(s_2-1)/2]}$$

$$= (-1)^{[\alpha_1(s_1^2-1)/8]}\left(\frac{s_0'}{s_1'}\right)(-1)^{[(s_1-1)/2][(s_2-1)/2]}$$

$$= (-1)^{[\alpha_1(s_1^2-1)/8]}(-1)^{[(s_1-1)/2][(s_2-1)/2]}\left(\frac{a'}{b'}\right)$$

where $a' = s_0' = s_1$ and $b' = s_1' = s_2$. It is easy to see that applying Theorem 3.111 to $a' = s_0' = s_1$ and $b' = s_1' = s_2$ we obtain $n' = m-1$, $s_i' = s_{i+1}$, $\alpha_i' = \alpha_{i+1}$, and $q_i' = q_{i+1}$. Applying the induction hypothesis for $k = n'$, we have

$$\left(\frac{a}{b}\right) = (-1)^{[\alpha_1(s_1^2-1)/8]}(-1)^{[(s_1-1)/2][(s_2-1)/2]}(-1)^{A'+B'}$$

where

$$A' = \frac{1}{8}\sum_{i=1}^{n'-1}\alpha_i'((s_i')^2-1) = \frac{1}{8}\sum_{j=2}^{m-1}\alpha_j(s_j^2-1)$$

$$B' = \frac{1}{4}\sum_{i=1}^{n'-2}(s_i'-1)(s_{i+1}'-1) = \frac{1}{4}\sum_{j=2}^{m-2}(s_j-1)(s_{j+1}-1)$$

The missing $j=1$ terms in A' and B' are found in the exponents of (-1) in the last expression for $\left(\dfrac{a}{b}\right)$ and the formula of the theorem holds for $n=m$. By mathematical induction, the formula of the theorem holds for all positive integers n with $n \ge 2$. ∎

Example 3.113 *It was shown in Example 3.110 that* $\left(\dfrac{231}{1993}\right) = \left(\dfrac{1993}{231}\right).$

If we let $a = s_0 = 1993$ and $b = s_1 = 231$ so that $a > b > 1$, then we have

$$
\begin{aligned}
1993 &= 231 \cdot 8 + 145 \\
&= 231 \cdot 8 + 2^0 \cdot 145 \\
231 &= 145 \cdot 1 + 86 \\
&= 145 \cdot 1 + 2^1 \cdot 43 \\
145 &= 43 \cdot 3 + 16 \\
&= 43 \cdot 3 + 2^4 \cdot 1
\end{aligned}
$$

So $n = 4$ and

$$
\begin{aligned}
s_0 &= 1993 \\
s_1 &= 231 & \alpha_1 &= 0 \\
s_2 &= 145 & \alpha_2 &= 1 \\
s_3 &= 43 & \alpha_3 &= 4 \\
s_4 &= 1
\end{aligned}
$$

$$
\begin{aligned}
A &= \frac{1}{8} \sum_{i=1}^{3} \alpha_i (s_i^2 - 1) \\[2mm]
&= \frac{1}{8} \left[0 \cdot (231^2 - 1) + 1 \cdot (145^2 - 1) + 4 \cdot (43^2 - 1) \right] \\[2mm]
&= 3552 \equiv 0 \ (\mathrm{mod}\ 2)
\end{aligned}
$$

$$
\begin{aligned}
B &= \frac{1}{4} \sum_{i=1}^{2} (s_i - 1)(s_{i+1} - 1) \\[2mm]
&= \frac{1}{4} \left[(231 - 1)(145 - 1) + (145 - 1)(43 - 1) \right] \\[2mm]
&= 9792 \equiv 0 \ (\mathrm{mod}\ 2)
\end{aligned}
$$

Finally, we have

$$
\left(\frac{231}{1993} \right) = \left(\frac{1993}{231} \right) = (-1)^{A+B} = (-1)^0 = 1
$$

as before in Example 3.110. The application of Theorem 3.112 may appear more complicated than the methods of Example 3.110; however, Theorem 3.112 is straightforward to implement in a computer program.

The Jacobi symbol is named after Karl Gustav Jacob Jacobi (1804-1851), who was a contemporary of Legendre. Jacobi had the good fortune to be born into a wealthy German banking family in Potsdam and had an excellent education at home. He studied at the University of Berlin and received his doctorate there in 1825. In 1826 he became lecturer at the University of Königsburg and was made professor in 1831.

Jacobi was best known for his work on elliptic functions. His reputation was established by the publication of *Fundamenta Nova Theoriae Functionum*

Ellipticarum on elliptic functions. He was also know for his work in analysis, geometry, mechanics, and number theory. He made many contributions to the theory of determinants and developed the functional determinant called the "Jacobian." In number theory he proved that every integer was the sum of the square of four integers and determined in how many ways. He also developed the Jacobi symbol, which is the subject of this section.

Jacobi's family lost their money when he was 36, so he had less leisure time for research until the King of Prussia gave him an allowance. Jacobi made the mistake of getting into politics and lost the office he ran for as well as his allowance from the King although it was later restored. He did not enter politics again.

Like Legendre, Jacobi had problems with proving theorems only to have Gauss claim that he had already proved them. In one case he apparently took a paper to Gauss only to have Gauss thumb through his notes and produce the proof. Jacobi is said to have asked why he had not published the paper since he had published so many poorer ones.

Exercises

1. Use Theorems 3.106 to 3.109 to evaluate these Jacobi symbols:

    (a) $\left(\dfrac{195}{977}\right)$ (Note that 977 is prime.)

    (b) $\left(\dfrac{44}{85}\right)$

    (c) $\left(\dfrac{-168}{715}\right)$

2. Prove these parts of Theorem 3.107:

    (a) Part (b)   (b) Part (c)   (c) Part (d)

    (d) Part (e)   (e) Part (f)

3. Show that $x^2 \equiv 2 \pmod{33}$ has no solution.

4. Prove these parts of Lemma 3.108:

    (a) Part (b)   (b) Part (c)   (c) Part (d)

5. Prove these parts of Theorem 3.109:

    (a) Part (b)   (b) Part (c)

6. Complete the details of Example 3.110 using methods of Section 3.9 by showing that:

(a) $\left(\dfrac{7}{1993}\right) = -1$ (b) $\left(\dfrac{11}{1993}\right) = -1$

7. Use Theorem 3.112 to evaluate these Jacobi symbols:

   (a) $\left(\dfrac{5863}{155}\right)$ (b) $\left(\dfrac{17575}{707}\right)$ (c) $\left(\dfrac{-168}{715}\right)$

8. Prove Theorem 3.111.

9. Show that $x^2 \equiv 21 \pmod{65}$ has no integer solutions.

10. What positive integers Q such that $\gcd(Q, 21) = 1$ have the property that $\left(\dfrac{21}{Q}\right) = 1$?

3.11   Application: Unit Orthogonal Matrices

It is often necessary to construct matrices which have certain special properties and whose entries are restricted in some way. Number theoretic methods can occasionally help in this construction. This section is about one such type of matrix. R. E. A. C. Paley's paper "On Orthogonal Matrices" [60] is the basis of much of the discussion below.

The term "orthogonal" is used in describing both vectors and matrices. The following definitions should help clarify the usages as well as how we will interpret the term.

Definition 3.114 *Let V be a row or column matrix with n entries and W be a row or column matrix with n entries so that*

$$V = \begin{bmatrix} V_1 \\ V_2 \\ \vdots \\ V_n \end{bmatrix} \quad or \quad V = \begin{bmatrix} V_1 & V_2 & \cdots & V_n \end{bmatrix}$$

and

$$W = \begin{bmatrix} W_1 \\ W_2 \\ \vdots \\ W_n \end{bmatrix} \quad or \quad W = \begin{bmatrix} W_1 & W_2 & \cdots & W_n \end{bmatrix}$$

Given V and W, both nonzero, we say that V and W are orthogonal *if and only if the dot product of V and W is zero, that is, if and only if*

$$V \bullet W = V_1 W_1 + V_2 W_2 + \cdots + V_n W_n = 0$$

Definition 3.115 *A square $n \times n$ matrix A whose entries are real numbers is said to be* orthogonal *provided that $AA^t = A^tA = I_n$, where I_n is the $n \times n$ identity matrix and A^t is the transpose of A.*

Recall that $AA^t = I_n$ means that the dot product of the i-th row of A with the j-th column of A^t produces the number 1 when $i = j$ and produces the number 0 when $i \neq j$; that is, the i-th row of A is orthogonal with the j-th column of A^t when $i \neq j$ and gives the dot product value 1 when $i = j$. Inasmuch as the rows of A contain the same entries as the columns of A^t, to say that A is orthogonal means that different rows of A are orthogonal and the dot product of a row of A with itself gives the value 1. Note that no row of A can be zero because the dot product of each row with itself must be 1. Because $A^tA = I_n$, similar statements may be made about the columns of A.

For the applications in this section, we will want to relax the properties of an orthogonal matrix somewhat by modifying the requirement that the dot product of a column or a row of A with itself be 1 and allow it to be any nonzero number. We will also want further to restrict the components of A.

Definition 3.116 *A given $n \times n$ real matrix $A = [a_{ij}]$ is a U-matrix provided that*

(a) $a_{ij} = \pm 1$ *for all i, j.*

(b)

$$AA^t = A^tA = D = \begin{bmatrix} d_1 & 0 & 0 & \cdots & 0 \\ 0 & d_2 & 0 & \cdots & 0 \\ 0 & 0 & d_3 & \cdots & 0 \\ \vdots & & & & \vdots \\ 0 & 0 & 0 & \cdots & d_n \end{bmatrix}$$

where $D = [d_{ij}]$ is a diagonal matrix with $d_i \neq 0$ for all i. A diagonal matrix $D = [d_{ij}]$ has the property that $d_{ij} = 0$ when $i \neq j$. Since only the "diagonal" elements of a diagonal matrix D may be nonzero, we sometimes write $d_i = d_{ii}$ and $D = \mathrm{diag}\{d_1, d_2, \ldots, d_n\}$.

Note that the only differences between a U-matrix and an orthogonal matrix are that the components of a U-matrix are required to be either 1 or -1 and the dot product of a row or column with itself is only required to be nonzero. Because A is a real matrix, $d_i > 0$ for all i.

Example 3.117 *Consider the matrix*

$$A = \begin{bmatrix} 1 & 1 & 1 & 1 \\ 1 & -1 & 1 & -1 \\ 1 & 1 & -1 & -1 \\ 1 & -1 & -1 & 1 \end{bmatrix}$$

It easy to verify by direct computation that

$$A^t A = AA^t = \begin{bmatrix} 4 & 0 & 0 & 0 \\ 0 & 4 & 0 & 0 \\ 0 & 0 & 4 & 0 \\ 0 & 0 & 0 & 4 \end{bmatrix}$$

Thus, A is a U-matrix. A is not an orthogonal matrix because the dot product of every row or column with itself is 4 and not 1.

Example 3.118 *If we can construct a U-matrix, we can construct an orthogonal matrix; for if A is a U-matrix of dimension $n \times n$, then*

$$AA^t = A^t A = D = \begin{bmatrix} d_1 & 0 & 0 & \cdots & 0 \\ 0 & d_2 & 0 & \cdots & 0 \\ 0 & 0 & d_3 & \cdots & 0 \\ \vdots & & & & \vdots \\ 0 & 0 & 0 & \cdots & d_n \end{bmatrix}$$

with $d_i \neq 0$. Let

$$E = diag \left\{ \frac{1}{\sqrt{d_1}}, \frac{1}{\sqrt{d_2}}, \cdots, \frac{1}{\sqrt{d_n}} \right\}$$

and let $B = EA$. The effect of multiplying A on the left by E is to multiply each entry in the k-th row of A by $\frac{1}{\sqrt{d_k}}$. Let B_j and B_k be the matrices consisting of the j-th and k-th rows, respectively, of B. Similarly, let A_j and A_k be the matrices consisting of the j-th and k-th rows, respectively, of A. Then the dot product of these rows is

$$B_j \bullet B_k = \left(\frac{1}{\sqrt{d_j}} A_j \right) \bullet \left(\frac{1}{\sqrt{d_k}} A_k \right) = \left(\frac{1}{\sqrt{d_j}\sqrt{d_k}} \right) (A_j \bullet A_k)$$

Thus, we obtain

$$B_j \bullet B_k = \begin{cases} \left(\dfrac{1}{\sqrt{d_j}\sqrt{d_k}} \right) 0 = 0 & if \ \ j \neq k \\[3mm] \left(\dfrac{1}{\sqrt{d_j}\sqrt{d_k}} \right) d_k = 1 & if \ \ j = k \end{cases}$$

So $BB^t = I_n$. Similarly, one may show that $B^t B = I_n$ by working with columns of B. Thus, $B = EA$ is orthogonal.

Examples of U-matrices with dimensions $n = 1$ and 2 are

$$A_1 = [1] \quad and \quad A_2 = \begin{bmatrix} 1 & 1 \\ 1 & -1 \end{bmatrix}$$

We next consider if there exist a U-matrix of dimension $n \times n$ with orthogonal rows and columns for every positive integer n? If the answer is yes for a particular positive integer n, then how can such a matrix A be constructed? A partial answer is given by the next theorem.

Theorem 3.119 *If A is $n \times n$ and A is a U-matrix, then $n = 1$, 2, or $n \equiv 0 \pmod 4$.*

Proof. If $n > 2$, then $A = [a_{ij}]$ has at least three rows so that

$$\sum_{j=1}^{n}(a_{1j} + a_{2j})(a_{1j} + a_{3j}) = \sum_{j=1}^{n}a_{1j}^2 + \sum_{j=1}^{n}a_{2j}a_{1j} + \sum_{j=1}^{n}a_{2j}a_{3j} + \sum_{j=1}^{n}a_{1j}a_{3j}$$

$$= \sum_{j=1}^{n}a_{1j}^2 = n$$

where three of the sums are zero because different rows of A are orthogonal and where the remaining sum is n because $a_{ij} = \pm 1$.

But in the first summation, an examination of the 8 possible combinations of cases for a_{1j}, a_{2j}, and a_{3j} shows that the product $(a_{1j} + a_{2j})(a_{1j} + a_{3j})$ can be only 0 or 4 so that

$$n = \sum_{j=1}^{n}(a_{1j} + a_{2j})(a_{1j} + a_{3j}) \equiv 0 \pmod 4 \quad \blacksquare$$

In order to construct U-matrices, we need a means of generating 1's and -1's. Such a vehicle is provided by the Legendre symbol $\left(\dfrac{b}{p}\right)$. We recall that if p is prime and $p > 2$, then

$$\left(\frac{b}{p}\right) = \begin{cases} 1 & \text{if } b \text{ is a quadratic residue of } p \\ -1 & \text{if } b \text{ is a quadratic nonresidue of } p \\ 0 & \text{if } p \mid b \end{cases}$$

The properties of the Legendre symbol are given in Section 3.9. The most pertinent of these are summarized in the next theorem for easy reference.

Theorem 3.120 *If p is an odd prime and each of a and b is an integer, then*

(a) $\displaystyle\sum_{k=0}^{p-1}\left(\frac{k}{p}\right) = 0$ *(Exercises of Section 3.9),*

(b) $\left(\dfrac{a}{p}\right)\left(\dfrac{b}{p}\right) = \left(\dfrac{ab}{p}\right)$ *(Theorem 3.96),*

(c) $\left(\dfrac{-1}{p}\right) = \begin{cases} 1 & \text{if } p \equiv 1 \pmod 4 \\ -1 & \text{if } p \equiv 3 \pmod 4 \end{cases}$ *(Theorem 3.96),*

(d) $\left(\dfrac{a}{p}\right) = \left(\dfrac{b}{p}\right)$ *if* $a \equiv b \pmod{p}$ (*Theorem 3.96*), *and*

(e) $\left(\dfrac{2}{p}\right) = (-1)^{(p^2-1)/8}$ (*Theorem 3.100*),

where the parenthetical expressions are Legendre symbols.

We note that because $\left(\dfrac{b}{p}\right) = 0$ when $p \mid b$, parts (b) to (d) hold when $p \mid b$.

In addition to these properties, a useful summation is evaluated in the next theorem. For a proof, see Plackett and Burman [63].

Theorem 3.121 *If p is an odd prime and $0 \le a, b \le p-1$ with $a \ne b$, then*

$$\sum_{k=0}^{p-1} \left(\frac{k-a}{p}\right)\left(\frac{k-b}{p}\right) = -1$$

where the parenthetical expressions are Legendre symbols.

In the context that follows, it is convenient to index the subscripts of a_{ij} in a square matrix $A = [a_{ij}]$ with $i = 0, 1, 2, \ldots, n-1$ and $j = 0, 1, 2, \ldots, n-1$ instead of with $1 \le i, j \le n$. The next theorem shows how to generate certain U-matrices.

Theorem 3.122 *For a prime p such that $p \equiv 3 \pmod 4$ and $n = p + 1$, define the $n \times n$ matrix $A = [a_{ij}]$ as follows:*

$$\begin{aligned}
a_{ij} &= 1 & &\text{if } i = 0 \text{ or } j = 0 \\
a_{ij} &= \left(\frac{j-i}{p}\right) & &\text{if } 1 \le i, j \le p \text{ and } i \ne j \\
a_{ii} &= -1 & &\text{if } 1 \le i \le p
\end{aligned}$$

where the parenthetical expression is the Legendre symbol. Then A is a U-matrix.

Proof. Clearly, $a_{ij} = \pm 1$ for all i and j so that the dot product of the i-th row of A with itself gives

$$\sum_{k=0}^{p} a_{ik} a_{ik} = \sum_{k=0}^{p} 1 = p + 1$$

If $i \ne 0$, then

$$\sum_{k=0}^{p} a_{0k} a_{ik} = \sum_{k=0}^{p} a_{ik} = a_{i0} + a_{ii} + \sum_{\substack{k=1 \\ k \ne i}}^{p} a_{ik}$$

$$= \sum_{k=1}^{p} \left(\frac{k-i}{p}\right) \text{ since } a_{i0} = 1, \ a_{ii} = -1, \text{ and } \left(\frac{0}{p}\right) = 0$$

$$= \sum_{j=0}^{p-1} \left(\frac{j}{p} \right)$$

where the last equality holds because

$$\{[[k-i]]_p : 1 \le k \le p\} = \{[[j]]_p : 0 \le j \le p-1\}$$

and $\left(\dfrac{0}{p} \right) = 0$. Thus, the row of A with index zero is orthogonal with all other rows of A.

If $0 \ne i \ne j \ne 0$, then

$$\sum_{k=0}^{p} a_{ik}a_{jk} = a_{ii}a_{ji} + a_{ij}a_{jj} + a_{i0}a_{j0} + \sum_{w=1}^{p} \left(\frac{w-i}{p} \right) \left(\frac{w-j}{p} \right)$$

since $\left(\dfrac{w-i}{p} \right) \left(\dfrac{w-j}{p} \right) = 0$ when $w = i$ or $w = j$. But $a_{i0}a_{j0} = 1 \cdot 1 = 1$ and

$$a_{ii}a_{ji} + a_{ij}a_{jj} = (-1)\left(\frac{i-j}{p} \right) + \left(\frac{j-i}{p} \right)(-1) = 0$$

because

$$\left(\frac{i-j}{p} \right) = \left(\frac{-(j-i)}{p} \right) = \left(\frac{-1}{p} \right)\left(\frac{j-i}{p} \right) = -\left(\frac{j-i}{p} \right)$$

where the last equality holds because $p \equiv 3 \pmod 4$. Thus

$$\sum_{k=0}^{p} a_{ik}a_{jk} = 1 + \sum_{w=0}^{p-1} \left(\frac{w-i}{p} \right) \left(\frac{w-j}{p} \right) = 1 - 1 = 0$$

because $p \equiv 0 \pmod p$ allows the summation to run from 0 to $p-1$ and because of Theorem 3.121. So any two different rows of A, neither of which has index zero, are orthogonal. The argument we have just applied to the rows of A may be applied to the columns of A; however, the modifications are left to the reader. ∎

Example 3.123 *If $p = 11 \equiv 3 \pmod 4$ so that $n = p + 1 = 12$, we can obtain a U-matrix as follows. We note that 2, 6, 7, 8, and 10 are quadratic nonresidues of 11 and that 1, 3, 4, 5, and 9 are quadratic residues of 11.*
To obtain the sixth and ninth rows of $A = [a_{ij}]$, that is, the rows with $i = 5$

and $i = 8$, *we generate the following table:*

j	$j-5$	$\left(\dfrac{j-5}{p}\right)$	$j-8$	$\left(\dfrac{j-8}{p}\right)$
0	$a_{50} = 1$		$a_{80} = 1$	
1	$-4 \equiv 7$	-1	$-7 \equiv 4$	1
2	$-3 \equiv 8$	-1	$-6 \equiv 5$	1
3	$-2 \equiv 9$	1	$-5 \equiv 6$	-1
4	$-1 \equiv 10$	-1	$-4 \equiv 7$	-1
5	$a_{55} = -1$		$-3 \equiv 8$	-1
6	1	1	$-2 \equiv 9$	1
7	2	-1	$-1 \equiv 10$	-1
8	3	1	$a_{88} = -1$	
9	4	1	1	1
10	5	1	2	-1
11	6	-1	3	1

The entire 12×12 *U-matrix is*

$$
A = \begin{bmatrix}
1 & 1 & 1 & 1 & 1 & 1 & 1 & 1 & 1 & 1 & 1 & 1 \\
1 & -1 & 1 & -1 & 1 & 1 & 1 & -1 & -1 & -1 & 1 & -1 \\
1 & -1 & -1 & 1 & -1 & 1 & 1 & 1 & -1 & -1 & -1 & 1 \\
1 & 1 & -1 & -1 & 1 & -1 & 1 & 1 & 1 & -1 & -1 & -1 \\
1 & -1 & 1 & -1 & -1 & 1 & -1 & 1 & 1 & 1 & -1 & -1 \\
1 & -1 & -1 & 1 & -1 & -1 & 1 & -1 & 1 & 1 & 1 & -1 \\
1 & -1 & -1 & -1 & 1 & -1 & -1 & 1 & -1 & 1 & 1 & 1 \\
1 & 1 & -1 & -1 & -1 & 1 & -1 & -1 & 1 & -1 & 1 & 1 \\
1 & 1 & 1 & -1 & -1 & -1 & 1 & -1 & -1 & 1 & -1 & 1 \\
1 & 1 & 1 & 1 & -1 & -1 & -1 & 1 & -1 & -1 & 1 & -1 \\
1 & -1 & 1 & 1 & 1 & -1 & -1 & -1 & 1 & -1 & -1 & 1 \\
1 & 1 & -1 & 1 & 1 & 1 & -1 & -1 & -1 & 1 & -1 & -1
\end{bmatrix}
$$

At this point, we know there are U-matrices of orders $n = 1$ and 2 and of order $n = p + 1$ when p is prime and $p \equiv 3 \pmod 4$. Next consider the following table of n with $n \equiv 0 \pmod 4$ and $4 \le n \le 50$ and of k with $n = k + 1$, which covers all possible cases with $4 \le n \le 50$ according to Theorem 3.119.

$n = k+1$		k	k Is Prime	$n = k+1$		k	k Is Prime
4		3	Yes	28	**	27	No
8		7	Yes	32		31	Yes
12		11	Yes	36	**	35	No
16	**	15	No	40	**	39	No
20		19	Yes	44		43	Yes
24		23	Yes	48		47	Yes

The unstarred values of n correspond to cases for which Theorem 3.122 implies that there is a U-matrix of order $n \times n$. The starred values of n

indicate those cases that are not addressed by Theorem 3.122; however, this fact does not mean that there are no U-matrices having these dimensions. There are other ways of constructing U-matrices in these cases.

Definition 3.124 *Let $A = [a_{ij}]$ be $n \times n$ and $B = [b_{ij}]$ be $m \times m$. We define the product $A \otimes B$ to be the $nm \times nm$ matrix*

$$A \otimes B = \begin{bmatrix} a_{11}B & a_{12}B & \cdots & a_{1n}B \\ a_{21}B & a_{22}B & \cdots & a_{2n}B \\ \vdots & & & \vdots \\ a_{n1}B & a_{n2}B & \cdots & a_{nn}B \end{bmatrix}$$

where $a_{ij}B$ is the $m \times m$ matrix obtained when B is multiplied by the scalar a_{ij}.

Note that $A \otimes B$ is not ordinary matrix multiplication. It is straightforward but tedious to show the following theorem.

Theorem 3.125 *If A, of dimension $n \times n$, and B, of dimension $m \times m$, are U-matrices, then $A \otimes B$ is also a U-matrix. Further, if the components in both the first rows and first columns of A and B are 1, then those of $A \otimes B$ are also.*

Example 3.126 *It was noted in the table following Example 3.123 that $n = 16$ was not covered by Theorem 3.122; however, we know that*

$$A_2 = \begin{bmatrix} 1 & 1 \\ 1 & -1 \end{bmatrix}$$

is a U-matrix. Then by Theorem 3.125,

$$A_4 = A_2 \otimes A_2 = \begin{bmatrix} A_2 & A_2 \\ A_2 & -A_2 \end{bmatrix} = \begin{bmatrix} 1 & 1 & 1 & 1 \\ 1 & -1 & 1 & -1 \\ 1 & 1 & -1 & -1 \\ 1 & -1 & -1 & 1 \end{bmatrix}$$

and

$$A_8 = A_2 \otimes A_4 = \begin{bmatrix} A_4 & A_4 \\ A_4 & -A_4 \end{bmatrix} = \begin{bmatrix} 1 & 1 & 1 & 1 & 1 & 1 & 1 & 1 \\ 1 & -1 & 1 & -1 & 1 & -1 & 1 & -1 \\ 1 & 1 & -1 & -1 & 1 & 1 & -1 & -1 \\ 1 & -1 & -1 & 1 & 1 & -1 & -1 & 1 \\ 1 & 1 & 1 & 1 & -1 & -1 & -1 & -1 \\ 1 & -1 & 1 & -1 & -1 & 1 & -1 & 1 \\ 1 & 1 & -1 & -1 & -1 & -1 & 1 & 1 \\ 1 & -1 & -1 & 1 & -1 & 1 & 1 & -1 \end{bmatrix}$$

are U-matrices. The process can be extended another stage, giving

$$A_{16} = A_2 \otimes A_8 = \begin{bmatrix} A_8 & A_8 \\ A_8 & -A_8 \end{bmatrix}$$

Also, $A_4 \otimes A_4$ is a 16×16 U-matrix.

Paley also proved the next two theorems using methods similar to those of Theorem 3.122; however, you will need to consult Paley's paper to obtain the actual definitions of the U-matrices.

Theorem 3.127 *If $n \equiv 0 \pmod 4$, $n = 2^k(p+1)$, p is prime, and $k \geq 1$, then there is a U-matrix, A, of dimension $n \times n$. Further, the components of both the first row and column of A may be chosen to be 1.*

Theorem 3.128 *If $n \equiv 0 \pmod 4$, $n = 2^k(p^s + 1)$, p is an odd prime, $k \geq 1$, and $s \geq 1$, then there is a U-matrix, A, of dimension $n \times n$. Further, the components of both the first row and column of A may be chosen to be 1.*

A discussion that applies the methods of this section to experimental design may be found in "The Design of Optimum Multifactorial Experiments" by R. L. Plackett and J. P. Burman [63]. These designs are referred to as "Plackett-Burman designs" in the statistical literature. Thus, U-matrices are used in the construction of experimental designs where the 1 and -1 values indicate whether an experimental factor will be at a low level or a high level when the experiment is done. Thus, some of the columns of a U-matrix are associated with experimental factors such as temperature, pressure, amount of catalyst, and so on, if one is interested in experimenting with a chemical reaction. For example, if temperature corresponds to the third column, a 1 as an entry in this column may mean to use a high temperature of $80°$ C and a -1 may mean to use a low temperature of $60°$ C. Each row of the matrix corresponds to a particular way of running the experiment by describing exactly what condition to use for each experimental factor for that trial. One of the values of using such experimental designs is that the effect of changing each separate experimental factor may be extracted and estimated mathematically even though several factors change simultaneously in each trial (row). In addition, the data from all the trials may be used in each estimate, improving the precision of the result greatly over that of a single trial.

J. A. Todd in "A Combinatorial Problem" [87] used U-matrices to solve a combinatorial problem involving subsets of a set. For example, suppose we have a committee of 7 members. We want to form subcommittees of three people each; however, in order to maintain communication among the subcommittees, we would like to have one from each subcommittee on every other subcommittee to function as a liaison. Is this procedure possible? If so, how can the subcommittees be constructed? To begin our attack on the problem, we need some characteristics of a U-matrix.

Theorem 3.129 *If R_1 and R_2 are different rows of a U-matrix of dimension $4n \times 4n$, then there are exactly n components that are 1 in both R_1 and R_2, exactly n components that are -1 in both R_1 and R_2, and exactly $2n$ components that are different in R_1 and R_2.*

Proof. The product $R_1 \bullet R_2 = 0$ because R_1 and R_2 are orthogonal. In the sum defining $R_1 \bullet R_2$, assume that there are

A terms of the form $(-1)(1)$
B terms of the form $(1)(1)$
C terms of the form $(-1)(-1)$

Since $A(-1)(1) + B(1)(1) + C(-1)(-1) = R_1 \bullet R_2 = 0$, we must have

$$A = B + C$$
$$A + B + C = 4n$$

Therefore, $A = 2n$. Assume that there are k terms of type $(-1)(1)$ that result from a 1 from R_1. The counts of the sources of 1 and -1 among the $A = 2n$ terms of type $(-1)(1)$ are given in the following table:

Source	1	-1
R_1	k	$2n - k$
R_2	$2n - k$	k

A table of types of components available for terms of the form $(-1)(-1)$ and $(1)(1)$ is therefore

Remaining in	1	-1
R_1	$2n - k$	k
R_2	k	$2n - k$

In the last table, there must be as many 1's from R_1 as there are 1's from R_2. Thus $k = 2n - k$ implying that $k = n$. We then have $A = 2n$ and $B = C = n$. ■

The U-matrices constructed earlier had $4n$ rows and columns and could be obtained with the first rows and columns consisting only of 1's. Assume that we have such a U-matrix, T, and we form a submatrix S from T by deleting the first row and first column of T. S has dimensions $(4n-1) \times (4n-1)$; and by Theorem 3.129, any two different rows of S have exactly $n - 1$ instances such that corresponding components of both rows are 1 and exactly n instances such that corresponding components of both rows are -1. We will use the matrix S to select subsets.

Let W be a set having $4n - 1$ objects. Next, associate each object of W with one and only one column of S. Each row of S will describe the subset of W obtained by including only those elements of W corresponding to a component that is 1. Thus, there will be $4n - 1$ subsets of W, each associated with a row of S and each containing $2n - 1$ elements. Further, each such subset has exactly $n - 1$ elements in common with any other such subset.

Example 3.130 *We return to the problem posed initially. Suppose that we have a committee of 7 members. We want to form subcommittees of three people each; however, we would like to have a member from each subcommittee on every other subcommittee. Let $W = \{a, b, c, d, e, f, g\}$ be the set of 7* committee members. W has $p = 7 = 4n - 1$ elements. $p = 7$ is prime and $n = 2$. We want $4n - 1 = 7$ subsets each of size $2n - 1 = 3$ and each having $n - 1 = 1$ members in common with every other committee.

A U-matrix A_8 of size $4n$ with first rows and columns all 1's was obtained in Example 3.126. Let S be the submatrix obtained by omitting the first row and first column of A_8. Thus,

Row of S	a	b	c	d	e	f	g	Subset
1	-1	1	-1	1	-1	1	-1	$\{b, d, f\}$
2	1	-1	-1	1	1	-1	-1	$\{a, d, e\}$
3	-1	-1	1	1	-1	-1	1	$\{c, d, g\}$
4	1	1	1	-1	-1	-1	-1	$\{a, b, c\}$
5	-1	1	-1	-1	1	-1	1	$\{b, e, g\}$
6	1	-1	-1	-1	-1	1	1	$\{a, f, g\}$
7	-1	-1	1	-1	1	1	-1	$\{c, e, f\}$

where a member is in a subset provided that there is a 1 in that column for that row. By inspection we observe that the subsets of W in the table have the required properties.

We summarize the discussion above in the following theorem.

Theorem 3.131 *If W is a set containing $4n - 1$ elements and if there is a U-matrix of dimension $4n \times 4n$ having the first row and first column consisting of only 1's, then there are $4n - 1$ subsets of W each containing $2n - 1$ elements of W and the intersection of each pair of such subsets contains exactly $n - 1$ elements.*

Exercises

1. Use Theorem 3.122 to generate $n \times n$ unit matrices with orthogonal rows for $n = 4$ and $n = 8$.

2. Complete the details of constructing the 12×12 unit matrix with orthogonal rows discussed in Example 3.123.

3. Prove Theorem 3.125.

4. Complete the construction of A_{16} of Example 3.126. Also, construct a 16×16 U-matrix in a different way.

5. Use Exercise 1 and Example 3.123 to construct $n \times n$ orthogonal matrices for $n = 4$, 8, and 12.

6. Use Theorems 3.122, 3.125, 3.127, and 3.128 to show that U-matrices can be constructed (but do not perform the construction) for every n with $4 \le n \le 100$ and $n \equiv 0 \pmod 4$ except perhaps for $n = 92$.

7. Complete the proof of Theorem 3.119 by showing that the product $(a_{1j} + a_{2j})(a_{1j} + a_{3j})$ can be only 0 or 4.

8. Construct a 24×24 U-matrix.

9. Generate 11 subsets of $W = \{a, b, c, d, e, f, g, h, i, j, k\}$, each of size 5, such that each pair of subsets intersects in exactly 2 elements of W.

10. Let W be a set with $4n-1$ elements such that there are $4n-1$ subsets of W each containing exactly $2n-1$ elements and such that the intersection of each pair contains exactly $n-1$ members of W. Prove there exists a U-matrix of dimension $4n \times 4n$.

11. Consider these two problems:

    (a) Use the U-matrix of Example 3.130 to generate $4n - 1 = 7$ subsets of W each containing exactly $2n = 4$ elements and such that the intersection of each pair contains exactly $n = 2$ members of W.

    (b) Prove that if W is a set containing $4n - 1$ elements and there is a U-matrix of dimension $4n \times 4n$ having the first row and column consisting only of 1's, then there are $4n - 1$ subsets of W each containing exactly $2n$ elements of W and the intersection of each pair of such subsets contains exactly n elements.

3.12   Application: Random Number Generation II

In Section 1.8 we introduced the multiplicative congruential method of generating (pseudo) random integers. We revisit such generators again since we now have more number theoretic tools available to assess some of their characteristics. Thus, if $m > 1$ and a are positive integers, we recall that $L(x) = [[a \cdot x]]_m$ produces the primary residue or remainder when $a \cdot x$ is divided by m. For $S = \{1, 2, 3, \ldots, m - 1\}$, L generates the integer sequence $x_0, x_1, x_2, \ldots$, where if x_0 is any integer in S, then

$$x_{i+1} = L(x_i) = [[a \cdot x_i]]_m$$

for any integer $i \geq 0$. Consequently, $x_{i+1} \equiv a \cdot x_i \pmod{m}$ for each i. Further, in order to ensure that $L(x_i) \neq 0$ and $L(x_i) \in S$, we also require that the multiplier a be such that $\gcd(a, m) = 1$.

In Chapter 1 we showed that

$$x_k \equiv a^k \cdot x_0 \pmod{m}$$

for each positive integer k. The period is the least positive integer k such that $x_k = x_0$ or such that

$$x_0 \equiv a^k \cdot x_0 \pmod{m}$$

The larger the period, the better, as far as random integer generation is concerned. If the initial seed x_0 is chosen with $\gcd(x_0, m) = 1$, then the period k must be the least positive integer k satisfying

$$a^k \equiv 1 \ (\text{mod } m)$$

Since $\gcd(a, m) = 1$, Euler's Theorem (Theorem 3.53) gives

$$a^{\phi(m)} \equiv 1 \ (\text{mod } m)$$

so that the period is no larger than $\phi(m)$. The largest value of $\phi(m)$ relative to m is when m is prime, giving $\phi(m) = m - 1$. For this reason, the modulus m of a multiplicative congruential generator is usually selected to be prime. The modulus m will be assumed to be prime for the remainder of this section unless stated otherwise.

For some choices of j, $1 \leq j < \phi(m) = m - 1$, it is possible that $a^j \equiv 1 \ (\text{mod } m)$. If such a j exists, then $j \mid \phi(m)$. Equivalently, there is an integer $t > 1$ such that $j \cdot t = \phi(m) = m - 1$, giving

$$a^{\phi(m)/t} = a^{(m-1)/t} \equiv 1 \ (\text{mod } m)$$

If a is also chosen to be a primitive root of m, then $t = 1$ and $m - 1 = \phi(m)$ is the least positive integer with $a^{m-1} \equiv 1 \ (\text{mod } m)$, making the period as large as possible with the modulus m.

We return again to Example 1.77, where we considered the prime

$$m = 2^{31} - 1 = 2147483647$$

and multiplier $a = 16807$. Recall that $m = 2^{31} - 1$ is also the largest positive integer representable as a signed integer in a 32-bit computer word size and that if the multiplicative congruential generator is implemented using Theorem 1.76, the results of all intermediate operations will remain within the 32-bit word size used by many computers. In this sense, the $m = 2^{31} - 1$ is an optimal prime with respect to 32-bit integer computer arithmetic. It is also the case that $a = 16807$ is a primitive root of $m = 2^{31} - 1$.

Although one can show directly that 16807 is a primitive root of $m = 2^{31} - 1$, the result may be obtained from knowing that 7 is also a primitive root of m and that $7^5 = 16807$. By Theorem 3.78, if a is a primitive root of $m > 1$ and $\gcd(s, m - 1) = 1$, then a^s is also a primitive root of m. The reason for choosing the larger primitive root is because if $a = 7$, then the random integers generated from a small x_i will still be small. There are other measures of performance described in "An Exhaustive Analysis of Multiplicative Congruential Random Generators with Modulus $2^{31} - 1$" by George S. Fishman and Louis R. Moore III [28]. Other potentially useful multipliers for modulus $m = 2^{31} - 1$ are evaluated in this reference.

The only part remaining is to show that 7 is a primitive root of $m = 2^{31} - 1$. We saw above that if 7 is a primitive root, we must have

$$7^{(m-1)/t} \equiv 1 \ (\text{mod } m)$$

for no positive divisor t of $m - 1$ except $t = 1$. One can show that the prime factorization of $m - 1$ is

$$\begin{aligned} m - 1 &= (2^{31} - 1) - 1 = 2^{31} - 2 = 2147483646 \\ &= 2 \cdot 3^2 \cdot 7 \cdot 11 \cdot 31 \cdot 151 \cdot 331 \end{aligned}$$

The factorization can be verified by direct computation; however, the special form of $m - 1$ allows the factors to be obtained by algebraic manipulation. Only values of t that are prime divisors of $m - 1$ need to be checked; for example,

$$7^{(m-1)/151} = 7^{2 \cdot 3^2 \cdot 7 \cdot 11 \cdot 31 \cdot 331} \equiv 535044134 \not\equiv 1 \pmod{m}$$

This result can be obtained readily using the algorithm described in Section 3.6. Thus, since

$$\begin{aligned} 2 \cdot 3^2 \cdot 7 \cdot 11 \cdot 31 \cdot 331 &= 14221746_{ten} \\ &= 1101\,1001\,0000\,0001\,1011\,0010_{binary} \end{aligned}$$

we have the following evaluation of $\left[\left[7^{14221746}\right]\right]_m$:

k	b_k	$[[p_k]]_m$
23	1	7
22	1	$7^2 \cdot 7 = 343$
21	0	$(343)^2 \equiv 117649$
20	1	$(117649)^2 \cdot 7 \equiv 252246292$
$\vdots$	$\vdots$	$\vdots$
1	1	$(504066171)^2 \cdot 7 \equiv 21925355$
0	0	$(21925355)^2 \equiv 535044134$

where b_k is a binary digit for the binary representation of 14221746 and p_k is the k-th product in the algorithm.

Exercises

1. Determine the number of primitive roots of $m = 2^{31} - 1$.

2. Show that the prime factorization of $m - 1 = 2^{31} - 2$ is $2 \cdot 3^2 \cdot 7 \cdot 11 \cdot 31 \cdot 151 \cdot 331$ using algebraic factoring. Note that $m - 1 = 2^{31} - 2 = 2 \cdot (2^{15} - 1)(2^{15} + 1)$, and so on.

3. Complete the proof that 7 is a primitive root of $m = 2^{31} - 1$ by showing that $7^{(m-1)/p} \not\equiv 1 \pmod{m}$ for every prime divisor p of $m - 1$. (Access to extended integer precision computer software makes the solution for this problem more tractable.)

4. Justify the procedure of Exercise 3 by proving that if $a^{(m-1)/p} \not\equiv 1 \pmod{m}$ for every prime divisor p of $m-1$, then $a^{(m-1)/t} \not\equiv 1 \pmod{m}$ for every positive divisor $t > 1$ of $m-1$.

5. Show that 7 is the smallest primitive root of $m = 2^{31}-1$ (see the comment in Exercise 3).

6. Let g be the smallest primitive root of a prime m. Show that every primitive root h of m has the form $h \equiv g^s \pmod{m}$, where $\gcd(s, m-1) = 1$.

7. If m is prime and $\gcd(a, m) = 1$ for the multiplicative congruential generator $x_{i+1} = [[a \cdot x_i]]_m$, show that every choice of x_0 in the set $S = \{1, 2, \ldots, m-1\}$ has the same period. If, further, a is a primitive root of m, show that the finite integer sequence $x_0, x_1, x_2, \ldots, x_{m-1}$ is a permutation (rearrangement) of the sequence $1, 2, \ldots, m-1$.

3.13   Application: Hashing Functions

In computer applications the situation frequently occurs in which there are n sets of information $I_1, I_2, \ldots, I_n$ where each set is "named" with a key, say $k_1, k_2, \ldots, k_n$, respectively. It is desirable to be able to find the information I_j quickly when the key k_j is given. The keys may be considered to be integers even though they may be finite sequences of alphabetic characters. Searching lists of keys may require too much time; moreover, if a search for information using the keys must be conducted many times, the computational work may be considerable. A compiler is a computer program that converts another computer program written in a "high-level" computer language such as FORTRAN or Pascal into a machine language program. During the process of "compiling," the compiler must produce tables and find items with the associated information in the tables many times. The important idea here is that the "look-up" table is created once but is used many times. Since there may be many such tables, one also needs for them to use the least storage feasible.

One solution to the "location" problem is to use a *hash function* $h : K \to \{0, 1, \ldots, m\}$ where $K = \{k_1, k_2, \ldots, k_n\}$ and where $m + 1 \geq n$. We then provide $m+1$ computer storage locations where the n sets of information are kept. Thus, to find I_j, we calculate $h(k_j)$ and go the location $h(k_j)$ where I_j is kept. It is generally difficult to find an ideal hash function, that is, one where $h(k_r) \neq h(k_s)$ when $k_r \neq k_s$ and where $m + 1 = n$. Hash functions that are often used map K into a set much larger than n elements and are not one to one. For h not one to one, provision must be made for what to do when there is collision, that is, when k_r and k_s are distinct keys but $h(k_r) = h(k_s)$. Donald E. Knuth in Vol. 3 of *The Art of Computer Programming* [43] discusses various choices for hash functions and methods of dealing with collisions. G. Jaeschke in "Reciprocal Hashing: A Method for Generating Minimal Perfect Hashing Functions" [37] describes a particular type of minimum hashing function, that is, a hash function that is one to one and uses minimum storage (*minimum perfect hashing function*). Jaeschke's hashing function requires that one be able to map the keys $\{k_1, k_2, \ldots, k_n\}$, which may not be relatively prime to one another, to a set of integers $\{f(k_1), f(k_2), \ldots, f(k_n)\}$ that are pairwise relatively prime. He proved the following theorem.

Theorem 3.132 *Let $K = \{k_1, k_2, \ldots, k_n\}$ be a set of n distinct positive integers. There exist two integers D and E such that if $f(x) = Dx + E$ is the polynomial thereby defined, then the members of the set of positive integers $\{f(k_1), f(k_2), \ldots, f(k_n)\}$ are pairwise relatively prime. Thus, the integers $Dk_1 + E$, $Dk_2 + E$, ..., and $Dk_n + E$ are pairwise relatively prime.*

Proof. The proof is longer than we can present here (see Jaeschke's 1981 paper [37]); however, we will consider a special case later that will be adequate for our purposes. ∎

Jaeschke's main result is the next theorem, where $\lfloor w \rfloor$ is the greatest positive integer less than or equal to w.

Theorem 3.133 *If the set $K = \{k_1, k_2, \ldots, k_n\}$ has distinct positive integer keys, then there exist integers C, D, and E such that*

$$h(x) = \left[\left[\left\lfloor \frac{C}{Dx + E} \right\rfloor\right]\right]_n = \left\lfloor \frac{C}{Dx + E} \right\rfloor \pmod{n}$$

where $x \in K$, is a minimal perfect hashing function.

Proof. Let $k_1 < k_2 < \cdots < k_n$ and let the integers D and E be given by Theorem 3.132 so that $f(k_j) = Dk_j + E$ and $\gcd(f(k_i), f(k_j)) = 1$ for $i \neq j$. Since D and E may be chosen so that $f(k_j) > n$ for each j, it is possible to choose n integers $a_1, a_2, \ldots, a_n$ such that $a_i \not\equiv a_j \pmod{n}$ and such that

$$(i - 1) \cdot (Dk_i + E) \leq a_i < i \cdot (Dk_i + E)$$

Then, by the version of the Chinese Remainder Theorem applicable to non-relatively prime moduli (see the Exercises of Section 3.2), there is a number C such that

$$\begin{aligned} C &\equiv a_1 \pmod{n(Dk_1 + E)} \\ C &\equiv a_2 \pmod{n(Dk_2 + E)} \\ &\vdots \\ C &\equiv a_n \pmod{n(Dk_n + E)} \end{aligned}$$

Therefore, there are integers q_i such that $C = q_i[n(Dk_i + E)] + a_i$ for all i. Consequently,

$$\left\lfloor \frac{C}{Dk_i + E} \right\rfloor = q_i n + (i - 1) \equiv i - 1 \pmod{n}$$

This result implies that the function h defined in the theorem has the property $h(k_i) \neq h(k_j)$ when $i \neq j$ so that h is one to one. Further, the range of h is $\{0, 1, \ldots, (n-1)\}$. ∎

Jaeschke gives algorithms for computing C, D, and E; however, the methods are exhaustive in nature. On the other hand, C. C. Chang and J. C. Shieh in "Pairwise Relatively Prime Generating Polynomials and Their Applications" [15] give a way to calculate a required polynomial of the form $f(x) = Dx + 1$ where $E = 1$. The proof of the next lemma is left to the reader.

Lemma 3.134 *If a and b are positive integers, $a > b$, and d is a multiple of $a - b$, then $d(a - b)$ and $da + 1$ are relatively prime.*

Theorem 3.135 *Let $K = \{k_1, k_2, \ldots, k_n\}$ be a set of n distinct positive integers with $k_i < k_{i+1}$. If $\{t_1, t_2, \ldots, t_s\} = \{k_i - k_j : 1 \leq j < i \leq n\}$ is the set of $s = n(n-1)/2$ differences, then $D = w \cdot \mathrm{lcm}(t_1, t_2, \ldots, t_s)$, where w is any positive integer, has the property that $Dk_1 + 1, Dk_2 + 1, \ldots, Dk_n + 1$ are pairwise relatively prime.*

Proof. Since $\gcd(a, b) = \gcd(a - b, a)$ when a and b are integers for which both $\gcd(a, b)$ and $\gcd(a - b, b)$ are defined, we obtain $\gcd(Dk_i + 1, Dk_j + 1) = \gcd(D(k_i - k_j), Dk_i + 1)$ for $i > j$, where D is as given in the theorem. By definition of D, D is a multiple of $(k_i - k_j)$; therefore, Lemma 3.134 gives $\gcd(Dk_i + 1, Dk_j + 1) = 1$ for $i > j$. ∎

We note that the proof of Theorem 3.135 requires only that D be a multiple of all of $t_1, t_2, \ldots, t_s$. Ordinarily, one would probably choose as small a D as possible ($w = 1$); however, the proof of Theorem 3.133 requires a D such that $f(k_i) = Dk_i + 1 > n$. Such a D can always be obtained by selecting w large enough.

Example 3.136 *Suppose that $K = \{3, 6, 7, 12\}$. Then*

$$\begin{aligned} \{k_i - k_j : 1 \leq j < i \leq 4\} \quad &= \quad \{6 - 3, 7 - 3, 12 - 3, 7 - 6, 12 - 6, 12 - 7\} \\ &= \quad \{3, 4, 9, 1, 6, 5\} \end{aligned}$$

so that $D = \mathrm{lcm}(3, 4, 9, 1, 6, 5) = 180$. Thus, $f(x) = 180x + 1$ and

i	k_i	$f(k_i) = Dk_i + 1$
1	3	$180 \cdot 3 + 1 = 541$
2	6	$180 \cdot 6 + 1 = 1081$
3	7	$180 \cdot 7 + 1 = 1261$
4	12	$180 \cdot 12 + 1 = 2161$

By inspection, any two of 541, 1081, 1261, and 2161 are relatively prime.

In order to implement Jaeschke's reciprocal hashing we must be able to compute the integer C. Fortunately, C. C. Chang and J. C. Shieh in "A Fast

Algorithm for Constructing Reciprocal Hashing Functions" [14] provide an algorithm for computing the integer C.

In the rest of this section we use both the greatest integer function $\lfloor w \rfloor$ described earlier and the ceiling function $\lceil w \rceil$, which is the least positive integer greater than or equal to w. Thus, $\lceil 5.23 \rceil = 6$ and $\lceil 8 \rceil = 8$. See Section 6.2 for a more complete discussion of these two functions.

Theorem 3.137 *If*

(a) $m_1, m_2, \ldots, m_n$ *are pairwise relatively prime integers,*

(b) $m_i > n$ *for all* i, $1 \le i \le n$,

(c) $M = \displaystyle\prod_{j=1}^{n} m_j$,

(d) $M_i = n \displaystyle\prod_{j \neq i} m_j = \dfrac{nM}{m_i}$ *for all* i,

(e) b_i *is such that* $M_i b_i \equiv n \pmod{n m_i}$, *and*

(f) $N_i = \left\lceil \dfrac{(i-1)m_i}{n} \right\rceil$ *for each* i,

then

1.

$$\left\lfloor \dfrac{\left| \displaystyle\sum_{j=1}^{n} M_j b_j N_j \right|}{m_i} \right\rfloor \equiv (i-1) \pmod{n} \text{ for all } i$$

2.

$$C = \left\lceil \left\lfloor \displaystyle\sum_{j=1}^{n} M_j b_j N_j \right\rfloor \right\rceil_{nM}$$

is the smallest positive integer such that $\left\lfloor \dfrac{C}{m_i} \right\rfloor \equiv (i-1) \pmod{n}$ *for all* i.

Proof. Let $a_i = N_i \cdot n$. Then a_i is the smallest multiple of n such that $(i-1) \cdot m_i \le a_i < i \cdot m_i$ for all i and $a_i \not\equiv a_j \pmod{n}$ when $i \neq j$. The sum

$\sum_{j=1}^{n} M_j b_j N_j$ is a solution of the n congruences

$$
\begin{aligned}
x &\equiv a_1 \ (\text{mod } nm_1) \\
x &\equiv a_2 \ (\text{mod } nm_2) \\
&\ \ \vdots \\
x &\equiv a_n \ (\text{mod } nm_n)
\end{aligned}
$$

because if $j \neq i$, then M_j contains the factor nm_i; so $M_j b_j N_j \equiv 0 \ (\text{mod } nm_i)$ and

$$
\begin{aligned}
\sum_{j=1}^{n} M_j b_j N_j &\equiv M_i b_i N_i \ (\text{mod } nm_i) \\
&\equiv nN_i = a_i \ (\text{mod } nm_i)
\end{aligned}
$$

Therefore,

$$
W_i = \frac{\sum\limits_{j=1}^{n} M_j b_j N_j}{m_i} = \frac{\sum\limits_{j \neq i}^{n} M_j b_j N_j}{m_i} + \frac{M_i b_i N_i}{m_i} = nJ_i + \frac{M_i b_i N_i}{m_i}
$$

for an appropriate integer J_i. Since $M_i b_i \equiv n \ (\text{mod } nm_i)$, there exists an integer t_i such that $M_i b_i = t_i(nm_i) + n$ so that

$$
W_i = nJ_i + t_i n N_i + \frac{nN_i}{m_i}
$$

Thus, because $(i - 1) \cdot m_i \leq a_i < i \cdot m_i$,

$$
\lfloor W_i \rfloor \equiv \left\lfloor \frac{nN_i}{m_i} \right\rfloor \equiv \left\lfloor \frac{a_i}{m_i} \right\rfloor \equiv (i - 1) \ (\text{mod } n)
$$

which is part 1.

Let $C = \left[\left[\sum\limits_{j=1}^{n} M_j b_j N_j \right] \right]_{nM}$. We need to show the congruence $\left\lfloor \dfrac{C}{m_i} \right\rfloor \equiv$
$(i-1) \ (\text{mod } n)$ for $1 \leq i \leq n$. By the Division Algorithm there is an integer J such that $\sum\limits_{j=1}^{n} M_j b_j N_j = J(nM) + C$. For $1 \leq i \leq n$, again using the Division Algorithm, there are integers q_i and r_i with $0 \leq r_i < m_i$ such that

$$
\sum_{j=1}^{n} M_j b_j N_j = q_i m_i + r_i
$$

so that

$$
\left\lfloor \frac{\sum\limits_{j=1}^{n} M_j b_j N_j}{m_i} \right\rfloor = q_i
$$

Therefore, by the result of part 1, there is an integer t_i such that

$$
\begin{aligned}
q_i &= (i-1) + t_i n \\
q_i m_i &= (i-1)m_i + t_i n m_i \\
q_i m_i + r_i &= (i-1)m_i + t_i n m_i + r_i \\
J(nM) + C &= (i-1)m_i + t_i n m_i + r_i
\end{aligned}
$$

so that

$$
\frac{C}{m_i} = (i-1) + \left(t_i - \frac{JM}{m_i}\right)n + \frac{r_i}{m_i}
$$

and

$$
\left\lfloor \frac{C}{m_i} \right\rfloor = (i-1) + \left(t_i - \frac{JM}{m_i}\right)n
$$

$$
\left\lfloor \frac{C}{m_i} \right\rfloor \equiv (i-1) \ (\mathrm{mod} \ n)
$$

It is left to the reader to show that if C' is any other positive integer such that $\left\lfloor \dfrac{C'}{m_i} \right\rfloor \equiv (i-1) \ (\mathrm{mod} \ n)$ for all i, $1 \le i \le n$, then $C' > C$. ∎

Finally, we combine the last several results into the following theorem.

Theorem 3.138 *Let $K = \{k_1, k_2, \ldots, k_n\}$ be a set of n distinct positive integers; and let D and E be the integers, and $f(x) = Dx + E$ the function, given by Theorem 3.133 or 3.135. Also, assume that C is given by part 2 of Theorem 3.137 where $m_i = f(k_i)$. Then*

$$
h(x) = \left[\!\!\left[\left\lfloor \frac{C}{Dx + E} \right\rfloor \right]\!\!\right]_n
$$

is a minimum perfect hashing function.

Proof. If $m_i = f(k_i)$, Theorem 3.137 implies that

$$
h(k_i) = \left[\!\!\left[\left\lfloor \frac{C}{f(k_i)} \right\rfloor \right]\!\!\right] = (i-1) \text{ for } 1 \le i \le n
$$

Thus, h is one to one and onto $\{0, 1, \ldots, (n-1)\}$. ∎

Example 3.139 *We now continue Example 3.136 where $K = \{k_1, k_2, k_3, k_4\} = \{3, 6, 7, 12\}$ and $f(x) = 180x + 1$. We found that $\{f(k_1), f(k_2), f(k_3), f(k_4)\} = \{541, 1081, 1261, 2161\}$. Let $m_i = f(k_i)$. The m_i's are pairwise relatively prime and $m_i > n = 4$. Using Theorem 3.137, we obtain the following table:*

i	m_i	M_i	b_i	N_i	$M_i b_i N_i$	$h(k_i)$
1	541	11782990804	235	0	0	0
2	1081	5896945444	12	273	19318393274544	1
3	1261	5055192724	172	631	548650176721168	2
4	2161	2949837124	1303	1621	6230536829339212	3

For example, to find b_3 we solve $5055192724 \cdot b_3 \equiv 4 \pmod{5044}$, which is equivalent to solving $1263798181 \cdot b_3 \equiv 1 \pmod{1261}$ or to solving $22 \cdot b_3 \equiv 1 \pmod{1261}$. Therefore, we want b_3 and y such that $22b_3 + 1261y = 1$. Using the Euclidean Algorithm and back substitution, we obtain the result $22 \cdot 172 + (-3) \cdot 1261 = 1$. Thus

$$ C = \left[\!\left[\sum_{j=1}^{n} M_j b_j N_j \right]\!\right]_{nM} = [[6798505399334924]]_{nM} = 3183904723300 $$

Next we use C, f, and h to hash $k_3 = 7$.

$$ h(k_3) = h(7) = \left[\!\left[\left\lfloor \frac{3183904723300}{1261} \right\rfloor \right]\!\right]_4 = [[2524904618]]_4 = 2 $$

Jaeschke's reciprocal hashing function generates large integers. The computational effort necessary to implement it should be compared to the resources required for using a nonminimal and nonperfect hashing function: (1) possible large memory requirements, (2) dealing with collisions, and (3) using search algorithms to find items in a table.

Exercises

1. If a and b are integers for which both $\gcd(a, b)$ and $\gcd(a - b, a)$ are defined, prove that $\gcd(a, b) = \gcd(a - b, a)$.

2. For each of these sets, find a polynomial $f(x) = Dx + 1$ that maps the set one to one and onto a set of pairwise relatively prime integers:

   (a) $\{12, 15, 18, 24\}$

   (b) $\{5, 15, 20, 25\}$

3. Let $a_1, a_2, \ldots, a_n$ be n distinct positive integers. Prove or disprove: $\gcd(a_1, a_2, \ldots, a_n) = 1$ if and only if $\gcd(a_i, a_j) = 1$ for all $i \neq j$.

4. In the proof of Theorem 3.137, show that $a_i = N_i n$ is the smallest multiple of n such that $(i - 1) \cdot m_i \leq a_i < i \cdot m_i$.

5. In the proof of Theorem 3.137, show that if C' is any other positive integer such that $\lfloor C/m_i \rfloor \equiv (i - 1) \pmod{n}$ for $1 \leq i \leq n$, then $C' > C$.

6. Verify the numbers in the table of Example 3.139.

7. For

   (a) $K = \{12, 15, 18, 24\}$ of Exercise 2(a)

   (b) $K = \{5, 15, 20, 25\}$ of Exercise 2(b)

compute the integer C of Theorem 3.137 and, therefore, the minimal perfect reciprocal hashing function of Theorem 3.138. Verify that $h(k_i) = i - 1$ for each i.

8. For $K = \{5, 25, 30, 48\}$ of Exercise 2(b), compute the integer C of Theorem 3.137 and, therefore, the minimal perfect reciprocal hashing function of Theorem 3.138. Verify that $h(k_i) = i - 1$ for each i.

3.14   Application: Indices

In Section 3.8, the concept of an index was introduced wherein if n has a primitive root r, then for any integer y relatively prime to n, there is a unique exponent x of r such that $1 \le x < n$ and $y \equiv r^x \pmod{n}$. The integer x is called the base r index of y modulo n or the modulo n logarithm of y with base r. We then write $x = \operatorname{ind}_r y$ or sometimes $x = \log_r y$ to emphasize the analogy with real logarithms. Thus

$$y \equiv r^x \pmod{n} \text{ if and only if } x = \operatorname{ind}_r y$$

For this section we will assume $n = p$ is prime so that p will always have a primitive root. Since the set of reduced residues of a prime p is $\{1, 2, \ldots, (p-1)\}$, any integer not divisible by p is congruent to a power of r modulo p.

Once a prime p and a primitive root r of p are selected, an important question is: Given any y not divisible by p, what is its index? If p is a small prime, then one could, as we did in Section 3.8, simply calculate $[[r^x]]_p$ for every x, $1 \le x < p$ creating a table of all indices or stopping when $[[r^x]]_p = y$. On the other hand, if p is large prime, this procedure of exhaustively listing all cases is impractical and computationally infeasible.

The import of easily finding indices is related to the fact that several cryptographic methods for enciphering messages rely upon the infeasibility of computing indices for the security of the enciphered message. If one has a message M, then, using a scheme of encipherment, an altered unreadable message called the ciphertext C is produced (Figure 10). The ciphertext C is available for any one to read because, say, it is transmitted via radio or over telephone lines to the intended recipient.

The intended recipient knows how to restore or decipher the original message. On the other hand, an hostile entity may intercept the ciphertext and through mathematical analysis be able to restore the message without prior knowledge of the decipherment scheme. The process of breaking the cipher without being privy to the secret keys of the method is known as cryptanalysis. Security here refers to the difficulty confronting a cryptanalysis attack to reveal the original message. Several enciphering methods use large primes or composite integers with only large prime factors and are secure to the extent that indices are difficult to calculate. We will discuss such a scheme in Section 3.15.

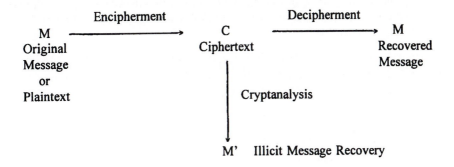

Figure 10: Encipherment and decipherment.

Steven Pohlig and Martin Hellman in "An Improved Algorithm for Computing Logarithms over $GF(p)$ and Its Cryptographic Significance" [64] give a method of computing indices requiring of the order of $(\log p)^2$ operations and storage complexity when $p - 1$ has no large prime factors rather than the order of $\sqrt{p}$ operations and space complexity required by previous methods. If $p \approx 10^{20}$, then $\sqrt{p} \approx 10^{10}$ but $(\log p)^2 \approx 2 \times 10^3$ giving a work reduction by a factor of about 5×10^6.

Theorem 3.56 will be useful in developing Pohlig and Hellman's algorithm. Let n be a positive integer and $\gcd(n, r) = 1$; and let $k = \text{ord}_n r$, the smallest positive integer such that $r^k \equiv 1 \pmod{n}$. Then $r^s \equiv r^t \pmod{n}$ if and only if $s \equiv t \pmod{k}$. If n is a prime p, then $k = \text{ord}_p r = p - 1$; so we have

$$r^s \equiv r^t \pmod{p} \text{ if and only if } s \equiv t \pmod{(p - 1)}$$

Thus, computation in the exponent is done modulo $(p - 1)$, but computation in the left congruence is done modulo p. The theorem above holds whether r is a primitive root of p or not.

Let p be an odd prime, r be a primitive root of p (the case of p an even prime is clear), and y be an integer not divisible by p. We want to find a positive integer x such that $1 \leq x < p$ and $y \equiv r^x \pmod{p}$ so that $x = \text{ind}_r y$.

Since p is prime and odd, $p - 1$ is even and composite. Let the prime factorization of $p - 1$ be

$$p - 1 = q_1^{\alpha_1} q_2^{\alpha_2} \cdots q_k^{\alpha_k}$$

where $q_1, \ldots, q_k$ are distinct primes and $\alpha_i \geq 1$. The method is summarized as follows:

1. For each i:

    (a) Express x in base q_i so that

    $$x = b_n q_i^n + b_{n-1} q_i^{n-1} + \cdots + b_1 q_i + b_0$$

where $0 \le b_j < q_i$ and $b_n \ne 0$ and where for notational simplicity we use a different set of b_j's for each i instead of using the notation $b_{i,n}$ or the like.

(b) Using that $y \equiv r^x \pmod{p}$, calculate

$$[[x]]_{q_i^{\alpha_i}} = b_{\alpha_i-1} q_i^{\alpha_i-1} + b_{\alpha_i-2} q_i^{\alpha_i-2} + \cdots + b_1 q_i + b_0$$

by determining the "digits" $b_0, b_1, \ldots, b_{\alpha_i-1}$.

2. We need an x such that

$$
\begin{aligned}
x &\equiv [[x]]_{q_1^{\alpha_1}} \pmod{q_1^{\alpha_1}} \\
x &\equiv [[x]]_{q_2^{\alpha_2}} \pmod{q_2^{\alpha_2}} \\
&\vdots \\
x &\equiv [[x]]_{q_k^{\alpha_k}} \pmod{q_k^{\alpha_k}}
\end{aligned}
$$

Application of the Chinese Remainder Theorem to this system of k congruences produces a solution x_0 such that $1 \le x_0 < p$ and x_0 is unique modulo $(p-1)$. From the properties of p prime, primitive root r of p, and $p \nmid y$, we know that x with $y \equiv r^x \pmod{p}$ exists and is unique. Since the exponent x satisfies the congruences, we obtain $x = x_0$.

The crux of the algorithm resides in part 1(b), where we calculate $b_0, b_1,$ $\ldots, b_{\alpha_i-1}$ for each i. Let

$$t_i = r^{(p-1)/q_i}$$

and let

$$[[x]]_{q_i^{\alpha_i}} = b_{\alpha_i-1} q_i^{\alpha_i-1} + b_{\alpha_i-2} q_i^{\alpha_i-2} + \cdots + b_1 q_i + b_0$$

Compute b_0 as follows:

Let $y_0 = y$. Then we note that

$$y_0^{(p-1)/q_i} \equiv (r^x)^{(p-1)/q_i} \equiv \left(r^{(p-1)/q_i}\right)^x \equiv (t_i)^{[[x]]_{q_i}} \equiv t_i^{b_0} \pmod{p}$$

because $\mathrm{ord}_p\left(r^{(p-1)/q_i}\right) = q_i$ implies that $x \equiv [[x]]_{q_i} \pmod{p}$ by Theorem 3.56 and because $[[x]]_{q_i} \equiv b_0 \pmod{q_i}$. Since by Theorem 3.56, t_i is a q_i-th root of unity, t_i^u can have only q_i distinct values modulo p. Next

(a) Compute explicitly $\left[\left[y_0^{(p-1)/q_i}\right]\right]_p$.

(b) Compute $[[t_i^u]]_p$ for every choice of u, $0 \le u < q_i$ until an integer equal to that obtained in (a) is found. Alternatively, use the Knuth method described below.

(c) The u giving $y_0^{(p-1)/q_i} \equiv t_i^u \pmod{p}$ is b_0.

Compute b_1 as follows:

Let $y_1 \equiv y_0 r^{-b_0} \equiv r^{x-b_0} \pmod{p}$. Then we see that

$$y_1^{(p-1)/q_i^2} \equiv \left(r^{x-b_0}\right)^{(p-1)/q_i^2} \equiv (t_i)^{(x-b_0)/q_i} \equiv t_i^{b_1} \pmod{p}$$

because

$$\frac{x - b_0}{q_i} \equiv b_1 \equiv \left[\left[\frac{x - b_0}{q_i}\right]\right]_{q_i} \pmod{q_i}$$

by Theorem 3.56 since $\text{ord}_r \, t_i = q_i$. Again t_i^u can have only q_i distinct values modulo p. Next

(a) Compute explicitly $\left[\left[y_1^{(p-1)/q_i}\right]\right]_p$.

(b) Compute $[[t_i^u]]_p$ for every choice of u, $0 \leq u < q_i$ until an integer equal to that obtained in (a) is found. Alternatively, use the Knuth method below.

(c) The u giving $y_1^{(p-1)/q_i} \equiv t_i^u \pmod{p}$ is b_1.

Compute b_2 as follows:
   Let $y_2 \equiv y_1 r^{-b_1 q_i} \equiv r^{x-b_0-b_1 q_i} \pmod{p}$. Then

$$y_2^{(p-1)/q_i^3} \equiv (t_i)^{(x-b_0-b_1 q_i)/q_i^2} \equiv t_i^{b_2} \pmod{p}$$

Next

(a) Compute explicitly $\left[\left[y_2^{(p-1)/q_i}\right]\right]_p$

(b) Compute $[[t_i^u]]_p$ for every choice of u, $0 \leq u < q_i$ until an integer equal to that obtained in (a) is found or use the Knuth method below.

(c) The u giving $y_2^{(p-1)/q_i} \equiv t_i^u \pmod{p}$ is b_2.

   This procedure is continued until $b_{\alpha_i - 1}$ is obtained.
   The Pohlig-Hellman algorithm is more efficient because in computing b_j in part (b), the choices for u involve only the integers $0 \leq u < q_i$ and not $0 \leq x < p$ of the original index problem. For very small q_i, the generation of a table of $[[t_i^u]]_p$ for $0 \leq u < q_i$ is efficient and for this reason is mentioned explicitly in part (b). For moderate and larger-sized q_i, we still need to reduce the computational complexity of this step, say, by using the method of Knuth below. First we consider an example.

Example 3.140 *The prime $p = 73$ has primitive root $r = 5$. Suppose that we want to find $x = \text{ind}_5 \, 13$; that is, we want an x, $1 \leq x < 72$ such that $5^x \equiv 13 \pmod{73}$. Then*

$$p - 1 = 72 = 2^3 \cdot 3^2 = q_1^3 \cdot q_2^2$$
$$t_1 \equiv r^{(p-1)/q_1} \equiv 5^{36} \equiv 72 \pmod{p}$$
$$t_2 \equiv r^{(p-1)/q_2} \equiv 5^{24} \equiv 8 \pmod{p}$$

First consider $q_1 = 2$. Represent x in base q_1 as

$$x = b_n q_1^n + b_{n-1} q_1^{n-1} + \cdots + b_1 q_1 + b_0$$

with $b_n \neq 0$ and $0 \leq b_j < q_1 = 2$. We want to obtain

$$[[x]]_{q_1^3} = b_2 q_1^2 + b_1 q_1 + b_0$$

by determining b_0, b_1, and b_2.

Let $y_0 = 13$. Then $y_0^{(p-1)/q_1} \equiv 13^{36} \equiv 72 \pmod{p}$ and $t_1^{b_0} \equiv (72)^{b_0} \pmod{p}$. We want u such that $0 \leq u < q_1 = 2$ and $72 \equiv 72^u \pmod{p}$. Clearly, by inspection, we have $u = 1 = b_0$.

Let $y_1 \equiv y_0 r^{-b_0} \equiv 13 \cdot 5^{-1} \equiv 13 \cdot 44 \equiv 61 \pmod{p}$. Then $y_1^{(p-1)/q_1^2} \equiv 61^{18} \equiv 72 \pmod{p}$ and $t_1^{b_1} \equiv (72)^{b_1} \pmod{p}$. We want u such that $0 \leq u < q_1 = 2$ and $72 \equiv 72^u \pmod{p}$. Again, by inspection, we have $u = 1 = b_1$.

Let $y_2 \equiv y_1 r^{-b_1 q_1} \equiv 61 \cdot 5^{-2} \equiv 61 \cdot 38 \equiv 55 \pmod{p}$. Then $y_2^{(p-1)/q_1^3} \equiv 55^9 \equiv 1 \pmod{p}$ and $t_2^{b_1} \equiv (72)^{b_2} \pmod{p}$. We want u such that $0 \leq u < q_1 = 2$ and $1 \equiv 72^u \pmod{p}$. By inspection, we have $u = 0 = b_2$. Therefore, for $q_1 = 2$ we have obtained:

$$[[x]]_{q_1^3} = b_2 q_1^2 + b_1 q_1 + b_0 = 0 \cdot 2^2 + 1 \cdot 2^1 + 1 = 3$$

Next, we consider $q_2 = 3$. Represent x in base q_2 as

$$x = b_n q_2^n + b_{n-1} q_2^{n-1} + \cdots + b_1 q_2 + b_0$$

with $b_n \neq 0$ and $0 \leq b_j < q_2 = 3$. We want to obtain

$$[[x]]_{q_2^2} = b_1 q_2 + b_0$$

by determining b_0 and b_1.

Let $y_0 = 13$. Then $y_0^{(p-1)/q_2} \equiv 13^{24} \equiv 64 \pmod{p}$ and $t_2^{b_0} \equiv 8^{b_0} \pmod{p}$. We want u such that $0 \leq u < q_2 = 3$ and $64 \equiv 8^u \pmod{p}$. By inspection, we have $u = 2 = b_0$.

Let $y_1 \equiv y_0 r^{-b_0} \equiv 13 \cdot 5^{-2} \equiv 13 \cdot 38 \equiv 56 \pmod{p}$. Then $y_1^{(p-1)/q_2^2} \equiv 56^8 \equiv 8 \pmod{p}$ and $t_1^{b_1} \equiv 8^{b_1} \pmod{p}$. We want u such that $0 \leq u < q_2 = 3$ and $8 \equiv 8^u \pmod{p}$. Again, by inspection, we have $u = 1 = b_1$. Therefore, for $q_2 = 3$ we have obtained

$$[[x]]_{q_2^2} = b_1 q_1 + b_0 = 1 \cdot 3^1 + 2 = 5$$

and we are finished with part 1 of the algorithm.

For part 2, we use the Chinese Remainder Theorem to solve

$$x \equiv [[x]]_{2^3} \equiv 3 \pmod{2^3}$$
$$x \equiv [[x]]_{3^2} \equiv 5 \pmod{3^2}$$

or

$$x \equiv 3 \pmod{8}$$
$$x \equiv 5 \pmod{9}$$

The solution of these congruences is $x = 59$; therefore, $59 = \mathrm{ind}_5\,13$ modulo 73. As a check, we have

$$5^{59} = (5^{10})^5 \cdot 5^9 \equiv (50)^5 \cdot 10 \equiv 67 \cdot 10 \equiv 13 \pmod{73}$$

The following algorithm for solving $w \equiv t^g \pmod{p}$ for g is a generalization of one by D. E. Knuth on page 9 of *The Art of Computer Programming*, Vol. 3 [43]. The problem solved by the Knuth method is essentially the same as the original problem $y \equiv r^x \pmod{p}$; however, t has order q_i and $0 \le g < q_i$ so that there are a smaller number of cases to consider. Therefore, the Pohlig-Hellman algorithm is used to break down the problem into smaller pieces and the Knuth method is used on these pieces.

Let p be a prime, r is a primitive root of p, and $p - 1$ has the prime factorization $p - 1 = q_1^{\alpha_1} q_2^{\alpha_2} \cdots q_k^{\alpha_k}$. Let $\mathrm{ord}_p\, t = q_i$. We recall that in the context of the original problem $t \equiv r^{(p-1)/q_i} \pmod{p}$. For a given w we want to determine g, $0 \le g < q_i$, such that $w \equiv t^g \pmod{p}$, in effect finding that $g = \mathrm{ind}_t\, w$ modulo p.

Let $0 \le r \le 1$ and let m be the least integer that is greater than or equal to q_i^r. Since g exists, the Division Algorithm provides for integers Q and S such that $g = Qm + S$ with $0 \le S < m$. It is straightforward to show that $0 \le Q < J$, where J is the least integer that is greater than or equal to q_i/m. We note that $m \approx q_i^r$ and $J \approx q_i^{1-r}$ and that Q and S exist even though g is unknown. At this point Q and S are unknown also. Substituting into $w \equiv t^g \pmod{p}$, we get

$$t^S \equiv w \cdot t^{-Qm} \pmod{p}$$

The algorithm proceeds by first calculating t^S, or more specifically, $\left[\left[t^S\right]\right]_p$, on the left side of the congruence above for every possible choice of S, namely, $S = 0, 1, \ldots, (m - 1)$. This process gives a look-up table of values congruent to t^S. Because we will look up values in this table many times, we need to sort the table so that a given value may be found quickly. The table generation and subsequent sorting occur only once, and there are $m \approx q_i^r$ entries in the table.

Next, we work with the right side of the congruence above, calculating successively $w \cdot t^{-Qm}$, or more specifically, $\left[\left[w \cdot t^{-Qm}\right]\right]_p$, for $Q = 0, 1, \ldots, J$. For each Q, look up the number $\left[\left[w \cdot t^{-Qm}\right]\right]_p$ in the table. If it is in the table for a given Q, then, using the corresponding S, calculate $g = Q \cdot m + S$. At most $J \approx q_i^{1-r} \cdot \left[\left[w \cdot t^{-Qm}\right]\right]_p$ calculations and look-ups will need to be done.

There is a constant $m \cdot J \approx q_i^r \cdot q_i^{1-r} = q_i$ of combination table storage and calculation/look-ups associated with the generalized Knuth method. If $r = 1$, then the complete table of all possible $[[t^u]]_p$ is generated and the method agrees with the one suggested in part (b) of the Pohlig-Hellman procedure for small q_i. If q_i is large, then there will not be enough storage and you will want to trade storage for computation by making r larger. We note that the worst computational case occurs when $p - 1 = 2 \cdot q$ for a prime q. The

Pohlig-Hellman algorithm works best when the prime divisors of $p - 1$ are all small; therefore, a "good" prime for a cryptographic enciphering-deciphering scheme relying upon a difficult index determination for its security should not use a prime p with all small prime divisors of $p - 1$.

Pohlig and Hellman provide a streamlined flowchart description of their algorithm in their 1978 paper. The algorithm itself is attributed to Roland Silver. For other more efficient methods see A. E. Western and J. C. Miller's *Tables of Indices and Primitive Roots* [89] and L. M. Adleman's "A Subexponential Algorithm for the Discrete Logarithm Problem with Applications to Cryptography" [1]. A general discussion is given by D. Coppersmith in "Cryptography" [19].

Exercises

1. For $p = 13$ and $r = 2$, use the Pohlig-Hellman algorithm to compute $\text{ind}_2 7$.

2. For $p = 13$ and $r = 6$, use the Pohlig-Hellman algorithm to compute $\text{ind}_6 5$.

3. For $p = 701$ and $r = 2$, use the Pohlig-Hellman algorithm to compute $\text{ind}_2 13$.

3.15   Application: Cryptography

Governments, companies, and individuals have a need to send messages in such a way that only the intended recipient is able to read the messages. Armies send battle orders, banks wire fund transfers, and individuals make purchases using credit or debit cards. The messages are sent over telephone lines, via radio and satellite, or through computer networks and are subject to being intercepted by, perhaps hostile, third parties. To prevent a third party from knowing or modifying the messages, they are scrambled in such a way that the original is obscured. The scrambling process is called encryption or encipherment and the unscrambling process is called decryption or decipherment. The original message is called the plaintext and the scrambled message is the ciphertext.

Of course, both the sender and recipient must have agreed upon the scrambling and unscrambling procedure. This procedure usually consists of a general method along with a "key." The key provides specific information that, with the general method, allows easy encryption by the sender and easy decryption by the recipient; however, without the key, anyone who intercepts the transmission finds it computationally impractical to recover the original message even though he knows the general method. For example, message recovery would require, say, a million years using the fastest computer avail-

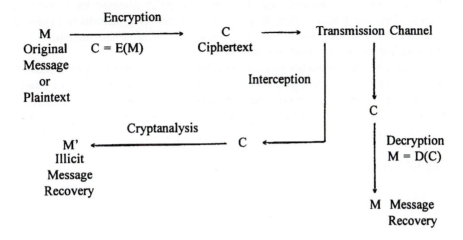

Figure 11: Cryptographic cycle.

able or would require examining all possible messages through an exhaustive trial-and-error search. An attempt to decrypt the plaintext from the cipher-text without benefit of the general method or the key is called cryptanalysis. The mechanism is illustrated in Figure 11 where E is the encryption function and D is the decryption function. Thus, $D(E(M)) = M$. Cryptography is the discipline that deals with the study of methods of encryption and the cryptanalysis of those methods.

Encryption and decryption are usually done by computer; therefore, a message M is first converted to an integer or a series of integers in order to allow manipulation using computer arithmetic. Thus, we will think of M as an integer. The integer message M is then usually divided into blocks $M_1, M_2, M_3, \ldots, M_k$ so that each block is an integer M_i with $0 \le M_i < n$, where n is previously chosen. The encryption is then done block by block as $E(M_i) = C_i$ and decryption proceeds similarly to $D(C_i) = M_i$. For example, one could use 8-bit ASCII computer codes for each alphabetic character as follows:

$$
\begin{array}{ll}
M & 01001101_{binary} = 77_{ten} \\
O & 01001111_{binary} = 79_{ten} \\
N & 01001110_{binary} = 78_{ten} \\
E & 01001010_{binary} = 69_{ten} \\
Y & 01011001_{binary} = 89_{ten}
\end{array}
$$

Then M_i could correspond to groups of two letters, giving

$$\text{``}MO\text{''} \rightarrow M_1 = 01001101\,01001111_{binary} = 19791_{ten}$$

so that $0 \le M_i < n = 2^{16} = 65536$. Of course, the entire message could be considered to be a single string of binary digits which could then be blocked into a sequence of blocks of any convenient size.

R. L. Rivest, A. Shamir, and L. Adleman in "A Method for Obtaining Digital Signatures and Public Key Cryptosystems" [73] introduce a general method of encryption, known as the RSA method after the authors. The method has many desirable characteristics.

Let p and q be two large primes, each having, say, 100 digits each and let $n = p \cdot q$. Two integers e and d that are related to p, q, and n will be used to encrypt and decrypt. Both e and d will be determined later. To encrypt a message $M = M_i$ we calculate

$$C = E(M) = [[M^e]]_n$$

and to decrypt we calculate

$$M = D(C) = [[C^d]]_n$$

Since the products M^e and C^d are reduced modulo n, both ciphertext and plaintext blocks are integers in the range 0 to $(n-1)$. The key to encrypt is the integer pair (n, e) and the key to decrypt is the integer pair (n, d).

First, choose d to be a "large" integer relatively prime to the product $(p-1) \cdot (q-1)$. Then determine the integer e, $0 \le e < n$, that is, the unique solution (Theorem 1.67) of the congruence

$$d \cdot x \equiv 1 \pmod{(p-1)(q-1)}$$

Therefore, since $\phi(n) = \phi(p)\phi(q) = (p-1)(q-1)$,

$$d \cdot e \equiv 1 \pmod{\phi(n)}$$

Theorem 3.141 *If*

(a) $n = p \cdot q$, *where p and q are distinct primes,*

(b) $\gcd(d, \phi(n)) = 1$,

(c) $d \cdot e \equiv 1 \pmod{\phi(n)}$,

(d) *for $0 \le J < n$, define functions*

$$E(J) = [[J^e]]_n \ \text{ and } D(J) = [[J^d]]_n$$

then, for $0 \le J < n$,

$$D(E(J)) = J \text{ and } E(D(J)) = J$$

Proof. Let $0 \le J < n$. Since, $A \equiv B \pmod{n}$ implies that $[[A]]_n = [[B]]_n$, we have

$$
\begin{aligned}
D(E(J)) &= D([[J^e]]_n) = \left[\left[[[J^e]]_n^d\right]\right]_n \\
&= [[(J^e)^d]]_n = [[J^{ed}]]_n \\
&= [[J^1]]_n = J
\end{aligned}
$$

because $d \cdot e \equiv 1 \pmod{\phi(n)}$ and because $r \equiv s \pmod{\phi(n)}$ if and only if $x^r \equiv x^s \pmod{n}$ by Theorem 3.56. The result $E(D(J)) = J$ is proved similarly. ■

Example 3.142 *Suppose that we decide to use 8-bit ASCII encoding for alphabetic characters with a block size of one character or letter. We will need for $n \geq 2^8 = 256$. Let $p = 41$ and $q = 73$ so that $n = p \cdot q = 2993$. $\phi(n) = 40 \cdot 72 = 2880 = 2^6 \cdot 3^2 \cdot 5$. Let $d = 217 = 31 \cdot 7$ so that $\gcd(217, 2880) = 1$. Using the Euclidean Algorithm we obtain $e = 1513$ as a solution to $217x \equiv 1 \pmod{2880}$. Then "MONEY" is encrypted in the following table using the RSA method with key $(n, e) = (2993, 1513)$:*

i	Block	M_i	$E(M_i) = [[M_i^e]]_n = C_i$
1	"M"	77	1683
2	"O"	79	79
3	"N"	78	2560
4	"E"	69	872
5	"Y"	89	2571

where we have used the algorithm of Section 3.6 to calculate M_i^e modulo n efficiently. One may show also that $D(C_i) = M_i$ in every case using the same algorithm with key $(n, d) = (2993, 217)$. The small size of n in this example does not yield a secure encryption function E because small n may be easily factored. In addition, secure blocking will have more than one letter in a block since otherwise we would have a simple substitution cipher which yields easily to direct cryptanalysis.

The RSA method is intended to be used as follows. Both the two primes p and q giving $n = pq$ and the integer d relatively prime to $\phi(n)$ are kept secret; however, the integers n and e are made public, and the general method is public. Therefore, anyone knowing the encryption key (n, e) may encrypt a message M, resulting in an encrypted result C that only someone with the key (n, d) may read. Everyone could have his own set of RSA cryptosystem keys. Each person's public key, (n, e), could be listed, say, in a register as telephone numbers now are. If a person, say Alice, wants to send a secure message M_A to Bob, then Alice would look up Bob's public key, (n_B, e_B) in the public registry. (In this literature, the two parties are often named Bob and Alice.) Next, Alice would encrypt her message M_A as follows:

$$E_B(M_A) = [[M_A^{e_B}]]_{n_B} = C_A$$

The ciphertext C_A is then transmitted to Bob by any convenient method. Anyone encountering C_A besides Bob would be unable to reconstruct M_A in a reasonable amount of time. When Bob receives the enciphered message C_A, he recovers the original using his private key (n_B, d_B) as follows:

$$D_B(C_A) = \left[\left[C_A^{d_B}\right]\right]_{n_B} = M_A$$

The feature of RSA that is different from typical cryptosystems is that part of the key is public. Such methods are called public-key cryptosystems. They were introduced by W. Diffie and M. E. Hellman in "New Directions in Cryptography" [24].

Because $D(E(J)) = J$ and $E(D(J)) = J$ in Theorem 3.141, the RSA cyptosystem has another very useful characteristic also proposed by Diffie and Hellman. Anyone can obtain Bob's public key (n_B, e_B) and send messages to Bob; therefore, if Alice sends Bob a message, how can Bob be sure that the message came from Alice and not someone else? Alice can compose a signature message S_A containing her name and other identifying documentation. Next, Alice uses her own *secret deciphering key* (n_A, d_A)

$$D_A(S_A) = \left[\left[S_A^{d_A}\right]\right]_{n_A} = T_A$$

and includes T_A along with or as part of the main ciphertext C_A of M_A. When Bob gets Alice's transmission, he isolates T_A and applies Alice's *public enciphering key* (n_A, e_A), giving

$$E_A(T_A) = E_A(D_A(S_A)) = S_A$$

Since, presumably, only Alice could have produced a message T_A that could be converted to S_A with Alice's public key (n_A, e_A), the associated message M_A must have been authentic. For such an application, the instrument of authentication must be message dependent and signer dependent. A simple way to implement this authentication for a message M from Alice to Bob consisting of blocks $M_1, M_2, \ldots, M_k$ would be for Alice to generate a signature S consisting of blocks $S_1, S_2, \ldots, S_k$ that are transformed to $T_i = D_A(S_i)$ using Alice's decryption key. We could even let $S_i = M_i$ so that $T_i = D_A(M_i)$. The super block pairs $T_1 M_1, T_2 M_2, \ldots, T_k M_k$ are enciphered using Bob's public key. The integers $T_1 M_1, T_2 M_2, \ldots, T_k M_k$ may need to be reblocked to fit Bob's modulus n_B, say as $J_1, J_2, \ldots, J_m$ where $0 \le J_i < n_B$. Thus, we have

$$U_i = E_B(J_i)$$

Upon receiving the sequence $U_1, U_2, \ldots, U_k$, Bob uses his secret key to obtain

$$J_i = D_B(U_i)$$

which may be unblocked to give again $T_1 M_1, T_2 M_2, \ldots, T_k M_k$. Then Bob checks the signatures with

$$E_A(T_i) = M_i$$

assuming that we used $S_i = M_i$. Since T_i giving $M_i = E_A(T_i)$ could have been produced only by Alice, the message is authentic.

The security of the RSA cryptosystem depends upon the difficulty of factoring $n = p \cdot q$, thereby obtaining p and q. Since e is public, d could then be determined if p and q were known by solving $ex \equiv 1 \pmod{(p-1)(q-1)}$. The security also depends upon the difficulty of finding indices. For a known

M, $C = [[M^e]]_n$ may be calculated using public knowledge. A cryptanalyst would need to solve $M \equiv C^d \pmod{n}$ for $d = \text{ind}_C M$ modulo n, which is also difficult. Also, p and q are randomly chosen large primes to make factoring n difficult. Moreover, d is randomly chosen relatively prime to $\phi(n)$ and large to make a quick exhaustive search over small d unproductive.

Traditional cryptosystems require that both parties be in possession of the same key but no one else. For example, the United States Data Encryption Standard (DES) uses a 56-bit key. (See *Cryptography: A New Dimension in Computer Data Security* by C. H. Meyer and S. M. Matyas [50].) The use of DES for high-volume message traffic may be preferable to using the RSA method since DES has less computational overhead. The question is: How can the parties communicate the key to one another? Diffie and Hellman in their 1976 paper [24] proposed the following method of exchanging keys.

Suppose that p is a prime and g is a primitive root of p. p and g may be publicly known. Suppose that Alice and Bob want to share the same key. Alice chooses a random integer a such that $1 \le a \le p - 2$ and Bob chooses a random integer b such that $1 \le b \le p - 2$. a and b are kept secret and are known only to Alice and Bob, respectively. Alice computes

$$A = [[g^a]]_p$$

and transmits A to Bob in any convenient way. On the other hand, Bob computes

$$B = [[g^b]]_p$$

and transmits B to Alice in any way. Alice computes

$$K_1 = [[B^a]]_p$$

and Bob computes

$$K_2 = [[A^b]]_p$$

But $K_1 = K_2$ because

$$[[A^b]]_p = [[(g^a)^b]]_p = [[(g^b)^a]]_p = [[B^a]]_p$$

Thus, Bob and Alice now have the same key, $K = K_1 = K_2$.

Since B, A, p, and g are perhaps all known, a cryptanalyst can break the cipher by solving $A \equiv g^a \pmod{p}$ for a or by solving $B \equiv g^b$ for b; that is, one must be able to compute $\text{ind}_g A$ or $\text{ind}_g B$ modulo p, both of which are computationally difficult. The complexity of this calculation is on the order of $\sqrt{p}$ according to the discussion of Section 3.14; therefore, p should be a large prime and $p - 1$ should have at least one large prime factor.

For general information about cryptography from ancient times, see David Kahn's *The Codebreakers: The Story of Secret Writing* [38]. See M. E. Hellman's "The Mathematics of Public-Key Cryptography" [34], for a discussion of public-key cryptography. The RSA method was patented by Rivest, Shamir, and Adleman in 1983 (U.S. Patent 4,405,829).

Exercises

1. In Example 3.142, verify that $D(J) = \left[\left[J^{217}\right]\right]_{2993}$ decrypts the message

$$1683, 79, 2560, 872, 2571$$

2. Generate an RSA cryptosystem using $n = 83 \cdot 107$ by finding a public key (n, e) and private key (n, d).

    (a) Use these keys to encipher the message "$HELP$."

    (b) Let the RSA keys of Example 3.142 belong to Alice and the keys for $n = 83 \cdot 107$ belong to Bob. Use the method of signatures described in this section to send the signed message "$MONEY$" from Alice to Bob.

    (c) Using Bob's private and Alice's public keys, decipher the message of part (b) and verify that it came from Alice.

3. In the RSA cryptosystem with (n, e) public, show that if $\phi(n)$ can be found, then the cryptosystem may be compromised by calculating d.

4. Consider the RSA cryptosystem where $n = p \cdot q$ and $\phi(n)$ are known but the primes p and q are not known.

    (a) Show that $p + q$ may be expressed in terms of n and $\phi(n)$.

    (b) Show that if $q > p$, then $q - p = \sqrt{(p+q)^2 - 4n}$.

    (c) Determine p and q.

    (d) Refer to Example 3.142 and assume only that $n = 2993$ and $\phi(n) = 2880$. Then use the method above to find p and q.

3.16   Application: Primality Testing

Thus far, several theorems have been given that may be used to assert that a positive integer n is prime. The exhaustive method of Theorem 2.11 requires checking every positive prime k such that $k^2 < n$ to determine if $k \mid n$. Corollary 3.46 implies that n is prime if $\phi(n) = n - 1$. Wilson's Theorem asserts that n is prime if $(n - 1)! \equiv -1 \pmod{n}$. This test is impractical for large n because $(n-1)!$ is very large for even moderately sized n.

Other tests are more suitable for large n. Fermat's Little Theorem (Theorem 3.54) implies that a necessary condition for n to be prime is that $a^{n-1} \equiv 1 \pmod{n}$ for every a, $0 < a < n$; therefore, if there is an integer a, $0 < a < n$ with $a^{n-1} \not\equiv 1 \pmod{n}$, then n cannot be prime. Also somewhat related to the criterion in Fermat's Little Theorem is Lucas' Primality Test (Theorem 3.61) which states that if there is an integer a with $a^{n-1} \equiv 1 \pmod{n}$ and $a^{(n-1)/p} \not\equiv 1 \pmod{n}$ for all primes p that divide $n - 1$, then n is prime. To apply these tests, we do not have to deal with computations involving large numbers of integers relative to n; we may just need to be lucky in choosing the integer a. In Section 3.6 it was found that there are integers a such that $a^{n-1} \equiv 1 \pmod{n}$ even when n is composite.

Definition 3.143 *If n is composite, $\gcd(a, n) = 1$, and $a^{n-1} \equiv 1 \pmod{n}$, then n is called a* pseudo-prime to base a.

We showed in Section 3.6 that $n = 91$ is a pseudo-prime to base 3 but not to base 2. If n is composite, we may anticipate that if we search long enough that we would be able to find an a such that $a^{n-1} \not\equiv 1 \pmod{n}$ or, equivalently, such that n is not a pseudo-prime to base a; however, for some composite n the search for an a with $\gcd(a, n) = 1$ and $a^{n-1} \not\equiv 1 \pmod{n}$ will be futile. Such a composite n is called a Carmichael number.

Definition 3.144 *If n is a composite positive integer and $a^{n-1} \equiv 1 \pmod{n}$ for all a with $\gcd(a, n) = 1$, then n is called a* Carmichael number.

Example 3.145 *Consider the composite $n = 1105 = 5 \cdot 13 \cdot 17$. Suppose that $\gcd(a, 1105) = 1$. Clearly, $\gcd(a, 5) = \gcd(a, 13) = \gcd(a, 17) = 1$ also. From Fermat's Little Theorem we obtain, since $n - 1 = 1104 = 3 \cdot 2^4 \cdot 23$,*

$$
\begin{aligned}
a^4 &\equiv 1 \pmod{5} &\text{and}\quad a^{1104} &= (a^4)^{276} \equiv 1 \pmod{5} \\
a^{12} &\equiv 1 \pmod{13} &\text{and}\quad a^{1104} &= (a^{12})^{92} \equiv 1 \pmod{13} \\
a^{16} &\equiv 1 \pmod{17} &\text{and}\quad a^{1104} &= (a^{16})^{69} \equiv 1 \pmod{17}
\end{aligned}
$$

Therefore,

$$a^{1104} \equiv 1 \pmod{5 \cdot 13 \cdot 17}$$

so that $n = 1105$ is a Carmichael number.

Carmichael numbers are named for R. D. Carmichael, who discovered them. (See R. D. Carmichael's "On Composite Numbers P Which Satisfy the Fermat Congruence $a^{P-1} \equiv 1 \pmod{P}$" [13].) Carmichael gave 16 such numbers in his 1912 paper. If there were only finitely many Carmichael numbers and all could be determined, then Fermat's Little Theorem would become more useful in determining compositeness. That there are infinitely many Carmichael numbers was established by W. R. Alford, Andrew Granville, and Carl Pomerance [2].

We next characterize Carmichael numbers.

Theorem 3.146 *Let n be a positive integer and $t \geq 2$, and let $p_1, p_2, \ldots, p_t$ be distinct primes with $n = \prod_{i=1}^{t} p_i$. If $(p_i - 1) \mid (n - 1)$ for all i, $1 \leq i \leq t$, then n is a Carmichael number.*

Proof. Clearly, n is composite. Let a be such that $\gcd(a, n) = 1$. Since $n = \prod_{i=1}^{t} p_i$, $\gcd(a, p_i) = 1$ for all i. By Fermat's Little Theorem,

$$a^{p_i - 1} \equiv 1 \pmod{p_i} \text{ for } 1 \leq i \leq t$$

Let b_i be such that $b_i \cdot (p_i - 1) = n - 1$ for each i. Therefore, for each i,

$$a^{n-1} = \left(a^{p_i - 1}\right)^{b_i} \equiv 1 \pmod{p_i}$$

and hence

$$a^{n-1} \equiv 1 \ \left(\operatorname{mod} \prod_{i=1}^{t} p_i \right) \ \blacksquare$$

Note that the method of this proof was used in Example 3.145 to show that $n = 1105$ is a Carmichael number.

Theorem 3.147 *If $p_1, p_2, \ldots, p_t$ are distinct odd primes, $n = \prod_{i=1}^{t} p_i$, and n is a Carmichael number, then $(p_i - 1) \mid (n - 1)$ for $1 \leq i \leq t$.*

Proof. We first show that for each i, $1 \leq i \leq t$, there is a primitive root g_i of p_i such that $\gcd(n, g_i) = 1$. Theorem 3.74 implies that p has a primitive root r_i. If $\gcd(n, r_i) = 1$, let $g_i = r_i$. If $\gcd(n, r_i) \neq 1$, consider integers of the form $r_i + kp_i$. Since $\gcd(r_i, p_i) = 1$, Dirichlet's Theorem (Theorem 2.18) implies that there are infinitely many integers k such that $r_i + kp_i$ is prime. Let g_i be a prime of the form $r_i + kp_i$ such that $r_i + kp_i > n$. Since $g_i \equiv r_i \pmod{p_i}$, g_i is also a primitive root of p_i.

Since n is a Carmichael number and $\gcd(n, g_i) = 1$ for $1 \leq i \leq t$, we obtain $g_i^{n-1} \equiv 1 \pmod{n}$; and since $p_i \mid n$, $g_i^{n-1} \equiv 1 \pmod{p_i}$. Thus, $(p_i - 1) \mid (n - 1)$ using Theorem 3.56 (a) because $\operatorname{ord}_{p_i} g_i = p_i - 1$. $\blacksquare$

In the next theorem it is shown that if $n = \prod_{i=1}^{t} p_i^{\alpha_i}$ is the prime factorization of a Carmichael number n, then $\alpha_i = 1$ for all i.

Theorem 3.148 *No Carmichael number has the square of a prime as a factor.*

Proof. Assume that n is a Carmichael number and p is a prime such that $n = p^e w$, $e > 1$, and $p \nmid w$.

First let p be odd. Theorem 3.74 implies that p^2 has a primitive root g. The following congruence holds modulo n because n is a Carmichael number and holds modulo p^2 because $p^2 \mid n$:

$$1 \equiv g^{n-1} = g^{p^e w - 1}$$

So $\phi(p^2) = p(p - 1)$ and $p(p - 1) \mid (p^e w - 1)$ by Theorem 3.56(a). Thus $p \mid (p^e w - 1)$, which is a contradiction.

Next let $p = 2$; and let q be a prime with the properties $q \equiv 3 \pmod{2^e}$ and $q > w$. Such a q exists because $q \equiv 3 \pmod{2^e}$ is equivalent to $q = 3 + k2^e$ for some k and because, by Dirichlet's Theorem, there will be a k such that $3 + k2^e$ is prime and greater than w. Since $q > w$ and q is prime, $\gcd(n, q) = 1$. Since n is a Carmichael number, $q^{n-1} \equiv 1 \pmod{n}$; and since $2^e \mid n$,

$$q^{2^e w - 1} = q^{n-1} \equiv 1 \pmod{2^e}$$

Because $\phi(2^e) = 2^{e-1}$, Theorem 3.53 implies that

$$q^{2^{e-1}} \equiv 1 \pmod{2^e}$$

But $2^{e-1} \mid 2^e w$ so that

$$q^{2^e w} \equiv 1 \pmod{2^e}$$

Hence,

$$1 \equiv q^{2^e w} \equiv q \cdot q^{2^e w - 1} \equiv q \pmod{2^e}$$

is a contradiction since $q \equiv 3 \pmod{2^e}$. ∎

Theorem 3.149 *A Carmichael number n is a product of at least three distinct odd primes.*

Proof. If n is a Carmichael number and $n = 2s$ for some integer s, then

$$2^{2s-1} \equiv 1 \pmod{n} \qquad \text{or} \qquad 2^{2s-1} - 1 = k \cdot 2 \cdot s$$

for some integer k. Thus, $2 \mid 1$, which is a contradiction.

Thus n is odd and so are all the prime factors of n. By Theorem 3.148, no prime factor of n occurs twice in the factorization of n. Assume that there are primes p_1 and p_2 such that $n = p_1 p_2$ and assume that $p_1 > p_2$. By Theorem 3.147 $(p_1 - 1) \mid (p_1 p_2 - 1)$ and there is an integer k such that $(p_1 - 1)k = p_1 p_2 - 1$. Hence,

$$k = \frac{p_1 p_2 - 1}{p_1 - 1} = p_2 + \frac{(p_2 - 1)}{(p_1 - 1)}$$

which is a contradiction because $(p_1 - 1) \nmid (p_2 - 1)$. ∎

Although the existence of Carmichael numbers prevents Fermat's Little Theorem from ferreting out compositeness of an integer n with certainty, we will see that some of those composite integers whose detection is missed by Fermat's Little Theorem are found by Lagrange's Theorem (Theorem 3.34). In Example 3.145 it was found that $n = 1105$ is a Carmichael number. But for $a = 2$,

$$
\begin{aligned}
2^{69} &\equiv 967 \pmod{1105} \\
2^{138} &= (2^{69})^2 \equiv 259 \pmod{1105} \\
2^{276} &= (2^{138})^2 \equiv 781 \pmod{1105} \\
2^{552} &= (2^{276})^2 \equiv 1 \pmod{1105} \\
2^{1104} &= (2^{552})^2 \equiv 1 \pmod{1105}
\end{aligned}
$$

Thus, because of Lagrange's Theorem, if $n = 1105$ is prime, the congruence $x^2 \equiv 1 \pmod{n}$ would have at most 2 incongruent solutions modulo n. But $x = 1$ and $x = -1$ are always solutions and are incongruent if $n > 2$. Since $x = 781$ is a third solution, n cannot be prime. A test comprised of a combination of Fermat's Little Theorem and Lagrange's Theorem is given in the following criterion.

Definition 3.150 *Let n be a positive integer and let $n - 1 = 2^k w$ where w is odd and $k \geq 1$. If $\gcd(n, a) = 1$, then n passes Miller's test for base a if either*

(a) $a^w \equiv 1 \pmod{n}$

or

(b) $a^{2^i w} \equiv -1$ *for some i, $0 \leq i \leq k - 1$.*

Miller's test is useful because every odd prime passes the test for any base, but the chances that a composite number passes the test for a base is less than $1/4$. Thus, if an integer n passes Miller's test for a base, we think of n as possibly prime; whereas if n fails Miller's test, it is certainly composite. These results are given in the next two theorems.

Theorem 3.151 *If p is an odd prime and $\gcd(a, p) = 1$, then p passes Miller's test for base a.*

Proof. Let $p - 1 = 2^e w$ such that $e \geq 1$ and w is odd. Also let a be relatively prime to p and

$$b_k = a^{2^{e-k} w} \text{ for } k = 0, 1, 2, \ldots, e$$

We want to show that either $b_k \equiv 1 \pmod{p}$ for $0 \leq k \leq e$ or there is a j, $0 \leq j \leq e - 1$ such that $b_j \equiv -1 \pmod{p}$. This result will establish the theorem since if $b_k \equiv 1 \pmod{p}$ for $0 \leq k \leq e$, then $1 \equiv b_e = a^w \pmod{p}$; and if not, then there is a j such that

$$a^{2^j w} = b_j \equiv -1 \pmod{p}$$

Assume that there is an i, $0 \leq i \leq e$, such that $b_i \not\equiv 1 \pmod{p}$. Let j be the least such integer i. $j > 0$ because Fermat's Little Theorem implies that $b_0 = a^{p-1} \equiv 1 \pmod{p}$. But $(b_j)^2 \equiv b_{j-1} \equiv 1 \pmod{p}$ so that $b_j \equiv 1 \pmod{p}$ or $b_j \equiv -1 \pmod{p}$, by Lagrange's Theorem. Since $b_j \not\equiv 1 \pmod{p}$, it follows that $b_j \equiv -1 \pmod{p}$. ∎

Example 3.152 *For the Carmichael number $n = 1105$ of Example 3.145, $n - 1 = 1104 = 2^4 \cdot 69$. In the discussion above it was shown that*

$$
\begin{aligned}
2^{69} &\equiv 967 \pmod{1105} \\
2^{2 \cdot 69} &\equiv 259 \pmod{1105} \\
2^{2^2 \cdot 69} &\equiv 781 \pmod{1105} \\
2^{2^3 \cdot 69} &\equiv 1 \pmod{1105}
\end{aligned}
$$

so that $n = 1105$ does not pass Miller's test and is composite.

Knowing that n is composite, in itself, does not reveal factors of n; however, because it was shown above that

$$2^{2^i \cdot 69} \equiv 1 \pmod{1105}$$

for $i = 3$, we can use this congruence to factor $n = 1105$. The congruence above is equivalent to

$$\left(2^{2^{i-1} \cdot 69} - 1\right)\left(2^{2^{i-1} \cdot 69} + 1\right) \equiv 0 \pmod{1105}$$

For $i = 3$, we have $2^{2^2 \cdot 69} \equiv 781 \pmod{1105}$ and so obtain

$$(781 + 1)(781 - 1) = s \cdot 1105$$

for some integer s; therefore, some factor of 782 or 780 divides 1105. If we find that $\gcd(782, 1105) > 1$ or $\gcd(781, 1105) > 1$, then we have a nontrivial divisor of $n = 1105$. But $\gcd(780, 1105) = 65$ and $\gcd(782, 1105) = 17$. These two factors of 1105 imply that $1105 = 5 \cdot 13 \cdot 17$.

The example illustrates that if in applying Miller's test for n and base a with $n - 1 = 2^k w$, one finds that

$$a^w \not\equiv 1 \pmod{n} \text{ and } a^{2^i w} \not\equiv \pm 1 \pmod{n} \text{ for } 1 \le i \le j < k - 1$$

but

$$a^{2^j w} \equiv 1 \pmod{n}$$

then

$$a^{2^i w} \equiv 1 \pmod{n} \text{ for } j \le i \le k$$

so that $a^{2^i w} \equiv -1 \pmod{n}$ will never occur and one may stop with $i = j$. As in Example 3.152, one can then look for factors of n by calculating the greatest common divisors

$$\gcd\left(\left[\left[a^{2^{j-1} w}\right]\right]_n + 1, n\right) \text{ and } \gcd\left(\left[\left[a^{2^{j-1} w}\right]\right]_n - 1, n\right)$$

for base a.

Michael Rabin in "Probabilistic Algorithm for Testing Primality" [69] determined the probability that a composite integer n would pass Miller's test for a base. This result enables one to test positive integers n for being prime so that the result is (a) n is composite or (b) n is probably prime.

Theorem 3.153 (Rabin) *Let n be an odd positive integer and choose at random k bases $a_1, a_2, \ldots, a_k$ with $0 < a_i < n$ and $\gcd(a_i, n) = 1$. If n is composite, then the probability that n passes Miller's test for base a_i for all i, $1 \le i \le k$, is less than $(1/4)^k$.*

Rabin's Theorem may be applied to test large odd integers n. Suppose that we choose at random 20 integers a_i, $1 \le i \le 20$ with $0 < a_i < n$ and

$\gcd(a_i, n) = 1$. There is no need to choose $a_i = 1$. This selection may be done in practice by considering $a_i = b_m 2^m + b_{m-1} 2^{m-1} + \cdots + b_1 2^1 + b_0$ to be a base two representation, by choosing the binary digits $b_j = 0$ or 1 at random, and by retaining the result if $1 < a_i < n$ and $\gcd(a_i, n) = 1$. (See Section 2.8 for methods of generating pseudo-random digits.) If any one of the a_i fails as a base in Miller's test for n, then we know, by Theorem 3.151, that n is not prime. On the other hand, if n passes Miller's test for all a_i, then we can say that "n is probably prime." In this case we are *not* asserting that n is prime, but we have applied a procedure that has a probability of being wrong less than one time in $4^{20} \approx 10^{12}$.

If we choose instead $k = 100$ integers a_i for bases and find that all a_i, $1 \le i \le 100$, pass as bases for Miller's test with n declaring n to be "probably prime," the method has a probability of being wrong less than one time in $4^{100} \approx 10^{60}$. This degree of uncertainty about primeness may be adequate for generating primes for cryptographic purposes.

Suppose that we want to generate a "random" prime n having 300 binary digits. Generate a random sequence of binary digits, $b_0, b_1, \ldots, b_{299}$ using methods of Section 1.8. In base two we have $n = b_{299} 2^{299} + b_{298} 2^{298} + \cdots + b_0$. Rabin suggests testing successively $n, n+2, n+4, \ldots$ if n is odd until a prime is "found" using Rabin's test. If n is even, test $n+1, n+3, n+5, \ldots$ instead. By the Prime Number Theorem, the density of primes near $n = 2^{300}$ is $1/\log(n) > 1/300$ so that, on average, about every 300-th integer will be prime. So test $n + 2i$ or $n + (2i - 1)$ for, say, $1 \le i \le 600$. If no prime is found among these 600, generate a new n having 300 binary digits and repeat the search.

Another probabilistic test for primality similar in spirit to Rabin's was introduced by R. Solovay and V. Strassen in "A Fast Monte-Carlo Test for Primality" [83]. It uses the Jacobi symbol and relies upon Euler's Criterion, Theorem 3.95, which is satisfied by odd primes.

Theorem 3.154 (Solovay and Strassen) *For a positive integer n, choose k integers $a_1, a_2, \ldots, a_k$ with $0 < a_i < n$ and $\gcd(a_i, n) = 1$ at random and for each determine whether*

$$a_i^{(n-1)/2} \equiv \left(\frac{a_i}{n} \right)_J \pmod{n}$$

If this congruence fails to hold for any i, then n is composite. The probability that n is composite but the congruence holds for every i, $1 \le i \le k$, is less than $(1/2)^k$.

If a positive integer passes Solovay-Strassen's test, then it is "probably prime" in the same sense as with Rabin's test. The method of computing the Jacobi symbol provided by Theorems 3.111 and 3.112 makes the Solovay-Strassen test computationally straightforward to apply.

Exercises

1. Show that the following integers are Carmichael numbers:

   (a) $n = 561 = 3 \cdot 11 \cdot 17$ (561 is the smallest Carmichael number).

   (b) $m = 15841 = 7 \cdot 31 \cdot 73$

2. Prove that these Carmichael numbers are composite using Theorem 3.151 and factor them using the method described in Example 3.152:

   (a) $n = 561$ (b) $n = 15841$

3. Prove that $n = 851$ is composite using Theorem 3.151 with $a = 2$.

4. Show that $n = 2625$ is composite using methods in this section.

5. Show $n = 4577$ is composite using Theorem 3.151.

6. Let p and q be primes such that $3 < p < q$ and $n = 3pq$ is a Carmichael number. Prove that $p = 11$ and $q = 17$ so that $n = 561$. (*Hint:* Using an argument similar to that in the proof of Theorem 3.149, show that

$$\frac{pq - 1}{2}, \quad \frac{3p - 1}{q - 1}, \quad \text{and} \quad \frac{3q - 1}{p - 1}$$

   must be integers.)

7. Let $n = 7 \cdot 19 \cdot q$.

   (a) Show that if $q = 67$, then n is a Carmichael number.

   (b) Show that there is only one Carmichael number of the form $n = 7 \cdot 19 \cdot q$ with q prime.

8. Apply the Solovay-Strassen primality test to prove that the following integers are composite:

   (a) $n = 91$ (b) $n = 117$

Chapter 4
ARITHMETIC
FUNCTIONS

4.1   Introduction

Each positive integer n has associated with it several numerical properties. For example, one may ask how many positive integer divisors n has, how many positive integers less than n are relatively prime to n, or what is the sum of the positive divisors of n. In each of these examples, the particular characterization of n is another positive integer; thus, the property of n under consideration is a function that assigns an integer to n. This integer, of course, is the measure of the characteristic in question. These functions are called arithmetic functions or number-theoretic functions.

Definition 4.1 *Let N be the set of positive integers and D an integral domain. A function $f : N \to D$ is called an* arithmetic function *or* number-theoretic function.

In most cases, we will consider the integral domain D to be the set of integers Z. Other contexts can be considered where D is the set of real or complex numbers. Many of the arithmetic functions will be related to the Unique Prime Factorization Theorem (Theorem 2.6) representation of a positive integers as a prime or as a product of primes:

$$n = p_1^{\alpha_1} p_2^{\alpha_2} \cdots p_t^{\alpha_t}$$

where the p_i are distinct primes and $\alpha_i > 0$.

An important arithmetic function is the Euler totient function $\phi : N \to N$, where $\phi(n)$ is defined to be the number of positive integers less than n but relatively prime to n. Several properties of ϕ were derived in Section 3.5. It was found that the order of an integer modulo n divides $\phi(n)$ and that ϕ was useful in counting primitive roots. For a positive integer n with the prime factorization given above, Theorem 3.48 implies that

$$\phi(n) = \prod_{i=1}^{t} \left[p_i^{\alpha_i - 1}(p_i - 1) \right] = n \prod_{i=1}^{t} \left(1 - \frac{1}{p_i} \right)$$

Theorem 3.42 states that

$$n = \sum_{d \mid n} \phi(d)$$

where the sum is over all positive divisors d of n; that is, a term in the sum is generated for every positive divisor d of n. In Theorem 3.44, it was proved that

$$\phi(m \cdot n) = \phi(m) \cdot \phi(n)$$

provided that the positive integers m and n are relatively prime.

In this chapter, analogs of the three properties of ϕ above will be obtained for several other arithmetic functions.

4.2   Multiplicative Functions

The properties of Euler's ϕ function described in Section 4.1 will be formalized in order to be able to characterize other arithmetic functions.

Definition 4.2 *An arithmetic function f is said to be* multiplicative *provided that $f(m \cdot n) = f(m) \cdot f(n)$ whenever m and n are relatively prime, that is, whenever $\gcd(m, n) = 1$. In addition, f is said to be* completely multiplicative *provided that $f(m \cdot n) = f(m) \cdot f(n)$ for all positive integers m and n.*

Let $U : N \to Z$ be defined by $U(n) = 1$ for all n in N. Clearly, U is an arithmetic function. For any positive integers n and m, $U(m \cdot n) = 1 = 1 \cdot 1 = U(m) \cdot U(n)$ so that U is completely multiplicative; therefore, U is also multiplicative. If $E : N \to N$ is defined by $E(n) = n$, then for any positive integers m and n, $E(m \cdot n) = m \cdot n = E(m) \cdot E(n)$. Thus, E is completely multiplicative. The zero function $Z(n) = 0$ for $n \in N$ is also completely multiplicative. The function $I : N \to Z$ defined by

$$I(n) = \begin{cases} 1 & \text{if } n = 1 \\ 0 & \text{if } n > 1 \end{cases}$$

is also completely multiplicative.

Definition 3.41 and Theorem 3.44 imply that Euler's totient function ϕ is multiplicative. But ϕ is not completely multiplicative since for $m = 6$ and $n = 9$, $\phi(6) = 2$, $\phi(9) = 6$, and $\phi(54) = \phi(2 \cdot 3^3) = (2 - 1)3^2(3 - 1) = 18 \neq \phi(6)\phi(9) = 12$.

Theorem 4.3 *If D is an integral domain and $f : N \to D$ is a multiplicative arithmetic function which is not identically zero, then $f(1) = 1$.*

Proof. Let n be an integer for which $f(n) \neq 0$. Then $\gcd(n, 1) = 1$ and $f(n) \cdot 1 = f(n) = f(n \cdot 1) = f(n)f(1)$. By cancellation of the nonzero $f(n)$, $f(1) = 1$. ∎

For a multiplicative arithmetic function, we can reduce the computation of $f(n)$ in terms of $f(p_i^{\alpha_i})$, where $n = p_1^{\alpha_1} p_2^{\alpha_2} \cdots p_t^{\alpha_t}$. Often the value of f for a power of a prime is easy to calculate. This procedure worked for ϕ to give the formula in Theorem 3.48. The same type of argument that was used for ϕ may be applied to other multiplicative arithmetic functions as shown in the following theorem. The proof of the theorem is left to the reader.

Theorem 4.4 *If f is a multiplicative function and $n = p_1^{\alpha_1} p_2^{\alpha_2} \cdots p_t^{\alpha_t}$ is the prime factorization of the positive integer n, then*

$$f(n) = f(p_1^{\alpha_1}) f(p_2^{\alpha_2}) \cdots f(p_t^{\alpha_t})$$

Similarly, a completely multiplicative arithmetic function is also completely determined by its action upon each prime since $f(n) = f(p_1)^{\alpha_1} f(p_2)^{\alpha_2} \cdots f(p_t)^{\alpha_t}$.

In the Exercises of Section 1.3 , we defined sums over consecutive integers as in $\sum_{i=1}^{n} a_i$. We shall generalize this sum. Earlier we used the notation $\sum_{d \mid n} \phi(d)$, which is a special case of the following definition.

Definition 4.5 *Let f be an arithmetic function. If T is a finite subset of the integers, $u : \{1, 2, \ldots, k\} \rightarrow T$ is one to one and onto, and f is an arithmetic function, then*

$$\sum_{t \in T} f(t) = \sum_{i=1}^{k} f(u(i))$$

and

$$\prod_{t \in T} f(t) = \prod_{i=1}^{k} f(u(i))$$

If $T = \varnothing$, then define $\sum_{t \in T} f(t) = 0$ and $\prod_{t \in T} f(t) = 1$.

Since the codomain of f is an integral domain, it is immaterial which bijection u from $\{1, 2, \ldots, k\}$ onto T is used. For a given integer n, let $T = \{d : d \text{ is a positive integer and } d \mid n\}$. We sometimes write $d \mid n$ instead of the less descriptive $d \in T$ so that

$$\sum_{d \in T} f(d) = \sum_{d \mid n} f(d)$$

The proof of the following theorem is left to the reader.

Theorem 4.6 *If f is an arithmetic function, then for each positive integer n,*

$$\sum_{d|n} f(d) = \sum_{d|n} f\left(\frac{n}{d}\right)$$

Summing ϕ over all positive divisors of an integer n gave an interesting and useful formula (Theorem 3.42). In fact, $\sum_{d|n} \phi(d)$ is the arithmetic function $E(n) = n$ just discussed. Applying this procedure more generally leads to the following definition which provides a way to generate new arithmetic functions from previously known ones.

Definition 4.7 *If f is an arithmetic function, then the arithmetic function*

$$F(n) = \sum_{d|n} f(d)$$

where the sum is over all divisors d of n, is called the summatory function of f.

If $E(n) = \sum_{d|n} \phi(d) = n$, we note that ϕ is multiplicative and so is its summatory function $E(n)$. The next theorem establishes that this relationship between a multiplicative arithmetic function and its summatory function is true in general.

Theorem 4.8 *If f is a multiplicative arithmetic function, then so is its summatory function*

$$F(n) = \sum_{d|n} f(d)$$

Proof. Let f be a multiplicative arithmetic function and assume that m and n are positive integers with $\gcd(m, n) = 1$. Since m and n are relatively prime, if d is a positive integer and $d \mid mn$, then $d = d_1 d_2$, where $d_1 \mid m$ and $d_2 \mid n$; and hence d_1 and d_2 are relatively prime. Let

$$
\begin{aligned}
D &= \{d : d \text{ is a positive integer and } d \mid mn\} \\
D_1 &= \{d_1 : d_1 \text{ is a positive integer and } d_1 \mid m\} \\
D_2 &= \{d_2 : d_2 \text{ is a positive integer and } d_2 \mid m\} \\
E &= \{e : e = d_1 d_2 \text{ for some } d_1 \in D_1 \text{ and } d_2 \in D_2\}
\end{aligned}
$$

From the statement above $D \subseteq E$. It is also clear that $E \subseteq D$ since if $d = d_1 d_2$

with $d_1 \in D_1$ and $d_2 \in D_2$, then $d \mid mn$.

$$F(mn) \;=\; \sum_{d \mid mn} f(d) = \sum_{d \in D} f(d) = \sum_{e \in E} f(e)$$

$$=\; \sum_{d_1 d_2 \in E} f(d_1 d_2) = \sum_{d_1 d_2 \in E} f(d_1) f(d_2)$$

$$=\; \sum_{d_2 \in D_2} \sum_{d_1 \in D_1} f(d_1) f(d_2)$$

$$=\; \sum_{d_2 \in D_2} \left[f(d_2) \sum_{d_1 \in D_1} f(d_1) \right]$$

$$=\; \left(\sum_{d_1 \in D_1} f(d_1) \right) \left(\sum_{d_2 \in D_2} f(d_2) \right)$$

$$=\; F(m) \cdot F(n)$$

so that F is multiplicative. ∎

We use Definition 4.7 to define several new arithmetic functions. Theorem 4.8 provides an easy path to establish that these new number-theoretic functions are multiplicative.

Definition 4.9 *In the following table, f is the generating arithmetic function, F is its summatory function, and k is a nonnegative integer.*

$f(m)$	$F(n) = \sum_{d \mid n} f(d)$
1	$\tau(n) = \sum_{d \mid n} 1$
m	$\sigma(n) = \sum_{d \mid n} d$
m^k	$\sigma_k(n) = \sum_{d \mid n} d^k$

$\tau(n)$ *is the number of positive divisors of n.*
$\sigma(n)$ *is the sum of the positive divisors of n.*
$\sigma_k(n)$ *is the sum of the k-th powers of the positive divisors of n.*

Clearly, $\sigma = \sigma_1$ and $\tau = \sigma_0$ are special cases of σ_k. The proof of the following theorem is left to the reader.

Theorem 4.10 *The three arithmetic functions of Definition 4.9 are multiplicative; and, therefore, τ, σ, and σ_k are multiplicative.*

We next find formulas that provide a means of calculating $\tau(n)$, $\sigma(n)$, and $\sigma_k(n)$ by first following the method of Theorem 4.4 where we obtained formulas for powers of primes. In the next two theorems, we have included the previously derived results for Euler's totient function ϕ for completeness.

Theorem 4.11 *If p is a prime and α is a positive integer, then*

(a) $\tau(p^\alpha) = \alpha + 1$.

(b) $\sigma(p^\alpha) = 1 + p + p^2 + \cdots + p^\alpha = \dfrac{p^{\alpha+1} - 1}{p - 1}$.

(c) $\sigma_k(p^\alpha) = \dfrac{p^{(\alpha+1)k} - 1}{p^k - 1}$.

(d) $\phi(p^\alpha) = p^{\alpha-1}(p - 1)$.

Proof. The positive divisors of p^α are $1, p, p^2, p^3, \ldots, p^\alpha$. There are $(\alpha + 1)$ of these divisors. Thus, $\tau(p^\alpha) = (\alpha + 1)$ and the sum of these divisors is $\sigma(p^\alpha) = 1 + p + p^2 + p^3 + \cdots + p^\alpha = \dfrac{p^{\alpha+1} - 1}{p - 1}$ since the sum is geometric. Part (c) is left to the reader. ∎

Theorem 4.12 *Let the positive integer n have the unique prime factorization $n = p_1^{\alpha_1} p_2^{\alpha_2} \cdots p_t^{\alpha_t}$. Then*

(a) $\tau(n) = \displaystyle\prod_{i=1}^{t} (\alpha_i + 1)$.

(b) $\sigma(n) = \displaystyle\prod_{i=1}^{t} \dfrac{p_i^{(\alpha_i+1)} - 1}{p_i - 1}$.

(c) $\sigma_k(n) = \displaystyle\prod_{i=1}^{t} \dfrac{p_i^{(\alpha_i+1)k} - 1}{p_i^k - 1} T$.

(d) $\phi(n) = \displaystyle\prod_{i=1}^{t} \left[p_i^{(\alpha_i-1)} (p_i - 1) \right]$.

Proof. The proof follows from Theorems 4.10 and 4.11. ∎

We note that the number of positive divisors of n is independent of the particular primes that are in the prime factorization of n. The number depends only upon the number of prime divisors and their exponents.

Example 4.13 Let $n = 12 = 2^2 \cdot 3$. We will first calculate $\tau(12)$, $\sigma(12)$, and $\sigma_4(12)$ directly from their definitions given in Definition 4.9. The integer d is *a positive divisor of n in the following table:*

d	$d^0 = 1$	d^4
1	1	1
2	1	16
3	1	81
4	1	256
6	1	1296
12	1	20736
28	6	22386

By direct calculation, $\tau(12) = 6$, $\sigma(12) = 28$, and $\sigma_4(12) = 22386$.
  For $n = 12 = 2^2 \cdot 3$ and $t = 2$, let $p_1 = 2$, $\alpha_1 = 2$, $p_2 = 3$, and $\alpha_2 = 1$.
Thus, using Theorem 4.12, we obtain

$$\tau(12) \quad = \quad (\alpha_1 + 1)(\alpha_2 + 1) = (2+1)(1+1) = 6$$

$$\sigma(12) \quad = \quad \frac{p_1^{(\alpha_1+1)} - 1}{p_1 - 1} \cdot \frac{p_2^{(\alpha_2+1)} - 1}{p_2 - 1} = \frac{2^3 - 1}{2 - 1} \cdot \frac{3^2 - 1}{3 - 1} = 28$$

$$\sigma_4(12) \quad = \quad \frac{p_1^{(\alpha_1+1) \cdot 4} - 1}{p_1^4 - 1} \cdot \frac{p_2^{(\alpha_2+1) \cdot 4} - 1}{p_2^4 - 1}$$

$$= \quad \frac{2^{3 \cdot 4} - 1}{2^4 - 1} \cdot \frac{3^{2 \cdot 4} - 1}{3^4 - 1} = 22386$$

Exercises

1. Calculate:

    (a) $\tau(41)$ **(b)** $\sigma(41)$ **(c)** $\sigma_2(41)$ **(d)** $\phi(41)$

2. Calculate:

    (a) $\tau(72)$ **(b)** $\sigma(72)$ **(c)** $\sigma_3(72)$ **(d)** $\phi(72)$

3. Calculate:

    (a) $\tau(1125)$ **(b)** $\sigma(1125)$ **(c)** $\sigma_2(1125)$ **(d)** $\phi(1125)$

4. Determine whether each of these arithmetic functions is multiplicative or completely multiplicative

    (a) $f(n) = 3$ **(b)** $f(n) = 3n$ **(c)** $f(n) = n + 3$

    (d) $f(n) = n^2$

5. Prove Theorem 4.11(c).

6. Prove Theorem 4.4.

7. Prove Theorem 4.6.

8. Prove Theorem 4.10.

9. Find all positive integers n such that $\phi(n) = 2^k$ for some $k \geq 1$.

10. Let n be a positive integer. Prove that $\prod_{d|n} d = n^{\tau(n)/2}$.

11. In Chapter 2 we defined a positive integer n to be perfect if it is the sum of its positive divisors other that itself.

    (a) Prove that a positive integer n is perfect if and only if $\sigma(n) = 2n$.

    (b) Assume that n is even. Provide an alternative proof of Theorem 2.21 using properties of σ to prove that if $n = 2^{p-1}(2^p - 1)$ where both p and $2^p - 1$ are prime, then n is perfect.

12. If n has prime factorization $\prod_{i=1}^{k} p_i^{\alpha_i}$, prove that $\tau(n) \geq 2^k$.

13. Prove the following when p is prime and α is a positive integer:

    (a) If $p \equiv 1 \pmod 4$, then $\sigma(p^\alpha) \equiv \alpha + 1 \pmod 4$.

    (b) If $p \equiv -1 \pmod 4$, then

    $$\sigma(p^\alpha) \equiv \begin{cases} 1 \pmod 4 \text{ when } \alpha \equiv 0 \pmod 2 \\ 0 \pmod 4 \text{ when } \alpha \equiv 1 \pmod 2 \end{cases}$$

14. Prove that if f and g are multiplicative arithmetic functions with integer values, then $h(n) = f(n) \cdot g(n)$ is multiplicative.

15. Prove that for $k > 1$, there are infinitely many integer n with $\tau(n) = k$.

16. Let n be a positive integer such that $\sigma(n)$ is prime. Prove that there is a prime q and an integer $k \geq 1$ such that $n = q^k$.

17. Find all positive integers n such that $\phi(n) \mid n$.

18. Prove that if n is an odd positive integer, then $\tau(n) \equiv \sigma(n) \pmod 2$.

19. Let n be a positive integer. Prove that $\tau(n)$ is odd if and only if n is the square of an integer.

20. If n is a positive integer, $\tau(n) = q$, and q is prime, prove that there exists a prime p such that $n = p^{q-1}$.

21. Prove that if m and n are positive integers and $m \mid n$, then $\phi(m) \mid \phi(n)$.

22. For a positive integer k, prove that there are only finitely many integers

n such that $\phi(n) = k$.

23. Find the least positive integer n such that:

   (a) $\phi(n) \geq 50$

   (b) $\phi(n) \geq 5000$

24. Prove that if $n \equiv 7 \pmod 8$, then $\sigma(n) \equiv 0 \pmod 8$.

25. For arithmetic functions f, g, and h, the Dirichlet product $f * g$ is defined by

$$(f * g)(n) = \sum_{d \mid n} f(d) g\left(\frac{n}{d}\right)$$

for each positive integer n. The function $f * g$ is also referred to as the Dirichlet convolution. Let U, E, and I be the arithmetic functions introduced at the beginning of Section 4.2.

   (a) Prove that for each positive integer n,

$$(f * g)(n) = \sum_{d \mid n} f\left(\frac{n}{d}\right) g(d)$$

   (b) Find a formula for $f * f$.

   (c) Prove that $f * g = g * f$.

   (d) Prove that $(f * g) * h = f * (g * h)$.

   (e) Prove that $f * I = f$.

   (f) Prove that $f * U = \sum_{d \mid n} f(d)$.

   (g) If f and g are multiplicative, prove that $f * g$ is multiplicative.

   (h) Use parts (f) and (g) to obtain an alternative proof of Theorem 4.8.

   (i) Prove that if $h = f * g$ and both g and h are multiplicative then f is multiplicative. (*Hint:* Assume that g is multiplicative and f is not. Then prove that h is not multiplicative. Do this by letting mn be the least product of positive integers such that $\gcd(m, n) = 1$ and such that $f(m \cdot n) \neq f(m) \cdot f(n)$. Then show that (i) $mn > 1$ and (ii) $h(mn) - h(m)h(n) = f(mn) - f(m)f(n) \neq 0$.)

   (j) If f is an arithmetic function and there is an arithmetic function g such that $f * g = I$, then g is called the Dirichlet inverse of f and is denoted by $g = f^{-1}$. Prove that every arithmetic function f with

$f(1) \neq 0$ has a Dirichlet inverse. (*Hint:* Let $g(1) = 1/f(1)$. Define g inductively by $g(n) = - \sum_{d|n, d \neq n} f\left(\frac{n}{d}\right) g(d)$.)

(k) Prove that if f is multiplicative and f is not the zero function, then:

(i) The Dirichlet inverse f^{-1} exists.

(ii) f^{-1} is multiplicative.

(iii) f^{-1} is not the zero function.

(l) Prove that for arithmetic functions f and g such that the Dirichlet inverses f^{-1} and g^{-1} exist, $(f * g)^{-1} = f^{-1} * g^{-1}$.

4.3   The Möbius Function μ

In Section 4.2, given an arithmetic function f, we defined another arithmetic function F by

$$F(n) = \sum_{d|n} f(d)$$

for each positive integer n. We found that the prime factorization of $n = p_1^{\alpha_1} p_2^{\alpha_2} \cdots p_t^{\alpha_t}$ was important in obtaining explicit computation formulas for F when f is multiplicative. We will now consider the special case of $n = p_1^2 \cdot p_2$.

There are $\tau(n) = (2+1)(1+1) = 6$ positive divisors of n. They are 1, p_1, p_1^2, p_2, $p_1 p_2$, and $p_1^2 p_2$. Thus

$$F(n) = \sum_{d|n} f(d) = f(1) + f(p_1) + f(p_1^2) + f(p_2) + f(p_1 p_2) + f(p_1^2 p_2)$$

We now find an explicit expression for $f(n)$ in terms of F. Solving for $f(n) = f(p_1^2 p_2)$, we have

$$f(n) = F(n) - [f(1) + f(p_1) + f(p_1^2) + f(p_2) + f(p_1 p_2)]$$

But,

$$F(p_1 p_2) = f(1) + f(p_1) + f(p_2) + f(p_1 p_2)$$

so that

$$f(n) = F(n) - F(p_1 p_2) - f(p_1^2)$$

By adding and subtracting $f(p_1) + f(1)$ we get

$$
\begin{aligned}
f(n) &= F(n) - F(p_1 p_2) - [f(p_1^2)) + f(p_1) + f(1)] + [f(p_1) + f(1)] \\
&= F(p_1^2 p_2) - F(p_1 p_2) - F(p_1^2) + F(p_1)
\end{aligned}
$$

This last expression contains several terms of the form $F(d)$, where d is a divisor of n. Consider the following table of divisors of $n = p_1^2 \cdot p_2$ and coefficients of $F(d)$ in the preceding equation.

d	$d^* = \dfrac{n}{d}$	coeff. of $F(d)$
1	$p_1^2 \cdot p_2$	0
p_1	$p_1 \cdot p_2$	$(-1)^2 = 1$
p_1^2	p_2	-1
p_2	p_1^2	0
$p_1 \cdot p_2$	p_1	-1
$p_1^2 \cdot p_2$	1	1

Thus, from the table,

$$f(n) = 0 \cdot F(1) + F(p_1) - F(p_1^2) + 0 \cdot F(p_2) - F(p_1 \cdot p_2) + F(p_1^2 \cdot p_2)$$

$$= \sum_{d \mid n} \mu\left(\frac{n}{d}\right) \cdot F(d)$$

$$= \sum_{d \mid n} \mu(d^*) \cdot F(d)$$

where $d^* = n/d$ for each positive divisor d of n and where μ is the function that gives the coefficients of $F(d)$ in the table. Each positive divisor d of n is associated with exactly one integer d^* so that we could specify the terms in this last sum by d and $d^* = n/d$. Given that $F(n)$ is defined in terms of a sum of values of f, we have "inverted" the sum and have expressed $f(n)$ as a sum of values of F for $n = p_1^2 p_2$. This example and other similar ones suggest the following definition.

Definition 4.14 *The arithmetic function μ is defined as follows:*

$$\mu(n) = \begin{cases} 1 & \text{if } n = 1 \\ (-1)^t & \text{if } \alpha_i = 1 \text{ for all } i \\ 0 & \text{if } \alpha_i > 1 \text{ for some } i \end{cases}$$

where the positive integer n has prime factorization $n = p_1^{\alpha_1} p_2^{\alpha_2} \cdots p_t^{\alpha_t}$. The function μ is called the Möbius function. *The summatory function of μ is denoted by*

$$I(n) = \sum_{d \mid n} \mu(d)$$

Theorem 4.15 *Let μ be the Möbius function. Then*

(a) *The function μ is multiplicative.*

(b) *The summatory function I of μ is multiplicative.*

(c) $I(n) = \begin{cases} 1 & \text{if } n = 1 \\ 0 & \text{if } n > 1. \end{cases}$

Proof. **(a)** Let m and n be positive integers and $gcd(m,n) = 1$. Let $m = p_1^{\alpha_1} p_2^{\alpha_2} \cdots p_s^{\alpha_s}$ and $n = q_1^{\beta_1} q_2^{\beta_2} \cdots q_t^{\beta_t}$. Then

$$mn = p_1^{\alpha_1} p_2^{\alpha_2} \cdots p_s^{\alpha_s} q_1^{\beta_1} q_2^{\beta_2} \cdots q_t^{\beta_t}$$

with $q_j \neq p_i$ for all i and j because m and n are relatively prime.

If one of m or n is 1, say $m = 1$, then $\mu(mn) = \mu(n) = 1 \cdot \mu(n) = \mu(m)\mu(n)$ since $\mu(m) = 1$. Consider the case $m > 1$ and $n > 1$. If $\beta_j = \alpha_i = 1$ for all i and j, then

$$\mu(m) = (-1)^s$$

and

$$\mu(n) = (-1)^t$$

so

$$
\begin{aligned}
\mu(m)\mu(n) &= (-1)^s(-1)^t \\
&= (-1)^{s+t} \\
&= \mu(mn)
\end{aligned}
$$

If some $\beta_j > 1$ or some $\alpha_i > 1$, say, $\alpha_k > 1$, then

$$\mu(mn) = 0 = 0 \cdot \mu(n) = \mu(m) \cdot \mu(n)$$

Thus μ is multiplicative.

(b) The function I is multiplicative by Theorem 4.8.

(c) By definition, $I(1) = \sum_{d|1} \mu(d) = \mu(1) = 1$. For any prime p and positive integer α,

$$
\begin{aligned}
I(p^\alpha) &= \sum_{d|p^\alpha} \mu(d) = \mu(1) + \mu(p) + \mu(p^2) + \cdots + \mu(p^\alpha) \\
&= \mu(1) + \mu(p) \\
&= 1 + (-1) = 0
\end{aligned}
$$

Thus, by Theorem 4.4, $I(n) = 0$ for $n > 1$. ∎

We are now able to state the main result of this section in the following theorem which was illustrated in the example at the beginning of this section. It is called the Möbius Inversion Theorem.

Theorem 4.16 (Möbius Inversion Theorem) *If f and F are arithmetic functions, then, for each positive integer n,*

$$F(n) = \sum_{d|n} f(d) \qquad \text{if and only if} \qquad f(n) = \sum_{d|n} \mu\left(\frac{n}{d}\right) \cdot F(d)$$

The theorem says that if F is the summatory function of the arithmetic function f, the function f can be generated from the values of F using the Möbius function. Further, the computation of $f(n)$ from F requires only knowledge of $F(t)$ for t's in the range $1 \leq t \leq n$. The theorem also answers the question: Given an arithmetic function F, for what arithmetic function f is F the summatory function? It should be noted that there is no need for either f or F to be multiplicative in Theorem 4.16. Before proving this theorem, we will consider an example.

Example 4.17 *Because the expressions of the equations in Theorem 4.16 use only values of functions in the range from 1 to n, the theorem applies to arithmetic-like functions defined upon finite sets of the form $\{1, 2, \ldots, k\}$. We will illustrate Theorem 4.16 with such finitely domained functions. The following table gives the definition of $f : \{1, 2, \ldots, 10\} \to Z$ in the first and third columns. The fourth column gives the summatory function F of f calculated from the formula $F(n) = \sum_{d|n} f(d)$. The value of μ is calculated for convenience in the second column only for reference in creating the next table and is not used to produce F.*

t	$\mu(t)$	$f(t)$	$F(t)$
1	1	-3	-3
2	-1	2	-1
3	-1	5	2
4	0	1	0
5	-1	7	4
6	$(-1)^2 = 1$	-4	0
7	-1	6	3
8	0	-2	-2
9	0	3	5
10	$(-1)^2 = 1$	1	7

In the next table, we use the values of F above and the formula of Theorem 4.16 to regenerate the original function f. Only $f(n)$ for $1 \le n \le 6$ will be calculated. The cases of $f(n)$ for $7 \le n \le 10$ are left for the reader.

n	d	$\dfrac{n}{d}$	$\mu\left(\dfrac{n}{d}\right)$	$F(d)$	$\mu\left(\dfrac{n}{d}\right) \cdot F(d)$	$\sum \mu\left(\dfrac{n}{d}\right) \cdot F(d)$
1	1	1	1	-3	-3	-3
2	1	2	-1	-3	3	
	2	1	1	-1	-1	2
3	1	3	-1	-3	3	
	3	1	1	2	2	5
4	1	4	0	-3	0	
	2	2	-1	-1	1	
	4	1	1	0	0	1
5	1	5	-1	-3	3	
	5	1	1	4	4	7
6	1	6	1	-3	-3	
	2	3	-1	-1	1	
	3	2	-1	2	-2	
	6	1	1	0	0	-4

The proof of Theorem 4.16 will be easier to organize if we use the following property of sums over elements of a set.

Lemma 4.18 *Assume that D is an integral domain. Let S and T be nonempty sets such that for every $s \in S$, there is a nonempty subset T_s of T and such that $\bigcup_{s \in S} T_s = T$. Let $W \subseteq S \times T$ be defined by $W = \{w : \text{there is an } s \in S \text{ and}$ a $t \in T_s$ such that $w = (s, t)\}$. Let $f : W \to D$; and for each $t \in T$, let $S_t = \{s : s \in S \text{ and } t \in T_s\}$. Then*

(a) $\bigcup_{t \in T} S_t = S$.

(b) $\displaystyle\sum_{s \in S} \left[\sum_{t \in T_s} f(s, t) \right] = \sum_{t \in T} \left[\sum_{s \in S_t} f(s, t) \right].$

Proof. **(a)** Clearly $\bigcup_{t \in T} S_t \subseteq S$. If $s \in S$, then there is a nonempty subset T_s of T corresponding to s. Thus, for any $t \in T_s$, $s \in S_t$ so that $s \in \bigcup_{t \in T} S_t$.

(b) If $s \in S$, $t \in T_s$, and $f(s, t)$ is an addend in $\displaystyle\sum_{s \in S} \left[\sum_{t \in T_s} f(s, t) \right]$, then $t \in T$ and $s \in S_t$ so that $f(s, t)$ is an addend in $\displaystyle\sum_{t \in T} \left[\sum_{s \in S_t} f(s, t) \right]$. Similarly, if $t \in T$, $s \in S_t$, and $f(s, t)$ is an addend in $\displaystyle\sum_{t \in T} \left[\sum_{s \in S_t} f(s, t) \right]$, then $s \in \bigcup_{t \in T} S_t = S$, $t \in T_s$, and $f(s, t)$ is an addend in $\displaystyle\sum_{s \in S} \left[\sum_{t \in T_s} f(s, t) \right]$. The proof is complete when we notice that for a given s and t, $f(s, t)$ occurs at most once in the sum on each side. ∎

Proof. (Theorem 4.16) If f is an arithmetic function and $F(m) = \displaystyle\sum_{d \mid m} f(d)$ for every positive integer m is the summatory function of f, then

$$\sum_{d \mid n} \mu\left(\frac{n}{d}\right) \cdot F(d) = \sum_{d \mid n} \mu(d) \cdot F\left(\frac{n}{d}\right)$$

$$= \sum_{d \mid n} \mu(d) \sum_{t \mid (d/n)} f(t)$$

$$= \sum_{d \mid n} \mu(d) \sum_{d \mid (n/t)} f(t)$$

$$= \sum_{d \mid n} \sum_{d \mid (n/t)} \mu(d) f(t)$$

The first equality holds because as d ranges over the divisors of n, then so does n/d. The second equality holds by the definition of $F(m)$ for $m = n/d$.

The third equality is true because $t \mid (n/d)$ if and only if there is an integer k such that $t \cdot k = (n/d)$ if and only if $d \mid (n/t)$. The distributive property was used in the fourth equality.

Let $S = \{d : d \in N \text{ and } d \mid n\}$. For $d \in S$, let $T_d = \{t : t \in N \text{ and } d \mid (n/t)\}$. Clearly, $T = \bigcup_d T_d = S$. For t in T, we have $S_t = \{d : d \in S \text{ and } t \in T_d\} = \{d : d \in N \text{ and } d \mid (n/t)\}$. These identifications allow Lemma 4.18 to be applied so that

$$
\sum_{d \mid n} \mu\left(\frac{n}{d}\right) \cdot F(d) = \sum_{d \in S} \sum_{t \in T_d} \mu(d) f(t)
$$

$$
= \sum_{t \in T} \sum_{d \in S_t} \mu(d) f(t)
$$

$$
= \sum_{t \mid n} \sum_{d \mid (n/t)} \mu(d) f(t)
$$

$$
= \sum_{t \mid n} f(t) \sum_{d \mid (n/t)} \mu(d)
$$

$$
= f(n)
$$

The last equality holds because $\sum_{d \mid (n/t)} \mu(d)$ is zero unless $n/t = 1$ or, equivalently, unless $t = n$ when the sum is 1 (Theorem 4.15(c)). ∎

The Möbius function is named after Augustus Ferdinand Möbius (1790-1868), who was born in Schulpforta in Prussia. As a youth, he was taught at home. He first considered studying law but soon came to his senses and switched to mathematics. Möbius studied at Liepzig, at Halle with Pfaff, and finally under Gauss at Göttingen. He became lecturer at Leipzig in 1815 and was named Professor of Higher Mathematics and Astronomy in 1844, which position he held until his death. His most important research was in geometry; but he also wrote in statics, astronomy, and number theory. He took a giant step toward a consistent algebra of rotations and vectors in space with his work on barycentric calculus, *Der Barycentrische Calcul*, which introduced homogeneous coordinates. His progress in this area was blocked by the fact that he was unable to accept a multiplication which was not commutative.

While Möbius is known in number theory for the Möbius function, he is far better known for two other accomplishments. He proposed the coloring of maps in 1840, which lead to the famous four-color problem. He presented this problem in a lecture in the form of a fairy tale. He also created the well-known Möbius strip, a surface with only one side.

Möbius's son was a neurologist who was much more widely known than his father because of his book dealing with his belief in the psychologically weaker minds of women.

Exercises

1. Calculate:

    (a) $\mu(30)$ (b) $\mu(50!)$ (c) $\mu(22)$ (d) $\mu(2 \cdot 5 \cdot 11 \cdot 13 \cdot 17)$

    (e) $\mu(101)$

2. Complete the second table of Example 4.17 for $7 \leq n \leq 10$.

3. Let f be an arithmetic function such that $f(1) = 5$, $f(2) = -3$, $f(3) = 7$, $f(4) = 5$, $f(5) = 10$, $f(6) = -2$, $f(7) = 3$, and $f(k) = 0$ for $k \geq 8$.

    (a) Calculate the summatory function $F(n)$ of f for every positive integer n, $1 \leq n \leq 10$.

    (b) Use the Möbius Inversion Theorem and $F(n)$ to compute $f(k)$ for $1 \leq k \leq 10$.

4. If n is a positive integer with prime factorization $\prod_{i=1}^{k} p_i$, prove that:

    (a) $\sum_{d|n} |\mu(d)| = 2^k$, where $|b|$ indicates absolute value of b

    (b) $\sum_{d|n} \mu(d) \cdot \tau(d) = (-1)^k$

5. If n is a positive integer, show that $\prod_{i=0}^{3} \mu(n+i) = 0$.

6. The Dirichlet convolution or Dirichlet product was defined in the Exercises of Section 4.2. Prove that:

    (a) $\mu * \tau = U$

    (b) $\mu * \mu = E$

    (c) $\mu * U = U * \mu$ so that $\mu^{-1} = U$ and $U^{-1} = \mu$

    (d) $U * U = \tau$

    (e) $\mu * E = \phi$

    (f) $U * E = \sigma$

    (g) $\phi * \tau = \sigma$

(h) $E * E = \sigma * \phi$

(i) $\sigma * \mu = E$

(j) $\sigma * U = E * \tau$

7. Using the Dirichlet Convolution of Exercise 6, show that the Möbius Inversion Theorem may be written as follows:

$$F = U * f \text{ if and only if } f = \mu * F$$

where F is the summatory function of f.

8. Show that, if $G = \{f : f \text{ is an arithmetic function and } f(1) \neq 0\}$, then G with the operation $*$, the Dirichlet Product of Exercise 6, is a group.

4.4   Generalized Möbius Function

For the set N of positive integers, "divides" is a partial order:

$$
\begin{array}{ll}
r \mid r \text{ for all } r \in N & \text{(reflexivity)} \\
r \mid s \text{ and } s \mid r \text{ imply that } r = s & \text{(antisymmetry)} \\
r \mid s \text{ and } s \mid t \text{ imply that } r \mid t & \text{(transitivity)}
\end{array}
$$

where partial order is defined in Chapter 0.

Definition 4.19 *Let S be a nonempty set and $\leq$ a partial order on S. If S contains a unique element, 0, with the property $0 \leq s$ for all $s \in S$, then 0 is called the* minimal *element of S. The set $[a, b] = \{s : s \in S, a \leq s \leq b\}$ is called the* segment *or* interval *from a to b. If, in addition, $[a, b]$ is finite for all $a, b \in S$, then S is said to be* locally finite *relative to the relation $\leq$. The intervals $[a, b)$, $(a, b]$, and (a, b) are defined by $[a, b) = \{s : s \in S, a \leq s < b\}$, $(a, b] = \{s : s \in S, a < s \leq b\}$, and $(a, b) = \{s : s \in S, a < s < b\}$.*

There is no loss of generality in requiring a minimal element; for if S with partial order $\leq$ has no such minimal element, then we can define $S^0 = S \cup \{0\}$ where 0 is assumed to be an object not in S and extend the ordering on S to the order $\leq^0$ on S^0 by defining $0 \leq^0 s$ for all $s \in S^0$ and for all $s, t \in S$ defining $s \leq^0 t$ if and only if $s \leq t$. The new extended relation $\leq^0$ will be a partial order on S^0 and will agree with $\leq$ when comparing elements of S.

We will assume that all the partially ordered sets considered in this section have a unique minimal element and are locally finite, unless stated otherwise.

Several partial orders are particularly useful. They are specified in the next theorem whose proof is left to the reader.

Theorem 4.20 *The following three statements hold:*

(a) *Let N be the set of positive integers. Then N with the relation divides, $|$, is a locally finite partially ordered set with minimal element 1.*

(b) *If T is a finite nonempty set and $S = \mathcal{P}(T)$ is the collection of all subsets of T, then S and the subset relation, $\subseteq$, on S is a locally finite partially ordered set with minimal element $\varnothing$.*

(c) *The set N with the relation $\leq$ ("less than or equal to" as defined in Chapter 1) is a locally finite partially ordered set with minimal element 1.*

For the partial order in Theorem 4.20(a), the interval $[a, b]$ is the set of all divisors of b that are also divided by a. Thus $[2, 9] = \varnothing$ and $[3, 24] = \{3, 6, 12, 24\}$. The interval $[1, b]$ is merely the positive integers that divide b since 1 divides every positive integer so that $[1, 12] = \{1, 2, 3, 4, 6, 12\}$.

For the partial order in Theorem 4.20(b), if A and B are nonempty subsets of a finite set T, then $[A, B]$ is the collection of subsets of B that contain A as a subset. Thus, $[\varnothing, B]$ is the collection of all subsets of B, namely, $\mathcal{P}(B)$. If $A \not\subseteq B$, then $[A, B] = \varnothing$.

In Theorem 4.20(c), if a and b are positive integers, then $[a, b] = \{a, a + 1, a + 2, \ldots, b\}$ provided that $a \leq b$. $[1, b] = \{1, 2, \ldots, b\}$. If $a > b$, then $[a, b] = \varnothing$. For this partial order, the summation $\sum_{i \in [1, b]}$ is equivalent to $\sum_{i=1}^{b}$.

We extend the definition of the Möbius function and the Möbius Inversion Theorem to generalized "partial order" sums such as $\sum_{t \in [a, b]}$, where $[a, b]$ is a segment or interval of the appropriate partially ordered set. Thus, applying Definition 4.5 to a locally finite partially ordered set S,

$$\sum_{t \in [a, b]} f(t) = \sum_{a \leq t \leq b} f(t)$$

is well defined for a function $f : S \to D$ for D an integral domain since $[a, b]$ is finite.

Referring to Theorem 4.15 and the definition of $I(n)$, the summatory function for μ , we see that

$$\sum_{d|n} \mu(d) = 0 \text{ for } n > 1$$

or

$$\sum_{d \in [1, n]} \mu(d) = 0 \text{ for } n > 1$$

Equivalently,

$$\mu(n) = - \sum_{d \in [1, n)} \mu(d) \text{ for } n > 1$$

Thus, $\mu(n)$ can be defined recursively if $\mu(1) = 1$ is given. If we explicitly include the lower "endpoint," 1, of $[1, n)$, then a reasonable generalization of μ to "partial order" sums would be the following.

Definition 4.21 *If S with the partial order $\leq$ is locally finite with a minimal element, then the* generalized Möbius function $\mu : S \times S \to N$ *is defined by*

(a) $\mu(s, s) = 1$ *for all $s \in S$.*

(b) $\mu(r, s) = - \displaystyle\sum_{t \in [r,s)} \mu(r,t) = - \sum_{r \leq t < s} \mu(r,t)$ *for $r < s$.*

Note that the summation over empty sets (Definition 4.5) implies that μ defaults to $\mu(r, s) = 0$ for $r > s$.

This definition of μ extends Theorem 4.16 in the following theorem which is also called the Möbius Inversion Theorem.

Theorem 4.22 (Mobius Inversion Theorem) *Let X be a locally finite partially ordered set with minimal element 0, D be an integral domain containing the integers, and $f : X \to D$ be a function. Define $F : X \to D$ by*

$$F(x) = \sum_{w \in [0,x]} f(w) = \sum_{0 \leq w \leq x} f(w)$$

Then, for all $x \in X$,

$$f(x) = \sum_{w \in [0,x]} \mu(w, x) F(w) = \sum_{0 \leq w \leq x} \mu(w, x) F(w)$$

Proof.

$$\sum_{0 \leq w \leq x} \mu(w, x) F(w) \;=\; \sum_{0 \leq w \leq x} \mu(w, x) \sum_{0 \leq t \leq w} f(t)$$

$$=\; \sum_{0 \leq w \leq x} \left[\sum_{0 \leq t \leq w} \mu(w, x) f(t) \right]$$

Let $S = \{w : 0 \leq w \leq x\}$. For $w \in S$, let $T_w = \{t : 0 \leq t \leq w\}$ so that $\bigcup_{w \in S} T_w = S$. Let $S_t = \{w : t \in T_w\} = \{w : t \leq w \leq x\}$.

$$\sum_{0 \leq w \leq x} \left[\sum_{0 \leq t \leq w} \mu(w, x) f(t) \right] \;=\; \sum_{w \in S} \sum_{t \in T_w} \mu(w, x) f(t)$$

$$=\; \sum_{t \in T} \sum_{w \in S_t} \mu(w, x) f(t) \quad \text{by Lemma 4.18}$$

$$=\; \sum_{t \in T} f(t) \sum_{w \in S_t} \mu(w, x)$$

$$=\; \sum_{0 \leq t \leq x} f(t) \sum_{t \leq w \leq x} \mu(w, x)$$

$$=\; f(x)$$

since $\sum\limits_{t \leq w \leq x} \mu(w, x) = 0$ unless $t = x$. ∎

Clearly, Theorem 4.22 reduces to Theorem 4.16(a) for $X = N$ and partial order $\leq$ defined by $a \leq b$ if and only if a divides b. Examples of integral domains containing the integers are the rationals and the reals. The next theorem shows how to calculate the value of the generalized Möbius function for the partial orders given in Theorem 4.20.

Theorem 4.23 *Let the nonempty set S with partial order $\leq$ be locally finite with a minimal element.*

(a) *Let $S = N$ and define $a \leq b$ if and only if $a \mid b$. If d and n are integers and $d \mid n$, then*

$$\mu(d, n) = \mu\left(\frac{n}{d}\right)$$

where the "μ" on the right is given by Definition 4.14.

(b) *Suppose that T is a nonempty finite set, $S = \mathcal{P}(T)$ with partial order $\subseteq$. If $A, B \in S$ (or equivalently $A, B \subseteq T$) and $A \subseteq B$, then*

$$\mu(A, B) = (-1)^{|B|-|A|}$$

where $|A|$ is the number of elements in the set A.

(c) *Let $S = N$ have the partial ordering $\leq$. If $i, n \in N$ and $i \leq n$, then*

$$\mu(i, n) = \begin{cases} 1 & if\ i = n \\ -1 & if\ i = n - 1 \\ 0 & if\ i \leq n - 2 \end{cases}$$

Before proving this theorem, we need to know for a finite set how many subsets there are having a given number of elements. The proof of this theorem is left to the reader.

Lemma 4.24 *If S is a finite set with n elements and k is an integer, $0 \leq k \leq n$. Then there are $\binom{n}{k} = \dfrac{n!}{k!(n-k)!}$ subsets of S containing k elements each.*

Proof. (Theorem 4.23(b)) The proofs of parts (a) and (c) are left to the reader. Let T be a nonempty finite set and $S = \mathcal{P}(T)$ and suppose that $A \subseteq B \subseteq S$. For a nonempty set , let A and B be subsets of a nonempty set $S = \mathcal{P}(T)$. If $A = B$, then $\mu(A, B) = \mu(A, A) = 1$ and $(-1)^{|B|-|A|} = (-1)^0 = 1$. Suppose that $A \subset B$ and let $\mu(A, W) = (-1)^{|W|-|A|}$ for any set

W with $A \subseteq W \subset B$ and $|W| < |B|$. Then

$$
\mu(A, B) \quad = \quad - \sum_{A \subseteq W \subset B} \mu(A, W)
$$

$$
= \quad - \sum_{A \subseteq W \subset B} (-1)^{|W|-|A|} \text{ by the induction hypothesis}
$$

$$
= \quad (-1)^{|B|-|A|} - \sum_{A \subseteq W \subseteq B} (-1)^{|W|-|A|}
$$

where for the last equality the quantity $(-1)^{|B|-|A|}$ was added and subtracted. But for $A \subseteq W$, $|W| - |A| = |W - A|$. Thus

$$
\sum_{A \subseteq W \subseteq B} (-1)^{|W|-|A|} \quad = \quad \sum_{(W-A) \subseteq (B-A)} (-1)^{|W-A|}
$$

$$
= \quad \sum_{X \subseteq (B-A)} (-1)^{|X|}
$$

$$
= \quad \sum_{k=0}^{|B-A|} \binom{|B-A|}{k} (-1)^k (1)^{|B-A|-k}
$$

$$
= \quad ((-1)+1)^{|B-A|} = 0
$$

The summation over $X \subseteq (B - A)$ is equivalent to the summation from $k = 0$ to $|B - A|$ because the sum may be grouped by $X's$ with $|X| = k$. For $0 \leq k \leq n$, there are $\binom{|B-A|}{k}$ such $X's$. Finally, the Binomial Theorem in invoked. Thus, by mathematical induction, part (b) is proved. ∎

We can now apply the Möbius Inversion Theorem to summations involving the binomial coefficient $\binom{n}{k}$ because the summation on segments for the subset partial order may be regrouped according to sets having a given number of elements.

Theorem 4.25 *If $a_0, a_1, a_2, \ldots$ is a sequence of elements of an integral domain which contains the integers and the sequence $b_0, b_1, b_2, \ldots$ is defined by*

$$
b_n = \sum_{k=0}^{n} \binom{n}{k} a_k
$$

for each nonnegative integer n, then

$$
a_n = \sum_{k=0}^{n} \binom{n}{k} (-1)^{n-k} b_k
$$

Proof. Let n be a nonnegative integer and S any set having n elements. If $A \subseteq S$, then define $f(A) = a_{|A|}$; that is, if A has k elements, $f(A) = a_k$.

Since there are $\binom{n}{k}$ subsets of S having k elements for $0 \leq k \leq n$,

$$F(S) = b_n = \sum_{k=0}^{n} \binom{n}{k} a_k = \sum_{\varnothing \subseteq A \subseteq S} f(A)$$

By the Möbius Inversion Theorem 4.22, we have

$$a_n = f(S) = \sum_{\varnothing \subseteq B \subseteq S} \mu(B, S) \cdot F(B)$$

$$= \sum_{\varnothing \subseteq B \subseteq S} (-1)^{|S|-|B|} F(B)$$

$$= \sum_{k=0}^{n} \binom{n}{k} (-1)^{n-k} b_k$$

where the last equality holds because there are $\binom{n}{k}$ subsets B of S having k elements and for each one, $F(B) = b_k$. $\blacksquare$

We often need to count the number of integers having various properties. In Section 4.2 we discussed arithmetic functions that counted the number of positive divisors of a positive integer and the number of positive integers relatively prime to a positive integer. We introduced a method of "sifting" through a set of integers including those having the desired property and excluding those that do not provides a widely applicable algorithm. Such algorithms are called sieve methods. We will obtain several sieve formulas and use the Möbius function to generate other ones.

For a finite set A, let $|A|$ denote the number of elements in A or the cardinality of A. For convenience and readability in this section, we will sometimes use the notation $n(A) = |A|$ instead. For example,

$$n(A) = \sum_{a \in A} n(\{a\}) = \sum_{a \in A} 1$$

We restate some results from Chapter 0 as follows.

Theorem 4.26 *For subsets A and B of universal set S:*

(a) *If $A \cap B = \varnothing$, then $n(A \cup B) = n(A) + n(B)$.*

(b) $n(A') = n(S - A) = n(S) - n(A).$

(c) $n\left((A \cup B)'\right) = n\left(A' \cap B'\right).$

(d) $n\left((A \cap B)'\right) = n\left(A' \cup B'\right).$

Consider the case of three sets A_1, A_2, and A_3 which are not necessarily disjoint and which are subsets of a universal set S. We are interested in counting the number of elements in $A_1 \cup A_2 \cup A_3$. If we write the sum $n(A_1) + n(A_2) + n(A_3)$ and some elements are members of several of the three sets, then we have counted them more than once. In that case, we could subtract those we counted twice as the sum $n(A_1 \cap A_2) + n(A_1 \cap A_3) + n(A_2 \cap A_3)$. But if an element happens to be in all three sets, we subtracted its contribution too many times and we need to add it back. There are $n(A_1 \cap A_2 \cap A_3)$ such elements. Thus,

$$n(A_1 \cup A_2 \cup A_3) \;=\; n(A_1) + n(A_2) + n(A_3) - [n(A_1 \cap A_2) + n(A_1 \cap A_3)$$

$$+ \, n(A_2 \cap A_3)] + n(A_1 \cap A_2 \cap A_3)$$

This formula may be restated using summation over sets, but first we need to develop some notation. Once the sets A_1, A_2, and A_3 are specified, then we can refer to $n(A_1 \cup A_3)$ by giving the set of subscripts $K = \{1, 3\} \subseteq \{1, 2, 3\}$ using the notation

$$n_\cup(K) = n\left(\bigcup_{i \in K} A_i\right) = n(A_1 \cup A_3)$$

When $K = \varnothing$, then the definition of the intersection gives

$$\bigcap_{i \in \varnothing} A_i = S$$

In the following definition it would seem natural to define

$$n_\cap\left(\bigcap_{i \in \varnothing} A_i\right) = n(S)$$

However, for convenience in the following theorems and proofs we define

$$n_\cap\left(\bigcap_{i \in \varnothing} A_i\right) = 0$$

Definition 4.27 *Suppose that t is a positive integer, $T = \{1, 2, \ldots, t\}$, S is finite, and $A_i \subseteq S$ for each $i \in T$. If $K \subseteq T$, then define*

$$n_\cup(K) = \begin{cases} n\left(\displaystyle\bigcup_{i \in K} A_i\right) & \text{if } K \neq \varnothing \\ 0 & \text{if } K = \varnothing \end{cases}$$

$$n_\cap(K) = \begin{cases} n\left(\displaystyle\bigcap_{i \in K} A_i\right) & \text{if } K \neq \varnothing \\ 0 & \text{if } K = \varnothing \end{cases}$$

Theorem 4.28 *If S is a finite set and $A_i \subseteq S$ for each $i \in T = \{1, 2, 3, \ldots, t\}$, then*

$$n_\cup(T) = \sum_{K \subseteq T} (-1)^{|K|+1} n_\cap(K)$$

Proof. The proof uses induction on t. If $T = \emptyset$, then $n_\cup(T) = 0$ and the sum has one addend, $(-1)^{|\emptyset|+1}n_\cap(\emptyset) = 0$. If $t = 1$, then $\emptyset$ and T are the only subsets of T and $n_\cup(T) = n_\cup(\{1\}) = n(A_1)$.

$$\sum_{K \subseteq T}(-1)^{|K|+1}n_\cap(K) = (-1)^{|\emptyset|+1}n_\cap(\emptyset) + (-1)^{|T|+1}n_\cap(T)$$
$$= 0 + n(A_1) = n(A_1)$$

and the theorem holds for $t = 1$.

Assume that the equality of the theorem holds for any T with $1 \leq t < r$. Let $t = r$ so that $T = \{1, 2, \ldots, r\}$. Let $W = T - \{r\} = \{1, 2, \ldots, r-1\}$ and $B_i = A_i \cap A_r$. Let $n'_\cap$ and $n'_\cup$ be $n_\cap$ and $n_\cup$ respectively, applied to the sets $B_1, B_2, \ldots, B_{r-1}$ instead of to $A_1, \ldots, A_r$. Then

$$n_\cup(T) = n\left(\bigcup_{i \in T} A_i\right) = n\left(\bigcup_{i \in W} A_i\right) + n(A_r) - n\left(\left(\bigcup_{i \in W} A_i\right) \cap A_r\right)$$

$$= n\left(\bigcup_{i \in W} A_i\right) + n(A_r) - n\left(\bigcup_{i \in W} B_i\right)$$

$$= n_\cup(W) + n(A_r) - n'_\cup(W)$$

$$= \sum_{K \subseteq W}(-1)^{|K|+1}n_\cap(K) + n(A_r) - \sum_{L \subseteq W}(-1)^{|L|+1}n_\cap(L)$$

where in the last equality we have used the induction hypothesis twice. But

$$n_\cap(L) = n\left(\bigcap_{i \in L}(A_i \cap A_r)\right) = n\left(\bigcap_{i \in L \cup \{r\}} A_i\right) = n_\cap(L \cup \{r\})$$

Thus,

$$\sum_{L \subseteq W}(-1)^{|L|+1}n_\cap(L) = -\sum_{\{r\} \subset K \subseteq T}(-1)^{|K|+1}n_\cap(K)$$

and

$$n_\cup(T) = \sum_{K \subseteq W}(-1)^{|K|+1}n_\cap(K) + n(A_r) + \sum_{\{r\} \subset K \subseteq T}(-1)^{|K|+1}n_\cap(K)$$

$$= \sum_{K \subseteq T}(-1)^{|K|+1}n_\cap(K)$$

Therefore, by mathematical induction, the theorem holds for all t. ∎

Example 4.29 *Let*

$$
\begin{aligned}
S &= \{a, b, c, d, e, f, g, h, i\} \\
A_1 &= \{a, b, c\} \\
A_2 &= \{b, c, d, f, g\} \\
A_3 &= \{a, g\} \\
T &= \{1, 2, 3\}
\end{aligned}
$$

and $t = 3$. The following table considers every subset K of T.

K	$|K|$	$n_\cap(K)$	$(-1)^{|K|+1} n_\cap(K)$
$\varnothing$	0	0	0
$\{1\}$	1	3	3
$\{2\}$	1	5	5
$\{3\}$	1	2	2
$\{1, 2\}$	2	2	-2
$\{1, 3\}$	2	1	-1
$\{2, 3\}$	2	1	-1
$\{1, 2, 3\}$	3	0	0

Thus,

$$n(A_1 \cup A_2 \cup A_3) = n\left(\bigcup_{i \in T} A_i\right) = n_\cup(T)$$

$$= \sum_{K \subseteq T}(-1)^{|K|+1} n_\cap(K) = 6$$

Also, by direct evaluation, $A_1 \cup A_2 \cup A_3 = \bigcup_{i \in T} A_i = \{a, b, c, d, f, g\}$ so that
$n(A_1 \cup A_2 \cup A_3) = 6$.

Theorem 4.28, which counts the number of elements in a finitely indexed union of sets in terms of intersections of these sets, has a dual, Theorem 4.30, in which the number of elements in a finitely indexed intersection is expressed in terms of the unions of these sets. For only three sets A_1, A_2, and A_3, the dual is

$$\begin{aligned} n(A_1 \cap A_2 \cap A_3) = {} & n(A_1) + n(A_2) + n(A_3) - [n(A_1 \cup A_2) + n(A_1 \cup A_3) \\ & + n(A_2 \cup A_3)] + n(A_1 \cup A_2 \cup A_3) \end{aligned}$$

where unions are interchanged with intersections in the sieve formula for unions. The Möbius Inversion Theorem provides an easy path to the proof of the next theorem.

Theorem 4.30 *If S is a finite set and $A_i \subseteq S$ for each $i \in T = \{1, 2, 3, \ldots t\}$, then*

$$n_\cap(T) = \sum_{K \subseteq T}(-1)^{|K|+1} n_\cup(K)$$

Proof.   From Theorem 4.28 we have

$$n_\cup(T) = \sum_{K \subseteq T}(-1)^{|K|+1} n_\cap(K)$$

By Theorem 4.23, $\mu(A, B) = (-1)^{|B|-|A|}$ for subsets A and B of T with $A \subseteq B$. The Möbius Inversion Theorem implies that

$$(-1)^{|T|+1} n_\cap(T) = \sum_{K \subseteq T}(-1)^{|K|-|T|} n_\cup(K)$$

or

$$n_\cap(T) = \sum_{K \subseteq T} (-1)^{|K|-1} n_\cup(K)$$

$$= \sum_{K \subseteq T} (-1)^{|K|+1} n_\cup(K)$$

where in the last equality, we have multiplied by $1 = (-1)^2$ to obtain a "symmetrical" version of the formula and its dual. ∎

We have just considered a collection of subsets $\{A_1, A_2, \ldots, A_t\}$ of a universal set S. Theorem 4.28 gave a means of computing the number of elements in the union of several sets. Theorem 4.30 gave a means of computing the number of elements in the intersection of several sets. We now see how to calculate the number of elements of S that are in none of the sets $A_1, A_2, \ldots, A_t$. For three sets A_1, A_2, and A_3, the next theorem provides a formula for $n((A_1 \cup A_2 \cup A_3)')$. This formula is referred as the Inclusion-Exclusion Principle or Sylvester's formula.

Theorem 4.31 *If S is a finite set and for each $i \in \{1, 2, \ldots, t\} = T$, $A_i \subseteq S$, then*

$$n\left(\left(\bigcup_{i \in T} A_i\right)'\right) = n\left(\bigcap_{i \in T} A_i'\right) = \sum_{k=0}^{t} (-1)^k \sum_{\substack{K \subseteq T \\ |K|=k}} n\left(\bigcap_{i \in K} A_i\right)$$

Proof.

$$n\left(\left(\bigcup_{i \in T} A_i\right)'\right) = n(S) - n(\bigcup_{i \in T} A_i)$$

$$= n(S) - \sum_{K \subseteq T} (-1)^{|K|+1} n_\cap(K) \quad \text{by Theorem 4.28}$$

$$= n(S) + \sum_{K \subseteq T} (-1)^{|K|} n_\cap(K)$$

Now regrouping the last sum by adding up by size or cardinality of K, i.e., by $k = |K|$, we obtain

$$n\left(\left(\bigcup_{i \in T} A_i\right)'\right) = n(S) + \sum_{k=1}^{t} (-1)^K \sum_{\substack{K \subseteq T \\ |K|=k}} n_\cap(K)$$

$$= \sum_{k=0}^{t} (-1)^k \sum_{\substack{K \subseteq T \\ |K|=k}} n\left(\bigcap_{i \in K} A_i\right)$$

since $\bigcap_{i \in \varnothing} A_i = S$. ∎

Example 4.32 *The number of positive integers less than or equal to* $n = 5000$ *that are relatively prime to* n *is* $\phi(n) = 5^3(5-1)2^2(2-1) = 2000$ *by Theorem 4.12. Sometimes we are interested in how many positive integers less than or equal to* n *are or are not divisible by various other positive integers.*

If $S = \{s : s \text{ is an integer and } 1 \le s \le n\}$, *one may ask, for example,* *(a) How many integers in* S *are divisible by 7?, (b) How many elements of* S *are not divisible by 7?, (c) How many are divisible by both 7 and 2?, and (d) How many are not divisible by 70?*

The possible positive multiples of 7 *are* $7 \cdot k$ *for* $k = 1, 2, 3, \ldots$. *Dividing* n *by 7, we have unique nonnegative integers* q *and* r *with* $n = 7 \cdot q + r$ *with* $0 \le r < n$. *Clearly,* q *is the largest* k *with* $7 \cdot k < n$; *and if* $1 \le k \le q$, *then* $7 \cdot k \le 7 \cdot q \le 7 \cdot q + r = n$. *Since* $n = 7 \cdot 714 + 2$, *there are* $q = 714$ *integers in* S *divisible by 7.*

Let $A_1 = \{s : s \in S \text{ and } 7 \mid s\}$. *We have just shown that* $|A_1| = n(A_1) = 714$. *Considering* S *to be the universal set,* A_1' *is the set of integers in* S *that are not divisible by 7. By Theorem 4.26(b),*

$$n(A_1') = n(S) - n(A_1) = 5000 - 714 = 4286$$

Because 2 and 7 are relatively prime, an integer is divisible by both 7 and 2 if and only if it is divisible by 14. Therefore, since $n = 14 \cdot 357 + 2$, *exactly 357 integers in* S *are divisible by both 2 and 7. If we let* $A_2 = \{s : s \in S \text{ and } 2 \mid s\}$, *then we have shown that* $n(A_1 \cap A_2) = 357$.

Consider $70 = 2 \cdot 5 \cdot 7$. *Let* $A_3 = \{s : s \in S \text{ and } 5 \mid s\}$. *If* $T = \{1, 2, 3\}$, *then* $A_1 \cup A_2 \cup A_3 = \bigcup_{i \in T} A_i$ *is the set of integers in* S *divisible by* $7, 2,$ *or* 5. *We want the number of integers in*

$$\left(\bigcup_{i \in T} A_i \right)' = \bigcap_{i \in T} A_i' = A_1' \cap A_2' \cap A_3'$$

which is the subset of S *containing those integers that are not divisible by* $2, 5,$ *or 7. Using the Inclusion-Exclusion Principle,*

$$n\left(\left(\bigcup_{i \in T} A_i \right)' \right) = \sum_{k=0}^{3} (-1)^k \sum_{\substack{K \subseteq T \\ |K| = k}} n\left(\bigcap_{i \in K} A_i \right)$$

k	$K \subseteq T$ $|K| = k$	$(-1)^k$	$n\left(\bigcap_{i \in K} A_i \right)$
0	$\varnothing$	1	5000
1	$\{1\}$	-1	714
1	$\{2\}$	-1	2500
1	$\{3\}$	-1	1000
2	$\{1,2\}$	1	357
2	$\{1,3\}$	1	142
2	$\{2,3\}$	1	500
3	$\{1,2,3\}$	-1	71

Thus, $n\left(\left(\bigcup_{i\in T} A_i\right)'\right) = 5000 - 4214 + 999 - 71 = 1714.$

Example 4.33 *For a positive integer n, we saw in Theorem 4.12 that Euler's totient function $\phi(n)$ could be calculated as*

$$\phi(n) = \prod_{i=1}^{t} \left[p_i^{(\alpha_i-1)}(p_i - 1)\right] = n\prod_{i=1}^{t}\frac{p_i - 1}{p_i}$$

where $n = p_1^{\alpha_1} p_2^{\alpha_2} \cdots p_t^{\alpha_t}$ is the prime factorization of n with $\alpha_i > 0$ for all i. This formula was derived using the multiplicative property of ϕ. An alternative route to proving this formula is offered by the Inclusion-Exclusion Principle. Let

$$S = \{s : s \in Z \text{ and } 1 \le s \le n\}$$

$$A_i = \{s : s \in S \text{ and } p_i \mid s\}$$

and

$$T = \{1, 2, \ldots, t\}$$

From the discussion in Example 4.29,

$$n(A_i) = \frac{n}{p_i}$$

Clearly,

$$n\left(\bigcap_{i\in K} A_i\right) = \frac{n}{\prod\limits_{i\in K} p_i}$$

for $K \subseteq S$ and $K \ne \varnothing$; and $n\left(\bigcap_{i\in K} A_i\right) = n(S) = n$ if $K = \varnothing$. Then

$$\phi(n) \;=\; n\left(\left(\bigcup_{i\in K} A_i\right)'\right)$$

$$=\; \sum_{k=0}^{t}(-1)^k \sum_{K\subseteq T} n\left(\bigcap_{i\in K} A_i\right)$$

$$=\; n + \sum_{k=1}^{t}(-1)^k \sum_{K\subseteq T} \frac{n}{\prod\limits_{i\in k} p_i}$$

$$=\; n - \sum_{i}\frac{n}{p_i} + \sum_{i<j}\frac{n}{p_i p_j} - \sum_{i<j<k}\frac{n}{p_i p_j p_k} + \cdots$$

$$=\; \prod_{i=1}^{t}\left[p_i^{(\alpha_i-1)}(p_i - 1)\right]$$

where the last equality is proved in the Exercises.

Exercises

1. Given a function f defined on the nonnegative integers and $F(n) = \sum_{k=0}^{n} f(k)$ for each integer $n \geq 0$, prove that $f(n) = F(n) - F(n-1)$ using Theorem 4.22 and Theorem 4.23.

2. Prove these parts of Theorem 4.20:

    (a) Part (a)   **(b)** Part (b)   **(c)** Part (c)

3. Prove Lemma 4.24.

4. Prove Theorem 4.23(a).

5. Prove Theorem 4.23(b).

6. How many positive integers less than or equal to $10,000$ are not divisible by $3, 11$, or 19?

7. How many positive integers less than or equal to $500,000$ are none of the following: the square of an integer, the cube of an integer, the fourth power of an integer?

8. Use Theorem 4.25 to show that $\sum_{k=0}^{n} \binom{n}{k} (-1)^{n-k} 2^k = 1$.

9. Assume that W is a set of n objects, say, $W = \{w_1, w_2, \ldots, w_n\}$. If $m \leq n$, the number of ways of distributing the n objects among m identical boxes (or sets) with no box being empty is denoted by S_n^m and called Stirling's number of the second kind. Each way of making such a distribution is called a partition of W. For example, if $W = \{w_1, w_2, w_3, w_4\}$ and if there are $m = 3$ boxes, two holding one item each and the other holding two items, then there six such ways of partitioning W. If $m = 2$, then there are two types of partitions: (1) both boxes have two objects each and (2) one box has one object and the other box has three objects. Note that the word "identical" means that if two boxes have the same number of objects, we cannot produce another partition by switching the two objects between the two boxes. One can show that

    $$n^r = \sum_{k=0}^{n} \binom{n}{k} (k! S_r^k)$$

    where we let $S_r^0 = 0$.

    (a) Show that an explicit formula for S_n^m is

    $$S_n^m = \frac{1}{m!} \sum_{k=0}^{m} (-1)^{m-k} \binom{m}{k} k^n$$

(b) Use this formula to show that $S_4^3 = 6$ and $S_4^2 = 7$.

(c) Write out the 6 partitions of $W = \{w_1, w_2, w_3, w_4\}$ into $m = 3$ identical boxes.

(d) Write out the 7 partitions of $W = \{w_1, w_2, w_3, w_4\}$ into $m = 2$ identical boxes.

10. Complete the derivation in Example 4.33 by proving that

$$\prod_{i=1}^{t} \left[p_i^{(\alpha_i - 1)} (p_i - 1) \right] = \left[\prod_{i=1}^{t} p_i^{(\alpha_i - 1)} \right] \left[\prod_{i=1}^{t} (p_i - 1) \right]$$

$$= \left[\frac{n}{\prod_{i=1}^{t} p_i} \right] \left[\prod_{i=1}^{t} (p_i - 1) \right]$$

$$= \left[\frac{n}{\prod_{i=1}^{t} p_i} \right] \cdot \left[p_1 p_2 \cdots p_k - \cdots + (-1)^{t-2} \sum_{i<j} p_i p_j \right.$$
$$\left. + (-1)^{t-1} \sum_i p_i + (-1)^t \right]$$

$$= n - \sum_i \frac{n}{p_i} + \sum_{i<j} \frac{n}{p_i p_j} - \sum_{i<j<k} \frac{n}{p_i p_j p_k} + \cdots + (-1)^t$$

where the terms in the last expression are in the same order as in the preceding expression.

4.5   Application: Inversions in Physics

In Theorem 4.16 it was shown that using the Möbius function μ, the function

$$F(n) = \sum_{d|n} f(n/d)$$

defined in terms of an arithmetic function f could be inverted giving f in terms of F by

$$f(n) = \sum_{d|n} \mu(d) \cdot F(n/d)$$

In Section 4.4 we found that the Möbius function could be generalized, providing an analog of the inversion theorem for sums over partially ordered sets. In this second formulation, the Möbius inversion theorem was used to solve several combinatorial problems.

Nan-xian Chen in "Modified Möbius Inversion Formula and Its Application to Physics" [16] gave a further generalization to series and functions of a real variable.

In place of the arithmetic function f defined on the set N of positive integers, we use a function B defined on the set of positive real numbers; and instead of integer division n/d, we consider real number division x/d. Thus, a generalization of

$$F(n) = \sum_{d|n} f(n/d)$$

is to define a real function $A(x)$ by

$$A(x) = \sum_{d=1}^{\infty} B(x/d)$$

where $x > 0$ since any positive integer d divides any real number x. The analogous inversion formula is

$$B(x) = \sum_{d=1}^{\infty} \mu(d) \cdot A(x/d)$$

Although the Möbius function μ still has only integer arguments, we will need to invoke the calculus concepts of series, integrals, and convergence. In preparation for the next theorem, recall that the series $\sum_{i=1}^{\infty} a_i$, for a_i real, is absolutely convergent if the series of absolute values, $\sum_{i=1}^{\infty} |a_i|$, converges. An important property of an absolutely convergent series is that the terms of the series may be rearranged and grouped in any way without affecting either the convergence or the sum of the series.

Theorem 4.34 *If the function B is defined for the positive real numbers and there are positive real numbers c and t such that $|B(x)| \le c \cdot x^{1+t}$ for $x > 0$, then*

(a) $A(x) = \sum_{k=1}^{\infty} B(x/k)$ *and* $\sum_{k=1}^{\infty} \mu(k) \cdot A(x/k)$ *are absolutely convergent.*

(b) $B(x) = \sum_{k=1}^{\infty} \mu(k) \cdot A(x/k)$ *for $x > 0$.*

Proof. **(a)** Using the integral test for series from calculus, it is straightforward to show that $J(t) = \sum_{k=1}^{\infty} 1/k^{t+1}$ is convergent for any $t > 0$. Since $|B(x/k)| \le c \cdot (x/k)^{1+t} = c \cdot x^{1+t} \cdot (1/k)^{1+t}$, $\sum_{k=1}^{\infty} |B(x/k)|$ converges by comparison with $J(t)$ for any $x > 0$. Thus $\sum_{k=1}^{\infty} B(x/k)$ converges absolutely for $x > 0$. Also,

$$\sum_{k=1}^{\infty}\sum_{m=1}^{\infty} c\left(\frac{x}{km}\right)^{1+t} = c\cdot x^{1+t}\sum_{k=1}^{\infty}\left[\left(\frac{1}{k}\right)^{1+t}\sum_{m=1}^{\infty}\left(\frac{1}{m}\right)^{1+t}\right]$$

$$= c\cdot x^{1+t}\sum_{k=1}^{\infty}\left[\left(\frac{1}{k}\right)^{1+t}J(t)\right]$$

$$= c\cdot x^{1+t}J(t)\sum_{k=1}^{\infty}\left(\frac{1}{k}\right)^{1+t}$$

$$= c\cdot x^{1+t}\left[J(t)\right]^2$$

Therefore, since $\left|\mu(k)\cdot B\left(\frac{x}{km}\right)\right| \leq c\left(\frac{x}{km}\right)^{1+t}$, then

$$\sum_{k=1}^{\infty}\sum_{m=1}^{\infty}\mu(k)\cdot B\left(\frac{x}{km}\right) = \sum_{k=1}^{\infty}\mu(k)\sum_{m=1}^{\infty}B\left(\frac{x}{km}\right)$$

converges absolutely, say to the real number S.

(b) Substituting for $A\left(x/k\right)$, we obtain

$$\sum_{k=1}^{\infty}\mu(k)\cdot A\left(\frac{x}{k}\right) = \sum_{k=1}^{\infty}\mu(k)\sum_{m=1}^{\infty}B\left(\frac{x}{km}\right)$$

$$= \sum_{k=1}^{\infty}\sum_{m=1}^{\infty}\mu(k)\cdot B\left(\frac{x}{km}\right) = S$$

Because of absolute convergence, we can rearrange the terms in any way we want. We will sum by all possible products $n = k\cdot m$. Thus,

$$S = \sum_{n=1}^{\infty}\left[\sum_{\substack{s=1\\ ds=n}}^{n}\mu(d)\right]B\left(\frac{x}{n}\right)$$

where, in the sum S on the left, the product $n = k\cdot m$ appears with exactly the coefficients $\mu(d)$ for each d such that $ds = n = km$ for some s, $1 \leq s \leq n$. But by Theorem 4.15, $\sum_{\substack{s=1\\ ds=n}}^{n}\mu(d) = \sum_{d|n}\mu(d)$ equals 1 if $n = 1$ and equals 0 if $n > 1$. Therefore, $S = B\left(x/1\right) = B(x)$. ∎

In physical problems measurable quantities may often be expressed theoretically in terms of other quantities that cannot be measured directly. Therefore, the objective of many investigations in physics is to express these unmeasurable quantities in terms of measurable ones. Chen gives the following application of the modified Möbius inversion theorem.

Example 4.35 *For a crystalline lattice of r atoms, the specific heat $C_v(T)$ of lattice vibration at temperature T is given by*

$$C_v(T) = rk \int_0^\infty \frac{(h\nu/kT)^2 e^{h\nu/kT}}{\left(e^{h\nu/kT} - 1\right)^2} g(\nu) d\nu$$

where k is the Boltzmann constant, h is Planck's constant, ν is the frequency, and $g(\nu)$ is the phonon density of states normalized so that $\int_0^\infty g(\nu) d\nu = 3N$. Using $\dfrac{x}{(x-1)^2} = \sum_{n=1}^\infty n x^n$, observing that $\dfrac{e^w}{(e^w - 1)^2} = \dfrac{e^{-w}}{(e^{-w} - 1)^2}$, and letting $u = \dfrac{h}{kt}$, we obtain

$$C_v\left(\frac{h}{ku}\right) = rk \int_0^\infty \sum_{n=1}^\infty n(u\nu)^2 e^{-nu\nu} g(\nu) d\nu$$

Again changing variables with $\omega = n\nu$, we get

$$\begin{aligned}
C_v\left(\frac{h}{ku}\right) &= rku^2 \int_0^\infty e^{-u\omega} \sum_{n=1}^\infty \left(\frac{\omega}{n}\right)^2 g\left(\frac{\omega}{n}\right) d\omega \\
&= rku^2 \int_0^\infty e^{-u\omega} G(\omega) d\omega \\
&= rku^2 L\{G(\omega)\}
\end{aligned}$$

where $G(\omega) = \sum_{n=1}^\infty \left(\frac{\omega}{n}\right)^2 g\left(\frac{\omega}{n}\right) d\omega$ and L is the Laplace transform. Next using Theorem 4.34, we invert to obtain

$$g(\omega) = \frac{1}{\omega^2} \sum_{n=1}^\infty \mu(n) G\left(\frac{\omega}{n}\right)$$

Therefore,

$$g(\omega) = \left(\frac{1}{rk\omega^2}\right) \sum_{n=1}^\infty \mu(n) L_n^{-1}\left\{ \frac{C_v\left(h/ku\right)}{u^2} \right\}$$

where $L_n^{-1}\{h(u)\} = H\left(\omega/n\right)$ is the inverse Laplace transform from the variable u to the variable ω/n. The method gives a practical solution to the phonon density because at low temperatures C_v may be determined experimentally:

$$\begin{aligned}
C_v(T) &= a_3 T^3 + a_5 T^5 + a_7 T^7 + \cdots \\
&= \sum_{n=2}^\infty a_{2n-1} (h/k)^{2n-1} u^{-(2n-1)}
\end{aligned}$$

Although Theorem 4.34 is equivalent to a previously known result (Theorem 270 of G. H. Hardy and E. M. Wright, *An Introduction to the Theory of Numbers* [33]), Chen's publication in *Physical Review Letters* made the result generally known in the physics community. Shang Y. Ren and John D. Dow in "Generalized Möbius Transforms for Inverse Problems" [71] use Hardy and Wright's result to prove the following theorem.

Theorem 4.36 *If α is a real number, $\tau(k)$ is the number of positive integer divisors of k, and $F(x)$ is a function defined for positive real numbers,*

$$\sum_{m=1}^{\infty} \sum_{n=1}^{\infty} |F(m^\alpha n^\alpha x)| = \sum_{k=1}^{\infty} \tau(k) |F(k^\alpha x)|$$

converges, and

$$G(x) = \sum_{n=1}^{\infty} F(n^\alpha x)$$

then

$$F(x) = \sum_{n=1}^{\infty} \mu(n) G(n^\alpha x)$$

Further, if G is a function defined for positive real numbers such that

$$\sum_{m=1}^{\infty} \sum_{n=1}^{\infty} |G(m^\alpha n^\alpha x)| = \sum_{k=1}^{\infty} \tau(k) |G(k^\alpha x)|$$

converges, and

$$F(x) = \sum_{n=1}^{\infty} \mu(n) G(n^\alpha x)$$

then

$$G(x) = \sum_{n=1}^{\infty} F(n^\alpha x)$$

We note that if $\alpha = -1$, then Theorem 4.36 gives Theorem 4.34.

Further treatments of series forms of the Möbius inversion may be found in R. P. Millane, "Möbius Transform Pairs" [52], in R. P. Millane, "A Product Form of the Möbius Transform" [51], and in Barry W. Ninham et al., "Möbius, Mellin, and Mathematical Physics" [56].

Exercises

1. In a one-dimensional lattice assume that the lattice points are occupied by atoms that interact so that the potential measured at any given atom is the Ewald summation $V(x) = \sum_{k=1}^{\infty} v(k \cdot x)$, where $v(x)$ is the pairwise interaction. Assuming that any necessary convergence criteria are satisfied, prove that we can write

$$v(x) = V(x) - V(2x) - V(3x) - V(5x) + V(6x) - V(7x) + V(10x) - V(11x) - \cdots$$

2. Use the geometric series $\dfrac{1}{2^x - 1} = \sum\limits_{n=1}^{\infty} 2^{-nx}$ for $x > 0$ to show that

$$2^{-x} = \sum_{n=1}^{\infty} \frac{\mu(n)}{2^{nx} - 1}.$$

3. Assuming convergence where necessary, show that if $G(x) = \prod\limits_{n=1}^{\infty} F(n^{\alpha}x)$,

then $F(x) = \prod\limits_{n=1}^{\infty} [G(n^{\alpha}x)]^{\mu(n)}$.

4. Using the result in Exercise 3 and the fact that

$$\sin(x) = x \prod_{n=1}^{\infty} \left[1 - \frac{x^2}{n^2 \pi^2}\right]$$

show that

$$\prod_{n=1}^{\infty} \left[\frac{n \sin(x/n)}{x}\right]^{\mu(n)} = 1 - \frac{x^2}{n^2}$$

when $x \neq k\pi$ for any integer k.

5. The power spectrum $P(\nu, T)$ radiated by a blackbody at T kelvin and of unit area is

$$P(\nu, T) = \frac{2h\nu^3}{c^2} \cdot \frac{1}{e^{h\nu/kT} - 1}$$

where ν is the frequency, c is the speed of light, k is Boltzmann's constant, and h is Planck's constant. The total radiated power spectrum is

$$W(\nu) = \frac{2h\nu^3}{c^2} \int_0^{\infty} \frac{a(t) dT}{e^{h\nu/kT} - 1}$$

where $a(T)$ is the area temperature distribution of the black body. Verify the following steps showing that $a(T)$ may be solved for in terms of $W(\nu)$.

(a) Change variables using $u = \dfrac{h}{kT}$ and $A(u) du = -a(T) dT$, giving

$$W(\nu) = \frac{2h\nu^3}{c^2} \int_0^{\infty} \frac{A(u) du}{e^{u\nu} - 1}$$

where $u\nu > 0$.

(b) Show that

$$W(\nu) = \frac{2h\nu^3}{c^2} \int_0^{\infty} e^{-w\nu} \sum_{n=1}^{\infty} \frac{1}{n} A\left(\frac{w}{n}\right) dw$$

by using the series expansion of $\dfrac{1}{1 - x}$ and changing variables with $u = \dfrac{w}{k}$.

(c) Let $f(w) = \sum\limits_{n=1}^{\infty} \dfrac{1}{w} A\left(\dfrac{w}{n}\right)$ and $g(\nu) = \dfrac{c^2}{2h\nu^3} W(\nu)$ so that

$$g(\nu) = \int_0^{\infty} e^{-w\nu} f(w) dw = L\{f(w)\}$$

and

$$f(w) = L^{-1}\{g(\nu)\}$$

where L is the Laplace transform. Apply Theorem 4.34 to

$$wf(w) = \sum_{n=1}^{\infty} \frac{w}{n} A\left(\frac{w}{n}\right)$$

to obtain

$$A(w) = \sum_{n=1}^{\infty} \frac{\mu(n)}{n} f\left(\frac{w}{n}\right)$$

Chapter 5
CONTINUED
FRACTIONS

5.1   Introduction

In Chapters 1 to 4, the logical development has been within the integers. In this chapter integers, rational numbers, and real numbers are used. This change allows one to treat continued fractions more fully. A brief construction of the rationals is presented below. We will not, however, construct the reals but will indicate the properties that are most immediately pertinent. The properties of the reals and rationals will not be developed fully here. It is assumed the reader is already familiar with them.

Let A be an integral domain. In particular, A could be the set of integers Z. Consider the set of ordered pairs

$$P = \{(a, b) : (a, b) \in A \times A \text{ and } b \neq 0\}$$

and define the relation $\sim$ on P as follows.

Definition 5.1 *If each of (a, b) and (c, d) is in P, then $(a, b) \sim (c, d)$ if and only if $ad = bc$.*

For example, if $A = Z$, the equivalence class $[(2, 3)]$ contains these ordered pairs: $(2, 3), (4, 6), (6, 9), \ldots, (-2, -3), (-4, -6), \ldots$, which correspond to the rational number representations $2/3, 4/6, 6/9, \ldots, (-2)/(-3), (-4)/(-6), \ldots$. All of these are different ways of expressing the same rational number, $[(2, 3)]$.

Theorem 5.2 *The relation $\sim$ on P is an equivalence relation.*

Proof. The proof is straightforward and is left to the reader. ∎

Notation: Let the equivalence class containing $(a, b) \in P$, namely $[(a, b)]$, be denoted by a/b. Let F denote the set of equivalence classes on P.

Definition 5.3 *Given elements a, b, c, and d of A, let addition be defined on F by $a/b + c/d = (ad + bc)/bd$, and let multiplication on F be defined by $(a/b)(c/d) = ac/bd$.*

The next theorem shows that the definitions of addition and multiplication on F are independent of the representatives of the equivalence classes used. When the definitions are independent of the representatives, the operations are well defined.

Theorem 5.4

(a) *Addition in F is well defined.*

(b) *Multiplication in F is well defined.*

Proof. Let $a/b = a'/b'$ and $c/d = c'/d'$. We have immediately that $ab' = a'b$ and $cd' = c'd$.

   (a) By definition of addition we have $a/b + c/d = (ad + bc)/bd$ and $a'/b' + c'/d' = (a'd' + b'c')/b'd'$. We need to show that $(ad + bc)/bd = (a'd' + b'c')/b'd'$ or, equivalently, that $b'd'(ad + bc) = bd(a'd' + b'c')$, but

$$
\begin{aligned}
b'd'(ad + bc) &= b'add' + cd'bb' \\
&= a'bdd' + c'dbb' \\
&= bd(a'd' + b'c')
\end{aligned}
$$

where the second equality holds because $ab' = a'b$ and $cd' = c'd$.

   (b) The definition of multiplication gives both $(a/b)(c/d) = ac/bd$ and $(a'/b')(c'/d') = a'c'/b'd'$. It is necessary to prove that $ac/bd = a'c'/b'd'$ or, equivalently, that $acb'd' = a'c'bd$. But $acb'd' = ab'cd' = a'bc'd = a'c'bd$. ∎

Theorem 5.5 *The set of equivalence classes F is a commutative ring with additive identity 0/1 and multiplicative identity 1/1.*

Proof. The proof is straightforward and is left to the reader. ∎

Lemma 5.6 *For a and b in A, $a/b = 0/1$ if and only if $a = 0$.*

Proof. Clearly, $a/b = 0/1$ if and only if $a = a(1) = b(0) = 0$. ∎

Theorem 5.7 *The commutative ring F is a field.*

Proof. If $a/b \neq 0$, then $a \neq 0$ and $b/a \in F$. But

$$(a/b)(b/a) = ab/ab = 1/1 = 1$$

Hence every nonzero element of F has an inverse, and F is a field. ∎

   Let $f : A \to F$ be defined by $f(a) = a/1$. $f(ab) = ab/1 = (a/1)(b/1) = f(a)f(b)$ and $f(a+b) = (a+b)/1 = (a(1)+b(1))/1 = a/1+b/1 = f(a)+f(b)$. Also, f is one to one for if $f(a) = f(b)$, then $a/1 = b/1$ and $a = a1 = b1 = b$.

Hence f is an monomorphism and the elements in A may be identified with the elements $\{a/1 : a \in A\}$ in F. Thus, A may be considered to be a subring of F.

Theorem 5.8 *The mapping $f : A \to F$ defined by $f(a) = a/1$ is a monomorphism and we say that the integral domain A is embedded in the field F or that F contains A.*

It is easily shown that F is the smallest field in which A can be embedded.

Definition 5.9 *F is called the* fraction field *of A. If A is the set of integers Z, then F is the set of* rational numbers *usually denoted by Q.*

Theorem 5.10 *The set of rational numbers Q is an ordered field.*

Proof. Q is ordered if and only if each a/b in Q has one and only one of the following three properties:

$$a/b \text{ is positive, } a/b \text{ is negative, or } a/b \text{ is zero}$$

We define a/b to be positive if a and b are both positive or both negative and a/b to be negative if a and b have opposite signs. It is straightforward to show that these definitions of positive, negative, and zero are independent of the representatives of the equivalence class. The following are also easily shown:

(a) The sum and product of two positive elements of Q is positive.

(b) If a/b is positive, then $-(a/b)$ is negative.

(c) If an integer is positive, then it is positive as an element of Q (that is, if a Z is positive, then $f(a) = a/1 \in Q$ is positive). If an integer is negative, then it is negative as an element of Q. The integer 0 is 0 as an element of Q.

(d) A rational number a/b is one and only one of the following: positive, negative, or zero. ■

At this point, we assume that the properties of the ordered field Q of rational numbers are known and available for use here.

We will assume the properties of reals are known. For example, we assume that the reals also form an ordered field containing both the integers and rationals and that the reals are complete, which means that a nonempty set of reals bounded from above has a least upper bound and one bounded from below has a greatest lower bound. Completeness also implies that a bounded nondecreasing or nonincreasing sequence $x_1, x_2, x_3, \ldots$ of real numbers converges to a real number x: $\lim\limits_{n \to \infty} x_i = x$, where x is the least upper bound or greatest lower bound, respectively, of $\{x_i : i \in N\}$. Completeness implies the Archimedean property; namely, if a and b are positive *real* numbers, then there is a positive *integer* n such that $n \cdot a > b$. From the Archimedean property it is easy to show that if x is any real number, then there is a unique integer n such that $n \leq x < n + 1$ confining x to be between two integers. Thus, the greatest integer function $\lfloor x \rfloor = n$, where $n \leq x < n + 1$, is well defined for each real x. We also use the absolute value function for reals: $|x| = x$ if $x \geq 0$ but $|x| = -x$ when $x < 0$. Another useful property of sequences of real numbers is the following: if $y_1, y_2, y_3, \ldots$ and $z_1, z_2, z_3, \ldots$ are sequences of real numbers such that $|y_i| \leq |z_i|$ for all i and $\lim_{n \to \infty} z_i = 0$, then $\lim_{n \to \infty} y_i = 0$ also.

For a detailed development of the real numbers, see Leon W. Cohen and Gertrude Ehrlich's *The Structure of the Real Number System* [17] and John M. H. Olmsted's *The Real Number System*, [58].

The Division Algorithm and the Euclidean Algorithm introduced in Chapter 1 will be applied frequently in this chapter. We consider first the rational a/b where a and b are integers and $b > 0$. If the Division Algorithm is applied to the integers a and b with $b > 0$, then there are unique integers t, the quotient, and r, the remainder, such that

$$a = b \cdot t + r \quad \text{and} \quad 0 \leq r < b$$

For $a = -124$ and $b = 35$ we have

$$-124 = (35)(-4) + 16$$

Rewriting these two equations using rational numbers gives

$$\frac{a}{b} = t + \frac{r}{b} \quad \text{and} \quad \frac{-124}{35} = -4 + \frac{16}{35}$$

Using the greatest integer function, the quotient may be written as

$$\left\lfloor \frac{a}{b} \right\rfloor = t \quad \text{and} \quad \left\lfloor \frac{-124}{35} \right\rfloor = -4$$

and the fractional remainder r/b has the property $0 \leq r/b < 1$. If $r \neq 0$, we can rewrite the equations as

$$\frac{a}{b} = t + \frac{1}{b/r} \quad \text{and} \quad \frac{-124}{35} = -4 + \frac{1}{35/16}$$

where $b/r > 1$. We can apply the Division Algorithm again to get $b = rt' + r'$ and $35 = (16)(2) + 3$. Thus, by inverting the fractional remainder and applying the Division Algorithm over and over, we obtain these ways of expressing the rational $-124/35$:

$$\frac{-124}{35} = -4 + \frac{16}{35} = -4 + \frac{1}{\left(\dfrac{35}{16}\right)}$$

$$= -4 + \frac{1}{2 + \dfrac{3}{16}} = -4 + \frac{1}{2 + \dfrac{1}{\left(\dfrac{16}{3}\right)}}$$

$$= -4 + \cfrac{1}{2 + \cfrac{1}{\left(5 + \cfrac{1}{3}\right)}}$$

Since the fraction $1/3$ has 1 as a numerator, we cannot reduce $3/1$ any more by applying the Division Algorithm. The expression

$$-4 + \cfrac{1}{2 + \cfrac{1}{5 + \cfrac{1}{3}}}$$

is called a continued fraction and is an alternative way of representing the rational number $-124/35$. Because of the regular structure of the continued fraction and because the numerators are always 1, we need to mention only the numbers -4, 2, 5, and 3 to specify the continued fraction; and we write $-124/35 = [-4; 2, 5, 3]$ instead. The numbers -4, 2, 5, and 3 are called the terms of the continued fraction or the partial quotients since they are produced by the Division Algorithm. The semicolon distinguishes the special nature of the first term -4. For example, $[3; 7] = 3 + 1/7 = 22/7$ but

$$[0; 3, 7] = 0 + \cfrac{1}{3 + \cfrac{1}{7}} = 7/22$$

Returning to the continued fraction $-124/35 = [-4; 2, 5, 3]$, we see that $3 = (3 - 1) + 1 = (3 - 1) + 1/1$. Thus, one can write

$$\frac{-124}{35} = -4 + \cfrac{1}{2 + \cfrac{1}{5 + \cfrac{1}{2 + \cfrac{1}{1}}}}$$

so that $-124/35 = [-4; 2, 5, 2, 1]$ also, but where the Division Algorithm is not used on the last step. This example suggests that every rational number has two distinct continued fraction representations with integers as terms: one with last term greater than 1 and the other with last term equal to 1.

   If x is real but not rational, we cannot apply the Division Algorithm; however, using the greatest integer function we can obtain a continued fraction representation for x of length n in a way that parallels the rational case. Thus, for x real but not rational, there is a unique integer t_0 with $t_0 \le x < t_0 + 1$ so that $t_0 = \lfloor x \rfloor$. Then

$$1 > y_1 = x - \lfloor x \rfloor = x - t_0 > 0$$

is the so-called fractional part of x. If $x_1 = 1/y_1$, then

$$x = \lfloor x \rfloor + y_1 = t_0 + y_1 = t_0 + \frac{1}{x_1}$$

and $x_1 > 1$. Let $t_1 = \lfloor x_1 \rfloor$, $y_2 = x_1 - \lfloor x_1 \rfloor$, and $x_2 = 1/y_2$ since $y_2 > 0$. Then $x_1 = t_1 + \dfrac{1}{x_2}$ with $x_2 > 0$. Combining these equalities gives

$$
\begin{aligned}
x &= t_0 + \frac{1}{x_1} \\
&= t_0 + \cfrac{1}{t_1 + \cfrac{1}{x_2}}
\end{aligned}
$$

Continuing in this way, we obtain

$$
x = t_0 + \cfrac{1}{t_1 + \cfrac{1}{t_2 + \cfrac{1}{t_3 + \cdots + \cfrac{1}{t_{k-1} + \cfrac{1}{x_k}}}}}
$$

where t_i is an integer for each i, $t_i > 0$ for $i \geq 1$, x_k is a nonrational real number, and $x_k > 1$. One could write as well

$$
x = [t_0; t_1, t_2, \ldots, t_{k-1}, x_k]
$$

For example, let $x = \sqrt{3} = 1.7320508\cdots$. We generate a continued fraction representation for x as follows:

$$
\begin{aligned}
t_0 &= \lfloor x \rfloor = \lfloor \sqrt{3} \rfloor = 1 \\
\text{where} \quad y_1 &= x - \lfloor x \rfloor = x - t_0 = \sqrt{3} - 1 \\
x_1 &= 1/y_1 = 1/(\sqrt{3} - 1) = (\sqrt{3} + 1)/2
\end{aligned}
$$

$$
\begin{aligned}
t_1 &= \lfloor x_1 \rfloor = 1 \\
\text{where} \quad y_2 &= x_1 - \lfloor x_1 \rfloor = x_1 - t_1 = (\sqrt{3} + 1)/2 - 1 \\
&= (\sqrt{3} - 1)/2 \\
x_2 &= 1/y_2 = 2/(\sqrt{3} - 1) = \sqrt{3} + 1
\end{aligned}
$$

$$
\begin{aligned}
t_2 &= \lfloor x_2 \rfloor = 2 \\
\text{where} \quad y_3 &= x_2 - \lfloor x_2 \rfloor = x_2 - t_2 = (\sqrt{3} + 1) - 2 \\
&= \sqrt{3} - 1 \\
x_3 &= 1/y_3 = 1/(\sqrt{3} - 1) = (\sqrt{3} + 1)/2
\end{aligned}
$$

Thus,

$$
\begin{aligned}
\sqrt{3} &= [t_0; x_1] = [1; (\sqrt{3} + 1)/2] \\
&= [t_0; t_1, x_2] = [t_0; [t_1; x_2]] = [1; 1, (\sqrt{3} + 1)] \\
&= [t_0; t_1, t_2, x_3] = [t_0; t_1, [t_2; x_3]] = [1; 1, 2, (\sqrt{3} + 1)/2]
\end{aligned}
$$

But because $x_3 = x_1$, the pattern repeats, giving

$$\sqrt{3} = [1; 1, 2, 1, 2, 1, 2, (\sqrt{3} + 1)/2]$$

and so on.

We now give a recursive definition of a continued fraction.

Definition 5.11 *For a finite sequence $t_0, t_1, t_2, \ldots, t_n$ of real numbers with $n \geq 0$ and $t_i > 0$ for $i \geq 1$, define the* finite continued fraction *$[t_0; t_1, t_2, \ldots, t_n]$ as follows:*

$$[t_0;] = t_0$$

$$[t_0; t_1] = t_0 + \frac{1}{t_1}$$

$$[t_0; t_1, t_2, \ldots, t_k] = [t_0; [t_1; t_2, \ldots, t_k]] \text{ for } 1 < k \leq n$$

The numbers $t_0, t_1, \ldots, t_n$ are called partial quotients *or* terms *of the continued fraction. The continued fraction $[t_0; t_1, t_2, \ldots, t_n]$ is said to be* simple *if t_i is an integer for each i; that is, every term in the continued fraction is an integer. The numbers $[t_0;]$, $[t_0; t_1]$, $[t_0; t_1, t_2]$, $\ldots$, $[t_0; t_1, t_2, \ldots, t_k]$, $\ldots$, $[t_0; t_1, t_2, \ldots, t_n]$ are called* convergents *of the continued fraction $[t_0; t_1, t_2, \ldots, t_n]$. $[t_0; t_1, t_2, \ldots, t_k]$ is the k-th convergent for $0 \leq k \leq n$. For convenience, if the notation $[t_0; t_1, t_2, \ldots, t_k]$ is used when $k = 0$, we will mean $[t_0;]$. We say that two continued fractions $[t_0; t_1, t_2, \ldots, t_n]$ and $[b_0; b_1, b_2, \ldots, b_m]$ are equal* term by term *provided that $n = m$ and $t_i = b_i$ for $0 \leq i \leq n$. If x is a real number and $x = [t_0; t_1, t_2, \ldots, t_n]$, then we say that $[t_0; t_1, t_2, \ldots, t_n]$ is a* continued fraction representation *of x. Two representations are the* same *or* equal *provided that they are equal term by term.*

The next theorem provides another way to decompose a continued fraction into convergents.

Theorem 5.12 *If n is a positive integer and $[t_0; t_1, t_2, \ldots, t_n]$ is a continued fraction, then, for each k, $1 \leq k \leq n$,*

$$[t_0; t_1, t_2, \ldots, t_n] = [t_0; t_1, t_2, \ldots, t_{k-1}, [t_k; t_{k+1}, \ldots, t_n]]$$

Proof. The equality is proved using mathematical induction. If $n = 1$, the definition gives $[t_0; t_1] = [t_0; [t_1;]]$. Assume the equality of the theorem is true for $n = m$; that is, for any continued fraction $[b_0; b_1, \ldots, b_m]$,

$$[b_0; b_1, b_2, \ldots, b_m] = [b_0; b_1, b_2, \ldots, b_{j-1}, [b_j; b_{j+1}, \ldots, b_m]]$$

for $1 \leq j \leq m$. Consider $n = m + 1$. If $[t_0; t_1, t_2, \ldots, t_{m+1}]$ is a continued fraction and $1 \leq k \leq m + 1$, then

$$\begin{aligned}
[t_0; t_1, \ldots, t_{m+1}] &= [t_0; [t_1; t_2, \ldots, t_{m+1}]] \text{ by definition} \\
&= [t_0; [t_1; t_2, \ldots, t_{k-1}, [t_k; t_{k+1}, \ldots, t_{m+1}]]]
\end{aligned}$$

by the induction hypothesis identifying $b_i = t_{i+1}$ and $j = k - 1$. Thus

$$[t_0; t_1, \ldots, t_{m+1}] = [t_0; t_1, t_2, \ldots, t_{k-1}, [t_k; t_{k+1}, \ldots, t_{m+1}]]$$

by using the definition of a continued fraction. Hence, by induction, the theorem holds. ∎

Definition 5.11, Theorem 5.12, and the results of Sections 5.2 and 5.3 hold for any ordered field containing the rational numbers. Properties of the real numbers are needed for the remainder of the chapter.

Exercises

1. Find two continued fraction representations, $[t_0; t_1, \ldots, t_n]$, for each of the following rational numbers:

   (a) $37/11$ **(b)** $48/1003$ **(c)** $-257/2003$

   (d) $11/37$ **(e)** $-5/44$ **(f)** 5

2. Compute these rational numbers and express them in the form p/q with p and q integers:

   (a) $[3; 5, 2]$ **(b)** $[0; 3, 5, 2]$ **(c)** $[-10; 1, 4, 3]$

   (d) $[6; 4, 7, 3, 5]$ **(e)** $[2; 5, 3]$ **(f)** $[5; 3, 7, 4, 6]$

3. Find the finite continued fraction representation of x of the form $x = [t_0; t_1, t_2, t_3, t_4, t_5, x_6]$ for:

   (a) $x = \sqrt{5}$

   (b) $x = \sqrt{2}$

   (c) $x = \pi$

   (d) $x = (1 + \sqrt{5})/2$

4. Let $[t_0; t_1, \ldots, t_n]$ be a finite continued fraction with $t_i > 0$ for $1 \leq i \leq n$. If $b > 0$, then prove that:

   (a) $[t_0; t_1, t_2, \ldots, t_n] > [t_0; t_1, t_2, \ldots, t_n + b]$ if n is odd

   (b) $[t_0; t_1, t_2, \ldots, t_n] < [t_0; t_1, t_2, \ldots, t_n + b]$ if n is even

5. Assuming the Archimedean property, prove that if x is any real number, then there is a unique integer n such that $n \leq x < n + 1$.

5.2   Convergents

Clearly, $[t_0; t_1, t_2, \ldots, t_k]$ is a real number for every k. We will write explicitly the first four convergents and simplify them by expressing each as a quotient of two real numbers with both the numerator and denominator given in terms of t_i. We will be particularly interested in the form of the numerator and denominator of each convergent.

$$[t_0;] = t_0 = \frac{t_0}{1} = \frac{p_0}{q_0}$$

where we have let $p_0 = t_0$ and $q_0 = 1$.

$$[t_0; t_1] = t_0 + \frac{1}{t_1} = \frac{t_0 t_1 + 1}{t_1} = \frac{p_1}{q_1}$$

where we have let $p_1 = t_0 t_1 + 1$ and $q_1 = t_1$.

$$
\begin{aligned}
[t_0; t_1, t_2] &= [t_0; [t_1; t_2]] \\
&= t_0 + \frac{1}{[t_1; t_2]} = t_0 + \cfrac{1}{t_1 + \cfrac{1}{t_2}} \\
&= t_0 + \frac{t_2}{t_1 t_2 + 1} = \frac{t_0 t_1 t_2 + t_0 + t_2}{t_1 t_2 + 1} \\
&= \frac{(t_0 t_1 + 1) t_2 + t_0}{t_1 t_2 + 1} \\
&= \frac{p_1 t_2 + p_0}{q_1 t_2 + q_0} = \frac{p_2}{q_2}
\end{aligned}
$$

where we have let $p_2 = p_1 t_2 + p_0$ and $q_2 = q_1 t_2 + q_0$. Similarly, we calculate

$$
\begin{aligned}
[t_0; t_1, t_2, t_3] &= \frac{t_0 t_1 t_2 t_3 + t_0 t_1 + t_0 t_3 + t_2 t_3 + 1}{t_1 t_2 t_3 + t_1 + t_3} \\
&= \frac{(t_0 t_1 t_2 + t_0 + t_2) t_3 + (t_0 t_1 + 1)}{(t_1 t_2 + 1) t_3 + t_1} \\
&= \frac{p_2 t_3 + p_1}{q_2 t_3 + q_1} = \frac{p_3}{q_3}
\end{aligned}
$$

where $p_3 = p_2 t_3 + p_1$ and $q_3 = q_2 t_3 + q_1$. In each case, the number $[t_0; t_1, t_2, \ldots, t_k]$ is "reassembled" from the continued fraction representation without any "cancellation" to form a quotient of two polynomials P and Q in the variables $t_0, t_1, \ldots, t_k$, that is,

$$[t_0; t_1, t_2, \ldots, t_k] = \frac{P(t_0, t_1, \ldots, t_k)}{Q(t_0, t_1, \ldots, t_k)}$$

It is also true that q_0, q_1, q_2, and q_3 are positive because they depend upon the multiplication and addition of positive numbers. These examples suggest the following theorem.

Theorem 5.13 *Let n be a nonnegative integer and $[t_0; t_1, t_2, \ldots, t_n]$ be a finite continued fraction with the finite sequences $p_0, p_1, \ldots, p_n$ and $q_0, q_1, \ldots, q_n$ defined recursively as follows:*

(a) $p_0 = t_0,$
$\quad\quad q_0 = 1.$

(b) $p_1 = t_0 t_1 + 1,$
$\quad\quad q_1 = t_1.$

(c) $p_k = p_{k-1} t_k + p_{k-2},$
$\quad\quad q_k = q_{k-1} t_k + q_{k-2} \quad$ *for $2 \le k \le n.$*

Then $q_k > 0$ and $[t_0; t_1, t_2, \ldots, t_k] = \dfrac{p_k}{q_k}$ for $0 \le k \le n.$

Proof. It is straightforward to show directly by mathematical induction that $q_k > 0$ for $0 \le k \le n$. The proof of this part is left to the reader. It was shown in the discussion leading up to this theorem that for $k = 0, 1,$ and 2 it is true that $[t_0;] = p_0/q_0$, $[t_0; t_1] = p_1/q_1$, and $[t_0; t_1, t_2] = p_2/q_2$. Assume that $2 \le k < n$ and, for any for any continued fraction $[b_0; b_1, \ldots, b_k]$ and any j, $0 \le j \le k$, it is true that $[b_0; b_1, \ldots, b_j] = p'_j/q'_j$ where the p'_j and q'_j are defined similar to (a) to (c) but for the continued fraction $[b_0; b_1, \ldots, b_k]$. Then

$$
\begin{aligned}
[t_0; t_1, \ldots, t_k, t_{k+1}] &= [t_0; t_1, \ldots, t_{k-1}, [t_k; t_{k+1}]] \text{ by Theorem 5.12} \\
&= [b_0; b_1, \ldots, b_{k-1}, b_k]
\end{aligned}
$$

where $b_i = t_i$ for $0 \le i \le k - 1$ and $b_k = [t_k; t_{k+1}] = t_k + \dfrac{1}{t_{k+1}}$. Thus, by the induction hypothesis,

$$
[t_0; t_1, \ldots, t_{k+1}] = \frac{p'_{k-1} b_k + p'_{k-2}}{q'_{k-1} b_k + q'_{k-2}}
$$

Since $b_i = t_i$ for $0 \le i \le k - 1$, we have that $p_i = p'_i$ and $q_i = q'_i$ for such i. Substituting for b_k, p'_i, and q'_i, we obtain

$$
\begin{aligned}
[t_0; t_1, \ldots, t_{k+1}] &= \frac{p_{k-1}\left(t_k + \dfrac{1}{t_{k+1}}\right) + p_{k-2}}{q_{k-1}\left(t_k + \dfrac{1}{t_{k+1}}\right) + q_{k-2}} \\
&= \frac{(p_{k-1} t_k + p_{k-2}) + p_{k-1}/t_{k+1}}{(q_{k-1} t_k + q_{k-2}) + q_{k-1}/t_{k+1}} \\
&= \frac{p_k t_{k+1} + p_{k-1}}{q_k t_{k+1} + q_{k-1}} = \frac{p_{k+1}}{q_{k+1}}
\end{aligned}
$$

Hence, $[t_0; t_1, \ldots, t_k] = p_k/q_k$ for $0 \le k \le n$ by mathematical induction. ∎

The numbers p_k and q_k of Theorem 5.13 are defined independently of whether they can be used as the numerator and denominator of an expression equal to the k-th convergent. The recurrence relation given by Theorem 5.13 provides a rapid means of calculating the convergents of a given continued fraction since the numerators p_i and denominators q_i may be computed simultaneously and easily. For example, for $x = [-4; 2, 5, 2, 1] = [t_0; t_1, t_2, t_3, t_4]$, the convergents may be calculated in a tableau:

k	t_k	p_k	q_k	p_k/q_k
0	-4	-4	1	$-4/1$
1	2	$(-4)(2) + 1 = -7$	2	$-7/2$
2	5	$(-7)(5) + (-4) = -39$	$(2)(5) + 1 = 11$	$-39/11$
3	2	$(-39)(2) + (-7) = -85$	$(11)(2) + 2 = 24$	$-85/24$
4	1	$(-85)(1) + (-39) = -124$	$(24)(1) + 11 = 35$	$-124/35$

So x is $-124/35$, the number that generated $[-4; 2, 5, 2, 1]$ in Section 5.1.

Because continued fractions are defined recursively, there are many relationships among the convergents in terms of the numerators and denominators p_k and q_k of Theorem 5.13. In order to allow some of these relationships to hold for $k = 0$, it will often be convenient to define p_k and q_k for $k = -1$ as

$$p_{-1} = 1$$
$$q_{-1} = 0$$

Thus, $p_1 = p_0 t_1 + p_{-1} = t_0 t_1 + 1$ and $q_1 = q_0 t_1 + q_{-1} = 1 \cdot t_1 + 0 = t_1$ as in Theorem 5.13.

Theorem 5.14 *If $[t_0; t_1, \ldots, t_n]$ is a finite continued fraction with t_i real and p_k and q_k are given by Theorem 5.13, then*

(a) $p_k q_{k-1} - p_{k-1} q_k = (-1)^{k-1}$ *for $0 \le k \le n$.*

(b) $\dfrac{p_k}{q_k} - \dfrac{p_{k-1}}{q_{k-1}} = \dfrac{(-1)^{k-1}}{q_k q_{k-1}}$ *for $1 \le k \le n$.*

(c) $\dfrac{p_k}{q_k} - \dfrac{p_{k-2}}{q_{k-2}} = \dfrac{(-1)^k t_k}{q_k q_{k-2}}$ *for $2 \le k \le n$.*

(d) *The even-numbered convergents form an increasing sequence, that is,*

$$\frac{p_0}{q_0} < \frac{p_2}{q_2} < \frac{p_4}{q_4} < \cdots$$

and the odd-numbered convergents form a decreasing sequence, that is,

$$\frac{p_1}{q_1} > \frac{p_3}{q_3} > \frac{p_5}{q_5} > \cdots$$

Further,

$$\frac{p_{2j}}{q_{2j}} \le [t_0; t_1, t_2, \ldots, t_n] \le \frac{p_{2i+1}}{q_{2i+1}}$$

for $0 \le j \le \lfloor n/2 \rfloor$ *and* $0 \le i \le \lfloor (n-1)/2 \rfloor$, *with the left equality holding when n is even and the right equality holding when n is odd.*

(e) *If* $[t_0; t_1, \ldots, t_n]$ *is simple, then* $q_k \ge k$ *for* $0 \le k \le n$.

(f) *If* $[t_0; t_1, \ldots, t_n]$ *is simple, then* $q_k < q_{k+1}$ *for* $1 \le k \le n-1$ *and* $q_0 \le q_1$.

Proof. (a) For $k = 1$, $p_1 q_0 - p_0 q_1 = (t_0 t_1 + 1)(1) - t_0 t_1 = 1 = (-1)^{1-1}$ by Theorem 5.13. Let $2 \le m < n$ and assume that the formula in part (a) is true for $k = m$. It is shown that the formula holds for $k = m + 1$. Again using Theorem 5.13, we have

$$
\begin{aligned}
p_{m+1} q_m - p_m q_{m+1} &= (p_m t_{m+1} + p_{m-1}) q_m - p_m (q_m t_{m+1} + q_{m-1}) \\
&= p_{m-1} q_m - p_m q_{m-1} \\
&= (-1)(p_m q_{m-1} - p_{m-1} q_m) \\
&= (-1)(-1)^{m-1} = (-1)^{(m+1)-1}
\end{aligned}
$$

Thus, the formula in part (a) holds for $1 \le k \le n$. Also $p_0 q_{-1} - q_0 p_{-1} = t_0 \cdot 0 - 1 \cdot 1 = -1$ and the formula holds for $k = 0$.

(b) This follows from part (a) by dividing by $q_k q_{k-1}$ when $k \ge 1$.

(c) If $k \ge 2$, then, by Theorem 5.13,

$$
\begin{aligned}
p_k &= p_{k-1} t_k + p_{k-2} \\
q_k &= q_{k-1} t_k + q_{k-2}
\end{aligned}
$$

Multiplying the first of these equations by q_{k-2} and the second by p_{k-2} gives

$$
\begin{aligned}
p_k q_{k-2} &= p_{k-1} q_{k-2} t_k + p_{k-2} q_{k-2} \\
q_k p_{k-2} &= q_{k-1} p_{k-2} t_k + q_{k-2} p_{k-2}
\end{aligned}
$$

Subtracting the last two equations produces

$$
p_k q_{k-2} - q_k p_{k-2} = (p_{k-1} q_{k-2} - p_{k-2} q_{k-1}) t_k
$$

Now, invoking part (a), we obtain

$$
p_k q_{k-2} - q_k p_{k-2} = (-1)^{(k-1)-1} t_k = (-1)^k (-1)^{-2} t_k = (-1)^k t_k
$$

Dividing by $q_k q_{k-2} > 0$ gives the desired result:

$$
\frac{p_k}{q_k} - \frac{p_{k-2}}{q_{k-2}} = \frac{(-1)^k t_k}{q_k q_{k-2}}
$$

(d) If $2 \le k \le n$ and k is even, then $k - 2$ is also even and $k - 2 \ge 0$. Thus $(-1)^k = 1$ and part (c) implies that

$$
\frac{p_k}{q_k} - \frac{p_{k-2}}{q_{k-2}} = \frac{t_k}{q_k q_{k-2}} > 0
$$

because t_k and $q_k q_{k-2}$ are positive for $k \geq 1$. If $3 \leq k \leq n$ and k is odd, then $k - 2$ is also odd and $k - 2 \geq 1$. In this case $(-1)^k = -1$ so that part (c) implies

$$\frac{p_k}{q_k} - \frac{p_{k-2}}{q_{k-2}} = \frac{(-1)t_k}{q_k q_{k-2}} < 0$$

because t_k and $q_k q_{k-2}$ are positive for $k \geq 3$.

If n is odd, then, by part (b), $\dfrac{p_n}{q_n} > \dfrac{p_{n-1}}{q_{n-1}}$ so that $[t_0; t_1, t_2, \ldots, t_n] = p_n/q_n$ is greater than the largest even convergent, p_{n-1}/q_{n-1}. If n is even, then $\dfrac{p_n}{q_n} < \dfrac{p_{n-1}}{q_{n-1}}$ so that $[t_0; t_1, t_2, \ldots, t_n] = p_n/q_n$ is less than the smallest odd convergent, p_{n-1}/q_{n-1}.

The proofs of parts (e) and (f) are left to the reader. ∎

The convergents for $x = [-4; 2, 5, 2, 1]$ were computed in the tableau earlier in this section. Summarizing those results relative to part (d), we see that

$$
\begin{array}{ccccccccc}
p_0/q_0 & < & p_2/q_2 & < & p_4/q_4 & = & x & < & p_3/q_3 & < & p_1/q_1 \\
-4/1 & < & -39/11 & < & -124/35 & = & x & < & -85/24 & < & -7/2
\end{array}
$$

Also, by direct calculation,

$$p_3 q_2 - p_2 q_3 = (-85)(11) - (-39)(24) = (-935) - (-936) = 1$$

The import of Theorem 5.14 is that it specifies that consecutive and alternate convergents can differ by no more than a quantity proportional to the reciprocal of the "denominators" of the convergents and that convergents are alternatively greater than and less than $[t_0; t_1, t_2, \ldots, t_n]$.

Several ratios of the numerators and denominators of the convergents p_k/q_k have some utility. They are given in the next theorem whose proof is left to the reader.

Theorem 5.15 *For the finite continued fraction* $[t_0; t_1, t_2, \ldots, t_n]$:

(a) $q_k/q_{k-1} = [t_k; t_{k-1}, \ldots, t_2, t_1]$ *for* $1 \leq k \leq n$.

(b) *If* $t_0 \neq 0$, *then* $p_k/p_{k-1} = [t_k; t_{k-1}, \ldots, t_1, t_0]$ *for* $1 \leq k \leq n$.

(c) *If* $t_0 = 0$, *then* $p_k/p_{k-1} = [t_k; t_{k-1}, \ldots, t_2, t_1]$ *for* $2 \leq k \leq n$.

Exercises

1. Complete the proof of Theorem 5.13 by proving that $q_k > 0$ for $0 \leq k \leq n$.

2. Compute the convergents for:

    (a) $[2; 5, 1, 2, 4]$ (b) $[2; 5, 1, 2, 3, 1]$ (c) $[5; 1, 2, 3]$

    (d) $[0; 5, 1, 2, 3]$ (e) $[-5; 3, 8, 2]$

3. Compute the convergents p_k/q_k by computing p_k and q_k for $0 \le k \le 5$ for $x = [t_0; t_1, t_2, t_3, t_4, t_5, x_6]$ for the following x (see the Exercises of Section 5.1):

   (a) $x = \sqrt{5}$

   (b) $x = \sqrt{2}$

   (c) $x = \pi$

   (d) $x = (1 + \sqrt{5})/2$

   Using a calculator, compute $x - p_k/q_k$ for each k.

4. Prove parts (e) and (f) of Theorem 5.14.

5. Prove Theorem 5.15. (*Hint:* Use Theorem 5.13.)

6. Let p_k and q_k be given by Theorem 5.13 for the continued fraction $[t_0; t_1, \ldots, t_n]$. Prove that

$$\begin{vmatrix} p_k & p_{k-1} \\ q_k & q_{k-1} \end{vmatrix} = (-1)^{k+1} \text{ for } 0 \le k \le n$$

   where the determinant is used on the left side of the equation.

7. In Theorem 5.13, we showed that if $p_{-1} = 1$ and $q_{-1} = 0$, part (c) held for $k = 1$. What would p_{-2} and q_{-2} have to be in order for part (c) to hold for $k = 0$?

8. For $k \ge 1$, let

$$\begin{bmatrix} p_0 & q_0 \\ p_{-1} & q_{-1} \end{bmatrix} = \begin{bmatrix} t_0 & 1 \\ 1 & 0 \end{bmatrix}$$

   and

$$\begin{bmatrix} p_k & q_k \\ p_{k-1} & q_{k-1} \end{bmatrix} = \begin{bmatrix} t_k & 1 \\ 1 & 0 \end{bmatrix} \begin{bmatrix} p_{k-1} & q_{k-1} \\ p_{k-2} & q_{k-2} \end{bmatrix}$$

   (a) Prove for $k \ge 0$ that

$$\begin{bmatrix} p_k & q_k \\ p_{k-1} & q_{k-1} \end{bmatrix} = \begin{bmatrix} t_k & 1 \\ 1 & 0 \end{bmatrix} \begin{bmatrix} t_{k-1} & 1 \\ 1 & 0 \end{bmatrix} \cdots \begin{bmatrix} t_0 & 1 \\ 1 & 0 \end{bmatrix}$$

   (b) Use the properties of determinants and part (a) to obtain an alternative proof of

$$p_k q_{k-1} - p_{k-1} q_k = (-1)^{k+1} \text{ for } k \ge 1$$

5.3   Simple Continued Fractions

The definitions and the fundamental recursive properties in the preceding sections of this chapter apply to all finite continued fractions $[t_0; t_1, t_2, \ldots, t_n]$ with t_i real. In this section, we consider simple finite continued fractions for which the terms t_i are integers. We first determine which real numbers are equal to simple finite continued fractions.

Theorem 5.16 *A real number x is rational if and only if there are integers t_i such that $x = [t_0; t_1, t_2, \ldots, t_n]$. Further, if $x = a/b$ for integers a and b with $b > 0$, then the integers $t_0, t_1, \ldots, t_n$ may be chosen such that $t_0, t_1, \ldots, t_n$ are the integer quotients obtained by applying the Euclidean Algorithm to calculate $\gcd(a, b)$:*

$$
\begin{array}{rclcccl}
a &=& bt_0 + r_0 & \quad 0 &<& r_0 &< b \\
b &=& r_0 t_1 + r_1 & \quad 0 &<& r_1 &< r_0 \\
r_0 &=& r_1 t_2 + r_2 & \quad 0 &<& r_2 &< r_1 \\
&\vdots& & & & \vdots & \\
r_k &=& r_{k+1} t_{k+2} + r_{k+2} & \quad 0 &<& r_{k+2} &< r_{k+1} \\
&\vdots& & & & \vdots & \\
r_{n-3} &=& r_{n-2} t_{n-1} + r_{n-1} & \quad 0 &<& r_{n-1} &< r_{n-2} \\
r_{n-2} &=& r_{n-1} t_n & & & &
\end{array}
$$

where $t_i > 0$ for $1 \le i \le n$ and $t_n > 1$ if $n > 0$.

Proof. Let $x = [t_0; t_1, t_2, \ldots, t_n]$ be a simple continued fraction so that t_i is an integer for each i and $t_i > 0$ for $1 \le i \le n$. We will show that x is rational by using the convergents given in Theorem 5.13. Specifically, we will show that p_k and q_k are integers for $0 \le k \le n$. It is obvious that p_k and q_k are integers for $k = 0$ and 1. If $1 \le k < n$ and p_j and q_j are integers for $1 \le j \le k$, then by Theorem 5.13,

$$
\begin{array}{rcl}
p_{k+1} &=& p_k t_{k+1} + p_{k-1} \\
q_{k+1} &=& q_k t_{k+1} + q_{k-1}
\end{array}
$$

are integers because $p_k, p_{k-1}, q_k, q_{k-1}$, and t_{k+1} are integers. Thus, by mathematical induction, p_i and q_i are integers for $0 \le i \le n$. Since $x = [t_0; t_1, t_2, \ldots, t_n] = p_n/q_n$, x is also rational.

Let x be a rational number and let a and b be integers such that $x = a/b$ and $b > 0$. Let $t_0, t_1, \ldots, t_n$ be the integers generated by the Euclidean Algorithm in calculating $\gcd(a, b)$ as described in the statement of the theorem. We first note that if $a = 0$, then $t_0 = 0$, $n = 0$, and $a/b = [0;] = [t_0;]$. Assume that $a \ne 0$. If $r_0 = 0$, then $a/b = t_0 = [t_0;]$. If $r_0 > 0$ and we define $r_{-1} = b$, then the equations produced by the Euclidean Algorithm may be rewritten as

$$
r_k/r_{k+1} = t_{k+2} + r_{k+2}/r_{k+1} = [t_{k+2}; r_{k+1}/r_{k+2}]
$$

for $-1 \leq k \leq n-3$ and

$$
\begin{aligned}
r_{n-3}/r_{n-2} &= t_{n-1} + r_{n-1}/r_{n-2} = [t_{n-1}; r_{n-2}/r_{n-1}] \\
&= [t_{n-1}; t_n]
\end{aligned}
$$

since $r_{n-2} = t_n r_{n-1}$.

We now show that $a/b = [t_0; t_1, t_2, \ldots, t_k, r_{k-1}/r_k]$ for each k, $0 \leq k \leq n-1$. The first of the equations above states that $a/b = [t_0; r_{-1}/r_0]$, so the claim is true for $k = 0$. If $0 \leq m < n-1$ and $a/b = [t_0; t_1, t_2, \ldots, t_m, r_{m-1}/r_m]$, then

$$
\begin{aligned}
a/b &= [t_0; t_1, t_2, \ldots, t_m, r_{m-1}/r_m] \\
&= [t_0; t_1, t_2, \ldots, t_m, [t_{m+1}; r_m/r_{m+1}]] \\
&= [t_0; t_1, t_2, \ldots, t_m, t_{m+1}, r_m/r_{m+1}] \text{ by Theorem 5.12}
\end{aligned}
$$

Thus, the claim is true for all k, $0 \leq k \leq n-1$. Letting $k = n-1$, we have

$$
\begin{aligned}
a/b &= [t_0; t_1, t_2, \ldots, t_{n-1}, r_{n-2}/r_{n-1}] \\
&= [t_0; t_1, t_2, \ldots, t_{n-1}, t_n]
\end{aligned}
$$

We note that $t_i > 0$ for $1 \leq i \leq n-1$ because $r_{i-2} = r_{i-1} t_i + r_i$ would imply $r_{i-2} = r_i$, which contradicts the fact that $r_i < r_{i-1} < r_{i-2}$. Also, $t_n > 0$ because if $t_n = 0$, then $r_{n-2} = r_{n-1} t_n = 0$, a contradiction. Further, $t_n > 1$ since if $t_n = 1$, we would have $r_{n-2} = r_{n-1} t_n = r_{n-1}$, a contradiction. ∎

Thus, only rational numbers are represented as simple finite continued fractions; and every finite simple continued fraction represents a rational number. The algorithm of Theorem 5.16 formalizes the procedure described in Section 5.1.

The next two theorems provide useful characterizations of simple continued fractions. The proofs of the next theorem and lemma are left to the reader.

Theorem 5.17 *If* $[t_0; t_1, t_2, \ldots, t_n]$ *is a simple continued fraction with* $n \geq 1$, *then*

(a) $[t_k; t_{k+1}, t_{k+2}, \ldots, t_n] > 1$ *for* $1 \leq k \leq n-1$.

(b) $[t_k; t_{k+1}, t_{k+2}, \ldots, t_n] \geq 1$ *for* $k = n$.

Lemma 5.18 *If* A *and* B *are real numbers with* $0 < A \leq 1$ *and* $0 < B \leq 1$, *then* $-1 < B - A < 1$.

The next theorem says how two finite simple continued fraction representations of the same rational number can differ.

Theorem 5.19 *If* $[a_0; a_1, a_2, \ldots, a_n]$ *and* $[b_0; b_1, b_2, \ldots, b_m]$ *are simple continued fractions that represent the same number so that* $[a_0; a_1, a_2, \ldots, a_n] = [b_0; b_1, b_2, \ldots, b_m]$ *and* $n \leq m$, *then*

(a) $[a_k; a_{k+1}, a_{k+2}, \ldots, a_n] = [b_k; b_{k+1}, b_{k+2}, \ldots, b_m]$ *for* $0 \le k \le n.$

(b) $a_k = b_k$ *for* $0 \le k \le n - 1.$

Proof.   **(a)** Clearly the equality holds for $k = 0$. If $0 \le j < n$ and $[a_j; a_{j+1}, a_{j+2}, \ldots, a_n] = [b_j; b_{j+1}, b_{j+2}, \ldots, b_m]$, then

$$a_j + \frac{1}{[a_{j+1}; a_{j+2}, \ldots, a_n]} = b_j + \frac{1}{[b_{j+1}; b_{j+2}, \ldots, b_m]}$$

or

$$a_j - b_j = \frac{1}{[b_{j+1}; b_{j+2}, \ldots, b_m]} - \frac{1}{[a_{j+1}; a_{j+2}, \ldots, a_n]}$$

Theorem 5.17 implies that

$$0 < \frac{1}{[a_{j+1}; a_{j+2}, \ldots, a_n]} \le 1 \quad \text{and} \quad 0 < \frac{1}{[b_{j+1}; b_{j+2}, \ldots, b_m]} \le 1$$

Lemma 5.18 gives $-1 < a_j - b_j < 1$; but, since a_j and b_j are integers, $a_j - b_j = 0$. Thus, $a_j = b_j$ and $[a_{j+1}; a_{j+2}, a_{j+3}, \ldots, a_n] = [b_{j+1}; b_{j+2}, b_{j+3}, \ldots, b_m]$. By mathematical induction, $[a_k; a_{k+1}, a_{k+2}, \ldots, a_n] = [b_k; b_{k+1}, b_{k+2}, \ldots, b_m]$ for $0 \le k \le n$.

   (b) Part (a) implies that $[a_k; a_{k+1}, a_{k+2}, \ldots, a_n] = [b_k; b_{k+1}, b_{k+2}, \ldots, b_m]$ for any k, $0 \le k \le n$. Certainly, if $0 \le k \le n - 1$, then

$$a_k + \frac{1}{[a_{k+1}; a_{k+2}, \ldots, a_n]} = b_k + \frac{1}{[b_{k+1}; b_{k+2}, \ldots, b_m]}$$

Again applying part (a), we obtain $a_k = b_k$. ∎

   It was shown in Section 5.1 that $-124/5$ had two finite simple continued fraction representations: $[-4; 2, 5, 3]$ and $[-4; 2, 5, 2, 1]$. The next theorem establishes that there are no others.

Theorem 5.20 *If x is rational and $x = [t_0; t_1, t_2, \ldots, t_{n-1}, t_n]$, where t_j is given by the Euclidean Algorithm in Theorem 5.16, then the only other simple finite continued fraction representation of x is $[t_0; t_1, t_2, \ldots, t_{n-1}, t_n - 1, 1]$.*

Proof. Let $x = [t_0; t_1, t_2, \ldots, t_n]$ be the simple continued fraction representation given by Theorem 5.16, where $t_n > 1$ if $n > 0$; and let $[b_0; b_1, b_2, \ldots, b_m]$ also be a simple continued fraction representation of x.

   For $n = 0$, if $m = 0$, then $t_0 = [t_0;] = [b_0;] = b_0$. If $m > 0$, then

$$a_0 = b_0 + \frac{1}{[b_1; b_2, \ldots, b_m]}$$

Since $[b_1; b_2, \ldots, b_m] \ge 1$, we have that

$$0 < a_0 - b_0 = \frac{1}{[b_1; b_2, \ldots, b_m]} \le 1$$

Thus, $a_0 - b_0 = 1$ or $b_0 = a_0 - 1$ and $[b_1; b_2, \ldots, b_m] = 1$. By Theorem 5.17, $m = 1$.

For $n > 0$, we first show that $n \leq m$. If $n > m$, then, by Theorem 5.19,

$$[t_m; t_{m+1}, \ldots, t_n] = [b_m;]$$

or

$$t_m + \frac{1}{[t_{m+1}; t_{m+2}, \ldots, t_n]} = b_m$$

If $m + 1 < n$, Theorem 5.17 implies that

$$0 < b_m - t_m = \frac{1}{[t_{m+1}; t_{m+2}, \ldots, t_n]} < 1$$

and $t_n > 1$ implies the same if $m + 1 = n$. Thus, $0 < b_m - t_m < 1$, which is a contradiction since $b_m - t_m$ is an integer.

Let $n \leq m$ and $0 \leq k \leq n$. Theorem 5.19 implies that the equalities $[t_k; t_{k+1}, t_{k+2}, \ldots, t_n] = [b_k; b_{k+1}, b_{k+2}, \ldots, b_m]$ and $a_k = b_k$ hold for $0 \leq k \leq n - 1$. For $k = n$ one obtains

$$[t_n;] = [b_n; b_{n+1}, \ldots, b_m]$$

If $n = m$, then

$$t_n = [t_n;] = [b_n;] = b_n$$

On the other hand, if $n < m$, then

$$[t_n;] = [b_n; b_{n+1}, \ldots, b_m]$$

or

$$t_n = b_n + \frac{1}{[b_{n+1}; b_{n+2}, \ldots, b_m]}$$

If $m \neq n + 1$, Theorem 5.17 implies that $[b_{n+1}; b_{n+2}, \ldots, b_m] > 1$ so that $0 < t_n - b_n < 1$, a contradiction. If $m = n + 1$, then

$$t_n = b_n + \frac{1}{[b_{n+1};]} = b_n + \frac{1}{b_{n+1}}$$

Since $b_{n+1} \geq 1$,

$$0 < t_n - b_n = \frac{1}{b_{n+1}} \leq 1$$

But because $t_n - b_n$ is an integer, $b_n = t_n - 1$ and $b_{n+1} = 1$. ∎

Theorem 5.21 *If $[t_0; t_1, t_2, \ldots, t_n]$ is a simple continued fraction with convergents p_k/q_k with p_k and q_k given by Theorem 5.13 for $0 \leq k \leq n$, then*

(a) *p_k and q_k are relatively prime, that is, $\gcd(p_k, q_k) = 1$ for $0 \leq k \leq n$.*

(b) *$\gcd(p_k, p_{k+1}) = 1$ and $\gcd(q_k, q_{k+1}) = 1$ for $0 \leq k \leq n - 1$.*

Proof. By Theorem 5.14, $p_k q_{k-1} - p_{k-1} q_k = (-1)^{k-1}$ for $1 \leq k \leq n$, which implies the results of both parts. ∎

By Theorem 5.21, the convergents p_k/q_k of a simple continued fraction $[t_0; t_1, t_2, \ldots, t_n]$ expressed as a ratio of p_k and q_k are always rationals in lowest terms; that is, the numerator and denominator are relatively prime. For the fraction $-124/35$ discussed in Section 5.2, we showed that the convergents as ratios of p_k and q_k were $-4/1$, $-7/2$, $-39/11$, $-85/24$, and $-124/35$, all of which are rationals in lowest terms. If the rational a/b is not in lowest terms, then $a/b = [t_0; t_1, t_2, \ldots, t_n] = p_n/q_n$ and p_n/q_n is in lowest terms. Part (b) implies that consecutive numerators and consecutive denominators of the convergents p_k/q_k are relatively prime.

Theorem 5.22 *If a and b are integers with $b > 0$, $a/b = [t_0; t_1, t_2, \ldots, t_n]$, and n odd, then $\gcd(a, b) = ax - by$ has the solution $x = q_{n-1}$ and $y = p_{n-1}$.*

Proof. Let $\gcd(a, b) = d$ and let p_k and q_k be the sequences given by Theorem 5.13 so that $p_k/q_k = [t_0; t_1, \ldots, t_k]$ are the convergents of a/b. Then $p_n/q_n = a/b$ and $a = d \cdot p_n$ and $b = d \cdot q_n$ by Theorem 5.21. Theorem 5.14 implies that

$$p_n q_{n-1} - p_{n-1} q_n = (-1)^{n-1}$$

so that multiplying by d gives

$$dp_n q_{n-1} - dp_{n-1} q_n = d(-1)^{n-1}$$

and since n is odd,

$$aq_{n-1} - bp_{n-1} = d$$

Thus, (q_{n-1}, p_{n-1}) is a solution of $ax - by = \gcd(a, b)$. ∎

Theorem 5.22 may at first appear unduly restrictive because of the requirement that n be odd; however, by Theorem 5.20, every rational has two simple continued fraction representations, one having one more term than the other. Thus, one of the two representations will be of the form $[t_0; t_1, t_2, \ldots, t_n]$ with n odd.

Example 5.23 *To solve $26x - 55y = 1$, we first apply the Euclidean Algorithm and find that $26/55 = [0; 2, 8, 1, 2]$; however, this continued fraction representation has an even number of terms following the semicolon. By Theorem 5.20 we know that $26/55 = [0; 2, 8, 1, 1, 1]$ also where there is an odd number, i.e., $n = 5$, of terms following the semicolon. The convergents are calculated in the following table:*

k	t_k	p_k	q_k
0	0	0	1
1	2	1	2
2	8	8	17
3	1	9	19
4	1	17	36
5	1	26	55

$p_{n-1} = p_4 = 17$ and $q_{n-1} = q_4 = 36$. Thus $x = 36$ and $y = 17$ is a solution.

Exercises

1. Prove Theorem 5.17.

2. Prove Lemma 5.18.

3. Let $[t_0; t_1, \ldots, t_n]$ be a simple continued fraction and let $t_0 \geq 0$. Prove that $[t_0; t_1, \ldots, t_n]$ is an integer N if and only if $n = 0$, for which case $t_0 = N$, or $n = 1$, for which case $t_0 = N - 1$.

4. Let $[t_0; t_1, \ldots, t_n]$ be a simple continued fraction. Prove that

$$q_k \geq 2^{(k-1)/2}$$

for $2 \leq k \leq n$.

5. Apply Theorem 5.16 to obtain the simple continued fraction representation of:

   (a) $88/25$ (b) 7.241 (c) -8.11

   and verify Theorem 5.21(a) and (b) for these three rationals.

6. Use Theorem 5.22 to solve the following integer equations:

   (a) $88x - 25y = 1$

   (b) $88x + 25y = 1$

   (c) $385x - 30y = 5$

   (d) $385x + 30y = 65$

5.4   Infinite Simple Continued Fractions

In Section 5.1, examples of continued fraction representations of rationals and reals were shown. By Theorem 5.16, a rational number x is represented by a continued fraction $[t_0; t_1, t_2, \ldots, t_n]$ where t_i is an integer for each i. If x is real but not rational, the discussion in Section 5.1 suggested that x is represented by a continued fraction $[t_0; t_1, t_2, \ldots, t_n, x_{n+1}]$ where t_i is an integer for each i and x_{n+1} is irrational. In fact, it appeared that for every $n \geq 0$ there is such a representation. So, in some sense, for x irrational, there seems to be a sequence of integers $t_0, t_1, t_2, \ldots$ associated with x. For example, in Section 5.1 it was found that the sequence $1, 1, 2, 1, 2, \overline{1, 2}, \ldots$, where the bar indicates that the finite sequence 1, 2 is repeated, is associated with $\sqrt{3}$.

Definition 5.24 *If $t_0, t_1, t_2, \ldots$ is a sequence of integers with $t_i > 0$ for $i \geq 1$, then the expression $[t_0; t_1, t_2, \ldots]$ is called an* infinite simple continued fraction *and is said to represent the real number x provided that*

$$\lim_{n \to \infty} [t_0; t_1, t_2, \ldots, t_n] = x$$

We write $x = [t_0; t_1, t_2, \ldots]$ and say that $[t_0; t_1, t_2, \ldots]$ converges to x. The integers $t_0, t_1, t_2, \ldots$ are called terms *of the infinite continued fraction. Two infinite simple continued fractions $[t_0; t_1, t_2, \ldots]$ and $[b_0; b_1, b_2, \ldots]$ are equal* term by term *provided that $t_i = b_i$ for each i. Two representations are the* same *or* equal *if they are equal term by term.*

To make this definition useful, we need both to ensure that infinite simple continued fractions represent real numbers and to find out what real numbers are so represented.

Theorem 5.25 *If $[t_0; t_1, t_2, \ldots]$ is an infinite simple continued fraction, then*

(a) *There is a real number x such that $x = [t_0; t_1, t_2, \ldots]$.*

(b) *x is irrational.*

Proof. **(a)** If $[t_0; t_1, t_2, \ldots, t_n] = p_n/q_n$ is the n-th convergent and p_n and q_n are given by Theorem 5.13, then, by Theorem 5.14(d), p_0/q_0, p_2/q_2, p_4/q_4, $\ldots$ is an increasing sequence and is bounded above by p_1/q_1. Further, by the completeness of the real number system, the sequence converges to the least upper bound of the set $\{p_{2i}/q_{2i} : i \geq 0\}$. Let a be the real number such that

$$\lim_{n \to \infty} \frac{p_{2n}}{q_{2n}} = a$$

Similarly, p_1/q_1, p_3/q_3, p_5/q_5, $\ldots$ is a decreasing sequence of numbers and is bounded from below by p_0/q_0; therefore, the sequence converges to the greatest lower bound of the set $\{p_{2i+1}/q_{2i+1} : i \geq 0\}$. Let b be the real number such that

$$\lim_{n \to \infty} \frac{p_{2n+1}}{q_{2n+1}} = b$$

Then

$$
\begin{aligned}
a - b &= \lim_{n \to \infty} \frac{p_{2n}}{q_{2n}} - \lim_{n \to \infty} \frac{p_{2n+1}}{q_{2n+1}} \\
&= \lim_{n \to \infty} \left(\frac{p_{2n}}{q_{2n}} - \frac{p_{2n+1}}{q_{2n+1}} \right) \\
&= \lim_{n \to \infty} \frac{(-1)^{2n+1}}{q_{2n} q_{2n+1}} \quad \text{by Theorem 5.14 (b)} \\
&= \lim_{n \to \infty} \frac{(-1)}{q_{2n} q_{2n+1}} = 0
\end{aligned}
$$

because, since $q_k \geq k$ for any $k \geq 0$,

$$\left| \left(\frac{-1}{q_{2n} q_{2n+1}} \right) \right| = \frac{1}{q_{2n} q_{2n+1}} \leq \frac{1}{(2n)(2n+1)}$$

and because $\lim_{n \to \infty} \dfrac{1}{(2n)(2n+1)} = 0$.

(b) If $[t_0; t_1, t_2, \ldots]$ is a simple infinite continued fraction and $x = [t_0; t_1, t_2, \ldots]$, then no convergent $[t_0; t_1, t_2, \ldots, t_k]$ equals x because the sequence of convergents with n even is strictly increasing and the sequence with n odd is strictly decreasing by Theorem 5.14(d). Let p_j/q_j be the j-th convergent of $[t_0; t_1, t_2, \ldots]$ where p_j and q_j are given by Theorem 5.13 so that $p_j/q_j = [t_0; t_1, t_2, \ldots, t_j]$. Then, by Theorem 5.14(d),

$$\frac{p_{2k}}{q_{2k}} < x < \frac{p_{2k+1}}{q_{2k+1}}$$

$$0 < x - \frac{p_{2k}}{q_{2k}} < \frac{p_{2k+1}}{q_{2k+1}} - \frac{p_{2k}}{q_{2k}}$$

Using Theorem 5.14(b) for the rightmost expression gives

$$0 < x - \frac{p_{2k}}{q_{2k}} < \frac{1}{q_{2k+1}q_{2k}}$$

because $(-1)^{(2k+1)-1} = 1$. Now Let x be rational and let $x = a/b$ with a and b integers and with $b > 0$. Then substituting in the last inequality for x and multiplying by bq_{2k}, we obtain

$$0 < aq_{2k} - bp_{2k} < \frac{b}{q_{2k+1}}$$

The quantity $aq_{2k} - bp_{2k}$ must be a positive integer so that $1 \le aq_{2k} - bp_{2k}$ for every k; therefore,

$$1 < \frac{b}{q_{2k+1}}$$

for every k, which is impossible because, by Theorem 5.14(e), $q_{2k+1} \ge 2k+1$. Thus, x cannot be rational. ■

Theorem 5.25 is the main reason why the definition of infinite continued fraction is restricted to have integral terms; that is, if the t_i are integers, then the infinite continued fraction is guaranteed to converge. A general theorem is: If $t_i > 0$ for $i \ge 1$, then $[t_0; t_1, t_2, \ldots]$ converges if and only if $\sum_{i=1}^{\infty} t_i$ diverges (see A. Ya Khinchin, *Continued Fractions* [40]).

At this point in our development, we still do not know whether every irrational number has an infinite simple continued fraction representation. The rationals were considered in Theorems 5.16 and 5.20. We now consider irrational real numbers.

Theorem 5.26 *For an irrational real number x, there is an integer sequence $t_0, t_1, t_2, \ldots$ with $t_i > 0$ for $i \ge 1$ such that $x = [t_0; t_1, t_2, \ldots]$. Further, the sequence $t_0, t_1, t_2, \ldots$ is defined as follows:*

(i) *Define the sequence $x_0, x_1, x_2, \ldots$ recursively by*

$$\begin{aligned} x_0 &= x \\ x_{k+1} &= \frac{1}{x_k - \lfloor x_k \rfloor} \text{ for } k \geq 1 \end{aligned}$$

(ii) *Let $t_k = \lfloor x_k \rfloor$ for $k \geq 0$.*

The numbers x_k have the properties:

(a) *x_k is irrational for $k \geq 1$.*

(b) *$x_k > 1$ for $k \geq 1$.*

(c) *$x = [t_0; t_1, t_2, \ldots, t_n, x_{n+1}]$ for $n \geq 0$.*

(d) *$x = (p_n x_{n+1} + p_{n-1})/(q_n x_{n+1} + q_{n-1})$ for $n \geq 0$, where p_i and q_i are given by Theorem 5.13.*

Proof. Let $x_0, x_1, x_2, \ldots$ be defined according to condition (i). We use induction for parts (a) and (b). Clearly, $x_0 = x$ and $y_1 = x_0 - \lfloor x_0 \rfloor$ are irrational. Further, $0 < x_0 - \lfloor x_0 \rfloor < 1$. So $x_1 = 1/(x_0 - \lfloor x_0 \rfloor) > 1$ and x_1 is irrational.

If $k \geq 1$, x_k is irrational, and $x_k > 1$, then $y_{k+1} = x_k - \lfloor x_k \rfloor$ is irrational and $0 < x_k - \lfloor x_k \rfloor < 1$ as before. Thus, $x_{k+1} = 1/(x_k - \lfloor x_k \rfloor) > 1$ and x_{k+1} is irrational. Hence, x_n is irrational and $x_n > 1$ for $n \geq 1$.

Let $t_k = \lfloor x_k \rfloor$ for each $k \geq 0$. We show that $x = [t_0; t_1, t_2, \ldots, t_n, x_{n+1}]$ for $n \geq 0$. For $n = 0$, let $x_0 = x$. Then $x = \lfloor x_0 \rfloor + (x - \lfloor x_0 \rfloor) = t_0 + 1/x_1 = [t_0; x_1]$. If $k \geq 0$ and $x = [t_0; t_1, t_2, \ldots, t_k, x_{k+1}]$, then

$$x_{k+1} = \lfloor x_{k+1} \rfloor + (x_{k+1} - \lfloor x_{k+1} \rfloor) = t_{k+1} + 1/x_{k+2} = [t_{k+1}; x_{k+2}]$$

so that

$$\begin{aligned} x &= [t_0; t_1, \ldots, t_k, x_{k+1}] \\ &= [t_0; t_1, \ldots, t_k, [t_{k+1}, x_{k+2}]] \\ &= [t_0; t_1, \ldots, t_k, t_{k+1}, x_{k+2}] \text{ by Theorem 5.12} \end{aligned}$$

Therefore, by mathematical induction, $x = [t_0; t_1, t_2, \ldots, t_n, x_{n+1}]$ for $n \geq 0$.

Showing that $\lim\limits_{n \to \infty} [t_0; t_1, t_2, \ldots, t_n] = x$ is equivalent to showing that

$$\lim_{n \to \infty} (x - [t_0; t_1, t_2, \ldots, t_n]) = 0$$

Consider

$$|w_n| = |x - [t_0; t_1, \ldots, t_n]| = \left| \frac{p'_{n+1}}{q'_{n+1}} - \frac{p_n}{q_n} \right|$$

where $w_n = x - [t_0; t_1, t_2, \ldots, t_n]$, where p_k and q_k are the numerator and denominator of the convergents of $[t_0; t_1, \ldots, t_n]$ given by Theorem 5.13, and where p'_k and q'_k are the numerator and denominator of the convergents of $[t_0; t_1, \ldots, t_n, x_{n+1}]$ also as given by Theorem 5.13. Then

$$|w_n| = \left| \frac{p'_n x_{n+1} + p'_{n-1}}{q'_n x_{n+1} + q'_{n-1}} - \frac{p_n}{q_n} \right| = \left| \frac{p_n x_{n+1} + p_{n-1}}{q_n x_{n+1} + q_{n-1}} - \frac{p_n}{q_n} \right|$$

because $p_k = p'_k$ and $q_k = q'_k$ for $k \leq n$. Theorem 5.14(b) and (e) imply

$$|w_n| = \left| \frac{(-1)^n}{q_n(q_n x_n + q_{n-1})} \right| \leq \frac{1}{n(n \cdot 1 + n - 1)} = \frac{1}{n(2n - 1)}$$

Since $\lim\limits_{n \to \infty} \dfrac{1}{n(2n - 1)} = 0$ and $|w_n| \leq 1/(n(2n - 1))$,

$$\lim_{n \to \infty} (x - [t_0; t_1, t_2, \ldots, t_n]) = \lim_{n \to \infty} w_n = 0 \;\blacksquare$$

Theorem 5.27 *If $[t_0; t_1, t_2, \ldots]$ is an infinite simple continued fraction, then for each $k \geq 1$,*

$$[t_0; t_1, t_2, \ldots] = [t_0; t_1, t_2, \ldots, t_{k-1}, [t_k; t_{k+1}, \ldots]]$$

Proof. By Theorem 5.25, $[t_0; t_1, t_2, \ldots]$ converges to a real number x. If $k = 1$, then

$$
\begin{aligned}
x &= \lim_{n \to \infty} [t_0; t_1, t_2, \ldots, t_n] = \lim_{n \to \infty} [t_0; [t_1; t_2, \ldots, t_n]] \\
&= \lim_{n \to \infty} \left(t_0 + \frac{1}{[t_1; t_2, \ldots, t_n]} \right) \\
&= t_0 + \frac{1}{\lim\limits_{n \to \infty} [t_1; t_2, \ldots, t_n]} \\
&= t_0 + \frac{1}{[t_1; t_2, t_3, \ldots]}
\end{aligned}
$$

since $x = \lim\limits_{n \to \infty} (t_0 + y_n)$ implies that $\lim\limits_{n \to \infty} y_n$ exists and equals $x - t_0$. If $k = m$, $m \geq 1$, and $x = [t_0; t_1, t_2, \ldots, t_{m-1}, [t_m; t_{m+1}, \ldots]]$, then

$$
\begin{aligned}
x &= [t_0; t_1, t_2, \ldots] = [t_0; t_1, t_2, \ldots, t_{m-1}, [t_m; t_{m+1}, \ldots]] \\
&= [t_0; t_1, t_2, \ldots, t_{m-1}, [t_m; [t_{m+1}, \ldots]]]
\end{aligned}
$$

because the result of the theorem holds for $k = 1$. Thus, by Theorem 5.12,

$$x = [t_0; t_1, t_2, \ldots, t_{m-1}, t_m, [t_{m+1}; \ldots]]$$

and the equation of the theorem holds for all $k \geq 1$. $\blacksquare$

Example 5.28 *Theorem 5.27 may be used to determine what numbers to which some infinite simple continued fractions converge. For example, if $t_i = 1$ for all i, then let*

$$
\begin{aligned}
x &= [t_0; t_1, t_2, \ldots] = [1; 1, 1, 1, \ldots] \\
&= [1; [1; 1, 1, 1, \ldots]] = [1; x] = 1 + 1/x
\end{aligned}
$$

Thus, x is a positive solution to the equation

$$x = 1 + \frac{1}{x}$$

or

$$x^2 - x - 1 = 0$$

Since x is positive, we discard the negative solution, giving

$$x = \frac{1 + \sqrt{5}}{2}$$

Similarly, for $y = [1; 2, 1, 2, 1, 2, \ldots]$,

$$y = [1; 2, [1; 2, 1, 2, 1, 2, \ldots]] = [1; 2, y]$$

Thus,

$$y = 1 + \cfrac{1}{2 + \cfrac{1}{y}} = \frac{3y + 1}{2y + 1}$$

The number y satisfies the following quadratic equation:

$$2y^2 - 2y - 1 = 0$$

which has positive solution

$$y = \frac{1 + \sqrt{3}}{2}$$

The particular feature of the representations of x and y that allowed us to compute x and y was that eventually there was a repeating pattern in the terms of $[t_0; t_1, t_2, \ldots]$. Since the infinite continued fraction converged to some real number, we then used Theorem 5.27 to generate a quadratic equation whose solutions included the real number sought. We will return to this topic again.

The main unresolved question about infinite simple continued fraction representations of real numbers is that of uniqueness which is resolved in the next theorem.

Theorem 5.29 *If the infinite simple continued fractions $[t_0; t_1, t_2, \ldots]$ and $[b_0; b_1, b_2, \ldots]$ are equal, then $t_k = b_k$ and $[t_k; t_{k+1}, t_{k+2}, \ldots] = [b_k; b_{k+1}, b_{k+2}, \ldots]$ for $k \geq 0$ so that the continued fractions are equal term by term.*

Proof. First, it is straightforward to show that if $y = [c_0; c_1, c_2, \ldots]$ is any infinite simple continued fraction, then $\lfloor y \rfloor = c_0$. We show that for every $k \geq 0$, $t_k = b_k$ and $[t_{k+1}; t_{k+2}, t_{k+3}, \ldots] = [b_{k+1}; b_{k+2}, b_{k+3}, \ldots]$. If $x = [t_0; t_1, t_2, \ldots] = [b_0; b_1, b_2, \ldots]$, then $t_0 = \lfloor x \rfloor = b_0$ so that $t_0 = b_0$. By Theorem 5.27,

$$[t_0; [t_1, t_2, \ldots]] = [b_0; [b_1, b_2, \ldots]]$$

Thus

$$t_0 + \frac{1}{[t_1; t_2, t_3, \ldots]} = b_0 + \frac{1}{[b_1; b_2, b_3, \ldots]}$$

Because $t_0 = b_0$, we can conclude that $[t_1; t_2, t_3, \ldots] = [b_1; b_2, b_3, \ldots]$.

If $j \geq 0$, $t_j = b_j$, and $[t_{j+1}; t_{j+2}, t_{j+3}, \ldots] = [b_{j+1}; b_{j+2}, b_{j+3}, \ldots] = y$, then $t_{j+1} = \lfloor y \rfloor = b_{j+1}$. Also, by Theorem 5.27,

$$[t_{j+1}; [t_{j+2}; t_{j+3}, \ldots]] = [b_{j+1}; [b_{j+2}; b_{j+3}, \ldots]]$$

or

$$t_{j+1} + \cfrac{1}{[t_{j+2}; t_{j+3}, t_{j+4}, \ldots]} = b_{j+1} + \cfrac{1}{[b_{j+2}; b_{j+3}, b_{j+4}, \ldots]}$$

because $t_{j+1} = b_{j+1}$, $[t_{j+2}; t_{j+3}, t_{j+4}, \ldots] = [b_{j+2}; b_{j+3}, b_{j+4}, \ldots]$. The claim is true by mathematical induction. ∎

Since an irrational number x has at most one infinite simple fraction representation, it must be the one produced by Theorem 5.26 so that

$$x = [t_0; t_1, \ldots] = [t_0; t_1, \ldots, t_n, x_{n+1}]$$

Since, by Theorem 5.27,

$$x = [t_0; t_1, \ldots, t_n, [t_{n+1}; t_{n+2}, t_{n+3}, \ldots]] = [t_0; t_1, \ldots, t_n, y]$$

with y irrational, we want to conclude that

$$x_{n+1} = [t_{n+1}; t_{n+2}, t_{n+3}, \ldots] = y$$

Theorems 5.19, 5.20, and 5.21 do not apply because they are stated only for finite simple continued fractions. It is easy to apply similar arguments to prove that if

$$[t_0; t_1, \ldots, t_n, x_{n+1}] = [t_0; t_1, \ldots, t_n, y]$$

with x_{n+1} and y irrational, then $y = x_{n+1}$.

Theorem 5.30 *If x has the infinite simple continued fraction representation $[t_0; t_1, t_2, \ldots]$ and x_n is given by Theorem 5.26 so that $x = [t_0; t_1, \ldots, t_{n-1}, x_n]$ for $n \geq 0$, then $x_n = [t_n; t_{n+1}, t_{n+2}, \ldots]$. Further, if $x = [t_0; t_1, \ldots, t_{n-1}, y]$, then $y = x_n$.*

Proof. The proof is left to the reader. ∎

The real number x represented by an infinite simple continued fraction $[t_0; t_1, t_2, \ldots]$ is the limit of the sequence p_0/q_0, p_1/q_1, p_2/q_2, $\ldots$ of convergents, where p_i and q_i are given by Theorem 5.13. We can also say how rapidly the convergents p_k/q_k converge to x.

Theorem 5.31

(a) *If $[t_0; t_1, t_2, \ldots]$ is an infinite simple continued fraction converging to x, then*

$$\left| x - \frac{p_k}{q_k} \right| < \frac{1}{q_k q_{k+1}} < \frac{1}{q_k^2}$$

for $k \geq 0$, where p_k and q_k are given by Theorem 5.13.

(b) *If* $[t_0; t_1, t_2, \ldots, t_n]$ *is an finite simple continued fraction equal to* x, *then*

$$\left| x - \frac{p_k}{q_k} \right| < \frac{1}{q_k q_{k+1}} < \frac{1}{q_k^2}$$

for $0 \leq k \leq n - 1$, *where* p_k *and* q_k *are given by Theorem 5.13. The equality holds when* $k = n - 1$.

Proof. (a) By Theorem 5.14(d), either $p_k/q_k < x < p_{k+1}/q_{k+1}$ or $p_{k+1}/q_{k+1} < x < p_k/q_k$ depending upon whether k is even or odd. Thus,

$$\left| x - \frac{p_k}{q_k} \right| < \left| \frac{p_{k+1}}{q_{k+1}} - \frac{p_k}{q_k} \right| = \frac{1}{q_k q_{k+1}} < \frac{1}{q_k^2}$$

by Theorem 5.14(a) and (f).

Part (b) is similarly proved. ∎

We want to characterize the repeating infinite simple continued fractions discussed in Example 5.28 and in Section 5.1. For, example, we computed that $[1; 1, 1, 1, 1, \overline{1}] = [1; \overline{1}] = (1+\sqrt{5})/2$ and $[\overline{1; 2}] = (1+\sqrt{3})/2$. The terms of these continued fractions repeat and both converge to irrational real numbers of the form $(e + \sqrt{d})/f$. It will be shown that only infinite simple continued fractions with terms that eventually repeat converge to such irrational numbers and that every such irrational number is represented by such a "repeating" continued fraction.

Definition 5.32 *A quadratic irrational number is an irrational real number that is a solution to a quadratic equation* $ax^2 + bx + c = 0$ *whose coefficients* a, b, *and* c *are integers and* $a \neq 0$.

We note that according to the quadratic formula, the solutions of $ax^2 + bx + c = 0$ are

$$x = \frac{-b \pm \sqrt{b^2 - 4ac}}{2a}$$

If x is irrational, these solutions are of the form $x = r \pm s\sqrt{t}$, where r and s are rational and t is a positive integer that is not the square of an integer. We can also rewrite the expression for x in terms of integers instead of rationals, a result given in the following theorem whose proof is left to the reader.

Theorem 5.33 *The number* x *is a quadratic irrational if and only if there are integers* d, e, *and* f *such that* $d > 0$ *is not the square of an integer and*

$$x = \frac{e + \sqrt{d}}{f}$$

For example, $w = (1 + \sqrt{5})/2$ is a quadratic irrational. The quadratic irrational $w' = (1 - \sqrt{5})/2$ has a special relationship with w; namely, both are solutions of the same quadratic with integer coefficients. Thus,

$$(x - w)(x - w') = 0$$

simplifies to

$$x^2 - x - 1 = 0$$

Definition 5.34 *Let $w = (e + \sqrt{d})/f$ and $w' = (e - \sqrt{d})/f$ for integers e, d, and f with $d > 0$ not the square of an integer. The numbers w and w' are said to be* conjugates *of one another.*

Definition 5.35 *An infinite simple continued fraction $[t_0; t_1, t_2, \ldots]$ is said to be* periodic *or* repeating *provided that there are nonnegative integers K and s with $s \geq 1$ such that if $k \geq K$, then $t_k = t_{k+s}$. The least such positive integer s is called the* period. *$[t_0; t_1, t_2, \ldots]$ is said to be* purely periodic *provided that there is a positive integer s such that $t_k = t_{k+s}$ for $k \geq 0$.*

For example, for $[1; 1, 1, 1, \overline{1}, \ldots] = [1; \overline{1}]$, let $K = 1$ and $s = 1$. For $[1; 2, 1, 2, 1, \overline{2, 1}] = [1; \overline{2, 1}]$, let $K = 1$ and $s = 2$. Both of these continued fractions are also purely periodic, with $K = 0$ and with $s = 1$ and 2, respectively. $\sqrt{3} = [1; 1, 2, \overline{1, 2}]$ is periodic with $K = 1$ and $s = 2$ but is not purely periodic. It will be convenient to have an alternative characterization of periodic that is not in terms of the t_i.

Theorem 5.36 *Let $x = [t_0; t_1, t_2, \ldots]$ be an infinite simple continued fraction and x_n be the irrational such that $x = [t_0; t_1, \ldots, t_{n-1}, x_n]$ for $n \geq 0$. If $[t_0; t_1, t_2, \ldots]$ is periodic with nonnegative integer K and positive integer s such that $t_k = t_{k+s}$ for $k \geq K$, then $x_k = x_{k+s}$ for $k \geq K$. Conversely, if there is a nonnegative integer K and a positive integer s such that $x_K = x_{K+s}$, then $t_k = t_{k+s}$ for $k \geq K$ and also $x_k = x_{k+s}$ for $k \geq K$.*

Proof. For a nonnegative integer K and a positive integer s such that $t_k = t_{k+s}$ for $k \geq K$, Theorem 5.30 implies that $x_k = [t_k; t_{k+1}, t_{k+2}, \ldots]$ and $x_{k+s} = [t_{k+s}; t_{k+1+s}, t_{k+2+s}, \ldots]$. Because these two continued fractions are equal term by term, they are the same so that $x_k = x_{k+s}$.

Conversely, if K is a nonnegative integer and s is a positive integer such that $x_K = x_{K+s}$, then, by Theorem 5.26, $t_K = \lfloor x_K \rfloor = \lfloor x_{K+s} \rfloor = t_{K+s}$. If $j \geq K$, $x_j = x_{j+s}$, and $t_j = t_{j+s}$, then

$$x_{j+1} = \frac{1}{x_j - \lfloor x_j \rfloor} = \frac{1}{x_{j+s} - \lfloor x_{j+s} \rfloor} = x_{j+1+s}$$

and $t_{j+1} = \lfloor x_{j+1} \rfloor = \lfloor x_{j+1+s} \rfloor = t_{j+1+s}$. So by mathematical induction, $x_k = x_{k+s}$ and $t_k = t_{k+s}$ for $k \geq K$. ∎

Theorem 5.37 *An infinite simple continued fraction $[t_0; t_1, t_2, \ldots]$ is periodic if and only if it represents a quadratic irrational number.*

Proof. Part 1. Assume that $[t_0; t_1, t_2, \ldots]$ is periodic and $x = [t_0; t_1, t_2, \ldots]$. Let K and s be positive integers such that $t_k = t_{k+s}$ for $k \geq K$. Let integers p_i and q_i be the numerator and denominator of the convergent p_i/q_i given by Theorem 5.13 for $[t_0; t_1, t_2, \ldots]$. Let the irrational x_i be given by Theorem 5.26 so that $x = [t_0; t_1, t_2, \ldots, t_i, x_{i+1}]$. Then, if $k \geq K$,

$$x = \frac{p_{k-1}x_k + p_{k-2}}{q_{k-1}x_k + q_{k-2}}$$

and also

$$x = \frac{p_{k+s-1}x_{k+s} + p_{k+s-2}}{q_{k+s-1}x_{k+s} + q_{k+s-2}} = \frac{p_{k+s-1}x_k + p_{k+s-2}}{q_{k+s-1}x_k + q_{k+s-2}}$$

because $x_k = x_{k+s}$. Setting these two expressions for x equal, we get

$$(p_{k-1}x_k + p_{k-2})(q_{k+s-1}x_k + q_{k+s-2}) = (p_{k+s-1}x_k + p_{k+s-2})(q_{k-1}x_k + q_{k-2})$$

which may be rearranged into an equation of the form

$$ax_k^2 + bx_k + c = 0$$

where a, b, and c are integers and $a = p_{k-1}q_{k+s-1} - p_{k+s-1}q_{k-1}$. Hence $a \neq 0$ because if $a = 0$, then we would have $p_{k-1}/q_{k-1} = p_{k+s-1}/q_{k+s-1}$, which contradicts Theorem 5.14(d). Thus, x_k, which is known to be irrational, is the solution of a quadratic equation. Thus, x_k is a quadratic irrational.

Since $x = (p_{k-1}x_k + p_{k-2})/(q_{k-1}x_k + q_{k-2})$, we substitute for x_k and obtain

$$x = \frac{p_{k-1}\left(\dfrac{e + \sqrt{d}}{f}\right) + p_{k-2}}{q_{k-1}\left(\dfrac{e + \sqrt{d}}{f}\right) + q_{k-2}}$$

which, by "rationalizing" the denominator may be rewritten in the form

$$x = \frac{A + B\sqrt{d}}{C}$$

where A, B, and C are integers and

$$B = (p_{k-1}q_{k-2} - q_{k-1}p_{k-2})f$$

If we can show that $B \neq 0$, then x can further be written in the form $(E + \sqrt{D})/F$, where D, E, and F are integers and D is a positive nonsquare integer, and so x must be a quadratic irrational. But

$$B = (p_{k-1}q_{k-2} - q_{k-1}p_{k-2})f = (-1)^{k-2}f \neq 0$$

by Theorem 5.14(b) since $f \neq 0$.

Part 2. Let $x = [t_0; t_1, t_2, \ldots]$ be a quadratic irrational. Then there are integers a, b, and c with $a \neq 0$ such that $ax^2 + bx + c = 0$. Let the sequence $x_0, x_1, x_2, \ldots$ be given by Theorem 5.26 so that for $n \geq 1$

$$x = [t_0; t_1, t_2, \ldots, t_{n-1}, x_n] = \frac{p_{n-1}x_n + p_{n-2}}{q_{n-1}x_n + q_{n-2}}$$

where p_k and q_k are given by Theorem 5.13. If we substitute the last expression for x into the equation $ax^2 + bx + c = 0$, we obtain

$$a(p_{n-1}x_n + p_{n-2})^2 + b(p_{n-1}x_n + p_{n-2})(q_{n-1}x_n + q_{n-2}) + c(q_{n-1}x_n + q_{n-2})^2 = 0$$

If we expand all products and collect coefficients of x_n^2 and x_n, we obtain the equation

$$A_n x_n^2 + B_n x_n + C_n = 0$$

where

$$
\begin{aligned}
A_n &= a p_{n-1}^2 + b p_{n-1} q_{n-1} + c q_{n-1}^2 \\
B_n &= 2a p_{n-1} p_{n-2} + b(p_{n-2} q_{n-1} + p_{n-1} q_{n-2}) + 2c q_{n-1} q_{n-2} \\
C_n &= a p_{n-2}^2 + b p_{n-2} q_{n-2} + c q_{n-2}^2
\end{aligned}
$$

are integers.

We will show below that there are only finitely many distinct integer triples: A_n, B_n, and C_n. Thus there are only finitely many distinct quadratic equations $A_n x_n^2 + B_n x_n + C_n = 0$ having x_n as a solution. Since a quadratic equation can have at most two real solutions, then there are only finitely many distinct real numbers x_n in the infinite sequence $x_1, x_2, x_3, \ldots$. Therefore, there are i and j such $x_i = x_j$ and, consequently, positive integers K and s such that $x_K = x_{K+s}$. Then, using Theorem 5.26, we have

$$
\begin{aligned}
x_{K+1} &= \frac{1}{x_K - \lfloor x_K \rfloor} \\
x_{K+1+s} &= \frac{1}{x_{K+s} - \lfloor x_{K+s} \rfloor}
\end{aligned}
$$

Since $x_K = x_{K+s}$, we get $x_{K+1} = x_{K+1+s}$. It is an easy induction to show that

$$x_k = x_{k+s} \text{ for } k \geq K$$

Therefore, by Theorem 5.36, $[t_0; t_1, t_2, \ldots]$ is periodic; and we are finished.

To complete the proof of part 2, it just remains for us to establish that there are only finitely many triples A_n, B_n, and C_n for $n \geq 1$. We show this fact by proving that the integers A_n, B_n, and C_n are bounded; that is, there are positive integers R, S, and T such that for all $n \geq 1$, $|A_n| \leq R$, $|B_n| \leq S$, and $|C_n| \leq T$. For example, there are exactly $2R + 1$ integers r such that $|r| \leq R$, namely, $-R, -(R-1), \ldots, -1, 0, 1, 2, \ldots, (R-1)$, and R.

By direct computation, the discriminant of $A_n x_n^2 + B_n x_n + C_n = 0$ is

$$B_n^2 - 4A_n C_n = (b^2 - 4ac)(p_{n-1}q_{n-2} - q_{n-1}p_{n-2})^2 = b^2 - 4ac$$

where the last equality holds because of Theorem 5.14(a). So $B_n^2 - 4A_nC_n$ equals the constant $b^2 - 4ac$ for every n. By Theorem 5.31,

$$\left| x - \frac{p_{n-1}}{q_{n-1}} \right| < \frac{1}{q_{n-1}^2}$$

or

$$|xq_{n-1} - p_{n-1}| < \frac{1}{q_{n-1}}$$

so that there is a real number θ_{n-1} such that $|\theta_{n-1}| < 1$ and

$$|xq_{n-1} - p_{n-1}| = \frac{|\theta_{n-1}|}{q_{n-1}}$$

Choose θ_{n-1} to have the same sign as $xq_{n-1} - p_{n-1}$ so that

$$p_{n-1} = xq_{n-1} - \frac{\theta_{n-1}}{q_{n-1}}$$

Substituting this expression for p_{n-1} into the one for A_n gives

$$
\begin{aligned}
A_n &= a\left(xq_{n-1} - \frac{\theta_{n-1}}{q_{n-1}} \right)^2 + b\left(xq_{n-1} - \frac{\theta_{n-1}}{q_{n-1}} \right)q_{n-1} + cq_{n-1}^2 \\
&= (ax^2 + bx + c)q_{n-1}^2 - 2ax\theta_{n-1} + a\left(\frac{\theta_{n-1}^2}{q_{n-1}^2} \right) - b\theta_{n-1}
\end{aligned}
$$

But $ax^2 + bx + c = 0$ so that

$$
\begin{aligned}
|A_n| &= \left| -2ax\theta_{n-1} + a\left(\frac{\theta_{n-1}^2}{q_{n-1}^2} \right) - b\theta_{n-1} \right| \\
&\leq 2|ax|\,|\theta_{n-1}| + |a|\left(\frac{\theta_{n-1}^2}{q_{n-1}^2} \right) + |b|\,|\theta_{n-1}| \\
&< 2|ax| + |a| + |b|
\end{aligned}
$$

Similarly, we can show that

$$|C_n| < 2|ax| + |a| + |b|$$

By the Archimedean property, there is a positive integer greater than $2|ax| + |a| + |b|$. Therefore, A_n and B_n are bounded by an integer; and there are only finitely many distinct integers A_n and C_n. Further, since $B_n^2 = 4A_nC_n + (b^2 - 4ac)$, there are only finitely many different B_n (B_n is bounded too). The proof of part 2 is now complete. ∎

If x is irrational, then, by definition of the infinite simple continued fraction $[t_0; t_1, t_2, \ldots]$ representing x, we know that

$$x = \lim_{n \to \infty} \frac{p_n}{q_n}$$

where $p_n/q_n = [t_0; t_1, t_2, \ldots, t_n]$ is the n-th convergent of x and p_n and q_n are given by Theorem 5.13. We saw in Theorem 5.31 that p_n/q_n is nearer to

x than the distance $1/q_n^2$. In fact, every convergent is closer to x than the preceding convergents.

Theorem 5.38 *If* $[t_0; t_1, t_2, \ldots]$ *is an infinite simple continued fraction representing* x, *then for each* $n \geq 1$,

(a) $|xq_n - p_n| < |xq_{n-1} - p_{n-1}|$

(b) $\left| x - \dfrac{p_n}{q_n} \right| < \left| x - \dfrac{p_{n-1}}{q_{n-1}} \right|$

where p_n *and* q_n *are given by Theorem 5.13.*

Proof. Let p_k and q_k be given by Theorem 5.13, let $n \geq 1$, and let x_n be given by Theorem 5.26 so that

$$x = [t_0; t_1, t_2, \ldots t_k, x_{k+1}] = \frac{p_k x_{k+1} + p_{k-1}}{q_k x_{k+1} + q_{k-1}}$$

where we define $p_{-1} = 1$ and $q_{-1} = 0$ so the formula holds when $k \geq 0$.

(a)

$$
\begin{aligned}
|xq_{n-1} - p_{n-1}| &= q_{n-1}\left| x - \frac{p_{n-1}}{q_{n-1}} \right| \\
&= q_{n-1}\left| \frac{p_{n-1}x_n + p_{n-2}}{q_{n-1}x_n + q_{n-2}} - \frac{p_{n-1}}{q_{n-1}} \right| \\
&= q_{n-1}\left| \frac{q_{n-1}p_{n-2} - q_{n-2}p_{n-1}}{q_{n-1}(q_{n-1}x_n + q_{n-2})} \right| \\
&= q_{n-1}\left| \frac{-(-1)^{n-2}}{q_{n-1}(q_{n-1}x_n + q_{n-2})} \right| \quad \text{by Theorem 5.14} \\
&= \frac{q_{n-1}}{q_{n-1}(q_{n-1}x_n + q_{n-2})} \\
&= \frac{1}{q_{n-1}x_n + q_{n-2}}
\end{aligned}
$$

But, by Theorem 5.26,

$$t_n = \lfloor x_n \rfloor \leq x_n < \lfloor x_n \rfloor + 1 = t_n + 1$$

so that

$$
\begin{aligned}
q_{n-1}x_n + q_{n-2} &< q_{n-1}(t_n + 1) + q_{n-2} \\
&= (q_{n-1}t_n + q_{n-2}) + q_{n-1} \\
&= q_n + q_{n-1} \text{ by Theorem 5.13} \\
&\leq q_n t_{n+1} + q_{n-1} \text{ since } t_{n+1} \geq 1 \\
&= q_{n+1} \text{ by Theorem 5.13}
\end{aligned}
$$

Thus

$$|xq_{n-1} - p_{n-1}| > \frac{1}{q_{n+1}} > |xq_n - p_n|$$

where the last inequality is implied by Theorem 5.31(a).

(b)

$$\left| x - \frac{p_n}{q_n} \right| = \frac{1}{q_n} |x q_n - p_n|$$

$$< \frac{1}{q_n} |x q_{n-1} - p_{n-1}| \text{ by part (a)}$$

$$\leq \frac{1}{q_{n-1}} |x q_{n-1} - p_{n-1}| \text{ by Theorem 5.14(f)}$$

$$= \left| x - \frac{p_{n-1}}{q_{n-1}} \right| \blacksquare$$

The values of both the forms $|x - a/b|$ and $|xb - a|$ measure the nearness of a/b to x. They are different measures since, for example, for the first form, $|x - (2a)/(2b)| = |x - a/b|$; however, for the second form $|xb - a| \neq |x(2b) - (2a)| = 2|xb - a|$. Thus two fractional representations of a rational may be equally "close" to x in the first sense but not in the second.

Using a continued fraction is an efficient method in representing an irrational number x because it provides rational number approximations to x that are the nearest to x of all rational number approximations having a given size denominator. The next three theorems address this efficiency of approximating irrationals.

Theorem 5.39 *Let the infinite simple continued fraction $[t_0; t_1, t_2, \ldots]$ represent the irrational number x and let p_k and q_k be given by Theorem 5.13. If a and b are any integers with $b > 0$ and n is a positive integer such that*

$$|bx - a| < |q_n x - p_n|$$

then $b \geq q_{n+1}$.

Proof. Let $x = [t_0; t_1, t_2, \ldots]$ and the integers $b > 0$, a, and n have the property that $|bx - a| < |q_n x - p_n|$ but that $b < q_{n+1}$. The key to the proof is to find a way to relate the integers a and b to x, or equivalently, to p_k and q_k. Since p_n and p_{n+1} are relatively prime and q_n and q_{n+1} are relatively prime, each of the following equations is separately solvable for u and v in integers (Theorem 1.45):

$$p_n u + p_{n+1} v = a$$
$$q_n u + q_{n+1} v = b$$

We need to find simultaneous solutions u and v, that is, solutions satisfying both equations.

If we multiply the two equations by q_n and p_n, respectively, and subtract, we obtain

$$(p_{n+1}q_n - p_n q_{n+1})v = a p_n - b q_n$$

so that $v = (-1)^n (a p_n - b q_n)$, using Theorem 5.14, and v is an integer. Similarly, using multipliers q_{n+1} and p_{n+1}, respectively, we obtain the integer solution $u = (-1)^n(bp_{n+1} - aq_{n+1})$. Thus there are integers u and v satisfying both equations. Substituting for a and b using the two equations for a and b above, we get

$$
\begin{aligned}
|bx - a| &= |(q_n u + q_{n+1}v)x - (p_n u + p_{n+1}v)| \\
&= |(q_n x - p_n)u + (q_{n+1}x - p_{n+1})v|
\end{aligned}
$$

In order to simplify this last expression, we need to know when the two terms $(q_n x - p_n)u$ and $(q_{n+1}x - p_{n+1})v$ are positive, negative, or zero; for, if A and B are real numbers having the same sign, then $|A + B| = |A| + |B|$. For the moment we will just state that neither of the two terms is zero and that they are both positive or both negative and that neither u nor v is zero, but we will briefly postpone proving it. Thus,

$$
\begin{aligned}
|bx - a| &= |(q_n x - p_n)u + (q_{n+1}x - p_{n+1})v| \\
&= |(q_n x - p_n)u| + |(q_{n+1}x - p_{n+1})v| \\
&\geq |q_n x - p_n| \cdot |u| \geq |q_n x - p_n|
\end{aligned}
$$

because u is a nonzero integer and must have the property that $|u| \geq 1$. But the string of inequalities above contradicts the assumption $|bx - a| < |q_n x - p_n|$. Consequently, we must have $b \geq q_{n+1}$.

Consider the two missing pieces of the argument. First show that (1) one of $(q_n x - p_n)$ and $(q_{n+1}x - p_{n+1})$ is positive and the other is negative, and (2) one of u and v is positive and the other is negative.

Since x is irrational, Theorem 5.14(d) implies that one of $x - p_n/q_n$ and $x - p_{n+1}/q_{n+1}$ is positive and the other is negative. Which expression is positive depends upon whether n is odd or even. Thus, multiplying the first expression by q_n and the second by q_{n+1}, we get that one of $(q_n x - p_n)$ and $(q_{n+1}x - p_{n+1})$ is positive and the other is negative. This result means that neither is zero.

Recall from above that

$$
\begin{aligned}
u &= (-1)^n(bp_{n+1} - aq_{n+1}) \\
v &= (-1)^n(ap_n - bq_n)
\end{aligned}
$$

If $u = 0$, then $bp_{n+1} = aq_{n+1}$; and therefore, $q_{n+1} \mid b$ because $\gcd(p_{n+1}, q_{n+1}) = 1$. This result contradicts $b < q_{n+1}$. So $u \neq 0$. If $v = 0$, then the two original equations imply that $p_n u = a$ and $q_n u = b$ so that

$$
\begin{aligned}
|bx - a| &= |q_n u x - p_n u| = |u| \cdot |q_n x - p_n| \\
&\geq |q_n x - p_n|
\end{aligned}
$$

which is a contradiction of the choice of a and b. So $v \neq 0$ also. Since $q_n u + q_{n+1}v = b < q_{n+1}$, one of u and v must be negative. Both u and v cannot be negative because $b > 0$. Thus one of u and v is positive and the other is negative.

Because one of $(q_n x - p_n)$ and $(q_{n+1} x - p_{n+1})$ is positive and the other is negative and because one of u and v is positive and the other is negative, both of $(q_n x - p_n)u$ and $(q_{n+1} x - p_{n+1})v$ are positive or both are negative and the proof is complete. ∎

The next theorem states a similar result wherein the nearness criterion is given in terms of the rational itself instead of in terms of the numerator and denominator of the representation.

Theorem 5.40 *Let the infinite simple continued fraction $[t_0; t_1, t_2, \ldots]$ represent the irrational number x and let p_k and q_k be given by Theorem 5.13. Let a/b represent a rational with integer a and positive integer b. Then, for every positive integer n, if*

$$\left| x - \frac{a}{b} \right| < \left| x - \frac{p_n}{q_n} \right|$$

then $b > q_n$.

Proof. If $n \geq 1$, $|x - a/b| < |x - p_n/q_n|$, and $b \leq q_n$, then

$$b \left| x - \frac{a}{b} \right| < q_n \left| x - \frac{p_n}{q_n} \right|$$

or

$$|bx - a| < |q_n x - p_n|$$

But $b \leq q_n < q_{n+1}$ and we have a contradiction of Theorem 5.39. Thus, $b > q_n$. ∎

We note that an equivalent form of the conclusion is: If $b \leq q_n$, then $|x - a/b| \geq |x - p_n/q_n|$.

Example 5.41 *We showed earlier that*

$$g = (1 + \sqrt{5})/2 = [1; 1, 1, \overline{1}] = 1.618033 \cdots$$

The convergents of this representation are given in the following table:

n	t_n	p_n	q_n	p_n/q_n			$g - p_n/q_n$
0	1	1	1	1/1	=	1.0	$0.6180\cdots$
1	1	2	1	2/1	=	2.0	$-0.3819\cdots$
2	1	3	2	3/2	=	1.5	$0.1180\cdots$
3	1	5	3	5/3	=	$1.666\cdots$	$-0.0486\cdots$
4	1	8	5	8/5	=	1.6	$0.0180\cdots$
5	1	13	8	13/8	=	$1.625\cdots$	$-0.0069\cdots$
6	1	21	13	21/13	=	$1.615\cdots$	$0.0026\cdots$

As expected, the convergents p_n/q_n are alternately greater than and less than g. Each convergent is nearer the irrational $(1 + \sqrt{5})/2$ than the preceding one.

Next consider those rationals a/b in the range $1 \leq a/b \leq 2$ with denominator $b \leq q_4 = 5$ as given in the following table:

b	a/b	$g - a/b$
5	5/5, 6/5, 7/5, **8/5**, 9/5, 10/5	.61, $-.41$, .21, **.01**, $-.18$, $-.38$
4	4/4, 5/4, 6/4, 7/4, 8/4	.61, $-.36$, .11, $-.13$, $-.38$
3	3/3, 4/3, **5/3**, 6/3	.61, .28, $-.04$, $-.38$
2	2/2, **3/2**, 4/2	.61, .11, $-.38$
1	1/1, **2/1**	.61, $-.38$

where in the section on the right only the first two digits to the right of the decimal are specified. The rationals in the table that are convergents p_n/q_n with $n \geq 1$ are highlighted in boldface.

Each of the rationals, a/b, in the table below a row containing a convergent is no closer to g than the convergent is. Note that $p_0/q_0 = 1/1$ is a special case not addressed in Theorems 5.39 and 5.40.

The convergents for $[1; 1, 1, \overline{1}] = (1 + \sqrt{5})/2$ are the most slowly converging convergents for all irrationals because $t_i = 1$, the minimum possible positive integer, for all i. The convergents p_n/q_n for g are ratios of consecutive Fibonacci numbers: $f_1 = 1$, $f_2 = 1$, and $f_n = f_{n-1} + f_{n-2}$ for $n \geq 2$. Thus, the sequence $f_1, f_2, f_3, \ldots$ is $1, 1, 2, 3, 5, 8, 13, 21, \ldots$ and every term is the sum of the preceding two. The number $g = (1 + \sqrt{5})/2$ is called the Golden Ratio. The identification of p_n/q_n with f_{n+2}/f_{n+1} establishes immediately that

$$\lim_{n \to \infty} \frac{f_{n+1}}{f_n} = \frac{1 + \sqrt{5}}{2}.$$

Another way to illustrate the closeness of convergents p_n/q_n to g is to consider all real numbers y and x such that $y/x = (1 + \sqrt{5})/2$, or

$$y = \left(\frac{1 + \sqrt{5}}{2}\right) x$$

which is the equation of a line in the plane. Because $(1 + \sqrt{5})/2$ is irrational, the only integers x and y satisfying $y = ((1 + \sqrt{5})/2)x$ are $x = 0 = y$, which do not define a rational number y/x. This line is graphed in Figure 12 in which only the first quadrant for $x \geq 0$ and $y \geq 0$ is shown.

The integral choices of x and y are the lattice points; but except for the point $(0,0)$, the line passes through none of the lattice points. We will show that every lattice point (a, b), except $(0,0)$, corresponds to a rational approximation a/b to g. If x_0 is a positive integer, then $y_0 = ((1 + \sqrt{5})/2)x_0$ guarantees that the point (x_0, y_0) is on the line. The number y_0, of course, is irrational. Let $y_1 = \lfloor y_0 \rfloor$ and $y_2 = \lfloor y_0 \rfloor + 1$. The integers y_1 and y_2 correspond to heights of lattice points just below and just above the height y_0 when $x = x_0$ so that $y_1 < y_0 < y_2$. The vertical distance between (x_0, y_2) and (x_0, y_0) is $y_2 - gx_0$ and between (x_0, y_0) and (x_0, y_1) is $-(y_1 - gx_0)$. But $|y_2 - gx_0| < |y_1 - gx_0|$ if and only if $|y_2/x_0 - g| < |y_1/x_0 - g|$ so that the vertical distance from the line to a lattice point is a measure of how close the rational y_i/x_0 is to g for $i = 1, 2$. The lattice points corresponding to the first

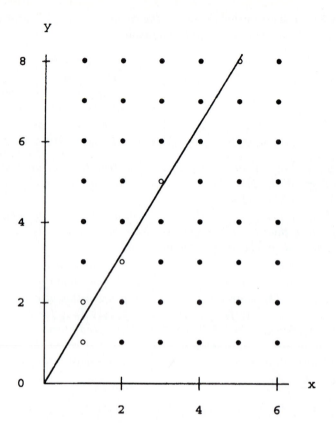

Figure 12: $y = \left(\dfrac{1 + \sqrt{5}}{2} \right) x.$

5 convergents are circled in Figure 12 and the corresponding convergents are highlighted in the first table above. In each case, any lattice point corresponding to a nearer rational approximation to $g = (1 + \sqrt{5})/2$ is to the right of the circled points (except for p_0/q_0). See the second table for the quantities $|a/b - g|$ for some lattice points (a, b) near to the line.

In Theorem 5.31 we saw that if x is irrational and p_k/q_k is the k-th convergent of x with p_k and q_k given by Theorem 5.13, then the distance from p_k/q_k to x is related to the denominator of the rational p_k/q_k: $|x - p_k/q_k| < 1/q_k^2$. Theorems 5.39 and 5.40 say that the best rational approximations with a given size denominator is a convergent. The next theorem gives sufficient conditions for determining whether a rational represented by a/b is so close to an irrational x that it must be a convergent of x. The condition depends only upon x, a, and b.

Theorem 5.42 *Let x be irrational, a/b be a rational determined by integers* *a and b with b > 0, and gcd(a, b) = 1. If*

$$\left| x - \frac{a}{b} \right| < \frac{1}{2b^2}$$

then a/b is a convergent of the infinite simple continued fraction of x; that is,
a/b = p_n/q_n, a = p_n, and b = q_n for some integer n ≥ 0 and for p_n and q_n
given by Theorem 5.13.

Proof. Let p_k and q_k be given by Theorem 5.13 for the irrational number
x; and let a and b be integers with $b > 0$ such that $|x - a/b| < 1/(2b^2)$, or
equivalently, $|bx - a| < 1/(2b)$. Also assume that a/b is not a convergent of x
so that $a/b \neq p_n/q_n$ for all n. This last nonequality is equivalent to $aq_n \neq bp_n$
for all n. Recall that $q_k \geq k$ for all $k \geq 0$ and $q_k < q_{k+1}$ for all $k \geq 0$ except
for $k = 0$, when only $q_0 \leq q_1$. Then there is a positive integer n such that
$q_n \leq b < q_{n+1}$, where we choose $n = 1$ if $q_0 = q_1$.

We want to compare a/b and p_n/q_n; therefore,

$$\left| \frac{a}{b} - \frac{p_n}{q_n} \right| = \left| \frac{aq_n - bp_n}{bq_n} \right| = \frac{|aq_n - bp_n|}{bq_n} \geq \frac{1}{bq_n} \qquad (*)$$

because $aq_n \neq bp_n$. Also, adding and subtracting x, we obtain

$$\left| \frac{a}{b} - \frac{p_n}{q_n} \right| \leq \left| \frac{a}{b} - x \right| + \left| x - \frac{p_n}{q_n} \right| < \frac{1}{2b^2} + \left| x - \frac{p_n}{q_n} \right| \qquad (**)$$

But because $b < q_{n+1}$, Theorem 5.39 implies that $|bx - a| \geq |q_n x - p_n|$.
Combining this inequality with the assumption $|bx - a| < 1/(2b)$, we get

$$\left| x - \frac{p_n}{q_n} \right| < \frac{1}{2bq_n} \qquad (***)$$

Combining the inequalities (*), (**), and (***), we have

$$\frac{1}{bq_n} \leq \left| \frac{a}{b} - \frac{p_n}{q_n} \right| < \frac{1}{2b^2} + \frac{1}{2bq_n}$$

If we multiply by $2b^2 q_n$ and simplify, this last sequence of inequalities implies that $b < q_n$, which is a contradiction. Therefore, $a/b = p_n/q_n$. Since
$\gcd(a, b) = 1$ and $\gcd(p_n, q_n) = 1$, $a = p_n$ and $b = q_n$. ∎

The requirement of $\gcd(a, b) = 1$ in Theorem 5.42 so that the rational
representation a/b is in lowest terms is a convenience only; for let $\gcd(a, b) =$
$d > 1$ and $|x - a/b| < 1/(2b^2)$ and let $a = da'$ and $b = db'$. Then

$$\left| x - \frac{da'}{db'} \right| < \frac{1}{2(db')^2}$$

and

$$\left| x - \frac{a'}{b'} \right| < \frac{1}{2d^2(b')^2} < \frac{1}{2(b')^2}$$

By Theorem 5.42, $a' = p_n$ and $b' = q_n$ for some positive integer n so that $a = dp_n$ and $b = dq_n$. Trivially, $a/b = p_n/q_n$ but $a \neq p_n$ and $b \neq q_n$. Thus, we have the following:

Corollary 5.43 *Let x be irrational and a/b be a rational determined by integers a and b with $b > 0$. If*

$$\left| x - \frac{a}{b} \right| < \frac{1}{2b^2}$$

then, a/b is a convergent of the infinite simple continued fraction of x; that is, $a/b = p_n/q_n$ for some integer $n \geq 0$ and for p_n and q_n given by Theorem 5.13.

The proof of the following theorem is left to the reader.

Theorem 5.44 *Let x be irrational and $x > 1$. If p_k/q_k and p'_k/q'_k are the convergents given by Theorem 5.13 for x and $1/x$, respectively, then $p'_k = q_{k-1}$ and $q'_k = p_{k-1}$ for $k \geq 1$. Further, if $x = [t_0; t_1, t_2, \ldots]$, then $1/x = [0; t_0, t_1, \ldots]$.*

Example 5.45 *If $x = [t_0; t_1, t_2, \ldots]$ is purely periodic, we can use Theorem 5.27 as was done in Example 5.28 to show that for $x = [\overline{1; 1}]$ we obtain $x = (1 + \sqrt{5})/2$ and for $y = [1; 2, \overline{1, 2}]$ we obtain $y = (1 + \sqrt{3})/2$. If $\tau = [t_0; t_1, t_2, \ldots]$ is not purely periodic, we can still use Theorem 5.27 in a modified manner. For example, for $w = [4; \overline{1, 3, 1, 8}]$, we can determine w explicitly as a quadratic irrational. Thus, $w = [4; \overline{1, 3, 1, 8}] = [4; z]$ where $z = [\overline{1; 3, 1, 8}]$ is purely periodic. So*

$$z = [1; 3, 1, 8, \overline{1, 3, 1, 8}] = [1; 3, 1, 8, z]$$

We next calculate the third convergent of z.

k	t_k	p_k	q_k
0	1	1	1
1	3	4	3
2	1	5	4
3	8	44	35

So

$$z = \frac{p_3 z + p_2}{q_3 z + q_2} = \frac{44z + 5}{35z + 4}$$

which, upon simplification, gives the quadratic equation

$$7z^2 - 8z - 1 = 0$$

The two real solutions are

$$z = \frac{4 \pm \sqrt{23}}{7}$$

where for our purposes we take z as $z = \left(4 + \sqrt{23}\right)/7$. Finally, $w = [4; z] = 4 + 1/z = \sqrt{23}$. Thus, the continued fraction representation of the square root of an integer, w, is not purely periodic; but those of $(1 + \sqrt{3})/2$ and $(1 + \sqrt{5})/2$ are purely periodic. If $w = [w_0; w_1, w_2, \ldots] = [4; 1, 3, 1, 8, 1, 3, 1, 8, \ldots]$, the only feature that needs modifying is the initial term $w_0 = 4$. If b is an integer, clearly, $w + b = [w_0; w_1, w_2, \ldots] + b = [w_0 + b; w_1, w_2, \ldots]$. Then

$$\sqrt{23} + 4 = [8; 1, 3, 1, 8, 1, 3, \ldots] = [\overline{8; 1, 3, 1}]$$

As we will soon see, it is not coincidental that $\lfloor \sqrt{23} \rfloor = 4 = w_0$ so that

$$\sqrt{23} + 4 = [2w_0; w_1, w_2, \ldots, w_{s-1}, \overline{2w_0, w_1, w_2, \ldots, w_{s-1}}]$$

The next theorem characterizes those quadratic irrationals that have purely periodic simple continued fractions representations.

Theorem 5.46 *If $\tau = [t_0; t_1, t_2, \ldots]$ is a quadratic irrational, then $[t_0; t_1, t_2, \ldots]$ is purely periodic if and only if $\tau > 1$ and the conjugate τ' of τ has the property $-1 < \tau' < 0$. Further, if $\tau = \overline{[t_0; t_1, \ldots, t_s]}$, then $-\dfrac{1}{\tau'} = \overline{[t_s; t_{s-1}, \ldots, t_1, t_0]}$.*

Proof. Let $x = [t_0; t_1, t_2, \ldots]$ and let x' be the conjugate of x. Assume that $x > 1$ and $-1 < x' < 0$, and let $x_0, x_1, x_2, \ldots$ be the sequence of irrationals given by Theorem 5.26. Since $x_0 = x > 1$, $t_0 = \lfloor x_0 \rfloor \geq 1$, and by the same theorem, $t_k = \lfloor x_k \rfloor \geq 1$ because $x_k > 1$ for $k \geq 1$. But for $k \geq 0$,

$$x_{k+1} = \frac{1}{x_k - t_k}$$

It is straightforward to show that x_k is a quadratic irrational for all k and that

$$x'_{k+1} = \frac{1}{x'_k - t_k}$$

Clearly, $-1 < x'_k < 0$ for $k \geq 0$; for since $-1 < x'_0 = x' < 0$, we have $x'_0 - t_0 < -1$ so that

$$-1 < \frac{1}{x'_0 - t_0} = x'_1 < 0$$

By induction we can easily obtain $-1 < x'_k < 0$ for $k \geq 0$. This last inequality implies, since $x'_k = \dfrac{1}{x'_{k+1}} + t_k$, that

$$-1 < \frac{1}{x'_{k+1}} + t_k < 0$$

and

$$t_k < \frac{-1}{x'_{k+1}} < t_k + 1$$

Thus, $\lfloor -1/x'_{k+1} \rfloor = t_k$ for $k \geq 0$.

By Theorem 5.37, $[t_0; t_1, \ldots]$ is periodic; and by Theorem 5.36, there are nonnegative integers K and s with $s \geq 1$ such that $x_K = x_{K+s}$. Then the conjugates of x_K and x_{K+s} are also equal and we obtain

$$\left\lfloor \frac{-1}{x'_K} \right\rfloor = \left\lfloor \frac{-1}{x'_{K+s}} \right\rfloor$$

Consequently, $t_{K-1} = t_{K-1+s}$, and by definition of the sequence $x_0, x_1, \ldots$ in Theorem 5.26,

$$x_{K-1} = \frac{1}{x_K} + \lfloor x_{K-1} \rfloor = \frac{1}{x_K} + t_{K-1}$$

$$= \frac{1}{x_{K+s}} + t_{K-1+s} = \frac{1}{x_{K+s}} + \lfloor x_{K-1-s} \rfloor$$

$$= x_{K-1+s}$$

By induction we can show similarly that $x_{K-j} = x_{K-j+s}$ for $1 \leq j \leq K$. Therefore, for $j = K$ we have

$$x_0 = x_s$$

But $x_0 = x = [t_0; x_1] = [t_0; t_1, \ldots, t_{s-1}, x_s]$ so that since $x_0 = x_s$, Theorem 5.36 implies that

$$t_k = t_{k+s} \text{ for } k \geq 0$$

and $[t_0; t_1, t_2, \ldots]$ is purely periodic.

If $x = [t_0; t_1, \ldots]$ is a purely periodic infinite simple continued fraction, then there is a positive integer s such that $t_k = t_{k+s}$ for $k \geq 0$. So $t_0 = t_s \geq 1$. Since $x = t_0 + 1/[t_1; t_2, \ldots]$, $x > 1$. Thus, $x = [t_0; t_1, t_2, \ldots t_{s-1}, [t_s; t_{s+1}, \ldots]] = [t_0; t_1, \ldots, t_{s-1}, x]$. Let p_k and q_k be given by Theorem 5.13 so that

$$x = \frac{p_s x + p_{s-1}}{q_s x + q_{s-1}} \text{ by Theorem 5.26} \tag{†}$$

Consider the irrational

$$y = [t_s; t_{s-1}, t_{s-2}, \ldots, t_1, t_0]$$

which is a purely periodic quadratic irrational whose simple continued fraction representation has the same repeating terms as for x except that the terms appear in reversed order. In a manner similar to our treatment of x, we have $y > 1$ and

$$x = \frac{p'_s x + p'_{s-1}}{q'_s x + q'_{s-1}} \text{ by Theorem 5.26} \tag{††}$$

where p'_k and q'_k are given by Theorem 5.13 so that p'_k/q'_k is the k-th convergent of y.

Rewriting (†) and (††) just above, one finds that x satisfies

$$q_s x^2 + (q_{s-1} - p_s)x - p_{s-1} = 0$$

and that y satisfies

$$q_s' y^2 + (q_{s-1}' - p_s')y - p_{s-1}' = 0$$

These two quadratic equations are closely related. Theorem 5.15 showed that ratios of the numerators and denominators of consecutive convergents are equal to continued fractions with terms written in reverse order. Thus,

$$\frac{p_s}{p_{s-1}} = [t_s; t_{s-1}, \ldots, t_1, t_0] = \frac{p_s'}{q_s'}$$

and

$$\frac{q_s}{q_{s-1}} = [t_s; t_{s-1}, \ldots, t_1] = \frac{p_{s-1}'}{q_{s-1}'}$$

By Theorem 5.21,

$$\gcd(p_s, p_{s-1}) = \gcd(q_s, q_{s-1}) = \gcd(p_s', q_s') = \gcd(p_{s-1}', q_{s-1}') = 1$$

so that

$$\begin{aligned} p_s &= p_s' \\ p_{s-1} &= q_s' \\ q_s &= p_{s-1}' \\ q_{s-1} &= q_{s-1}' \end{aligned}$$

Using these equivalents of the primed p's and q's in the quadratic equation for y gives

$$p_{s-1} y^2 + (q_{s-1} - p_s)y - q_s = 0$$

or, equivalently,

$$q_s\left(-\frac{1}{y}\right)^2 + (q_{s-1} - p_s)\left(-\frac{1}{y}\right) - p_{s-1} = 0$$

Thus, the equation

$$q_s w^2 + (q_{s-1} - p_s)w - p_{s-1} = 0$$

is satisfied by $w = x$ and $w = -1/y$. Since x is a quadratic irrational satisfying the quadratic equation above in w with integer coefficients, x', the conjugate of x, also satisfies the equation. Since a quadratic equation has no more than two solutions, $x' = -1/y$. But $y > 1$ implies that $0 < 1/y < 1$ and, therefore, $-1 < -1/y < 0$. Thus, $-1 < x' < 0$. ■

We can now completely characterize the simple continued fraction representation of $\sqrt{N}$ whenever N is not the square of an integer.

Theorem 5.47 *If N is a positive integer that is not the square of an integer and $\sqrt{N} = [t_0; t_1, t_2, \ldots]$, then there is a positive integer s such that*

(a) $\sqrt{N} = [t_0; \overline{t_1, t_2, \ldots, t_s}]$.

(b) $\sqrt{N} = [t_0; \overline{t_1, t_2, \ldots, t_{s-1}, 2t_0}]$ so that $t_s = 2t_0$.

(c) $\lfloor \sqrt{N} \rfloor + \sqrt{N} = \overline{[t_0; t_1, \ldots, t_{s-1}]}$.

(d) $t_i = t_{s-i}$ for $1 \le i \le s - 1$.

Proof. By Theorem 5.26, $t_0 = \lfloor \sqrt{N} \rfloor$ so that

$$
\begin{aligned}
\lfloor \sqrt{N} \rfloor + \sqrt{N} &= t_0 + [t_0; t_1, t_2, \ldots] \\
&= t_0 + [t_0; [t_1; t_2, \ldots]] \\
&= 2t_0 + 1/[t_1; t_2, \ldots] \\
&= [2t_0; [t_1; t_2, \ldots]] \\
&= [2t_0; t_1, t_2, \ldots]
\end{aligned}
$$

Let $x = \lfloor \sqrt{N} \rfloor + \sqrt{N}$. Then x is a quadratic irrational, $x > 1$, and the conjugate, x', of x is $x' = \lfloor \sqrt{N} \rfloor - \sqrt{N}$. Since $\lfloor \sqrt{N} \rfloor < \sqrt{N} < \lfloor \sqrt{N} \rfloor + 1$, we have $-1 < x' = \lfloor \sqrt{N} \rfloor - \sqrt{N} < 0$ and Theorem 5.46 implies that x is purely periodic. Let n be the least nonnegative integer such that

$$
\begin{aligned}
x &= \lfloor \sqrt{N} \rfloor + \sqrt{N} = \overline{[2t_0; t_1, t_2, \ldots, t_n]} \\
&= [2t_0; t_1, t_2, \ldots, t_n, \overline{2t_0, t_1, t_2, \ldots, t_n}]
\end{aligned}
$$

when $n \ge 1$ and $x = \overline{[2t_0;]}$ when $n = 0$. The case of $n = 0$ will be left to the reader. If we subtract $\lfloor \sqrt{N} \rfloor = t_0$ from both sides, we have

$$
\sqrt{N} = [t_0; \overline{t_1, t_2, \ldots, t_n, 2t_0}]
$$

If we let $s = n + 1$, then we have proved parts (a) to (c).

We will later show two ways of expressing $-1/x'$ as an infinite simple continued fraction; namely,

$$
-\frac{1}{x'} = \overline{[t_1; t_2, \ldots, t_n, 2t_0]}
$$

and

$$
-\frac{1}{x'} = \overline{[t_n; t_{n-1}, \ldots, t_1, 2t_0]}
$$

Assuming for the moment that these two representations are valid, consider part (d). Since the representation of an irrational as an infinite simple continued fraction is unique,

$$
t_i = t_{n-i+1} \text{ for } 1 \le i \le n
$$

Letting $s = n + 1$, we obtain that $t_i = t_{s-i}$ for $1 \le i \le s - 1$.

To complete the proof, we have only to establish the foregoing two continued fraction representations of $-1/x'$. Recall that $x' = \lfloor \sqrt{N} \rfloor - \sqrt{N}$ and $\lfloor \sqrt{N} \rfloor = t_0$. Thus,

$$
\begin{aligned}
-x' &= \sqrt{N} - \lfloor \sqrt{N} \rfloor = [t_0; \overline{t_1, t_2, \ldots, t_n, 2t_0}] - t_0 \\
&= [0; \overline{t_1, t_2, \ldots, t_n, 2t_0\}]] \\
&= 0 + \cfrac{1}{[\overline{t_1, t_2, \ldots, t_n, 2t_0}]}
\end{aligned}
$$

which gives $-1/x' = [\overline{t_1; t_2, \ldots, t_n, 2t_0}]$. Also, since $x = \lfloor \sqrt{N} \rfloor + \sqrt{N} = [\overline{2t_0; t_1, \ldots, t_n}]$ is purely periodic, Theorem 5.46 implies that

$$
-\frac{1}{x'} = [\overline{t_n; t_{n-1}, \ldots, t_1, 2t_0}] \qquad \blacksquare
$$

Theorem 5.47 says that square roots of nonsquare integers have a rather regular simple continued fraction representation, namely, a zeroth term followed by a repeating segment of terms where the last term of the segment is twice the zeroth. We showed in Example 5.45 that for the nonsquare $N = 23$

$$
\begin{aligned}
\sqrt{23} &= [4; \overline{1, 3, 1, 8}] \\
&= [t_0; \overline{t_1, t_2, t_3, 2t_0}]
\end{aligned}
$$

In the repeating pattern, the terms form a palindrome, that is, the same sequence in forward or reverse order. So for $N = 23$, $s = 4$ and $s - 1 = 3$ in Theorem 5.47. If $N = 2$, we have $\sqrt{2} = [1; \overline{2}] = [t_0; \overline{2t_0}]$, where $s = 1$ and $s - 1 = 0$ in Theorem 5.46. Thus, the square root of a nonsquare positive integer N is the sum of an integer and the reciprocal of a purely periodic quadratic irrational:

$$
\begin{aligned}
\sqrt{N} &= [t_0; \overline{t_1, t_2, \ldots, t_{s-1}, 2t_0}] \\
&= t_0 + \cfrac{1}{[\overline{t_1; t_2, \ldots, t_{s-1}, 2t_0}]}
\end{aligned}
$$

In Section 5.1 we illustrated for $x = \sqrt{3}$ how to generate the continued fraction representation $[t_0; t_1, t_2, \ldots] = [1; \overline{1, 2}]$. This algorithm was formalized for any irrational x in Theorem 5.26 by way of the sequence $x = x_0, x_1, x_2, \ldots$. The procedure can be streamlined further for x a quadratic irrationals via a recursion similar to that in the algorithm of Theorem 5.13 to generate p_k and q_k and convergents p_k/q_k.

To illustrate, suppose that x is a quadratic irrational. From Theorem 5.26, $x_0 = x$, $x_{k+1} = 1/(x_k - \lfloor x_k \rfloor)$, and $t_k = \lfloor x_k \rfloor$ so that $x = [t_0; t_1, t_2, \ldots]$. Suppose that x is expressed as

$$
x = x_0 = \frac{c_0 + \sqrt{d}}{e_0}
$$

where c_0, d, and e_0 are chosen as integers with $e_0 \neq 0$ and $e_0 \mid (d - c_0^2)$ and with $d > 0$ not the square of an integer (see the Exercises). Then $t_0 = \lfloor x_0 \rfloor = \left\lfloor (c_0 + \sqrt{d})/e_0 \right\rfloor$ and

$$
\begin{aligned}
x_1 \;&=\; \frac{1}{x_0 - \lfloor x_0 \rfloor} \;=\; \frac{1}{\left(\dfrac{c_0 + \sqrt{d}}{e_0}\right) - t_0} \\[2mm]
&=\; \frac{e_0}{(c_0 - t_0 e_0) + \sqrt{d}} \;=\; \frac{e_0\left[(c_0 - t_0 e_0) - \sqrt{d}\right]}{(c_0 - t_0 e_0)^2 - d} \\[2mm]
&=\; \frac{(t_0 e_0 - c_0) + \sqrt{d}}{\left[\dfrac{-(c_0 - t_0 e_0)^2 + d}{e_0}\right]}
\end{aligned}
$$

But $-(c_0 - t_0 e_0)^2 + d = e_0(2 t_0 c_0 - t_0^2 e_0) + (d - c_0^2)$, which is divisible by e_0. Therefore, we define the integers

$$
\begin{aligned}
c_1 \;&=\; t_0 e_0 - c_0 \\
e_1 \;&=\; \left(d - c_1^2\right)/e_0
\end{aligned}
$$

so that

$$
x_1 = \frac{c_1 + \sqrt{d}}{e_1} \quad \text{and } e_1 \mid \left(d - c_1^2\right)
$$

We have shown the first step of an inductive proof of the following theorem whose proof is left to the reader.

Theorem 5.48 *Let x be a quadratic irrational represented as follows:*

$$
x = x_0 = \frac{c_0 + \sqrt{d}}{e_0}
$$

where c_0, d, and e_0 are integers, $e_0 \neq 0$, and $d > 0$ is not the square of an integer and where $e_0 \mid \left(d - c_0^2\right)$. Then the irrational sequence x_k and integer sequence t_k of Theorem 5.26 and the integer sequences c_k and e_k are given recursively by

$$
\begin{aligned}
x_k &= \frac{c_k + \sqrt{d}}{e_k} \\
t_k &= \lfloor x_k \rfloor \\
c_{k+1} &= t_k e_k - c_k \\
e_{k+1} &= \frac{d - c_{k+1}^2}{e_k}
\end{aligned}
$$

for $k \geq 0$ so that $x = [t_0; t_1, t_2, \ldots]$.

Example 5.49 *Reconsider $x = \sqrt{23}$.*

$$x = \sqrt{23} = \frac{0 + \sqrt{23}}{1}$$

so that $c_0 = 0$, $e_0 = 1$, $d = 23$, and $e_0 \mid (d - c_0^2)$. Applying the algorithm of Theorem 5.48 yields

k	c_k	e_k	x_k	t_k
0	0	1	$(0 + \sqrt{23})/1$	4
1	$(4 \cdot 1 - 0) = 4$	$(23 - 4^2)/1 = 7$	$(4 + \sqrt{23})/7$	1
2	$(1 \cdot 7 - 4) = 3$	$(23 - 3^2)/7 = 2$	$(3 + \sqrt{23})/2$	3
3	$(3 \cdot 2 - 3) = 3$	$(23 - 3^2)/2 = 7$	$(3 + \sqrt{23})/7$	1
4	$(1 \cdot 7 - 3) = 4$	$(23 - 4^2)/7 = 1$	$(4 + \sqrt{23})/1$	8
5	$(8 \cdot 1 - 4) = 4$	$(23 - 4^2)/1 = 7$	$(4 + \sqrt{23})/7$	1
6	$(1 \cdot 7 - 4) = 3$	$(23 - 3^2)/7 = 2$	$(3 + \sqrt{23})/2$	3
7	$(3 \cdot 2 - 3) = 3$	$(23 - 3^2)/2 = 7$	$(3 + \sqrt{23})/7$	1
8	$(1 \cdot 7 - 3) = 4$	$(23 - 4^2)/7 = 1$	$(4 + \sqrt{23})/1$	8

Thus, $\sqrt{23} = [4; \overline{1, 3, 1, 8}]$.

Exercises

1. Let a be a positive integer. Prove that:

   (a) $\sqrt{a^2 + 1} = [a; \overline{2a}]$

   (b) $\sqrt{a^2 + 2} = [a; \overline{a, 2a}]$

   (c) $\frac{1}{2}\left(a + \sqrt{a^2 + 4}\right) = [\overline{a};]$

   (d) $\sqrt{a^2 - 1} = [a - 1; \overline{1, 2(a - 1)}]$ for $a \geq 2$

   (e) $\sqrt{a^2 - 2} = [a - 1; \overline{1, a - 2, 1, 2(a - 1)}]$ for $a \geq 3$

   (f) $\sqrt{a^2 + 4} = [a; \overline{a/2, 2a}]$ if a is even

2. Make a table of the continued fraction representations of $\sqrt{N}$ for $2 \leq N \leq 30$ when N is not the square of an integer. Note that the results of Exercise 1 can be applied in many of these cases.

3. Prove Theorem 5.30.

4. If $x = [t_0; t_1, \ldots]$ is a finite or an infinite simple continued fraction, then prove that for $0 \leq i < j$,

$$\left| x - \frac{p_i}{q_i} \right| > \left| x - \frac{p_j}{q_j} \right|$$

5. If

    (1) the integer $d > 0$ is not the square of an integer,

    (2) a, b, r, and s are integers, and

    (3) $a + b\sqrt{d} = r + s\sqrt{d}$,

    then prove that $a = r$ and $b = s$.

6. Let N be a positive integer that is not the square of an integer. What positive integers k have the property that

$$\frac{\left\lfloor \sqrt{N} \right\rfloor + \sqrt{N}}{k}$$

is purely periodic.

7. Prove Theorem 5.33.

8. Let v and w be either rational numbers or quadratic irrationals, n be a nonzero integer, and $'$ denote the conjugate.

    (a) Give an example of quadratic irrationals v and w such that $v + w$ is not a quadratic irrational.

    (b) Give an example of quadratic irrationals v and w such that $v \cdot w$ is not a quadratic irrational.

    (c) Prove that:

        (i) $(v \pm w)' = v' \pm w'$

        (ii) $(v \cdot w)' = v' \cdot w'$

        (iii) $(v/w)' = v'/w'$ if $w \neq 0$

        (iv) $(v^n)' = (v')^n$ if n is an integer

9. Let $f_1, f_2, f_3, \ldots$ be the Fibonacci sequence defined in Example 5.41 by

$$\begin{aligned} f_1 &= 1 \\ f_2 &= 1 \\ f_n &= f_{n-1} + f_{n-2} \text{ for } n \geq 2 \end{aligned}$$

Let $g = (1 + \sqrt{5})/2$. It was shown earlier that $g = \lfloor \overline{1}; \rfloor$. Let p_k and q_k be given by Theorem 5.13 for g. Prove that:

    (a) $p_n = f_{n+2}$ and $q_n = f_{n+1}$ for $n \geq 0$.

(b) $\displaystyle\lim_{n\to\infty}\frac{f_{n+1}}{f_n}=g$

(c) $\gcd(f_n,f_{n+1})=1$ for $n\geq 1$

(d) $f_{n+1}f_{n-1}-f_n^2=(-1)^n$ for $n\geq 1$

(e) $f_n\geq g^{n-2}$ for $n\geq 1$

(f) $f_n\leq g^{n-1}$ for $n\geq 1$

(g) $q_n\geq g^{n-1}$ for $n\geq 0$

10. Complete the proof of Theorem 5.46(a) by showing that:

(a) x_k is a quadratic irrational for all k.

(b) $-1<x_k'<0$ for $k\geq 0$.

11. Let C, D, and E be integers. Assume that $E\neq 0$ and $D>0$ is not the square of an integer and let $x=(C+\sqrt{D})/E$ be a quadratic irrational. Prove that there are integers c_0, d, and e_0 such that $e_0\neq 0$, $d>0$ is not the square of an integer, and $e_0\mid(d-c_0^2)$ such that $x=(c_0+\sqrt{d})/e_0$. (*Hint:* Multiply numerator and denominator of $x=(C+\sqrt{D})/E$ by $|E|$ and note that $\sqrt{E^2}=|E|$.)

12. Complete the proof of Theorem 5.48.

13. Express the following quadratic irrationals in the form $(c+\sqrt{d})/e$, where c, d, and e are integers:

(a) $[\overline{3;1,7}]$ **(b)** $[3;\overline{1,7}]$ **(c)** $[\overline{2;1,1}]$

(d) $[5;\overline{2,1,3}]$ **(e)** $[6;\overline{1,5,1,12}]$

14. In the elementary school grades, $22/7$ is sometimes used as an approximation to π. Show that this approximation is a convergent of π and argue that there is no better rational approximation with positive denominator less than 7.

15. Let $x=[t_0;t_1,\ldots]$ be irrational and let p_k and q_k be given by Theorem 5.13 so that p_k/q_k is the k-th convergent of x. For $k\geq 2$ and $0\leq j\leq t_k$, define $C_{k,j}$ as the fraction

$$\frac{j\cdot p_{k-1}+p_{k-2}}{j\cdot q_{k-1}+q_{k-2}}$$

(a) Prove that $C_{k,j}$ as defined above is in lowest terms for $0\leq j\leq t_k$ and $k\geq 2$.

(b) For $k \geq 2$, prove that for each j, $0 \leq j < t_k$, we have

$$C_{k,j+1} - C_{k,j} = \frac{(-1)^k}{[(j+1)q_{k-1} + q_{k-2}][jq_{k-1} + q_{k-2}]}$$

Further prove that

$$C_{k,0} < C_{k,1} < C_{k,2} < \cdots < C_{k,t_k}$$

when k is even and that

$$C_{k,0} > C_{k,1} > C_{k,2} > \cdots > C_{k,t_k}$$

when k is odd.

Since $C_{k,0} = \dfrac{p_{k-2}}{q_{k-2}}$ and $C_{k,t_k} = \dfrac{p_k}{q_k}$, the fractions $C_{k,j}$ are called *intermediate fractions* because they are intermediate between the convergents p_{k-2}/q_{k-2} and p_k/q_k. Note that p_{k-2}/q_{k-2} and p_k/q_k are either both greater than x or both less than x.

(c) If $k \geq 0$, prove that

$$\left| x - \frac{p_k}{q_k} \right| > \left| \frac{p_k + p_{k+1}}{q_k + q_{k+1}} - \frac{p_k}{q_k} \right| = \frac{1}{q_k(q_k + q_{k+1})}$$

A rational number a/b with $b > 0$ is a *best rational approximation* of the irrational number x provided that if c/d is any other rational with $d > 0$ and $a/b \neq c/d$ and if $0 < d \leq b$, then $|x - c/d| > |x - a/b|$.

(d) Prove that every best approximation of irrational x is either a convergent of x or an intermediate fraction of x; that is, if

(i) $x = [t_0; t_1, \ldots]$ is irrational,

(ii) c/d is a fraction with $d > 0$, and

(iii) $|x - c/d| \leq |x - C_{k,j}|$ for $k \geq 1$ and $0 < j < t_k$,

then $d > jq_{k-1} + q_{k-2}$, the denominator of $C_{k,j}$ or $c/d = p_{k-1}/q_{k-1}$.

(e) Find the intermediate fractions for $x = \sqrt{3}$ and $x = \sqrt{5}$ for $k = 2$, 3, and 4.

(f) Find all rational approximations to π that are nearer to π than $22/7$ and having a denominator less than 106. Why is the number 106 of significance here?

16. Determine if the following numbers have purely periodic simple periodic continued fraction representations:

(a) $3 + \sqrt{5}$

(b) $2 + \sqrt{5}$

(c) $5 + \sqrt{47}$

(d) $(5 + \sqrt{47})/2$

17. If $\sqrt{47} = [6; \overline{1, 5, 1, 12}]$, find the continued fraction representation for

(a) $6 + \sqrt{47}$

(b) $(6 + \sqrt{47})/11$

5.5 Pell's Equation

In Chapter 1 we solved the linear equation $ax + by = c$. We next consider the quadratic equation in two variables x and y:

$$ax^2 + bxy + cy^2 + dx + ey + f = 0$$

where the coefficients a, b, c, d, e, and f are integers and $a \cdot c \neq 0$. We seek solutions x and y that are integers. We proceed along the lines of Section 3.9, where we solved the quadratic congruence $ax^2 + bx + c = 0 \pmod{p}$.

Rewriting the quadratic equation as though it were a quadratic in x alone gives

$$ax^2 + (by + d)x + (cy^2 + ey + f) = 0$$

If there is a real solution x and y, then from the quadratic formula we must have

$$x = \frac{-(by + d) \pm \sqrt{(by + d)^2 - 4a(cy^2 + ey + f)}}{2a}$$

If y is an integer and x is rational, then the discriminant must be the square of an integer so that

$$(by + d)^2 - 4a(cy^2 + ey + f) = w^2$$

for some integer w. Rewriting as a quadratic equation in y, we obtain

$$Ay^2 + By + C = 0$$

where

$$
\begin{aligned}
A &= b^2 - 4ac \\
B &= 2bd - 4ae \\
C &= d^2 - 4af - w^2 = D - w^2 \\
D &= d^2 - 4af
\end{aligned}
$$

In order for y to be an integer, the discriminant $B^2 - 4AC$ must be the square of an integer so that

$$z^2 = B^2 - 4A(D - w^2)$$

for some integer z, or, equivalently,

$$z^2 - 4Aw^2 = B^2 - 4AD$$

where $4A$ and $B^2 - 4AD$ are integers.

Thus, if the original equation is solvable for integers x and y, then for $m = 4A$ and $n = B^2 - 4AD$

$$z^2 - mw^2 = n$$

is solvable for integers z and w. We note that the existence of integral solutions z and w is a necessary condition only. We still must be able to have x and y be integers and not just rational numbers.

The discussion above motivates wanting to solve an equation of the form

$$x^2 - dy^2 = n$$

with d and n integers. This equation is known as *Pell's equation*.

A procedure which was essentially a method for solving Pell's equation occurs as early as 400 B.C. in India and Greece. Diophantus used special Pell's equations to solve other problems. A problem which became intertwined with Pell's equation was the "Cattle Problem of Archimedes" (see p. 59 of [12]). Bháscara Achárya in the twelfth century gave a method for finding new sets of solutions of $x^2 - dy^2 = 1$ when one set is found. Various solutions and specific solutions were discovered and rediscovered using several different methods.

Fermat posed the general Pell equation as a challenge problem to Lord Brouncker and John Wallis, two English mathematicians. Both Brouncker and Wallis obtained solutions. The association of John Pell's name with the equation is due to Euler who apparently mixed the contributions of Pell and Brouncker when reading an algebra by Wallis. Euler mistakenly credited the theorem to Pell, who had neither stated the problem nor solved it. Pell, however, did work on solving the equation.

Lagrange used continued fractions to give a direct method to find integral solutions to $x^2 - dy^2 = n$. He also proved that real roots of any quadratic equation with rational coefficients can be developed into a periodic continued fraction and conversely, any such fraction is a root of a quadratic equation with rational coefficients.

Dirichlet solved $x^2 - dy^2 = 1$ by use of trigonometric functions. He proved that if d is a complex integer (a complex number $a + bi$ where a and b are integers) that is not a square, then $x^2 - dy^2 = 1$ is solvable in complex integers and showed how to find all solutions. This method also works in the case where all the integers are real.

First we will address several special cases in which the integral solutions of $x^2 - dy^2 = n$ are routine. We note that it is sufficient to consider only positive integer solutions x and y of $x^2 - dy^2 = n$ because x and y are squared in Pell's equation. Also, the solutions when one of x or y is zero are obvious.

Case 1. If $d < 0$, then $n > 0$. Pell's equation is of the form

$$x^2 + |d| \, y^2 = n$$

so that any solution must have both $x^2 \leq n$ and $|d| \, y^2 \leq n$, or equivalently, both $|x| \leq \sqrt{n}$ and $|y| \leq \sqrt{n/|d|}$. Thus, only finitely many integral solutions x and y of $x^2 + |d| \, y^2 = n$ are possible. An exhaustive check of all possible such x and y will produce all solutions if there are any.

Case 2. If $d > 0$ and d is the square of an integer, let $d = r^2$ so that

$$n = x^2 - dy^2 = x^2 - (ry)^2 = (x - ry)(x + ry)$$

Then a possible solution exists only if

$$\begin{cases} x - ry = a \\ x + ry = b \end{cases}$$

for some integers a and b with $ab = n$. Hence

$$2ry = b - a \qquad \text{and} \qquad 2x = a + b$$

So for each choice of integers a and b with $ab = n$, there is at most one pair of integers x and y such that $x^2 - dy^2 = n$. Since there are only finitely many divisors of n, there are only finitely many integer pairs a and b with $ab = n$ to consider; and so there are at most finitely many integer solutions x and y of $x^2 - dy^2 = n$ in this case.

Case 3. If $d > 0$, $n = \pm 1$, and d is not the square of an integer, then there are infinitely many integral solutions. The proof of this statement is the subject of most of the rest of the section. We will not consider here systematically the situation of other integers n besides ± 1 in the case of $d > 0$; however, some theorems will be stated for n not necessarily equal to ± 1.

For $n = \pm 1$, Pell's equation may be written as

$$(x/y)^2 - d = \pm \frac{1}{y^2}$$

for any positive integer solutions x and y with $y \neq 0$. Such a solution of this equation is tantamount to having a rational number x/y that approximates $\sqrt{d}$ since the equation above is equivalent to requiring

$$\left| \frac{x}{y} - \sqrt{d} \right| = \frac{1}{y^2 \left(\dfrac{x}{y} + \sqrt{d} \right)} < \frac{1}{y^2}$$

It appears, therefore, that solutions x and y of Pell's equation are related to rational approximations to $\sqrt{d}$. It was shown in Section 5.4 that the convergents of the simple continued fraction representation of irrationals are good

approximations so that we might expect that the continued fractions representation of $\sqrt{d}$ could be helpful in constructing solutions to Pell's equation.

Theorem 5.50 *Let d be a positive integer that is not the square of an integer
and let n be an integer with $n^2 < d$. If x and y are positive integer solutions
of $x^2 - dy^2 = n$, then x/y is a convergent of the infinite simple continued
fraction for $\sqrt{d}$. Further, if x and y are relatively prime, then $x = p_k$ and
$y = q_k$ for x/y equal to the k-th convergent p_k/q_k for some k.*

Proof. If x and y are positive integer solutions of $x^2 - dy^2 = n$ and $n^2 < d$,
then $n < \sqrt{d}$. We first consider $n > 0$. We can then write

$$
\begin{aligned}
\frac{x}{y} - \sqrt{d} &= \frac{x - y\sqrt{d}}{y} \\
&= \frac{x^2 - y^2 d}{y\left(x + y\sqrt{d}\right)} \\
&= \frac{n}{y\left(x + y\sqrt{d}\right)}
\end{aligned}
$$

Since x, y, and n are positive, we have $x/y - \sqrt{d} > 0$. Thus,

$$
\begin{aligned}
\frac{x}{y\sqrt{d}} &> 1 \\
\frac{x}{y\sqrt{d}} + 1 &> 2 \\
x + y\sqrt{d} &> 2y\sqrt{d}
\end{aligned}
$$

and because $n < \sqrt{d}$, we obtain

$$
\left| \frac{x}{y} - \sqrt{d} \right| < \frac{\sqrt{d}}{y\left(x + y\sqrt{d}\right)} < \frac{1}{2y^2}
$$

By Theorem 5.42 and its corollary, x/y is a convergent of the infinite simple
continued fraction for $\sqrt{d}$.

On the other hand, if $n < 0$, we rewrite the equation $x^2 - dy^2 = n$ as

$$
y^2 - \left(\frac{1}{d}\right) x^2 = -\frac{n}{d}
$$

where $\sqrt{1/d}$ is also a quadratic irrational and $(-n/d)^2 < 1/d$. We can
apply the case proved above to give that y/x is a convergent of $\sqrt{1/d} =
1/\sqrt{d}$. Finally, Theorem 5.44 implies that $1/(y/x) = x/y$ is a convergent of
$1/\left(1/\sqrt{d}\right) = \sqrt{d}$.

We note that $x^2 - dy^2 = n$ has no solution in positive integers when
$n = 0$. ∎

We next consider the case when $n = \pm 1$. We will see later that knowing solutions when $n = \pm 1$ leads to solutions for general n. In order to make the proof of the main theorem easier, we prove the following.

Theorem 5.51 *If $d > 0$ is not the square of an integer, p_k/q_k is the k-th convergent of $\sqrt{d}$ as given by Theorem 5.13, and the sequence e_k is given by Theorem 5.48 for the quadratic irrational $\sqrt{d}$, then*

$$p_k^2 - dq_k^2 = (-1)^{k-1}e_{k+1}$$

for $k \geq 0$.

Proof. By Theorem 5.26

$$\sqrt{d} = \frac{p_k x_{k+1} + p_{k-1}}{q_k x_{k+1} + q_{k-1}}$$

By Theorem 5.48

$$\sqrt{d} = \frac{p_k(c_{k+1} + \sqrt{d}) + e_{k+1}p_{k-1}}{q_k(c_{k+1} + \sqrt{d}) + e_{k+1}q_{k-1}}$$

because $x_j = (c_j + \sqrt{d})/e_j$ for $j \geq 0$. Clearing the denominator from the second expression for $\sqrt{d}$ and simplifying, we get

$$p_k c_{k+1} + e_{k+1}p_{k-1} - dq_k = (q_k c_{k+1} + e_{k+1}q_{k-1} - p_k)\sqrt{d}$$

Since all terms are integers and $\sqrt{d}$ is irrational, both of the following equations hold:

$$\begin{cases} p_k c_{k+1} + e_{k+1}p_{k-1} - dq_k = 0 \\ q_k c_{k+1} + e_{k+1}q_{k-1} - p_k = 0 \end{cases}$$

Multiplying the first equation by q_k and the second by p_k and subtracting, we obtain

$$p_k^2 - dq_k^2 = e_{k+1}(p_k q_{k-1} - q_k p_{k-1})$$

Substituting for the parenthetical expression from Theorem 5.14(a) gives

$$p_k^2 - dq_k^2 = e_{k+1}(-1)^{k-1}$$

for $k \geq 0$. ∎

Theorem 5.52 *Given that $d > 0$ is not the square of an integer and that p_k/q_k is the k-th convergent of $\sqrt{d}$ where p_k and q_k are given by Theorem 5.13, let s be the period of the simple continued fraction representation of $\sqrt{d}$.*

(a) *The positive solutions of $x^2 - dy^2 = 1$ are*

$$\begin{cases} x = p_{is-1} \\ \quad\quad\quad\quad \text{for } i \geq 1 \quad \text{when } s \text{ is even} \\ y = q_{is-1} \end{cases}$$

$$\begin{cases} x = p_{2is-1} \\[2mm] y = q_{2is-1} \end{cases} \quad \text{for } i \geq 1 \quad \text{when } s \text{ is odd}$$

(b) *The positive solutions of* $x^2 - dy^2 = -1$ *are nonexistent when* s *is even and*

$$\begin{cases} x = p_{(2i-1)s-1} \\[2mm] y = q_{(2i-1)s-1} \end{cases} \quad \text{for } i \geq 1 \quad \text{when } s \text{ is odd}$$

Proof. We first notice that any solutions x and y of $x^2 - dy^2 = \pm 1$ must be relatively prime. Let p_k/q_k be the k-th convergent of $\sqrt{d}$ given by Theorem 5.13 and let x_k, t_k, c_k, and e_k be given by Theorem 5.48 for $x_0 = \sqrt{d}$ so that $\sqrt{d} = [t_0; t_1, t_2, \ldots]$. Theorem 5.51 implies that

$$p_k^2 - dq_k^2 = e_{k+1}(-1)^{k-1}$$

for $k \geq 0$. Let x_k^*, t_k^*, c_k^*, and e_k^* be the sequences given by Theorem 5.48 for $x_0^* = \left\lfloor \sqrt{d} \right\rfloor + \sqrt{d}$ that are analogous to the unstarred sequences in the theorem. It is straightforward to show that

$$\begin{aligned} x_k^* &= x_k \\ t_k^* &= t_k \\ c_k^* &= c_k \\ e_k^* &= e_k \end{aligned}$$

for all $k \geq 0$ except for the following cases:

$$\begin{aligned} x_0^* &= t_0 + \sqrt{d} \\ x_0 &= \sqrt{d} \end{aligned}$$

$$t_0^* = 2t_0$$

$$\begin{aligned} c_0^* &= t_0 \\ c_0 &= 0 \end{aligned}$$

$$\begin{aligned} c_1^* &= 2t_0 \\ c_1 &= t_0 \end{aligned}$$

By Theorem 5.48, $\left\lfloor \sqrt{d} \right\rfloor + \sqrt{d} = t_0 + \sqrt{d}$ is purely periodic of period s and we have that $x_0^* = x_{js}^*$ for $j \geq 1$. But $t_0 + \sqrt{d} = t_0 + x_0 = x_0^* = x_{js}^* = x_{js}$ for $j \geq 1$. Therefore,

$$t_0 + \sqrt{d} = \frac{c_{js} + \sqrt{d}}{e_{js}} \quad \text{for } j \geq 1$$

Since t_0, c_{js}, and e_{js} are integers, the coefficients of $\sqrt{d}$ must be equal so that $e_{js} = 1$ for $j \geq 1$. Thus, for $k = js - 1$,

$$p_{js-1}^2 - dq_{js-1}^2 = (-1)^{js-2} = (-1)^{js}$$

When s is even, p_{js-1} and q_{js-1} comprise a solution of $x^2 - dy^2 = 1$ and $x^2 - dy^2 = -1$ has no solution of the form above. When s is odd, then

$$js \text{ is even when } j = 2i \text{ for some } i \geq 1$$
$$js \text{ is odd when } j = 2i - 1 \text{ for some } i \geq 1$$

Thus, when s is odd,

$$p_{2is-1} \text{ and } q_{2is-1} \text{ comprise a solution of } x^2 - dy^2 = 1$$

and

$$p_{(2i-1)s-1} \text{ and } q_{(2i-1)s-1} \text{ comprise a solution of } x^2 - dy^2 = -1$$

We know by Theorem 5.50 that the only solutions to $x^2 - dy^2 = \pm 1$ are of the form $x = p_k$ and $y = q_k$. Also, it is true that

$$p_k^2 - dq_k^2 = e_{k+1}(-1)^{k-1}$$

for every k. There will be no solutions other that those found above provided that $e_k = 1$ if and only if $s \mid k$ and provided that $e_k \neq -1$ for all k.

We already know that if $s \mid k$, then $e_k = 1$. If $e_k = 1$ for some $k > 0$, then $x_k = (c_k + \sqrt{d})/e_k = c_k + \sqrt{d}$. Since $x_0 = \sqrt{d} = [t_0; \overline{t_1, t_2, \ldots, t_s}]$ and $k > 0$, x_k is purely periodic. By Theorem 5.46, we have $-1 < x_k' = c_k - \sqrt{d} < 0$ or $\sqrt{d} - 1 < c_k < \sqrt{d}$ so that $c_k = \lfloor \sqrt{d} \rfloor$. But $t_0 = \lfloor \sqrt{d} \rfloor$ also. Thus

$$
\begin{aligned}
x_k &= t_0 + \sqrt{d} \\
&= t_0 + [t_0; \overline{t_1, t_2, \ldots, t_s}], \text{ where } t_s = 2t_0 \\
&= [\overline{2t_0; t_1, t_2, \ldots, t_{s-1}}]
\end{aligned}
$$

But

$$
\begin{aligned}
x_k &= t_0 + \sqrt{d} = t_0 + [t_0; t_1, \ldots, t_{k-1}, x_k] \\
&= [\overline{2t_0; t_1, t_2, \ldots, t_{k-1}, x_k}]
\end{aligned}
$$

Thus, $s \mid k$.

If $e_k = -1$ for some $k \geq 1$, then $x_k = (c_k + \sqrt{d})/e_k = -c_k - \sqrt{d}$. But $x_k > 1$ is purely periodic so that by Theorem 5.44 we have $1 < x_k = -c_k - \sqrt{d}$ and $-1 < x_k' = -c_k + \sqrt{d} < 0$. Thus $\sqrt{d} < c_k < -\sqrt{d} - 1$, which is a contradiction. Thus $e_k \neq -1$ for all $k \geq 0$ since $e_0 = 1$. ∎

Corollary 5.53 *Let $d > 0$ not be the square of an integer. Then $x^2 - dy^2 = 1$ has infinitely many solutions; and $x^2 - dy^2 = -1$ has infinitely many solutions only if the period of the simple continued fraction for $\sqrt{d}$ is an odd integer.*

Example 5.54 *Consider the equations $x^2 - 23y^2 = \pm 1$. We showed in Example 5.49 that $\sqrt{23} = [4; \overline{1, 3, 1, 8}]$ so that the period is $s = 4$. The equation $x^2 - 23y^2 = -1$ has no positive integral solutions because the period of $\sqrt{23}$*

is even. The only other possibility is for x or y to be zero. Neither $x = 0$ nor $y = 0$ leads to a solution, so this equation has no integral solutions.

We now consider $x^2 - 23y^2 = 1$. The first few convergents are given by

k	t_k	p_k	q_k
0	4	4	1
1	1	5	1
2	3	19	4
3	1	24	5
4	8	211	44
5	1	235	49
6	3	916	191
7	1	1151	240

If we look for solutions to $x^2 - dy^2 = 1$, by Theorem 5.51, $x = p_{is-1} = p_{4i-1}$ and $y = q_{is-1} = q_{4i-1}$ for $i \geq 1$ are solutions. The first solution generated by the convergents of $\sqrt{23}$ are $x_1 = p_3 = 24$ and $y_1 = q_3 = 5$. Checking, we have

$$x_1^2 - dy_1^2 = (24)^2 - 23 \cdot (5)^2 = 576 - 575 = 1$$

as expected. Some solutions may have x and y neither positive or negative, but they can be at most four in number. For example, for $x = 0$, $23y^2 = 1$ for no integer y; however, for $y = 0$, $x = \pm 1$ is a solution of $x^2 = 1$.

Looking forward to the next theorem, we observe that

$$\left(x_1 + y_1\sqrt{d}\right)^2 = \left(24 + 5\sqrt{23}\right)^2$$
$$= 1151 + 240\sqrt{23} = x_2 + y_2\sqrt{d}$$

and

$$x_2^2 - dy_2^2 = (1151)^2 - 23 \cdot (240)^2 = 1324801 - 1324800 = 1$$

so that $x_2 = 1151$ and $y_2 = 240$ also comprise a solution. Comparing p_k and q_k from the table above we see that the next solution obtained is when $i = 2$, giving $p_{4i-1} = p_7 = 1151$ and $q_{4i-1} = q_7 = 240$.

Although the integer solutions to $x^2 - dy^2 = 1$ may be found among the integers p_k and q_k associated with the convergents of $\sqrt{d}$, if we obtain one particular positive solution it will be shown that we can generate all the other positive ones without calculating p_k and q_k.

Definition 5.55 *Given that $d > 0$ is not the square of an integer and that x_1 is the least positive integer such that there is a corresponding positive integer y_1 with $x_1{}^2 - dy_1{}^2 = 1$, we say that the pair (x_1, y_1) comprises the* least positive integral solution *to $x^2 - dy^2 = 1$.*

The least positive solution is well-defined since $\sqrt{d} = [t_0; t_1, t_2, \ldots]$ is positive implies that t_0 is positive also. One can easily show in this case that p_k and q_k as given by Theorem 5.13 for $\sqrt{d}$ have the property $p_0 < p_1 < p_2 < \cdots$

and $q_0 < q_1 < q_2 < \cdots$. If x_1 is the least p_k such that $x = p_k$ and $y = q_k$ satisfy $x^2 - dy^2 = 1$, then y_1 is also the least q_j such that $x = p_j$ and $y = q_j$ satisfy the same equation. If x^* and y^* are any positive integral solutions, then clearly $x_1 \leq x^*$ and $y_1 \leq y^*$ and we can talk about the least positive solution.

Theorem 5.56 *If $d > 0$ is not the square of an integer and x_1 and y_1 comprise the least positive integral solution to $x^2 - dy^2 = 1$, then all positive integral solutions are given by the sequence of integer pairs $(x_1, y_1), (x_2, y_2),$ $(x_3, y_3), \ldots,$ where*

$$x_n + y_n\sqrt{d} = (x_1 + y_1\sqrt{d})^n$$

Proof. Let x_1 and y_1 comprise the least positive solution of $x^2 - dy^2 = 1$. Define integers x_n and y_n as in the statement of the theorem. Then x_n and y_n are well defined because if a and b are integers and $d > 0$ is not the square of an integer, then $(a + b\sqrt{d})^n = a' + b'\sqrt{d}$, where a' and b' are integers. Clearly, x_n and y_n are positive and $x_n - y_n\sqrt{d} = (x_1 - y_1\sqrt{d})^n$ because if $a + b\sqrt{d} = r + s\sqrt{d}$ for integers a, b, r, and s, then $a = r$ and $b = s$. Thus

$$\begin{aligned} x_n^2 - dy_n^2 &= \left(x_n + y_n\sqrt{d}\right)\left(x_n - y_n\sqrt{d}\right) \\ &= \left(x_1 + y_1\sqrt{d}\right)^n \left(x_1 - y_1\sqrt{d}\right)^n \\ &= \left(x_1^2 - dy_1^2\right)^n = 1 \end{aligned}$$

and x_n and y_n do comprise a positive integral solution of $x^2 - dy^2 = 1$.

We now show that no positive solutions are omitted from the sequence $(x_1, y_1), (x_2, y_2), (x_3, y_3), \ldots$. Let u and v comprise a positive integer solution of $x^2 - dy^2 = 1$ that is not among the x_n and y_n in the theorem. One can show for real numbers a and b with $1 < a \leq b$ that there is a positive integer m such that $a^m > b$. Consequently, there is a positive integer k such that $a^k \leq b < a^{k+1}$. Since x_1 and y_1 comprise the least positive solution, if u and v form another positive solution, then $x_1 < u$, $y_1 < v$, and $1 < x_1 + y_1\sqrt{d} < u + v\sqrt{d}$. Let k be the positive integer such that

$$x_k + y_k\sqrt{d} = \left(x_1 + y_1\sqrt{d}\right)^k \leq u + v\sqrt{d} < (x_1 + dy_1)^{k+1}$$

where the first inequality is strict because u and v cannot be x_n and y_n, respectively. If we multiply by $\left(x_1 - y_1\sqrt{d}\right)^k$, we get

$$\begin{aligned} 1 &= x_k^2 - dy_k^2 \\ &< \left(u + v\sqrt{d}\right)\left(x_1 - y_1\sqrt{d}\right)^k \\ &< \left(x_1 + y_1\sqrt{d}\right)^{k+1}\left(x_1 - y_1\sqrt{d}\right)^k \\ &= x_1 + y_1\sqrt{d} \end{aligned}$$

The middle expression $\left(u + v\sqrt{d}\right)\left(x_1 - y_1\sqrt{d}\right)^k$ is a quadratic irrational of the form $r + s\sqrt{d}$ with r and s integers.

We show below that r and s are positive integers and satisfy $x^2 - dy^2 = 1$ so that $x_1 \leq r$ and $y_1 \leq s$; and therefore, $x_1 + y_1\sqrt{d} \leq r + s\sqrt{d}$, which is a contradiction, since we know that $r + s\sqrt{d} < x_1 + y_1\sqrt{d}$. Thus, u and v must be equal to x_n and y_n, respectively, for some positive integer n; and we are finished.

The integers r and s comprise a solution since

$$
\begin{aligned}
r^2 - ds^2 &= \left(r - s\sqrt{d}\right)\left(r + s\sqrt{d}\right) \\
&= \left(u - v\sqrt{d}\right)\left(x_1 - y_1\sqrt{d}\right)^k \left(u + v\sqrt{d}\right)\left(x_1 + y_1\sqrt{d}\right)^k \\
&= \left(u^2 - dv^2\right)\left(x_1^2 - dy_1^2\right)^k \\
&= \left(u^2 - dv^2\right)\left(x_k^2 - dy_k^2\right) = 1
\end{aligned}
$$

But $1 < r + s\sqrt{d}$ implies that $0 < 1/\left(r + s\sqrt{d}\right) < 1$ so that multiplying numerator and denominator by $r - s\sqrt{d}$ gives

$$
0 < \frac{r - s\sqrt{d}}{r^2 - s^2\sqrt{d}} = r - s\sqrt{d} < 1
$$

Then

$$
r = \frac{r + s\sqrt{d}}{2} + \frac{r - s\sqrt{d}}{2} > 1 \cdot (1/2) + 0 \cdot (1/2) > 0
$$

and

$$
s\sqrt{d} = \frac{r + s\sqrt{d}}{2} - \frac{r - s\sqrt{d}}{2} > 1 \cdot (1/2) - 1 \cdot (1/2) = 0
$$

implying that r and s are positive integers. ■

We now reconsider the more general Pell's equation $x^2 - dy^2 = n$, where $d > 0$ is not the square of an integer. Information about integral solutions is not as well developed as when $n = \pm 1$; however, if we can find one solution r and s of $x^2 - dy^2 = n$, then we can generate infinitely many others.

Let x^* and y^* comprise any integral solution of $x^2 - dy^2 = 1$ and let r and s provide an integral solution to $x^2 - dy^2 = n$. Then

$$
\begin{aligned}
n &= 1 \cdot n = \left[(x^*)^2 - d(y^*)^2\right]\left(r^2 - ds^2\right) \\
&= \left(x^*r \pm dy^*s\right)^2 - d\left(x^*s \pm y^*r\right)^2
\end{aligned}
$$

Thus, $r^* = x^*r \pm dy^*s$ and $s^* = x^*s \pm y^*r$ comprise an integral solution to $x^2 - dy^2 = n$. Since there are infinitely many positive integer solutions to $x^2 - dy^2 = 1$, there are infinitely many solutions to $x^2 - dy^2 = n$.

Exercises

1. Find all solutions in integers x and y for:

    (a) $x^2 + 4y^2 = 7$

    (b) $x^2 + 3y^2 = 31$

    (c) $x^2 + 3y^2 = -2$

    (d) $x^2 - 4y^2 = 6$

    (e) $x^2 - 4y^2 = 13$

    (f) $x^2 - 9y^2 = 25$

    (g) $x^2 - 9y^2 = 12$

    (h) $9x^2 - 25y^2 = 125$

2. Find three positive integral solutions for:

    (a) $x^2 - 11y^2 = 1$

    (b) $x^2 - 13y^2 = 1$

    (c) $x^2 - 13y^2 = -1$

3. Find three positive integral solutions for $x^2 - 13y^2 = 12$.

4. Prove that if $d \equiv 3 \pmod 4$ and d is not a square, then $x^2 - dy^2 = -1$ has no integral solutions x and y.

5.6   Application: Relative Rates

Suppose that two gears are on different rotating shafts. Suppose that gear A on shaft 1 has a teeth and gear B on shaft 2 has b teeth as in Figure 13. Suppose shaft 1 turns at the rate of ω_1 turns per unit time and shaft 2 turns at the rate of ω_2 turns per unit time. Suppose that shaft 1 is driven and shaft 2 is the "slave" which turns only because gear A moves it. If the gears do not slip, then the number of teeth passing by the point of contact per unit time is the same for both gears. Thus, since $a\omega_1$ is the number of teeth per unit time for gear A and $b\omega_2$ is that for gear B, we have $b\omega_2 = a\omega_1$, or

$$\omega_2 = \frac{a}{b}\omega_1$$

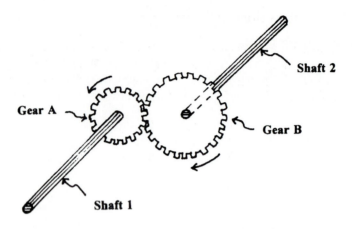

Figure 13: Two shafts and two gears.

Thus, if we are given ω_1 turns per unit time of shaft 1, then ω_2 is determined by the ratio of the number of teeth on the two gears. On the other hand, if two gears are on the same shaft, then the turns per unit time is the same for both gears; but the number of teeth per unit time depends the number of teeth on the respective gears.

Suppose that we have a shaft that turns at the rate of ω_1 turns per second and we want to run a device at the rate of $\sqrt{2}\omega_1$ turns per second. Since $\sqrt{2}$ is irrational, there will not be two gears with $\sqrt{2}$ as the ratio of their numbers of teeth. Thus, we need to approximate $\sqrt{2}$ with a rational number. We know that convergents of continued fractions for $\sqrt{2}$ will provide good rational approximations. Since it was shown that $\sqrt{2} = [1; \overline{2}]$ in the Exercises of Section 5.4, the first few convergents are

k	t_k	p_k	q_k	k	t_k	p_k	q_k
0	1	1	1	7	2	577	408
1	2	3	2	8	2	1393	985
2	2	7	5	9	2	3363	2378
3	2	17	12	10	2	8119	5741
4	2	41	29	11	2	19601	13860
5	2	99	70	12	2	47321	33461
6	2	239	169	13	2	114243	80782

Consider the convergent $p_9/q_9 = 3363/2378 \approx \sqrt{2}$. By Theorem 5.31, $|\sqrt{2} - p_9/q_9| < 1/q_9^2 < 2 \cdot 10^{-7}$. Thus, we could have the gear arrangement of Figure 13 with gear A on shaft 1 having $a = 3363$ teeth and gear B

on shaft 2 having $b = 2378$ teeth so that shaft 2 will turn at the rate of $\omega_2 = (3363/2378)\omega_1 \approx \sqrt{2}\omega_1$. The rate will be off by less than 1 part in a million. However, gears with thousands of teeth are impractical. The solution to this difficulty is to notice that 3363 and 2378 factor. Thus

$$\frac{p_9}{q_9} = \frac{3363}{2378} = \frac{3 \cdot 19 \cdot 59}{2 \cdot 29 \cdot 41} = \frac{57 \cdot 59}{58 \cdot 41}$$

where we have combined the primes 3 and 19 and the primes 2 and 29 because a gear with only 2 or 3 teeth is also impractical. A gear assembly that is equivalent to the foregoing "correct" but impractical solution is the three-shaft arrangement in Figure 14, where the gears have these number of teeth:

Gear	Number of Teeth
A	57
B	58
C	59
D	41

Thus,

$$\omega_2 = \frac{57}{58}\omega_1$$
$$\omega_3 = \frac{59}{41}\omega_2$$

so that

$$\omega_3 = \frac{59 \cdot 57}{41 \cdot 58}\omega_1 \approx \sqrt{2}\omega_1$$

The convergent $p_{10}/q_{10} = 8119/5741$ gives a better approximation to $\sqrt{2}$; however, 5741 is prime so that a multistage, multishaft configuration cannot be devised. Of course, a gear with 5741 teeth is probably unreasonable.

The system above used to implement irrational rates was mechanical in nature. A digital analog to this use of gears is to use a computer register or counter that is updated regularly according to an oscillating signal. For example, an 8-bit register can count 256 items before "rolling over" to zero; and a 16-bit register can count 65536 items. Regularly occurring electrical signals commonly available are the 60-hertz (i.e., cycles per second) U.S. household alternating current and the several timing signals available as part of a computer's circuitry, which are typically in the megahertz range, say 100 MHz. We describe next an application of continued fractions in creating a digital circuit that converts ordinary time to sidereal time (see "A Digital Clock For Sidereal Time" by Frank Reid and Kent Honeycutt [70]).

In the example above, the object was to convert from a rotation rate of ω_1 turns per unit time to say ω_2 turns per unit time. A similar situation occurs when converting from one time measure to another. For example, in astronomy, the mean solar day is the average time for the earth to rotate one time relative to the sun, that is, the average time between noons when the sun is highest in the sky. The sidereal day is the time required for the earth

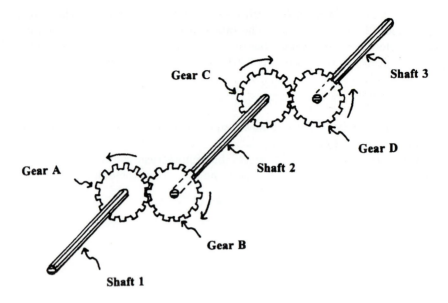

Figure 14: Three shafts and four gears.

to rotate one time relative to the fixed stars. These two times are different because the earth moves in its orbit around the sun while it is rotating. Since the earth rotates in the same direction as it orbits the sun, once it completes a turn relative to the stars, it must continue turning for about 4 more minutes until the same point on the earth faces the sun again. Thus, the time required for a sidereal day is shorter than for a mean solar day.

A sidereal minute is 60 sidereal seconds, and a sidereal hour is 60 sidereal minutes, and so forth; but nearly all timing mechanisms, including watches, household alternating current, and computer timing circuits, are based upon the ordinary second. In this example, we want a way to measure sidereal time using ordinary electrical or electronic timing circuits.

The ordinary second is sometimes defined to be 1/86400 of a mean solar day. Since the mean solar day changes slightly over time, an atomic clock defines a second to be 9192631770 cycles of the radiation associated with a certain transition between two energy levels of the cesium-133 atom (see p. 158 of *The Science of Measurement: A Historical Survey* by Herbert Arthur Klein [41]); however, we will ignore this distinction and other fine ones here.

Thus, the ordinary second, minute, hour, and day are based upon the mean solar time and are the "natural" time measures for normal earth activities. Astronomy, on the other hand, deals mostly with the earth's orientation relative to the stars so that sidereal time is the "natural" time for astronomy. We will take the ratio of the mean solar day to the mean sidereal day to be 1.00273790934. (See Guy Ottewell's *The Astronomical Companion* [59] for a detailed discussion of these issues.)

The meaning of this ratio is that every "tick" of an ordinary clock must be converted to 1.00273790934 ticks of a sidereal clock. This conversion can be implemented in digital form using a digital register of adequate size. The idea is to increment sidereal time according to the contents of this register; that is, if this register is incremented every 1/60 of a sidereal second, then when the register value is congruent to 0 modulo 60, we would have an additional sidereal second. So suppose that the rational number 1.00273790934 is approximated with continued fraction convergents:

k	t_k	p_k	q_k
0	1	1	1
1	365	366	365
2	4	1465	1461
3	7	10621	10592
4	1	12086	12053
5	3	46879	46751
6	47	2215399	2209350

Reid and Honeycutt used the approximation $1465/1461 \approx 1.00273785$ to construct their digital circuit to produce sidereal time from ordinary 60-Hz alternating house-current timing. The circuit is designed so that the counting register is incremented by 1 each cycle (that is, each 1/60 of an ordinary second); however, when the count is congruent to 1461, an extra 4 counts are added to the register. Thus, the sidereal time is updated or corrected about every 24 seconds. According to Reid and Honeycutt, the theoretical error is only -0.005 second per day or -1.85 seconds per year.

This circuit was dependent upon the frequency of the alternating house circuit. The faster frequencies available from computer timing circuits would allow the same mechanism to be used either with more frequent updating or with convergents having a larger denominator than in $1465/1461$, both of which will improve the accuracy of the approximation. Note, however, a precision beyond the number of significant digits in 1.00273790934 is unnecessary.

Exercises

1. Determine which convergents from among p_k/q_k for $k = 0$ to 13 may be used to approximate a gearing assembly for the rate $\sqrt{2}\omega_1$ provided that a gear must have at least 12 teeth but no more than 100 teeth.

2. Show that $70226/40545$ is a convergent for $\sqrt{3}$ and devise a gear assembly to implement the turn rate $\sqrt{3}\omega_1$.

3. Roger W. Sinnott [80] describes an adaptation of a motor that turns once in seven days to drive a moon-phase figure as is sometimes found on a grandfather clock. The circular-disk moon dial contains two moon

symbols on opposite sides of the diameter and should make half a turn per month. There are two moon symbols because just as one moon disappears or "sets," the other is "rising" on the opposite side of the dial disk. The synodic month is 29.53058868 days and is the average time from one new moon to the next. (See [59] for details.) Thus, there should be $(1/7)$ motor turns/day $\times$ 29.5$\cdots$ days/month $\times$ 1 month/$(1/2)$ dial turn $=$ $2 \cdot 29.53058868/7$ motor turns per dial turn. Thus

$$\omega_2 = 8.437311051\omega_1$$

where ω_1 is the turn rate of the motor and ω_2 is the turn rate of the moon dial.

Use continued fractions to design a three-shaft gear train so that the motor can turn the dial at nearly the proper rate, say, so that the ratio is at least within 0.01 % of the amount 8.437311051 according to Theorem 5.31.

4. Show that

$$1.00273790934$$
$$= [1; 365, 4, 7, 1, 3, 47, 1, 2, 1, 4, 1, 1, 3, 1, 1, 1, 2, 2, 1, 4]$$

5. Determine using Theorem 5.31 how much p_k/q_k for $k = 0$ to 6 at most differs from the ratio 1.00273790934 and compare to the actual differences.

5.7   Application: Factoring

Theorem 2.13 gave an algorithm, called Fermat's Method, of factoring odd integers. An odd integer n is not prime if and only if $n = x^2 - y^2$, the difference of two squares. In this case n factors as $n = (x + y)(x - y)$. The method proceeds by trial and error by calculating $n + j^2$ for $j = 1, 2, 3, \ldots$ and looking for $n + j^2$ to be a perfect square, that is, the square of another integer. If $n = ab$ and $1 < b < n$ is the smallest such factor, then we obtain a perfect square when $j = (a - b)/2$. Thus, Fermat's Method will find success when two factors of n are nearly equal.

The requirement that $n = x^2 - y^2$, exactly, may be relaxed somewhat by seeking nonzero integers x, y, and z such that $nz = x^2 - y^2$, which implies that n divides $x^2 - y^2$ or, equivalently, $x^2 - y^2 \equiv 0 \pmod{n}$. This result means that any factor of n must also divide $x^2 - y^2$. In this case, certainly some divisor d of n must divide either $x + y$ or $x - y$; and if that divisor d is not 1 or n, then we have found a nontrivial divisor of n and are on the way to a complete factorization of n. The crux of the method is that once we have found particular integers x and y such that $n \mid (x^2 - y^2)$, we can quickly and easily determine whether n and, say, $x + y$ have a nontrivial common divisor by calculating $d = \gcd(x + y, n)$ using the Euclidean algorithm. Similarly, we could do the same for n and $x - y$. Thus, given that $nz = x^2 - y^2$ has positive integer solutions x, y, and z, we just compute $\gcd(n, x + y) = d_1$ and $\gcd(n, x - y) = d_2$. Both $d_1 \mid n$ and $d_2 \mid n$ and we will have at least partially factored a composite n provided that $1 < d_1 < n$ or $1 < d_2 < n$. Of course, if n is prime, d_1 and d_2 will be either 1 or n; however, just because d_1 and d_2 are both 1 or n does not mean that n is prime.

What is needed to make this modified Fermat method work is a way to generate differences of two squares that are divisible by n. The theory of continued fractions provides such a method. If $n > 0$ is not the square of an integer, Theorem 5.51 implies that

$$p_k^2 - nq_k^2 = (-1)^{k-1}e_{k+1}$$

for $k \geq 1$ where p_k and q_k give the k-th convergent p_k/q_k of the quadratic irrational $\sqrt{n}$ and e_{k+1} is given by Theorem 5.48. This equality may be rearranged as

$$p_k^2 - (-1)^{k-1}e_{k+1} = nq_k^2$$

If k is odd so that $(-1)^{k-1} = 1$ and if e_{k+1} happens to be a perfect square, say $e_{k+1} = y^2$ for some positive integer y, then

$$p_k^2 - y^2 = nq_k^2$$

so that p_k and $\sqrt{e_{k+1}}$ fulfill the roles of x and y, respectively, in the discussion above.

This method is viable because Theorem 5.48 provides an efficient recursive algorithm for generating t_i for $\sqrt{n} = [t_0; t_1, t_2, \ldots]$ and e_i and because Theorem 5.13 provides an efficient algorithm for generating p_i (note that we do not actually have to compute q_i for this application). Once we have p_k and $y = \sqrt{e_{k+1}}$ for k odd, we calculate

$$\gcd(p_k + y, n) \text{ and } \gcd(p_k - y, n)$$

looking for nontrivial divisors of n. This method of seeking factors of n is called the Continued Fraction Factoring Method.

Consider $n = 893$. Applying Theorem 5.48 and, for the last column, Theorem 5.13, we obtain

k	c_k	e_k	x_k	t_k	p_k
0	0	1	$29.8\ldots$	29	29
1	29	52	$1.1\ldots$	1	30
2	23	7	$7.5\ldots$	7	239
3	26	31	$1.8\ldots$	1	269
4	5	28	$1.2\ldots$	1	508
5	23	13	$4.0\ldots$	4	$2301 \equiv 515 \pmod{n}$
6	29	4	$14.7\ldots$	14	$32722 \equiv 574 \pmod{n}$

Since $e_6 = 4 = (2)^2$ is the square of an integer, we have $p_5^2 - nq_5^2 = (-1)^4 \cdot e_6$ or $p_5^2 - e_6 \equiv 0 \pmod{n}$. Thus,

$$x^2 - y^2 \equiv 0 \pmod{893}$$

where $x = p_5 = 2301$ and $y = \sqrt{e_4} = 2$. So $(x+y)(x-y)$ is divisible by n.

$$
\begin{aligned}
x + y &= 2301 + 2 = 2303 \\
x - y &= 2301 - 2 = 2299
\end{aligned}
$$

$$
\begin{aligned}
\gcd(x+y, n) &= \gcd(2303, 893) = 47 \\
\gcd(x-y, n) &= \gcd(2299, 893) = 19
\end{aligned}
$$

Since $n = 19 \cdot 47$ and since both 19 and 47 are prime, n factors completely as $n = 893 = 19 \cdot 47$.

We note that in this application, we only want to know $x^2 - y^2 = p_k^2 - e_{k+1}$ up to modulo $n = 893$; therefore, p_k may be reduced modulo $n = 893$ at every stage of the computation in the table above. We could have used the form $p_5^2 - e_6 \equiv 515^2 - 2^2 \equiv 0 \pmod{893}$ instead. Thus,

$$
\begin{aligned}
\gcd(515 + 2, 893) &= \gcd(517, 893) = 47 \\
\gcd(515 - 2, 893) &= \gcd(513, 893) = 19
\end{aligned}
$$

also factors n.

Not every occurrence of e_j that is a perfect square when j is even leads to a nontrivial factorization of n. For example, consider $n = 17819$. Computing e_j using Theorem 5.48 gives $e_4 = 169 = (13)^2$ and Theorem 5.51 gives $p_3 = 142565 \equiv 13 \pmod{17819}$. So

$$
\begin{aligned}
x + y &= p_3 + 13 \equiv 13 + 13 = 26 \pmod{17819} \\
x - y &= p_3 - 13 \equiv 13 - 13 = 0 \pmod{17819}
\end{aligned}
$$

$\gcd(26, n) = 1$ and $\gcd(0, n) = n$, and this method yields only the trivial divisors of n. This fact does not mean that $n = 17819$ is prime but just that the particular perfect square e_4 did not yield nontrivial factors. If we continue computing e_j for $j > 4$ we find that $e_{10} = 49 = (7)^2$ and that $p_9 \equiv 7100 \pmod{17819}$. For this case,

$$
\begin{aligned}
x + y &= p_9 + \sqrt{e_{10}} \equiv 7100 + 7 = 7112 \pmod{17819} \\
x - y &= p_9 - \sqrt{e_{10}} \equiv 7100 - 7 = 7093 \pmod{17819}
\end{aligned}
$$

$$
\begin{aligned}
\gcd(x+y, n) &= \gcd(7112, 17819) = 103 \\
\gcd(x-y, n) &= \gcd(7093, 17819) = 173
\end{aligned}
$$

Since $n = 17819 = 103 \cdot 173$ and both 103 and 173 are primes, we have found the complete prime factorization of 17819.

Exercises

1. Verify the details of factoring $n = 17819$ by the continued fraction method.

2. Apply the continued fraction method to factor these integers completely:

   (a) $n = 343097$

   (b) $n = 265043$

   (c) $n = 57671$ (Computational aid may be helpful.)

3. If, in attempting to factor n using the continued fraction method, one repeatedly produces perfect squares e_k for k even that lead only to trivial factors of n, then sometimes applying the method to $m \cdot n$ instead, where m is the product of the first few primes, say $m = 2$, $6 = 2 \cdot 3$, $30 = 2 \cdot 3 \cdot 5$, and so on, will lead to nontrivial factors of n. Show that the continued fraction method can be applied to $m \cdot n$, with m described above, to produce integers x and y such that $x^2 - y^2 \equiv 0 \pmod{mn}$ and show how nontrivial $\gcd(x+y, m \cdot n)$ and $\gcd(x-y, m \cdot n)$ may also yield nontrivial factors of n. Apply this method to $n = 17819$ discussed above.

Chapter 6
BERTRAND'S POSTULATE

6.1 Introduction

When primes were introduced in Chapter 2, we discussed their distribution among the positive integers. For example, we found that there were infinitely many primes and that there were arbitrarily long sequences of consecutive integers that are composite. Bertrand's Postulate, namely, that for each positive integer $n > 1$ there is a prime p with $n < p < 2n$, was stated but not proved. This postulate is named after Joseph Bertrand (1822-1900), who verified it for all numbers less than three million. He reportedly also used the postulate to prove another theorem. Some sources say that Chebyshev proved it in trying to prove the Prime Number Theorem. Others indicate that his work led the way to a proof. E. Landau is also given credit for the proof. The objective of this chapter is to prove Bertrand's Postulate. The proof we give depends upon properties of rational and real numbers and so we delayed the treatment of this topic until now. Some properties about rational and real numbers are discussed in Chapter 5 on continued fractions.

6.2 Preliminaries

The greatest integer function was defined in Chapter 5 as follows: If x is a real number, then $\lfloor x \rfloor = n$, where n is the unique integer such that $n \leq x < n + 1$. The greatest integer function is sometimes called the *floor function* — thus, the notation $\lfloor x \rfloor$. The *ceiling function* is defined as follows: If x is a real number, then $\lceil x \rceil = m$, where m is the unique integer such that $m - 1 < x \leq m$. In this case $\lceil x \rceil$ is the least integer that is greater than or equal to x so that this function is sometimes called the *least integer function.*

These functions are very useful and the next few theorems summarize some of their properties.

Theorem 6.1 *Given real numbers x and y,*

(a) $\lfloor x \rfloor = x$ *if and only if x is an integer.*

(b) $x - 1 < \lfloor x \rfloor \le x < \lfloor x + 1 \rfloor$.

(c) $\lfloor x + m \rfloor = \lfloor x \rfloor + m$ *for any integer m.*

(d) *If $x \le y$, then $\lfloor x \rfloor \le \lfloor y \rfloor$.*

(e) $\lfloor x \rfloor + \lfloor y \rfloor \le \lfloor x + y \rfloor \le \lfloor x \rfloor + \lfloor y \rfloor + 1$.

(f) *If $x \ge 0$ and $y \ge 0$, then $\lfloor x \rfloor \cdot \lfloor y \rfloor \le \lfloor x \cdot y \rfloor$.*

(g) $\lfloor x \rfloor + \lfloor -x \rfloor = \begin{cases} 0 & \text{if } x \text{ is an integer} \\ -1 & \text{otherwise.} \end{cases}$

(h) $\lceil x \rceil = -\lfloor -x \rfloor$.

(i) $\left\lfloor \dfrac{x}{m} \right\rfloor = \left\lfloor \dfrac{\lfloor x \rfloor}{m} \right\rfloor$ *for any positive integer m.*

Proof. Part (a) is a restatement of the definition.

(c) Let n be the integer such that $n = \lfloor x \rfloor$ so that $n \le x < n + 1$. Then $n + m \le x + m < (n + m) + 1$. Since $n + m$ is an integer, $\lfloor x + m \rfloor = n + m = \lfloor x \rfloor + m$.

(d) If $x \le y$ and n and m are the unique integers such that $n \le x < n + 1$ and $m \le y < m + 1$, then $n \le x \le y < m + 1$ so that $n < m + 1$. Thus, $n \le m$ and $\lfloor x \rfloor = n \le m = \lfloor y \rfloor$.

(g) If $\lfloor x \rfloor = n$ so that $n \le x < n + 1$, then $-n \ge -x > -n - 1$. If x is an integer, then $x = n$ and $-x = -n$ so that $\lfloor x \rfloor + \lfloor -x \rfloor = n + (-n) = 0$. If x is not an integer, then $\lfloor x \rfloor = n$ but $\lfloor -x \rfloor = -n - 1$ so that $\lfloor x \rfloor + \lfloor -x \rfloor = n + (-n - 1) = -1$.

(i) Suppose that m is a positive integer. Let $\lfloor x \rfloor = n$ so that $n \le x < n + 1$ and let $y = x - n$ so that $0 \le y < 1$. Dividing n by m using the Division Algorithm we obtain integers q and r such that $n = mq + r$ and $0 \le r < m$. Then we have

$$\frac{x}{m} = \frac{n + y}{m} = \frac{mq + r + y}{m} = q + \frac{r + y}{m}$$

But $r < m$ implies that $r + 1 \le m$ and that $r + y < r + 1 \le m$. Thus, $q \le x/m < q + 1$ and $\lfloor x/m \rfloor = q$. Next we have

$$\frac{\lfloor x \rfloor}{m} = \frac{mq + r}{m} = q + \frac{r}{m}$$

Since $0 \le r < m$, $q \le \lfloor x \rfloor / m < q + 1$ and we obtain that $\lfloor \lfloor x \rfloor / m \rfloor = q$ also.

The proofs of parts (b), (e), (f), and (h) are left to the reader. ∎

The special case of $\lfloor x \rfloor$ when x is rational is given by the next theorem whose proof is left to the reader.

Theorem 6.2

(a) *If a, b, q, and r are integers and a = bq + r with 0 ≤ r < b, then*

(i) $q = \lfloor a/b \rfloor$.

(ii) $r = a - b \cdot \lfloor a/b \rfloor$.

(b) *If m and b are positive integers, then $\lfloor m/b \rfloor$ is the number of integers in the set $\{1, 2, 3, \ldots, m\}$ that are divisible by b.*

In Theorem 6.2(b), the integers of $\{1, 2, \ldots, m\}$ that are divisible by b are exactly $\{b, 2b, 3b, \ldots, \lfloor m/b \rfloor \cdot b\}$, which are the multiples of b up to the largest multiple that is less than or equal to m.

We will shortly use factorials, $n!$. The next theorem calculates the largest power of a prime dividing $n!$.

Theorem 6.3 *Let n and p be positive integers and p be prime. The largest integer s such that $p^s \mid n!$ is*

$$s = \sum_{j \geq 1} \left\lfloor \frac{n}{p^j} \right\rfloor$$

The sum in this theorem is over all positive integers j. Since $p > 1$, there is an integer k such that $p^j > n$ for $j \geq k$ so that $\left\lfloor \dfrac{n}{p^j} \right\rfloor = 0$ for $j \geq k$. Thus, the sum always has only a finite number of nonzero terms.

Proof. Let p be prime and n be a positive integer. Since $p > 1$, clearly there will always be a largest integer s such that $p^s \mid n!$. If $n = 1$, then $1! = 1$ and $p^0 = 1$ so that $p^0 \mid 1!$. But $p^j > 1$ for $j > 0$ so that $s = 0$ and $\left\lfloor \dfrac{1}{p^j} \right\rfloor = 0$ for $j \geq 1$. Thus, the sum in the theorem equals zero.

Assume that the conclusion of the theorem is true for $n = m \geq 1$. Let t be the largest positive integer such that $p^t \mid (m+1)$ so that, by the induction hypothesis, the largest integer s such that p^s divides $(m+1)! = (m+1) \cdot m!$ is

$$s = t + \sum_{j \geq 1} \left\lfloor \frac{m}{p^j} \right\rfloor$$

We need to show that this last expression equals

$$\sum_{j \geq 1} \left\lfloor \frac{(m+1)}{p^j} \right\rfloor$$

That is, we need to show that

$$\begin{aligned}
t &= \sum_{j \geq 1} \left\lfloor \frac{(m+1)}{p^j} \right\rfloor - \sum_{j \geq 1} \left\lfloor \frac{m}{p^j} \right\rfloor \\
&= \sum_{j \geq 1} \left[\left\lfloor \frac{(m+1)}{p^j} \right\rfloor - \left\lfloor \frac{m}{p^j} \right\rfloor \right]
\end{aligned}$$

For each j, we have

$$
\left\lfloor \frac{(m+1)}{p^j} \right\rfloor =
\begin{cases}
\left\lfloor \dfrac{m}{p^j} \right\rfloor + 1 & \text{if } p^j \mid (m+1) \\[3mm]
\left\lfloor \dfrac{m}{p^j} \right\rfloor & \text{if } p^j \nmid (m+1)
\end{cases}
$$

Since, if $p^j \mid (m+1)$, then $m+1 = J \cdot p^j$ for some integer J and $m = (J-1)p^j + (p^j - 1)$ so that $\left\lfloor \dfrac{(m+1)}{p^j} \right\rfloor = J$ and, by Theorem 6.2, $\left\lfloor \dfrac{m}{p^j} \right\rfloor = J - 1$. On the other hand, if $p^j \nmid (m+1)$, there are integers L and r such that $m+1 = Lp^j + r$ and $1 \le r < p^j$. But this last equation implies that $m = Lp^j + (r - 1)$ and $0 \le r - 1 < p^j$. So, by Theorem 6.2, $\left\lfloor \dfrac{(m+1)}{p^j} \right\rfloor = L = \left\lfloor \dfrac{m}{p^j} \right\rfloor$. Thus

$$
t = \sum_{j \ge 1} \left[\left\lfloor \frac{(m+1)}{p^j} \right\rfloor - \left\lfloor \frac{m}{p^j} \right\rfloor \right]
$$

because $\left\lfloor \dfrac{(m+1)}{p^j} \right\rfloor - \left\lfloor \dfrac{m}{p^j} \right\rfloor = 1$ for $j = 1, 2, \ldots, t$ and zero otherwise. $\blacksquare$

Example 6.4 *Suppose that we want to find the highest power of the prime $p = 11$ that divides 8335!. According to Theorem 6.3, we should evaluate*

$$
\left\lfloor \frac{8335}{11} \right\rfloor = 757
$$

$$
\left\lfloor \frac{8335}{11^2} \right\rfloor = 68
$$

$$
\left\lfloor \frac{8335}{11^3} \right\rfloor = 6
$$

$$
\left\lfloor \frac{8335}{11^4} \right\rfloor = 0
$$

Thus $757 + 68 + 6 = 831$, $11^{831} \mid 8335!$, and $11^{832} \nmid 8335!$.

Part (i) of Theorem 6.1 can conveniently be used to reduce the complexity of the above calculation because

$$
\left\lfloor \frac{n}{p^j} \right\rfloor = \left\lfloor \frac{\left(\dfrac{n}{p^{j-1}} \right)}{p} \right\rfloor = \left\lfloor \frac{\left\lfloor \dfrac{n}{p^{j-1}} \right\rfloor}{p} \right\rfloor
$$

Thus, instead of the method above, we have

$$\left\lfloor \frac{8335}{11^2} \right\rfloor = \left\lfloor \frac{\left(\dfrac{8335}{11} \right)}{11} \right\rfloor = \left\lfloor \frac{\left\lfloor \dfrac{8335}{11} \right\rfloor}{11} \right\rfloor = \left\lfloor \frac{757}{11} \right\rfloor = 68$$

$$\left\lfloor \frac{8335}{11^3} \right\rfloor = \left\lfloor \frac{\left(\dfrac{8335}{11^2} \right)}{11} \right\rfloor = \left\lfloor \frac{\left\lfloor \dfrac{8335}{11^2} \right\rfloor}{11} \right\rfloor = \left\lfloor \frac{68}{11} \right\rfloor = 6$$

and so on.

The following theorem about the binomial coefficients was given in the Exercises of Chapter 1 and is repeated here for easy reference.

Theorem 6.5 *If n and k are nonnegative integers and $0 \le k \le n$, the binomial coefficient is given by*

$$\binom{n}{k} = \frac{n!}{k!(n-k)!}$$

and, if $k > n$, then $\binom{n}{k} = 0$. The following properties hold:

(a) $\binom{n}{k} = \binom{n}{n-k}$ *for all k, $0 \le k \le n$.*

(b) $\binom{n}{k-1} + \binom{n}{k} = \binom{n+1}{k}$ *for all k, $1 \le k \le n$.*

(c) $\binom{n}{k}$ *is a positive integer for $0 \le k \le n$.*

(d)

$$\binom{n}{k} \le \begin{cases} \dbinom{n}{n/2} & \text{if } n \text{ is even} \\[2ex] \dbinom{n}{(n-1)/2} = \dbinom{n}{(n+1)/2} & \text{if } n \text{ is odd} \end{cases}$$

where the inequality is strict unless $k = n/2$ in the first case and unless $k = (n-1)/2$ or $k = (n+1)/2$ in the second case.

In the Exercises of Section 1.3, the Binomial Theorem was proved, namely,

$$(a+b)^n = \sum_{k=0}^{n} \binom{n}{k} a^{n-k} b^k$$

for integers a and b. By the Binomial Theorem,

$$2^n = (1+1)^n = \sum_{k=0}^{n} \binom{n}{k}$$

$$= \binom{n}{0} + \binom{n}{1} + \binom{n}{2} + \cdots + \binom{n}{n-1} + \binom{n}{n}$$

Thus, if $n > 1$, we obtain

$$2^n > \binom{n}{k} + \binom{n}{n-k} = 2\binom{n}{k}$$

for any k, $0 \le k \le n$. Consequently, $\binom{n}{k} < 2^{n-1}$ for $n > 1$ and any k, $0 \le k \le n$. Thus, the binomial coefficients are bounded above by a simple power of 2. This formula and the other characteristics of the binomial coefficients will be useful in the remainder of this section.

The next theorem shows that the product of consecutive primes which are no larger than a real number x is also bounded by a simple power of 4. Thus, if $x \ge 2$ and $p_1 < p_2 < \cdots < p_s$ are the primes that are less than or equal to x, we will show that $p_1 p_2 \cdots p_s < 4^x = 2^{2x}$.

Theorem 6.6 *If x is a real number, $x \ge 2$, and p indicates a prime, then*

$$\prod_{p \le x} p < 4^x$$

where the product on the left has one factor for each prime $p \le x$.

Proof. The inequality of the theorem holds if $2 \le x < 3$ since

$$\prod_{p \le x} p = 2 < 4^2 \le 4^x$$

We now consider the case of $x \ge 3$. Although x is real, it is sufficient to prove the inequality of the theorem for only odd integers; for if $n \ge 3$ is odd and $\prod_{p \le n} p < 4^n$ holds and if $n \le x < n+2$ so that x is between two consecutive odd integers, then

$$\prod_{p \le x} p = \prod_{p \le n} p < 4^n \le 4^x$$

The proof will now proceed by induction over the odd integers.

If $n = 3$, then $\prod_{p \le 3} p = 2 \cdot 3 < 4^3$. Suppose that $n > 3$ is odd and that $\prod_{p \le m} p < 4^m$ for any odd m with $3 \le m < n$. We will show that $\prod_{p \le n} p < 4^n$.

Since n is odd, either $n \equiv 1 \pmod 4$ or $n \equiv 3 \pmod 4$. If $n \equiv 1 \pmod 4$, let

$s = (n+1)/2$; otherwise, let $s = (n-1)/2$. These choices ensure that s is odd and that $s \geq 3$ because $n > 3$.

If we can show that any prime p with $s < p \leq n$ has the property that

$$p \mid \binom{n}{s}$$

then, certainly,

$$\prod_{s<p\leq n} p \leq \binom{n}{s}$$

Therefore, by the induction hypothesis,

$$\prod_{p\leq n} p = \left[\prod_{p\leq s} p\right]\left[\prod_{s<p\leq n} p\right] < 4^s \binom{n}{s} < 4^s \cdot 2^{n-1} = 2^{2s+n-1} \leq 2^{2n} = 4^n$$

Therefore, by induction, the inequality of the theorem holds for odd $n \geq 3$; and by virtue of the discussion above, the inequality of the theorem also hold for all real $x \geq 2$.

To complete the proof, we have only to establish that $p \mid \binom{n}{s}$ for any prime p with $s < p \leq n$. First we have $p \nmid (s!)$ because $p > s$.

Second, note that

$$n - s = \begin{cases} s - 1 & \text{if } s = (n+1)/2 \\ s + 1 & \text{if } s = (n-1)/2 \end{cases}$$

so that in either case $n - s \leq s + 1$. Since $p > s$, $p \geq s + 1$; but $p \neq s + 1$ because p is odd and $s + 1$ is even. Therefore, $p \nmid (n - s)!$.

Third, $p \mid (n!)$ because $p \leq n$. Therefore, since $p \mid n!$, $p \nmid s!$, and $p \nmid (n-s)!$,

$$p \mid \frac{n!}{s!(n-s)!}$$

and the proof is complete. ■

Exercises

1. Evaluate:

    (a) $\lfloor 19/5 \rfloor$ (b) $\lceil 19/5 \rceil$ (c) $\lfloor -29/7 \rfloor$ (d) $\lceil -29/7 \rceil$

2. Prove Theorem 6.1(c).

3. Prove Theorem 6.1(e).

4. Prove Theorem 6.1(f).

5. Prove Theorem 6.1(h).

6. Prove Theorem 6.2.

7. Prove the following statements:

   (a) If x is a real number, then $\lfloor x \rfloor + \lfloor x + 1/2 \rfloor = \lfloor 2x \rfloor$.

   (b) If x and y are real numbers, then $\lfloor x - y \rfloor \leq \lfloor x \rfloor - \lfloor y \rfloor \leq \lfloor x - y \rfloor + 1$.

8. Let x and y be real numbers. Prove that:

   (a) $\lceil x \rceil = x$ if and only if x is an integer.

   (b) $\lceil x - 1 \rceil < x \leq \lceil x \rceil < x + 1$.

   (c) $\lceil x + m \rceil = \lceil x \rceil + m$ for any integer m.

   (d) If $x \leq y$, then $\lceil x \rceil \leq \lceil y \rceil$.

   (e) $\lceil x \rceil + \lceil y \rceil \geq \lceil x + y \rceil \geq \lceil x \rceil + \lceil y \rceil - 1$.

   (f) $\lceil x \rceil + \lceil -x \rceil = \begin{cases} 0 & \text{if } x \text{ is an integer} \\ 1 & \text{if } x \text{ is not an integer.} \end{cases}$

9. For a real number x, prove that $\lfloor x + 1/2 \rfloor$ is the integer nearest to x unless x is equidistant from two integers when $\lfloor x + 1/2 \rfloor$ is the larger of the two integers.

10. Find the largest power of 7 that divides 10000!.

11. Find the number of trailing zeros in the base ten representation of 10000!.

12. Compute using the definition of binomial coefficient:

   (a) $\dbinom{50}{0}$ (b) $\dbinom{50}{50}$ (c) $\dbinom{50}{1}$

   (d) $\dbinom{50}{2}$ (e) $\dbinom{50}{3}$ (f) $\dbinom{50}{4}$

13. Compute using the definition of binomial coefficient:

   (a) $\dbinom{4}{k}$ for $0 \leq k \leq 4$

   (b) $\dbinom{5}{k}$ for $0 \leq k \leq 5$

   (c) Verify Theorem 6.5(a) to (d) for parts (a) and (b) of this exercise.

14. The following arrangement of binomial coefficients is called Pascal's Tri-

angle.

$$\binom{0}{0}$$

$$\binom{1}{0} \qquad \binom{1}{1}$$

$$\binom{2}{0} \qquad \binom{2}{1} \qquad \binom{2}{2}$$

$$\binom{3}{0} \qquad \binom{3}{1} \qquad \binom{3}{2} \qquad \binom{3}{3}$$

$$\binom{4}{0} \qquad \binom{4}{1} \qquad \binom{4}{2} \qquad \binom{4}{3} \qquad \binom{4}{4}$$

or

```
            1
         1     1
      1     2     1
   1     3     3     1
1     4     6     4     1
```

(a) Each row may be constructed from the previous row using Theorem 6.5(b) and the fact that $\binom{n}{0} = 1$ for $n \geq 0$. Use this theorem to continue the construction of Pascal's Triangle through the row of $\binom{n}{k}$ with $n = 10$. Compare your results with those of the Exercise 13.

(b) Verify that the row for $n = 4$ gives the coefficients of $x^j y^k$ when $(x + y)^4$ is "multiplied out."

15. Use Pascal's Triangle to help write out all the terms in the expansion of these expressions:

   (a) $(x + y)^6$ (b) $(2a + 3b)^6$ (c) $(3a - 2b)^6$ (d) $(4 + 5w)^8$

16. Prove the following for $n \geq 1$:

   (a) $\displaystyle\sum_{k=0}^{n} (-1)^k \binom{n}{k} = 0$

   (b) $\displaystyle\sum_{j=0}^{\lfloor n/2 \rfloor} \binom{n}{2j} = \sum_{k=0,\ k\ \text{even}} \binom{n}{k} = 2^{n-1}$

   (c) $\displaystyle\sum_{j=0}^{\lfloor (n-1)/2 \rfloor} \binom{n}{2j+1} = \sum_{k=0,\ k\ \text{odd}} \binom{n}{k} = 2^{n-1}$

17. Verify that the equations in Exercise 16 hold for the rows of binomial coefficients obtained in Exercise 14.

18. Verify that Theorem 6.5(a) to (d) hold for the rows of binomial coefficients obtained in Exercise 14.

19. Prove that $\binom{n}{k}\binom{k}{j} = \binom{n}{j}\binom{n-j}{k-j}$.

20. Prove that $\displaystyle\sum_{j=0}^{n}\binom{k+j}{k} = \binom{n+1}{k+1}$.

21. For $n! = m^k$ and $m > 1$, prove that $k = 1$.

22. Find the highest power s of a prime p such that p^s divides the product of the first n even positive integers: $2 \cdot 4 \cdot 6 \cdots (2n)$.

23. Find the highest power s of a prime p such that p^s divides the product of the first n odd positive integers $1 \cdot 3 \cdot 5 \cdots (2n - 1)$.

24. If n is a positive integer, then prove that the following formula holds:

$$n! = \prod_{p \text{ prime}} p^{\sum_{k=1}^{} \lfloor n/p^k \rfloor}$$

where there is one factor in the product for every prime.

25. A generalization of Theorem 6.3 is given in this exercise.

    (a) For a positive integer n, a prime p, and s the largest exponent of p such that $p^s \mid n!$, let

    $$n = a_m p^m + a_{m-1}p^{m-1} + \cdots + a_1 p + a_0 = [a_m a_{m-1} \cdots a_1 a_0]_p$$

    be the base p representation of n so that $0 \le a_i < p$ and $a_m > 0$. Let $d_p(n) = \displaystyle\sum_{i=0}^{m} a_i$ be the sum of the digits in the base p representation of n. Prove that

    $$s = \frac{\lfloor n - d_p(n) \rfloor}{p - 1}$$

    (b) Use the formula of part (a) to calculate the largest exponent s of 11 such that $11^s \mid 8335!$. (Compare Example 6.4.)

    (c) Prove that if k is the largest exponent of p such that $p^k \mid \binom{n}{r}$, then

    $$k = \frac{d_p(r) + d_p(n - r) - d_p(n)}{p - 1}$$

394 Chapter 6 BERTRAND'S POSTULATE

6.3 Bertrand's Postulate

Theorem 6.7 (Bertrand's Postulate) *For each positive integer $n > 1$ there is a prime p such that $n < p < 2n$.*

Proof. For each integer n with $2 \leq n \leq 127$ one can easily find a prime p with $n < p < 2n$ by inspection. For example, for $n = 2$, $p = 3$ will do. For $3 \leq n \leq 4$, $p = 5$ will do. Continuing in this manner, we obtain

Range of n	p
$n = 2$	3
$3 \leq n \leq 4$	5
$5 \leq n \leq 6$	7
$7 \leq n \leq 10$	11
$11 \leq n \leq 12$	13
$13 \leq n \leq 22$	23
$23 \leq n \leq 42$	43
$43 \leq n \leq 82$	83
$83 \leq n \leq 127$	131

At this point we need only to show that the result of the theorem holds for $n \geq 128$. Suppose not; that is, suppose that there is some $n \geq 128$ for which there is no prime p such that $n < p < 2n$. For this n, consider the binomial coefficient

$$\binom{2n}{n} = \frac{(2n)!}{n!n!}$$

where we will be interested in its prime factorization: $\binom{2n}{n} = p_1^{\alpha_1} p_2^{\alpha_2} \cdots p_k^{\alpha_k}$

in which α_i is the largest power of p_i such that $p_i^{\alpha_i} \left| \binom{2n}{n} \right.$. We will proceed more generally as follows: For each prime p, let s_p be the largest power of p such that

$$p^{s_p} \left| \binom{2n}{n} \right.$$

Of course, if p is one of the prime factors of $\binom{2n}{n}$, say $p = p_i$, then $s_p = \alpha_i$ and otherwise, $s_p = 0$. Also, for each prime p, let t_p be the unique integer such that

$$p^{t_p} \leq 2n < p^{t_p+1}$$

Theorem 6.3 implies that

$$s_p = \sum_{j \geq 1} \left(\left\lfloor \frac{2n}{p^j} \right\rfloor - 2 \left\lfloor \frac{n}{p^j} \right\rfloor \right) \tag{*}$$

because p with the largest exponent that divides $\binom{2n}{n}$ is p with the largest power that divides $(2n)!$ divided by the square of p with the largest power that divides $n!$. But

$$\left\lfloor \frac{2n}{p^j} \right\rfloor - 2\left\lfloor \frac{n}{p^j} \right\rfloor \begin{cases} = & 0 \quad \text{if } j > t_p \\ < & 2 \quad \text{if } 1 \le j \le t_p \end{cases}$$

The second case is true using Theorem 6.1(b) since $\left\lfloor \dfrac{2n}{p^j} \right\rfloor \le \dfrac{2n}{p^j}$ and $\left\lfloor \dfrac{n}{p^j} \right\rfloor >$
$\dfrac{n}{p^j} - 1$. Clearly,

$$\left\lfloor \frac{2n}{p^j} \right\rfloor - 2\left\lfloor \frac{n}{p^j} \right\rfloor \le 1 \text{ for all } j \ge 1$$

and when applied to the expression (*) gives

$$s_p \le \sum_{j=1}^{t_p} 1 = t_p$$

Also,

$$\binom{2n}{n} = \prod_{p \le 2n} p^{s_p} = \prod_{p \le n} p^{s_p}$$

where the products are over the primes p satisfying the conditions given. The first equality holds because if $p = p_i$ for some i, then $\alpha_i = s_p$; and if $p \ne p_i$ for all i, then $s_p = 0$. The second equality holds because, by assumption, there is no prime p with $n < p < 2n$.

We next divide the factors of $\prod_{p \le n} p^{s_p}$ into three disjoint groups:

$$p \le \sqrt{2n}, \quad \sqrt{2n} < p \le \frac{2n}{3}, \quad \text{and} \quad \frac{2n}{3} < p \le n$$

Case 1. Let $p \le \sqrt{2n}$. Then

$$p^{s_p} \le p^{t_p} \le 2n$$

The first inequality holds because $s_p \le t_p$ and the second holds by definition of t_p.

Case 2. Let $\sqrt{2n} < p \le \dfrac{2n}{3}$. Then $p < 2n < p^2$ so that $t_p = 1$ and $s_p \le t_p = 1$.

Case 3. Let $\dfrac{2n}{3} < p \le n$. Then $2np/3 < p^2 \le np$. Since $p \ge 3$, we obtain $p^2 > 2np/3 \ge 2n$. Also $(2n)/(3p) < 1 \le n/p$ so that $1 \le n/p < 3/2$ or equivalently, $2 \le 2n/p < 3$. In this case, using the expression (*) above for s_p, we obtain

$$\begin{aligned} s_p &= \sum_{j \ge 1} \left(\left\lfloor \frac{2n}{p^j} \right\rfloor - 2\left\lfloor \frac{n}{p^j} \right\rfloor \right) \\ &= \left\lfloor \frac{2n}{p^1} \right\rfloor - 2\left\lfloor \frac{n}{p^1} \right\rfloor = 2 - 2(1) = 0 \end{aligned}$$

Partitioning the product, we get

$$\binom{2n}{n} = \prod_{p \le n} p^{s_p}$$

$$= \left[\prod_{p \le \sqrt{2n}} p^{s_p} \right] \left[\prod_{\sqrt{2n} < p \le 2n/3} p^{s_p} \right] \left[\prod_{2n/3 < p \le n} p^{s_p} \right]$$

$$\le \left[\prod_{p \le \sqrt{2n}} (2n) \right] \left[\prod_{\sqrt{2n} < p \le 2n/3} p \right] \left[\prod_{2n/3 < p \le n} 1 \right]$$

$$\le \left[\prod_{p \le \sqrt{2n}} (2n) \right] \left[\prod_{p \le 2n/3} p \right] \text{ since } p > 1$$

Theorem 6.6 implies for the second product that

$$\prod_{p \le 2n/3} p < 4^{2n/3}$$

In the first product, every factor is the same, so we only have to determine the number of factors in the product or, equivalently, the number of primes less than or equal to $\sqrt{2n}$.

We will show that if $n \ge 128$, then the number of primes less than or equal to $\sqrt{2n}$ is no more than $\sqrt{n/2} - 1$. Recall that $\pi(x)$ is the number of primes less than or equal to x. For $x \ge 16$, the number of primes less than or equal to x does not exceed the number of odd positive integers less than or equal to x. In this case, the one even prime, 2, that is not included by this accounting is matched by the odd nonprime 15 that is counted. Thus

$$\pi(x) \le \frac{x+1}{2}$$

because the k-th odd positive integer is $2k - 1$. Since the odd integers 1 and 9 are also nonprimes less than or equal to 16, we could also write for any $x \ge 16$

$$\pi(x) \le \frac{x+1}{2} - 2 = \frac{x-3}{2} < \frac{x}{2} - 1$$

Since for $n \ge 128$, we have $\sqrt{2n} \ge \sqrt{2 \cdot 128} = \sqrt{256} = 16$ and

$$\pi\left(\sqrt{2n}\right) \le \frac{\sqrt{2n}}{2} - 1$$

then there are at most $\dfrac{\sqrt{2n}}{2} - 1$ factors in the first product. Thus,

$$\binom{2n}{n} \le \left[\prod_{p \le \sqrt{2n}} (2n) \right] \left[\prod_{p \le 2n/3} p \right] < (2n)^{(\sqrt{2n})/2 - 1} 4^{2n/3}$$

This inequality gives an upper bound for $\binom{2n}{n}$. We proceed to find a lower bound. The Binomial Theorem implies

$$2^{2n} = (1+1)^{2n} = \sum_{k=0}^{2n} \binom{2n}{k}$$

By Theorem 6.5, the largest of the $2n+1$ terms in the rightmost summation is $\binom{2n}{n}$ so that

$$2^{2n} \leq (2n)\binom{2n}{n}$$

since the sum of the first and last terms is $1 + 1 = 2 \leq \binom{2n}{n}$. Thus

$$\frac{2^{2n}}{2n} \leq \binom{2n}{n} < (2n)^{(\sqrt{2n})/2 - 1} 4^{2n/3}$$

which implies that

$$2^{2n/3} < (2n)^{\sqrt{n/2}}$$

Applying logarithms to both sides, we get

$$\sqrt{8n} \cdot \log(2) - 3 \cdot \log(2n) < 0 \qquad\qquad (**)$$

Consider the function

$$f(x) = \sqrt{8x} \cdot \log(2) - 3 \cdot \log(2x)$$

defined for real numbers $x > 0$. The derivative of $f(x)$ is

$$f'(x) = \frac{\sqrt{2x} \cdot \log(2) - 3}{x}$$

If $x \geq 128$, $f'(x) \geq (16 \cdot \log(2) - 3)/x > 0$ and by a well-known theorem of calculus, $f(x)$ is increasing for $x \geq 128$. But $f(128) = 8 \cdot \log(2) > 0$. Since $f(128) > 0$ and $f(n)$ is increasing for $n \geq 128$, the quantity

$$f(n) = \sqrt{8n} \cdot \log(2) - 3 \cdot \log(2n)$$

is never negative if $n \geq 128$. This last statement is a contradiction of inequality (**). Therefore, the assumption that there is no prime p with $n < p < 2n$ is false. ∎

Exercises

1. Let n be a positive integer, p be a prime, and t_p be the unique positive integer such that $p^{t_p} \leq 2n < p^{t_p+1}$. Show that $t_p = \left\lfloor \dfrac{\log(n)}{\log(p)} \right\rfloor$, where log is the natural logarithm.

2. For a real number x greater than one, prove that there is a prime p such that $x < p < 2x$.

Chapter 7
DIOPHANTINE
EQUATIONS

7.1 Linear Diophantine Equations

Definition 7.1 *A* Diophantine equation *is an equation such that only solutions in which all of the variables have integral values are considered. A* linear Diophantine equation *is a Diophantine equation of the form*

$$a_1 x_1 + a_2 x_2 + \cdots + a_n x_n = c$$

where the a_i and c are constants and the x_i are variables all raised to the first power.

Hence $ax^2 + bx = 2$ is not a linear Diophantine equation while $ax + by + cz = 3$ is linear and has a solution (x_0, y_0, z_0) only if x_0, y_0, and z_0 are integers such that $ax_0 + by_0 + cz_0 = 3$.

We have already shown that a Diophantine equation of the form $ax + by = c$ has solutions if and only if $\gcd(a, b)$ divides c. We showed in Chapter 1 how to find the solutions. We generalize the procedure in the next theorem.

Theorem 7.2 *The linear Diophantine equation $a_1 x_1 + a_2 x_2 + \cdots + a_n x_n = c$ such that $a_j \neq 0$ for some j has a solution if and only if $\gcd(a_1, a_2, \ldots, a_n)$ divides c.*

Proof. Assume that the equation $a_1 x_1 + a_2 x_2 + \cdots + a_n x_n = c$ has a solution. Since $\gcd(a_1, a_2, \ldots, a_n)$ divides a_i for all i, then $\gcd(a_1, a_2, \ldots, a_n)$ divides c. To prove the converse it is sufficient to prove that there exist $u_1, u_2, \ldots, u_n$ such that $a_1 u_1 + a_2 u_2 + \cdots + a_n u_n = \gcd(a_1, a_2, \ldots, a_n)$, which follows from Theorem 1.43. ∎

Example 7.3 *Consider the equation* $3x + 4y + 2z = 5$. *Since* $\gcd(3, 4, 2) = 1$, *there exist* u, v, *and* w *so that* $3u + 4v + 2w = 1$. *One example is* $u = w = 1$ *and* $v = -1$. *Hence* $3(1) + 4(-1) + 2(1) = 1$ *so that* $3(5) + 4(-5) + 2(5) = 5$.

Example 7.4 *Consider the equation* $252x + 576y + 42z = 18$. *In Example 1.39 it was shown that* $\gcd(252, 576) = 36$ *and that* $36 = (7)(252) + (-3)(576)$. *To find* $\gcd(36, 42)$, *we note that*

$$
\begin{aligned}
42 &= 36 \cdot 1 + 6 \\
36 &= 6 \cdot 6 + 0
\end{aligned}
$$

so that $\gcd(36, 42)$ *is* 6 *and*

$$
\begin{aligned}
6 &= (1)(42) + (-1)(36) \\
&= (1)(42) + (-1)((7)(252) + (-3)(576))
\end{aligned}
$$

Finally, a solution $x = -7$, $y = 3$, *and* $z = 1$ *is obtained.*

Diophantine equations are named after Diophantus of Alexandria, a mathematician in the third century A.D. He was probably a Hellenized Babylonian. It is ironic that Diophantine equations require integral solutions, while Diophantus accepted rational solutions. Indian mathematicians around A.D. 1100 accepted only integral solutions. Diophantus had no concept of negative numbers, so his solutions were restricted. He developed symbols for mathematics operations in the Greek language and had the first systematic notation to represent an unknown. He solved specific problems to get specific solutions rather than proving general theorems. Many of these problems involved solving equations having the form of Pell's equation. He restricted his mathematics to quadratic equations. Diophantus is considered the last of the great mathematicians of the Alexandrian School. His life is described by a puzzle: His boyhood was 1/6 of his life. His beard grew after 1/12 more. He married after 1/7 more. Five years later his son was born and lived to half the father's age. The father died four years after the son. Solving, one finds among other things that Diophantus died at age 84.

Diophantus wrote three books. His earliest and greatest work was *Arithmetica*, which was devoted to algebra. Only 6 of the 13 books of *Arithmetica* survive. *Arithmetica* became accessible to European mathematicians in 1621, when Claude Bachet published the Greek text along with the Latin translation. The other two works were *De Polygonis Numeris*, which also has a portion missing, and *Porisms*, a collection of propositions.

Exercises

1. Find an integral solution for each of the following linear Diophantine equations:

   (a) $9x + 12y + 16z = 13$

(b) $15x + 12y + 30z = 24$

(c) $15x + 9y + z = 300$

(d) $w - 2x + 3y + 5z = 25$

2. Find an integral solution for each of the following linear Diophantine equations:

   (a) $3x + 2y + z = 1$

   (b) $3x + 2y + z = 13$

   (c) $x + 5y + 7z = 14$

   (d) $2w + 3x - 2y - 7z = 16$

3. Using the method for finding all solutions for linear Diophantine equation in two variables, determine how to find all solutions to linear Diophantine equation in three variables.

4. Find all integral solutions for the linear Diophantine equations in Exercise 1.

5. Find all integral solutions for the linear Diophantine equations in Exercise 2.

6. Find all integral solutions for these Diophantine equations:

   (a) $6x + 15y + 22z = 38$

   (b) $25w - 35x + 14y + 11z = 101$

   (c) $15x + 35y + 6z = 105$

7.2   Pythagorean Triples

In this section solutions are found for equations of the form $x^2 + y^2 = z^2$, where x, y, and z are positive integers. Triples x, y, and z which have this property are called *Pythagorean triples*. Historically, this equation was of interest because of the Pythagorean Theorem, which states that in a right triangle the square of the hypotenuse is equal to the sum of the squares of the other two sides. With Pythagorean triples all three sides of the triangle have integral length. We note that there are infinitely many solutions to $x^2 + y^2 = z^2$ if $x = 0$ or $y = 0$; however, such solutions are considered trivial and correspond to degenerate triangles.

Lemma 7.5 *If $a^2 + b^2 = c^2$ for positive integers a, b, and c, then $\gcd(a, b) = \gcd(b, c) = \gcd(a, b, c)$.*

Proof. If $d = \gcd(a, b)$, then d^2 divides a^2 and d^2 divides b^2. Hence d^2 divides c^2 and d divides c so $\gcd(a, b)$ divides $\gcd(b, c)$. Similarly, $\gcd(b, c)$ divides $\gcd(a, b)$ and they are equal. From this it follows that $\gcd(a, b) = \gcd(a, b, c)$. ∎

If the equation $a^2 + b^2 = c^2$ for positive integers a, b, and c is divided by d^2 where $d = \gcd(a, b, c)$, then we have $(a')^2 + (b')^2 = (c')^2$ where $a' = a/d$, $b' = b/d$, and $c' = c/d$ are relatively prime in pairs. The triple (a', b', c') of relatively prime positive integers is then said to be a *primitive solution* of $x^2 + y^2 = z^2$. Obviously, every positive integer solution of $x^2 + y^2 = z^2$ is of the form $(da')^2 + (db')^2 = (dc')^2$ where (a', b', c') is a primitive solution of $x^2 + y^2 = z^2$. Hence to find positive integer solutions to the equation $x^2 + y^2 = z^2$, one need only find primitive solutions.

Theorem 7.6 *For any integer a, a^2 is congruent to 0 or 1 modulo 4. Further, for any integers a and b, $a^2 + b^2$ is congruent to 0, 1, or 2 modulo 4.*

Proof. If a is an integer, then a is even or odd. If a is even, then $a^2 \equiv 0 \pmod 4$. If a is odd, then $a \equiv 1 \pmod 4$ or $a \equiv 3 \pmod 4$. Then $a^2 \equiv 1 \cdot 1 = 1 \pmod 4$ or $a^2 \equiv 3 \cdot 3 = 9 \equiv 1 \pmod 4$. For integers a and b, we have, using the first part of this theorem, the four cases $a^2 + b^2 \equiv 0 + 0, 0 + 1, 1 + 0$, or $1 + 1 \pmod 4$; therefore, $a^2 + b^2 \equiv 0, 1,$ or $2 \pmod 4$. ∎

Corollary 7.7 *No positive integer a such that $a \equiv 3 \pmod 4$ is the sum of two squares.*

Assume that (a, b, c) is a primitive solution of $x^2 + y^2 = z^2$ so that $\gcd(a, b, c) = 1$. If c were even, then $c^2 \equiv 0 \pmod 4$. In order for $c^2 = a^2 + b^2 \equiv 0 \pmod 4$, Theorem 7.6 implies that both $a^2 \equiv 0 \pmod 4$ and $b^2 \equiv 0 \pmod 4$; therefore, both a and b must be even, a contradiction. Thus c is odd and only one of a and b is even. This result means that the parities assumed for a, b, and c in the next theorem are not restrictive.

Theorem 7.8 *Assume that a is even and both b and c are odd. Then (a, b, c) is a primitive solution of $x^2 + y^2 = z^2$ if and only if $a = 2rs$, $b = r^2 - s^2$ and $c = r^2 + s^2$ where r and s are relatively prime positive integers with $r > s$ and with $r + s$ odd.*

Proof. Assume (a, b, c) is a primitive solution. Then

$$\left(\frac{c+b}{2}\right)\left(\frac{c-b}{2}\right) = \frac{c^2 - b^2}{4} = \frac{a^2}{4} = \left(\frac{a}{2}\right)^2$$

$(c - b)/2$ and $(c + b)/2$ are relatively prime; for if d divides $(c + b)/2$ and d divides $(c - b)/2$, then d divides their sum c and their difference b. Thus, $d = 1$ because b and c are relatively prime. Since the product of $(c + b)/2$ and

$(c - b)/2$ is a square and the two are relatively prime, both of them must be squares, say $(c + b)/2 = r^2$ and $(c - b)/2 = s^2$. Hence $r^2 > s^2$ and $r > s > 0$. Since r^2 and s^2 are relatively prime, so are r and s. It is easy to show that $a = 2rs$, $b = r^2 - s^2$, and $c = r^2 + s^2$.

Conversely, let $a = 2rs$, $b = r^2 - s^2$, and $c = r^2 + s^2$ where $r > s > 0$, $r + s$ is odd, and $\gcd(r, s) = 1$. Thus, $a^2 + b^2 = 4r^2s^2 + r^4 - 2r^2s^2 + s^4 = r^4 + 2r^2s^2 + s^4 = c^2$, $c - b = 2s^2$, and $c + b = 2r^2$. Hence $\gcd(c - b, c + b) = 2$. If d divides b and c, then d divides $c - b$ and $c + b$ and hence d divides 2. But b and c are both odd. Hence $d = 1$, $\gcd(b, c) = 1$, and (a, b, c) is a primitive solution. ∎

For example, if $r = 8$ and $s = 3$, let $a = 2rs = 48$, $b = r^2 - s^2 = 55$, and $r^2 + s^2 = 73$. Since $48^2 + 55^2 = 5329 = 73^2$ and $\gcd(48, 55, 73) = 1$, the triple $(48, 55, 73)$ is a primitive solution of $x^2 + y^2 = z^2$.

There has been interest in finding integral solutions to $x^2 + y^2 = z^2$ for a long time. A Babylonian clay tablet (Plimpton 322) written in cuneiform was translated by O. Neugebauer. It contained columns of cuneiform numerals which formed lists of Pythagorean triples. Such interest may have been associated with the relationship between Pythagorean triples and right triangles given by the Pythagorean Theorem. A knotted rope formed into a triangle having x knotted sections on one side and y and z sections on the other two would provide a means of accurately laying out parcels of land with right angles. See B. L. van der Waerden's, *Science Awakening* [88], O. Neugebauer's *The Exact Sciences in Antiquity* [53], and Neugebauer and Sach's *Mathematical Cuneiform Texts* [54] for early uses of mathematics.

Exercises

1. Find a primitive Pythagorean triple (a, b, c) with:

    (a) $r = 8$ and $s = 7$

    (b) $r = 5$ and $s = 4$

2. Use Theorem 7.8 for the following:

    (a) Generate all primitive Pythagorean triples (a, b, c) with $0 < s < r \leq 9$.

    (b) For each such primitive Pythagorean triple with $c \leq 20$, generate two non-primitive Pythagorean triples.

3. Find all primitive Pythagorean triples (a, b, c) with $0 < c < 30$.

4. If (a, b, c) is a primitive Pythagorean triple generated by r and s, find the missing variable:

    (a) $r = 4$ and $a = 24$

(b) $r = 4$ and $b = 15$

(c) $r = 5$ and $c = 29$

5. Assume that (a, b, c) is a primitive Pythagorean triple with a even.

(a) Prove that the "representation" of a, b, and c in terms of r and s in Theorem 7.8 is unique.

(b) Show that the number of distinct primitive Pythagorean triples (a, b, c) (for example, $(3, 4, 5)$ is not counted as distinct from $(4, 3, 5)$), say, with a even generated by r and s of Theorem 7.8 is:

(i) $\phi(r)$ if $r \equiv 0 \pmod 2$

(ii) $\phi(r)/2$ if $r \equiv 1 \pmod 2$

6. Let (a, b, c) be a Pythagorean triple with $a^2 + b^2 = c^3$.

(a) Prove that one of a and b is divisible by 3.

(b) Prove that one of a, b, and c is divisible by 5.

7. Let T be a right triangle having legs of length x, y, and z so that $x^2 + y^2 = z^2$ and let R be the radius of the circle inscribed in T.

(a) Show that $R(x + y + z) = xy = 2 \cdot \{\text{area of } T\}$.

(b) Use Theorem 7.8 to prove that if (a, b, c) is a Pythagorean triple producing a right triangle T, then the radius R of the inscribed circle is $R = ks(r - s)$ for some positive integer k and that, therefore, R is always an integer.

8. According to Exercise 7, if R is the radius of the inscribed circle of a "primitive" right triangle (one whose legs are given by a primitive Pythagorean triple), then $R = s(r - s)$, where r and s are given by Theorem 7.8.

(a) Show that $r - s$ must be odd for each such R.

(b) Let R be any positive integer and u and v be positive integers such that $R = uv$ and v is odd. Prove that there is a "primitive" right triangle having an inscribed circle of radius R.

(c) Prove that each factoring uv of R in part (b) gives a distinct primitive right triangle having an inscribed circle of radius R.

(d) Let $R = 2^{\alpha_0} p_1^{\alpha_1} p_2^{\alpha_2} \cdots p_n^{\alpha_n}$ be the prime factorization of R, where $\alpha_i \geq 1$ for $i \geq 1$ and we allow $\alpha_0 = 0$ in case R is odd. If $n = 0$, then $R = 2^\alpha$. Show that the number of choices of factors u and v of part (b) is the same as the number of factors of $p_1 p_2 \cdots p_n$, namely, 2^n; therefore, there are exactly 2^n "primitive" triangles having an inscribed triangle of radius R.

(e) Find all primitive triangles having an inscribed circle of radius:

     (i) $R = 1$

     (ii) $R = 6$

     (iii) $R = 9$

     (iv) $R = 15$

9. Prove that if n is any integer, then the equation $x^2 - y^2 = n$ may be solved in integers x and y.

10. Prove that if n is any integer, then the equation $n = x^2 + y^2 - z^2$ can be solved in integers x, y, and z.

7.3   Integers as Sums of Two Squares

In this section, we explore integers that may be expressed as a sum of two squares. Equivalently we are also finding solutions to the Diophantine equation $z = x^2 + y^2$. We begin with the following theorem which shows that all nonnegative integers which may be expressed as the sum of two squares form a semigroup.

Theorem 7.9 *If m and n are integers which may be expressed as the sum of two squares, then their product mn may be expressed as the sum of two squares.*

Proof. Let $m = a_1^2 + a_2^2$ and $n = b_1^2 + b_2^2$; then

$$mn = (a_1^2 + a_2^2)(b_1^2 + b_2^2) = (a_1 b_1 + a_2 b_2)^2 + (a_1 b_2 - a_2 b_1)^2 \qquad \blacksquare$$

Theorem 7.10 *A prime p may be uniquely written as the sum of two squares if and only if $p \not\equiv 3 \pmod 4$.*

Proof. From the Corollary 7.7 it is known that a prime p with $p \equiv 3 \pmod 4$ cannot be written as the sum of two squares. Conversely, if a prime $p \not\equiv 3 \pmod 4$, then $p = 2$ or $p \equiv 1 \pmod 4$. If $p = 2$, then $p = 1^2 + 1^2$. Assume that $p \equiv 1 \pmod 4$. Using Theorem 3.96 to evaluate the Legendre symbol, it follows that $\left(\dfrac{-1}{p}\right) = 1$ so that $x^2 \equiv -1 \pmod p$ has a solution. Let r be such a solution. Consider all integers of the form $ar + b$ where a and b are nonnegative integers whose squares are less than p. Let s be the number of nonnegative integers whose squares are less than p. Thus, $s^2 > p$ because 0 is included in the count; and since there are s^2 ordered pairs (a, b) satisfying the conditions above, there are s^2 integers of the form $ar + b$ as described above. Hence two of these integers must be congruent modulo p, say $a_1 r + b_1 \equiv a_2 r + b_2 \pmod{p}$. Then $(a_1 - a_2)r \equiv b_2 - b_1 \pmod{p}$. Let $u = |a_1 - a_2|$ and $v = |b_2 - b_1|$. Then $(ur)^2 \equiv v^2 \pmod{p}$. Since $r^2 + 1 \equiv 0 \pmod{p}$, $u^2 + v^2 \equiv u^2 + (ur)^2 \equiv u^2(r^2 + 1) \equiv u^2 \cdot 0 = 0 \pmod{p}$. Therefore, $u^2 + v^2$ is a multiple of p. But $u^2 + v^2 < 2p$. Hence $u^2 + v^2 = p$ and p is expressed as the sum of two squares.

To show uniqueness, 2 is obviously uniquely expressed as $1^2 + 1^2$; so assume that $p > 2$ and $p = a^2 + b^2 = c^2 + d^2$. Then $a^2 \equiv -b^2 \pmod{p}$ and $c^2 \equiv -d^2 \pmod{p}$. Hence $a^2 c^2 \equiv b^2 d^2 \pmod{p}$ and

$$a^2 c^2 - b^2 d^2 = (ac - bd)(ac + bd) \equiv 0 \pmod{p}$$

Thus, p divides $ac - bd$ or p divides $ac + bd$. But a^2, b^2, c^2, and d^2 are all less than p. Hence $-p < ac - bd < p$ and $0 < ac + bd < 2p$. Hence p does not divide $ac - bd$ unless $ac - bd = 0$ and $ac = bd$. Then $a = bd/c$. Thus, c and d are relatively prime so that c divides b. Similarly, $d = ac/b$ and b divides c. Hence $b = c$ and $a = d$. If p divides $ac + bd$, then $p = ac + bd = a^2 + b^2$ giving $a(a - c) = -b(b - d)$; and since a and b are relatively prime, $a \mid (b - d)$ and $b \mid (a - c)$. Similarly, $ac + bd = c^2 + d^2$, $c(c - a) = d(b - d)$, $c \mid (b - d)$, and $d \mid (c - a)$. Hence

$$\begin{array}{cc} a \mid (b - d) & b \mid (a - c) \\ c \mid (b - d) & d \mid (c - a) \end{array}$$

Assume that a is the largest of the four integers (by symmetry it doesn't matter which is assumed largest); then $a \mid (b - d)$ only if $b - d = 0$ and $b = d$. Hence $a = c$. ∎

If $n \equiv 1 \pmod{4}$ and is prime, then we have shown that n is uniquely expressible as a sum of squares. This property does not hold in general since, for example, $745 = 27^2 + 4^2 = 24^2 + 13^2$ and 745 is not prime.

Theorem 7.11 *A positive integer n is expressible as a sum of two squares if and only if, when n is uniquely factored into primes, the power of each prime of the form $4m + 3$ is even.*

Proof. Factor n into $a^2 b$ where b is uniquely expressed as a product of primes and where the power of each such prime is 1. In this case, b is said to be "square free." If each of the primes in b is of the form $4m + 1$ or is even, then each can be expressed as a sum of squares and, by Theorem 7.10, b may be expressed as the sum of squares, say $c^2 + d^2$. Hence

$$n = a^2(c^2 + d^2) = (ac)^2 + (ad)^2$$

and n is expressed as the sum of two squares.

Assume that $n = a^2 + b^2$. Let p be a prime such that $p \mid n$ and such that p has an odd power in the prime factorization of n. Let $d = \gcd(a, b)$, $a = ud$ and $b = vd$. Then $n = d^2(u^2 + v^2)$. Let $r = n/d^2$ so $r = u^2 + v^2$. Hence, $p \mid r$. Since u and v are relatively prime, so are u and r; and there is an integer k such that $uk \equiv v \pmod r$. Hence $r \equiv u^2 + u^2k^2 \equiv u^2(1 + k^2) \equiv 0 \pmod r$. Since r and u are relatively prime, p and u are relatively prime and $p \mid (1+k^2)$ where $1 + k^2$ is square free (see the Exercises). Therefore, $k^2 \equiv -1 \pmod p$ and -1 is a quadratic residue of p. By Theorem 3.96, p must have the form $4m + 1$. ∎

Theorem 7.12 *An integer can be expressed as the difference of two squares if and only if it is not of the form* $4m + 2$.

Proof. A square must have the form $4m + 1$ or $4m$ or, equivalently, it is congruent to 1 or 0 modulo 4. Hence the difference of two squares must be congruent to 0, 1, or -1 modulo 4 and hence of the form $4m$, $4m + 1$, or $4m + 3$.

Conversely, if an integer n is odd then $n - 1$ and $n + 1$ are even and $n = ((n+1)/2)^2 - ((n-1)/2)^2$. On the other hand, if an integer n is divisible by 4, then $n = (n/4 + 1)^2 - (n/4 - 1)^2$. ∎

The expression of an integer as the sum of two squares was studied by Mohammed Ben Alhocain in the tenth century, by Fibonacci early in the thirteenth century, and by Bachet early in the seventeenth century. Albert Girard (1595-1632) first stated correct necessary and sufficient conditions for an integer to be the sum of two squares. Fermat independently stated these conditions and apparently had a proof which he did not publish. Euler gave the first published proof.

Fermat again formulated correct conditions for an integer to be the sum of three squares. The proof was given by Legendre. Gauss obtained a formula for finding the number of primitive representations of the sum of three squares.

Fermat again apparently had a proof that any integer can be written as the sum of four squares but did not publish it. Euler was unable to find a proof but made great progress. Lagrange proved the theorem. Euler was then able to find two easier proofs.

Integers as sums of three and four squares will be discussed in Sections 7.5 and 7.6.

Exercises

1. Express the following integers as the sum of two squares:

    (a) 105   (b) 221   (c) 1073

2. Express the following integers as the sum of two squares:

    (a) 401   (b) 450   (c) 520

3. Express the following integers as the difference of two squares:

    (a) 400   (b) 401   (c) 75

4. Express the following integers as the difference of two squares:

    (a) 200   (b) 105   (c) 163

5. Determine which of the following can be expressed as the sum of two squares:

    (a) 12!   (b) 1995^{1995}

6. Find a positive integer that can be written with at least 3 distinct representations as the sum of two squares.

7. Prove that the positive integer n has as many distinct representations as does the sum of two squares as the integer $2n$.

8. Prove that for any 4 consecutive integers, one of them cannot be written as the sum of two squares.

9. Prove that an integer is the difference of two squares if and only if it can be written as the product of two even integers or two odd integers.

10. If q is a positive divisor of n and $n = a^2 + b^2$, where a and b are relatively prime, prove that there exist relatively prime integers c and d such that $q = c^2 + d^2$.

11. Show that if p and q are distinct primes which are both congruent to 1 modulo 4, then $pq = a^2 + b^2$ where $0 < a < b$ has two distinct solutions.

12. Show that 45 is the smallest positive integer that has three distinct representations as the difference of two squares.

13. Prove that every prime of the form $4k+1$ divides the sum of two relatively prime squares.

14. Prove that the number of distinct representations of an integer m greater than 1 as a sum of two squares of positive relatively prime integers equals the number of solutions of the congruence $x^2 \equiv -1 \pmod{m}$.

15. Prove or disprove: if $a^2 + b^2$ is divisible by 27, then it is divisible by 81.

16. How many prime factors greater than m can the number $m^2 + 1$ have?

17. Prove that if every prime factor which divides a number of the form $n^2 + 1$ for $n < m$ is the sum of two squares, then every prime factor of $m^2 + 1$ which is less than m can be expressed as the sum of two squares.

18. Prove that if k is a positive integer, then $1 + k^2$ is not the square of any integer.

7.4 Quadratic Forms

Definition 7.13 *An expression is called a* polynomial of n variables $x_1, x_2,$
$\dots, x_n$ *if it is the sum of terms each of which is the product of an integer and positive integral powers of selected variables.*

For example, $xy + 8y + 5xyx$ is a polynomial of three variables. $x^2 + y^2$ is a polynomial of two variables.

Definition 7.14 *A polynomial of n variables is called* homogeneous *if the sums of the exponents of the variables in each term are the same. This common sum is called the* degree *of the polynomial.*

For example, $x^2 + 4xy + (-5)z^2$ is a homogeneous polynomial of degree 2 and $x^2y + 7xyz + z^3$ is a homogeneous polynomial of degree 3.

Definition 7.15 *If a polynomial is homogeneous of degree two, it is called a* quadratic form. *A quadratic form in the n variables $x_1, x_2, \dots, x_n$ is denoted by $Q(x_1, x_2, \dots, x_n)$.*

Evidently, a quadratic form $Q(x_1, x_2, \dots, x_n)$ has terms only of the form ax_ix_j for $i \neq j$ and of the form $bx_ix_i = bx_i^2$. Thus,

$$Q(x_1, x_2, x_3) = 2x_1^2 - 7x_2^2 + x_3^2 + 4x_1x_2 + 2x_1x_3 + 8x_2x_3$$

is a quadratic form in three variables. If we rewrite the form as

$$\begin{aligned} Q(x_1, x_2, x_3) &= B_{11}x_1^2 + B_{22}x_2^2 + B_{33}x_3^2 + (B_{12} + B_{21})x_1x_2 \\ &\quad + (B_{13} + B_{31})x_1x_3 + (B_{23} + B_{32})x_2x_3 \end{aligned}$$

then we can express the quadratic form as a matrix product:

$$[Q(x_1, x_2, x_3)] = x^t B x = \begin{bmatrix} x_1 & x_2 & x_3 \end{bmatrix} \begin{bmatrix} B_{11} & B_{12} & B_{13} \\ B_{21} & B_{22} & B_{23} \\ B_{31} & B_{32} & B_{33} \end{bmatrix} \begin{bmatrix} x_1 \\ x_2 \\ x_3 \end{bmatrix}$$

where $x = \begin{bmatrix} x_1 \\ x_2 \\ x_3 \end{bmatrix}$ and $B = [B_{ij}]$. The matrix product will always be a 1×1 matrix whose only component is $Q(x_1, x_2, x_3)$. Even though a 1×1 matrix with only one component is distinctly different from that component itself, in the application of this matrix product to quadratic forms, it is customary to write just $Q(x_1, x_2, x_3) = x^t B x$ rather than having to switch back and forth between $Q(x_1, x_2, x_3)$ and $[Q(x_1, x_2, x_3)]$. The components B_{ij} and B_{ji} for $i \neq j$ may be any numbers as long as the sum $B_{ij} + B_{ji}$ has the value of the coefficient of x_ix_j. Thus, the matrix B is not unique.

$$Q(x_1, x_2, x_3) = \begin{bmatrix} x_1 & x_2 & x_3 \end{bmatrix} \begin{bmatrix} 2 & -8 & 5 \\ 8 & -7 & 6 \\ -3 & 2 & 1 \end{bmatrix} \begin{bmatrix} x_1 \\ x_2 \\ x_3 \end{bmatrix}$$

$$= \begin{bmatrix} x_1 & x_2 & x_3 \end{bmatrix} \begin{bmatrix} 2 & 0 & 1 \\ 0 & -7 & 4 \\ 1 & 4 & 1 \end{bmatrix} \begin{bmatrix} x_1 \\ x_2 \\ x_3 \end{bmatrix}$$

If B_{ij} is chosen equal to B_{ji} when $i \neq j$, then the matrix B will be symmetric. This symmetric matrix is unique. Also if the coefficients of $Q(x_1, x_2, x_3)$ are integers and the coefficient of $x_i x_j$ is even for $i \neq j$, the corresponding symmetric matrix A such that $Q(x_1, x_2, x_3) = x^t A x$ will have integer components. This discussion leads to the following theorem whose proof is left to the reader.

Theorem 7.16 *Given any quadratic form*

$$Q(x_1, x_2, \ldots, x_n) = \sum_{i,j=1}^{n} B_{ij} x_i x_j = x^t B x \ \text{ with } x = \begin{bmatrix} x_1 \\ x_2 \\ \vdots \\ x_n \end{bmatrix}$$

where $B = [B_{ij}]$ and B_{ij} are integers, there is a unique symmetric matrix $A = [A_{ij}]$ such that

$$Q(x_1, x_2, \ldots, x_n) = x^t A x = \sum_{i,j=1}^{n} A_{ij} x_i x_j = \sum_{i=1}^{n} A_{ii} x_i^2 + \sum_{i,j=1 \ \text{and} \ i<j}^{n} 2 A_{ij} x_i x_j$$

If $B_{ij} + B_{ji}$ is even for $i \neq j$, then $A_{ij} = A_{ji}$ is an integer; otherwise, $A_{ij} = A_{ji}$ will be rational but not an integer.

Because matrix A of Theorem 7.16 is symmetric with $A_{ij} = A_{ji}$, it is often convenient to suppress the notational difference between each pair of symmetric components and write

$$\begin{bmatrix} A_{11} & A_{12} & A_{13} \\ A_{21} & A_{22} & A_{23} \\ A_{31} & A_{32} & A_{33} \end{bmatrix} = \begin{bmatrix} A_{11} & A_{12} & A_{13} \\ A_{12} & A_{22} & A_{23} \\ A_{13} & A_{23} & A_{33} \end{bmatrix}$$

in order to simplify proofs and make expressions more compact.

Definition 7.17 *If the quadratic form $Q(x_1, x_2, \ldots, x_n) = x^t A x$, where x is the $n \times 1$ matrix $\begin{bmatrix} x_1 \\ x_2 \\ \vdots \\ x_n \end{bmatrix}$, x^t is the transpose of x, and A is an $n \times n$ symmetric matrix, then the determinant of A is called the* discriminant *of Q and is denoted by $d(Q)$ so that $d(Q) = \det(A)$.*

For example, if $A = \begin{bmatrix} a & b \\ b & d \end{bmatrix}$, then

$$\begin{bmatrix} x_1 & x_2 \end{bmatrix} \begin{bmatrix} a & b \\ b & d \end{bmatrix} \begin{bmatrix} x_1 \\ x_2 \end{bmatrix} = ax_1^2 + 2bx_1 x_2 + dx_2^2$$

and $d(Q) = ad - b^2$. On the other hand, if $A = \begin{bmatrix} 2 & 5 & 7 \\ 5 & 9 & 2 \\ 7 & 2 & 9 \end{bmatrix}$, then

$$Q(x_1, x_2, x_3) = x^t A x = 2x_1^2 + 10x_1 x_2 + 14x_1 x_3 + 9x_2^2 + 4x_2 x_3 + 9x_3^2$$

and $d(Q) = \det(A) = -372$.

The proof of the next theorem is left to the reader.

Theorem 7.18 *Let M be an $n \times n$ matrix and x and y be $n \times 1$ matrices. Assume that M has a nonzero determinant and $x = My$. If*

$$Q(x_1, x_2, \ldots, x_n) = x^t A x = (My)^t A(My) = y^t M^t A M y$$

let $B = M^t A M$ so that $Q_1(y_1, y_2, \ldots, y_n) = y^t B y$. Then

$$d(Q_1) = \det(M^t A M) = \det(M^t) \det(A) \det(M) = \det(A)(\det(M))^2$$

Further, $d(Q)$ is positive if and only if $d(Q_1)$ is positive. In addition, B is symmetric if and only if A is symmetric.

Definition 7.19 *We say that A and B are* equivalent *matrices provided there is a matrix M with $\det(M) \neq 0$ such that $B = M^t A M$. We extend the term equivalent to say that the quadratic forms Q and Q_1 of Theorem 7.18 that were associated with matrices A and B, respectively, are equivalent.*

The requirement that $\det(M) \neq 0$ in Theorem 7.18 and Definition 7.19 is to ensure that the inverse M^{-1} exists (Theorem 3.19). Also, if M^{-1} exists, then $(M^t)^{-1} = (M^{-1})^t$.

For an example of equivalence, let

$$Q(x_1, x_2, x_3) = \begin{bmatrix} x_1 & x_2 & x_3 \end{bmatrix} \begin{bmatrix} 2 & -1 & 1 \\ -1 & 4 & 1 \\ 1 & 1 & 4 \end{bmatrix} \begin{bmatrix} x_1 \\ x_2 \\ x_3 \end{bmatrix}$$

and

$$\begin{bmatrix} x_1 & x_2 & x_3 \end{bmatrix} = \begin{bmatrix} 1 & 2 & 1 \\ 1 & 3 & 0 \\ 2 & 6 & 1 \end{bmatrix} \begin{bmatrix} y_1 \\ y_2 \\ y_3 \end{bmatrix}$$

Then

$$Q_1(y_1, y_2, y_3) = \begin{bmatrix} y_1 \\ y_2 \\ y_3 \end{bmatrix}^t \begin{bmatrix} 1 & 1 & 2 \\ 2 & 3 & 6 \\ 1 & 0 & 1 \end{bmatrix} \begin{bmatrix} 2 & -1 & 1 \\ -1 & 4 & 1 \\ 1 & 1 & 4 \end{bmatrix} \begin{bmatrix} 1 & 2 & 1 \\ 1 & 3 & 0 \\ 2 & 6 & 1 \end{bmatrix} \begin{bmatrix} y_1 \\ y_2 \\ y_3 \end{bmatrix}$$

$$= \begin{bmatrix} y_1 & y_2 & y_3 \end{bmatrix} \begin{bmatrix} 28 & 81 & 13 \\ 81 & 236 & 36 \\ 13 & 36 & 8 \end{bmatrix} \begin{bmatrix} y_1 \\ y_2 \\ y_3 \end{bmatrix}$$

The proof of the following theorem is left to the reader.

Theorem 7.20 *If S is the set of all $n \times n$ matrices (or equivalently, all quadratic forms with n variables), then equivalence as given in Definition 7.19 is an equivalence relation on S.*

Definition 7.21 *An $n \times n$ matrix A is a* diagonal *matrix if $A_{ij} = 0$ for $i \neq j$.*

The following matrix is diagonal:

$$A = \begin{bmatrix} 5 & 0 & 0 \\ 0 & -8 & 0 \\ 0 & 0 & 2 \end{bmatrix}$$

as are the identity matrices I_2 and I_3.

Definition 7.22 *A quadratic form $Q(x_1, x_2, \ldots, x_n)$ is* positive definite *if $Q(x_1, x_2, \ldots, x_n) > 0$ for all values of $x_1, x_2, \ldots, x_n$ that are not all zero. We extend the term positive definite to the unique symmetric matrix A such that $Q(x_1, x_2, \ldots, x_n) = x^t A x$.*

Theorem 7.23 *A diagonal matrix $A = [A_{ij}]$ is positive definite if and only if $A_{ii} > 0$ for all $1 \leq i \leq n$.*

Proof. The proof is left to the reader. ■

Theorem 7.24 *A binary (i.e., a two-variable) quadratic form $Q(x_1, x_2) = A_{11}x_1^2 + 2A_{12}x_1x_2 + A_{22}x_2^2$ is positive definite if and only if $A_{11} > 0$ and*

$$\det(A) = \begin{vmatrix} A_{11} & A_{12} \\ A_{12} & A_{22} \end{vmatrix} = A_{11}A_{22} - (A_{12})^2 > 0$$

Proof. If Q is positive definite, then $A_{11} = Q(1,0) > 0$. The quadratic form

$$A_{11}x_1^2 + 2A_{12}x_1x_2 + A_{22}x_2^2$$

is positive if and only if

$$A_{11}(A_{11}x_1^2 + 2A_{12}x_1x_2 + A_{22}x_2^2)$$

is positive. But completing the square we have

$$A_{11}x_1^2 + 2A_{11}A_{12}x_1x_2 + A_{11}A_{22}x_2^2 = (A_{11}x_1 + A_{12}x_2)^2 + (A_{11}A_{22} - A_{12}^2)x_2^2$$

which is positive for all x_1 and x_2 not both zero only if $A_{11}A_{12} - (A_{12})^2 > 0$. Conversely, since $A_{11}Q = (A_{11}x_1 + A_{12}x_2)^2 + (A_{11}A_{22} - (A_{12})^2)x_2^2$, Q is always positive if A_{11} and $A_{11}A_{22} - (A_{12})^2$ are positive and not both of x_1 and x_2 are zero. ■

Theorem 7.25 *For each equivalence class of binary positive definite quadratic forms there exists a "reduced" quadratic form in the class with the unique symmetric matrix $A = [A_{ij}]$ of the form having the properties*

$$2|A_{12}| \leq A_{11} \leq A_{22}$$

and

$$3A_{11}^2 \leq 4 \begin{vmatrix} A_{11} & A_{12} \\ A_{12} & A_{22} \end{vmatrix}$$

Proof. Let $Q(x_1, x_2) = B_{11}x_1^2 + 2B_{12}x_1x_2 + B_{22}x_2^2$ be a binary positive definite quadratic form. Let a be the smallest positive integer such that $Q(x_1, x_2) = a$ for some $x_1 = x_1^*$ and $x_2 = x_2^*$. Then $a = B_{11}(x_1^*)^2 + 2B_{12}(x_1^*)(x_2^*) + B_{22}(x_2^*)^2$. We know that $\gcd(x_1^*, x_2^*) = 1$ or a would not be the smallest positive integer represented by $Q(x_1, x_2)$. Hence, there exist u_0 and v_0 such that $u_0 x_1^* - v_0 x_2^* = 1$. The general solution is given by

$$u = u_0 + tx_2^* \quad \text{and} \quad v = v_0 + tx_1^* \quad \text{for } t \in Z$$

Let $x = \begin{bmatrix} x_1 \\ x_2 \end{bmatrix}$, $y = \begin{bmatrix} y_1 \\ y_2 \end{bmatrix}$, and $M = \begin{bmatrix} x_1^* & v \\ x_2^* & u \end{bmatrix}$; and let y be given by $x = My$. Direct calculation gives $\det(M) = 1$. Say that

$$Q_2(y_1, y_2) = A_{11}y_1^2 + 2A_{12}y_1y_2 + A_{22}y_2^2 = y^t A y = y^t M^t B M y$$

and $A = [A_{ij}]$. Then

$$A = \begin{bmatrix} A_{11} & A_{12} \\ A_{21} & A_{22} \end{bmatrix}$$

where

$$A_{11} = B_{11}(x_1^*)^2 + 2B_{12}x_1^*x_2^* + B_{22}(x_2^*)^2$$

$$A_{12} = x_1^*(B_{11}v + B_{12}u) + x_2^*(B_{21}v + B_{22}u)$$

$$A_{21} = v(B_{11}x_1^* + B_{12}x_2^*) + u(B_{21}x_1^* + B_{22}x_2^*)$$

$$A_{22} = u(B_{22}u + B_{21}v) + v(B_{11}v + B_{12}u)$$

We see that $a = A_{11}$. Letting $u = u_0 + tx_2^*$ and $v = v_0 + tx_1^*$, we have

$$A_{12} = v_0(B_{11}x_1^* + B_{12}x_2^*) + u_0(B_{12}x_1^* + B_{22}x_2^*) + ta$$

By choosing the proper value of t we can make $2|A_{12}| \leq a$. $A_{22} = Q_2(0, 1)$ and hence by definition of a, $a = A_{11} \leq A_{22}$. Also,

$$\begin{aligned} 4A_{11}^2 &\leq 4A_{11}A_{22} = 4A_{12}^2 + 4A_{11}A_{22} - 4A_{12}A_{21} \text{ since } A_{12} = A_{21} \\ &= 4A_{12}^2 + 4 \begin{vmatrix} A_{11} & A_{12} \\ A_{12} & A_{22} \end{vmatrix} \\ &\leq A_{11}^2 + 4 \begin{vmatrix} A_{11} & A_{12} \\ A_{12} & A_{22} \end{vmatrix} \end{aligned}$$

so that $3A_{11}^2 \leq 4 \begin{vmatrix} A_{11} & A_{12} \\ A_{12} & A_{22} \end{vmatrix}$. ■

Theorem 7.26 *Every positive definite binary quadratic form with discriminant 1 is equivalent to a quadratic form that is the sum of two squares.*

Proof. We use the notation of Theorem 7.25. Since $\begin{vmatrix} A_{11} & A_{12} \\ A_{12} & A_{22} \end{vmatrix} = 1$, then $3A_{11}^2 < 4$ and hence $A_{11} = 1$. Since $2|A_{12}| \leq A_{11} = 1$, we obtain $A_{12} = A_{21} = 0$. Solving for A_{22}, since $\det(A) = A_{11}A_{22}$ we find that $A_{22} = 1$. Hence $Q(x_1, x_2) = x_1^2 + x_2^2$. ■

Corollary 7.27 *For every positive definite binary quadratic form $Q(x, y)$ with discriminant 1, there exist integers x^* and y^* such that $Q(x^*, y^*) = 1$.*

Example 7.28 *Suppose that we are given the quadratic form*

$$Q(x_1, x_2) = x_1^2 + 4x_1x_2 + 5x_2^2 = \begin{bmatrix} x_1 & x_2 \end{bmatrix} \begin{bmatrix} 1 & 2 \\ 2 & 5 \end{bmatrix} \begin{bmatrix} x_1 \\ x_2 \end{bmatrix} = x^t Bx$$

where $B = \begin{bmatrix} 1 & 2 \\ 2 & 5 \end{bmatrix}$ is symmetric. We want to find a matrix $M = [M_{ij}]$ so that

$$\begin{bmatrix} x_1 \\ x_2 \end{bmatrix} = \begin{bmatrix} M_{11} & M_{12} \\ M_{21} & M_{22} \end{bmatrix} \begin{bmatrix} y_1 \\ y_2 \end{bmatrix}$$

and if B is the corresponding matrix of Q, then $Q_1(y_1, y_2) = y_1^2 + y_2^2 = y^t Ay$, where $A = M^t BM$. Thus, $M^t BM = \begin{bmatrix} 1 & 0 \\ 0 & 1 \end{bmatrix}$, the identity matrix.

In constructing M, we follow the proof of Theorem 7.25. Thus, we find values x_1^* and x_2^* so that $Q(x_1^*, x_2^*) = a$, the minimum positive value of Q. Clearly, $Q(1, 0) = 1$, the least positive integer, so that $a = 1$. M has the form $\begin{bmatrix} 1 & v \\ 0 & u \end{bmatrix}$. Since $1 \cdot u = \det(M) = 1$, we obtain $u = 1$ and $M = \begin{bmatrix} 1 & v \\ 0 & 1 \end{bmatrix}$. There exist u_0 and v_0 such that $1 \cdot u_0 - 0 \cdot v_0 = 1$, namely $u_0 = 1$ and $v_0 = 1$. The general solution is $u = 1 + 0 \cdot t$ and $v = 1 + 1 \cdot t$. From the proof to Theorem 7.25 we have

$$\begin{aligned} A_{12} &= v_0 (B_{11}x_1^* + B_{12}x_2^*) + u_0 (B_{12}x_1^* + B_{22}x_2^*) + ta \\ &= 1 \cdot [1 \cdot 1 + 2 \cdot 0] + 1 \cdot [2 \cdot 1 + 5 \cdot 0] + t \cdot 1 \\ &= 3 + t \end{aligned}$$

The value of t that makes $2 \cdot |3 + t| < 1$ is $t = -3$; therefore, $v = -2$. Finally,

$$M = \begin{bmatrix} 1 & -2 \\ 0 & 1 \end{bmatrix} \text{ so that } x_1 = y_1 - 2y_2 \text{ and } x_2 = y_2. \text{ Checking, we see that}$$

$$M^t B M = \begin{bmatrix} 1 & 0 \\ -2 & 1 \end{bmatrix} \begin{bmatrix} 1 & 2 \\ 2 & 5 \end{bmatrix} \begin{bmatrix} 1 & -2 \\ 0 & 1 \end{bmatrix}$$

$$= \begin{bmatrix} 1 & 2 \\ 0 & 1 \end{bmatrix} \begin{bmatrix} 1 & -2 \\ 0 & 1 \end{bmatrix} = \begin{bmatrix} 1 & 0 \\ 0 & 1 \end{bmatrix} = A$$

So $Q_1(y_1, y_2) = y^t A y = y_1^2 + y_2^2$.

Example 7.29 *Suppose that we are given the binary quadratic form*

$$Q(x_1, x_2) = 2x_1^2 + 10x_1 x_2 + 13x_2^2 = \begin{bmatrix} x_1 & x_2 \end{bmatrix} \begin{bmatrix} 2 & 5 \\ 5 & 13 \end{bmatrix} \begin{bmatrix} x_1 \\ x_2 \end{bmatrix} = x^t B x$$

We again want to find a matrix $M = [M_{ij}]$ such that

$$\begin{bmatrix} x_1 \\ x_2 \end{bmatrix} = \begin{bmatrix} M_{11} & M_{12} \\ M_{21} & M_{22} \end{bmatrix} \begin{bmatrix} y_1 \\ y_2 \end{bmatrix}$$

and such that $Q_1(y_1, y_2) = y_1^2 + y_2^2$, where the corresponding matrix is $A = M^t B M$. These properties imply that $M^t B M = \begin{bmatrix} 1 & 0 \\ 0 & 1 \end{bmatrix}$, the identity matrix. We will again follow the construction found in Theorem 7.25.

To construct M we find values x_1^* and x_2^* so that $Q(x_1^*, x_2^*) = 1$. This time x_1^* and x_2^* are not so easily found. Possible values are $(-2, 1)$, $(2, -1)$, $(-3, 1)$, and $(3, -1)$. We will use $(-2, 1)$ and leave the others to the reader. We know that M has the form $\begin{bmatrix} -2 & v \\ 1 & u \end{bmatrix}$. There exist u_0 and v_0 such that $-2 \cdot u_0 - 1 \cdot v_0 = 1$, namely $u_0 = -1$ and $v_0 = 1$. The general solution is $u = -1 + t$ and $v = 1 - 2t$. From the proof to Theorem 7.25 we have

$$\begin{aligned}
A_{12} &= v_0 \left(B_{11} x_1^* + B_{12} x_2^* \right) + u_0 \left(B_{12} x_1^* + B_{22} x_2^* \right) + ta \\
&= 1 \cdot [2(-2) + 5(1)] + (-1)[5(-2) + 13(1)] + t(1) \\
&= -2 + t
\end{aligned}$$

The value of t that makes $2 \cdot |-2 + t| < 1$ is $t = 2$ so that $v = -3$ and $u = 1$. These values give $M = \begin{bmatrix} -2 & -3 \\ 1 & 1 \end{bmatrix}$, so $x_1 = -2y_1 - 3y_2$ and $x_2 = y_1 + y_2$. It is a straightforward calculation to verify that $A = M^t B M = I_2$.

Theorem 7.30 *Let $Q(x_1, x_2, x_3) = x^t A x$ be a quadratic form, where*

$$A = \begin{bmatrix} A_{11} & A_{12} & A_{13} \\ A_{12} & A_{22} & A_{23} \\ A_{13} & A_{23} & A_{33} \end{bmatrix}$$

is symmetric. Then $Q(x_1, x_2, x_3)$ *is positive definite if and only if*

$$d(Q) > 0, \quad \begin{vmatrix} A_{11} & A_{12} \\ A_{12} & A_{22} \end{vmatrix} > 0, \quad and \quad A_{11} > 0$$

Proof. By direct computation, one can show that $A_{11}Q(x_1, x_2, x_3) = V^2 + Q_1(x_2, x_3)$ where $V = A_{11}x_1 + A_{12}x_2 + A_{13}x_3$ and

$$Q_1(x_2, x_3) = \begin{bmatrix} x_2 & x_3 \end{bmatrix} \begin{bmatrix} A_{11}A_{22} - A_{12}^2 & A_{11}A_{23} - A_{12}A_{13} \\ A_{11}A_{23} - A_{12}A_{13} & A_{11}A_{33} - A_{13}^2 \end{bmatrix} \begin{bmatrix} x_2 \\ x_3 \end{bmatrix}$$

We begin by showing that given $A_{11} > 0$, $Q_1(x_2, x_3)$ is positive definite if and only if $Q(x_1, x_2, x_3)$ is positive definite. It is obvious that if $Q_1(x_2, x_3)$ is positive definite, then $Q(x_1, x_2, x_3)$ is positive definite. Conversely, if there exist a_2 and a_3 not both zero such that $Q_1(a_2, a_3)$ is not positive, let $a_2' = A_{11}a_2$ and $a_3' = A_{11}a_3$. Then $Q_1(a_2', a_3') = A_{11}^2 Q_1(a_2, a_3)$ is also not positive. Choose $a_1' = -(A_{12}a_2' + A_{13}a_3')/A_{11}$. Then $V = 0$ and $Q(a_1', a_2', a_3')$ is not positive and hence $Q(x_1, x_2, x_3)$ is not positive definite. Thus, $Q(x_1, x_2, x_3)$ is positive definite if and only if $Q_1(x_2, x_3)$ is.

Assume that $Q(x_1, x_2, x_3)$ is positive definite. A_{11} is positive since

$$A_{11} = \begin{bmatrix} 1 & 0 & 0 \end{bmatrix} A \begin{bmatrix} 1 \\ 0 \\ 0 \end{bmatrix}$$

is positive making both $Q_1(x_2, x_3)$ and $Q(x_1, x_2, x_3)$ positive definite. By Theorem 7.24, $A_{11}A_{22} - A_{12}^2$ is positive since it is the upper left element of $d(Q_1)$. By direct computation we see that $d(Q_1) = A_{11}d(Q)$. By Theorem 7.24, $d(Q_1)$ is positive and so $d(Q)$ is positive as well.

Conversely, assume that $d(Q) > 0$, $\begin{vmatrix} A_{11} & A_{12} \\ A_{12} & A_{22} \end{vmatrix} > 0$, and $A_{11} > 0$. Since $A_{11} > 0$ and $d(Q)$ is positive, then $d(Q_1)$ is positive. Also, since $A_{11}A_{22} - A_{12}A_{21}$ is positive, then, by Theorem 7.24, $Q_1(x_2, x_3)$ is positive definite; and hence $Q(x_1, x_2, x_3)$ is positive definite. ■

Lemma 7.31 *Each equivalence class of positive definite quadratic forms containing* $Q(x_1, x_2, x_3)$ *contains a reduced quadratic form; that is, one such that*

$$Q_2(y_1, y_2, y_3) = \begin{bmatrix} y_1 & y_2 & y_3 \end{bmatrix} \begin{bmatrix} B_{11} & B_{12} & B_{13} \\ B_{21} & B_{22} & B_{23} \\ B_{31} & B_{32} & B_{33} \end{bmatrix} \begin{bmatrix} y_1 \\ y_2 \\ y_3 \end{bmatrix}$$

where $B = [B_{ij}]$ *is symmetric,* $0 < 27 \cdot B_{11}^3 \le 64 \cdot d(Q_2)$, $2 \cdot |B_{12}| \le B_{11}$, *and* $2 \cdot |B_{13}| \le B_{11}$.

Proof. Given $Q = x^t A x$, where

$$A = \begin{bmatrix} A_{11} & A_{12} & A_{13} \\ A_{21} & A_{22} & A_{23} \\ A_{31} & A_{32} & A_{33} \end{bmatrix}$$

is symmetric, let a be the smallest positive integer so that $a = Q(a_1, a_2, a_3)$ for some a_1, a_2, and a_3. As an intermediate step, we wish to construct $C = [C_{ij}]$ so that C has $\begin{bmatrix} C_{11} \\ C_{21} \\ C_{31} \end{bmatrix} = \begin{bmatrix} a_1 \\ a_2 \\ a_3 \end{bmatrix}$ as its first column and has determinant equal to 1. From Definition 3.6, the determinant of C is given by

$$C_{11}(C_{22}C_{33} - C_{32}C_{23}) - C_{21}(C_{12}C_{33} - C_{32}C_{13}) + C_{31}(C_{12}C_{23} - C_{13}C_{22})$$

Since $\gcd(C_{11}, C_{21}, C_{31}) = \gcd(a_1, a_2, a_3) = 1$ or a is not minimal, there exist u, v, and w so that $uC_{11} + vC_{21} + wC_{31} = 1$. Hence we need to solve the following equations by choosing C_{ij} not already defined:

$$\begin{aligned} C_{22}C_{33} - C_{32}C_{23} &= u \\ C_{12}C_{33} - C_{32}C_{13} &= -v \\ C_{12}C_{23} - C_{13}C_{22} &= w \end{aligned}$$

Let $e = \gcd(u, v)$, $u = eu'$, and $v = ev'$. Assigning the values $C_{22} = u'$, $C_{12} = -v'$, $C_{32} = 0$, and $C_{33} = e$, we obtain

$$\begin{aligned} C_{22}C_{33} - C_{32}C_{23} &= eu' &= u \\ C_{12}C_{33} - C_{32}C_{13} &= -ev' &= -v \end{aligned}$$

We then need to solve $C_{23}(-v') - C_{13}u' = w$ for C_{23} and C_{13}. Since u' and v' are relatively prime, such a solution exists; and we have constructed C with $d(C) = 1$.

Define $Q_1(z_1, z_2, z_3) = Q(Cz) = z^t M z$ where $M = C^t A C$ and $x = Cz$. Then $M_{11} = Q_1(1, 0, 0) = Q(C_{11}, C_{21}, C_{31}) = a$. Let

$$N = \begin{bmatrix} 1 & u & v \\ 0 & R_{11} & R_{12} \\ 0 & R_{21} & R_{22} \end{bmatrix}$$

where u and v will be specified later and where $R = \begin{bmatrix} R_{11} & R_{12} \\ R_{21} & R_{22} \end{bmatrix}$ has determinant 1 and will also be specified later. It is easily seen that for such N and R, $\det(N) = 1$. Next, let $Q_2(y) = Q_1(Ny) = y^t B y$ where $B = N^t M N$. By direct computation one finds that $B_{11} = M_{11}$; and hence, using the same equation as in the proof of Theorem 7.30, we have $M_{11}Q_1 = V_1^2 + Q_1'(z_2, z_3)$, where $V_1 = M_{11}z_1 + M_{12}z_2 + M_{13}z_3$, $z = Ny$, and

$$Q_1'(z_2, z_3) = \begin{bmatrix} z_2 & z_3 \end{bmatrix} \begin{bmatrix} M_{11}M_{22} - M_{12}^2 & M_{11}M_{23} - M_{12}M_{13} \\ M_{11}M_{23} - M_{12}M_{13} & M_{11}M_{33} - M_{13}^2 \end{bmatrix} \begin{bmatrix} z_2 \\ z_3 \end{bmatrix}$$

and $M_{11}Q_2 = V_2^2 + Q_2'(y_2, y_3)$, where $V_2 = B_{11}y_1 + B_{12}y_2 + B_{13}y_3$, and

$$Q_2'(y_2, y_3) = \begin{bmatrix} y_2 & y_3 \end{bmatrix} \begin{bmatrix} B_{11}B_{22} - B_{12}^2 & B_{11}B_{23} - B_{12}B_{13} \\ B_{11}B_{23} - B_{12}B_{13} & B_{11}B_{33} - B_{13}^2 \end{bmatrix} \begin{bmatrix} y_2 \\ y_3 \end{bmatrix}$$

By direct computation we see that

$$Q_2'(y_2, y_3) = \begin{bmatrix} y_2 \\ y_3 \end{bmatrix}^t R^t \begin{bmatrix} M_{11}M_{22} - M_{12}^2 & M_{11}M_{23} - M_{12}M_{13} \\ M_{11}M_{23} - M_{12}M_{13} & M_{11}M_{33} - M_{13}^2 \end{bmatrix} R \begin{bmatrix} y_2 \\ y_3 \end{bmatrix}$$

The coefficient of y_2^2 is $B_{11}B_{22} - B_{12}^2$. Let $b = B_{11}B_{22} - B_{12}^2$. We see in the proof of Theorem 7.25 that R may be selected so that $3b^2 < 4d(Q_2')$. But $d(Q_2') = M_{11}d(Q_2)$, so $3(B_{11}B_{22} - B_{12}^2)^2 < 4M_{11}d(Q_2)$. Also, by direct computation we see that

$$\begin{aligned} B_{12} &= uM_{11} + R_{11}M_{12} + M_{13}R_{21} \\ B_{13} &= vM_{11} + M_{12}R_{12} + M_{13}R_{22} \end{aligned}$$

Therefore, we can select u and v so that $2|B_{12}| < M_{11} = B_{11}$ and $2|B_{13}| < B_{11}$. Also, $B_{22} = Q_2(0, 1, 0)$ so $B_{11} = a < B_{22}$ by definition of a. We can then write

$$\begin{aligned} 3(4B_{11}^2 - B_{11}^2)^2 &< 3\left((4B_{11}B_{22} - B_{12}^2 + B_{12}^2) - B_{11}^2\right)^2 \\ &< 48(B_{11}B_{22} + B_{12}^2)^2 \text{ because } B_{11} > 2B_{12} \\ &< 64M_{11}d(Q_2) = 64B_{11}d(Q_2) \end{aligned}$$

Therefore, $3(3B_{11}^2)^2 < 64B_{11}d(Q_2)$ and $27B_{11}^3 < 64d(Q_2)$. ∎

Theorem 7.32 *Every positive definite quadratic form $Q(x_1, x_2, x_3)$ such that $d(Q) = 1$ is equivalent to a quadratic form $Q^*(z_1, z_2, z_3) = z_1^2 + z_2^2 + z_3^2$.*

Proof. For $Q(x_1, x_2, x_3) = x^t A x$ with A symmetric and $d(Q) = \det(A) = 1$, let $Q_1(y_1, y_2, y_3) = y^t B y$ be a reduced quadratic form equivalent to $Q(x_1, x_2, x_3)$ as given by Lemma 7.31, where $x = My$, $B = M^t A M$, $x = \begin{bmatrix} x_1 & x_2 & x_3 \end{bmatrix}^t$, and $y = \begin{bmatrix} y_1 & y_2 & y_3 \end{bmatrix}^t$. Let $B = [B_{ij}]$ and note that $d(Q_1) = \det(B) = 1$ and that B is symmetric also. Then $0 < 3B_{11} \le 4$ implies that $B_{11} = 1$. Since $2|B_{12}| \le B_{11} = 1$ and $2|B_{13}| \le B_{11} = 1$, we obtain $B_{12} = B_{13} = 0$. Thus

$$B = \begin{bmatrix} 1 & 0 & 0 \\ 0 & B_{22} & B_{23} \\ 0 & B_{23} & B_{33} \end{bmatrix}$$

Consider the submatrix

$$C = \begin{bmatrix} C_{11} & C_{12} \\ C_{12} & C_{22} \end{bmatrix} = \begin{bmatrix} B_{22} & B_{23} \\ B_{23} & B_{33} \end{bmatrix}$$

Since $1 = \det(B) = 1 \cdot \begin{vmatrix} B_{22} & B_{23} \\ B_{23} & B_{33} \end{vmatrix} = \det(C)$, Theorem 7.26 implies that there exist an equivalent 2×2 matrix D with $\det(D) = 1$ and a 2×2 matrix $N = [N_{ij}]$ with $\det(N) \ne 0$ such that

$$\bar{y} = \begin{bmatrix} y_2 \\ y_3 \end{bmatrix} = N \begin{bmatrix} w_2 \\ w_3 \end{bmatrix} = N\bar{w} \text{ and } D = N^t C N$$

If $\overline{w} = \begin{bmatrix} w_2 \\ w_3 \end{bmatrix}$, then

$$\overline{w}^t D \overline{w} = w_2^2 + w_3^2$$

Direct calculation gives

$$
\begin{aligned}
Q(y_1, y_2, y_3) &= t^t B y = y_1^2 + \begin{bmatrix} y_2 & y_3 \end{bmatrix} \begin{bmatrix} B_{22} & B_{23} \\ B_{23} & B_{33} \end{bmatrix} \begin{bmatrix} y_2 \\ y_3 \end{bmatrix} \\
&= y_1^2 + \overline{y}^t C \overline{y} = y_1^2 + \overline{w}^t D \overline{w} \\
&= y_1^2 + w_2^2 + w_3^2 = w_1^2 + w_2^2 + w_3^2
\end{aligned}
$$

where we have let $w_1 = y_1$. More explicitly, we have

$$
\begin{aligned}
Q(x_1, x_2, x_3) &= x^t A x \\
&= (My)^t A (My) = y^t (M^t A M) y \\
&= y^t B y \text{ where } B = M^t A M \\
&= (Rw)^t B (Rw)
\end{aligned}
$$

where $R = \begin{bmatrix} 1 & 0 & 0 \\ 0 & N_{11} & N_{12} \\ 0 & N_{12} & N_{22} \end{bmatrix}$ and $D = N^t C N$. Continuing, we obtain

$$
\begin{aligned}
Q(x_1, x_2, x_3) &= w^t (R^t B R) w \text{ with } w = \begin{bmatrix} w_1 \\ w_2 \\ w_3 \end{bmatrix} \\
&= w^t E w = w_1^2 + w_2^2 + w_3^2
\end{aligned}
$$

where $E = R^t B R$. ∎

Example 7.33 *Suppose that the quadratic form* $Q(x_1, x_2, x_3)$ *has the form* $x^t A x$, *where*

$$A = \begin{bmatrix} 1 & 5 & 1 \\ 5 & 26 & 0 \\ 1 & 0 & 27 \end{bmatrix}$$

so that $Q(x_1, x_2, x_3) = x_1^2 + 10x_1 x_2 + 2x_1 x_3 + 26x_2^2 + 27x_3^2$ *and* $\det(A) = 1$. *We want to find an equivalent quadratic form*

$$Q_2(y_1, y_2, y_3) = y^t B y = y_1^2 + y_2^2 + y_3^2$$

such that $B = N^t A N$ *and* $x = Ny$. *We will use the construction of Lemma 7.31 and Theorem 7.32.*

The quadratic form Q *has smallest positive value 1 and* $Q(1, 0, 0) = 1$. *In the construction of the proof in Lemma 7.31, we can skip the construction of matrix* C *since already* $Q(1, 0, 0) = 1$. *Thus, we can identify* M *with* A *and* z *with* x *and will seek matrices* N *and* B *such that*

$$N = \begin{bmatrix} 1 & T & Z \\ 0 & R_{11} & R_{12} \\ 0 & R_{21} & R_{22} \end{bmatrix}$$

where $x = Ny$ and $B = N^t MN$. Let $R = [R_{ij}]$ be the 2×2 submatrix in the lower right of N. Then

$$A_{11}Q(x_1, x_2, x_3) = V^2 + Q'(x_2, x_3)$$

where $V = A_{11}x_1 + A_{12}x_2 + A_{13}x_3$ and

$$Q'(x_2, x_3) = \begin{bmatrix} x_2 & x_3 \end{bmatrix} \begin{bmatrix} A_{11}A_{22} - A_{12}^2 & A_{11}A_{23} - A_{12}A_{13} \\ A_{11}A_{23} - A_{12}A_{13} & A_{11}A_{33} - A_{13}^2 \end{bmatrix} \begin{bmatrix} x_2 \\ x_3 \end{bmatrix}$$

Next, let

$$W = [W_{ij}] = \begin{bmatrix} A_{11}A_{22} - A_{12}^2 & A_{11}A_{23} - A_{12}A_{13} \\ A_{11}A_{23} - A_{12}A_{13} & A_{11}A_{33} - A_{13}^2 \end{bmatrix} = \begin{bmatrix} 1 & -5 \\ -5 & 26 \end{bmatrix}$$

so that $Q'(x_2, x_3) = x_2^2 - 10x_2x_3 + 26x_3^2$. We need an $R = [R_{ij}]$ such that $D = [D_{ij}] = R^t WR$ is given by Theorem 7.25 and $\begin{bmatrix} x_2 \\ x_3 \end{bmatrix} = R\begin{bmatrix} y_2 \\ y_3 \end{bmatrix}$.

$Q'(x_2, x_3)$ has a positive minimum value of 1 when $x_2^* = 1$ and $x_3^* = 0$. Note that $D_{11} = W_{11} = 1$. We next consider

$$u_0 \cdot x_2^* - v_0 \cdot x_3^* = 1$$
$$u_0 \cdot 1 - v_0 \cdot 0 = 1$$

which can be solved by inspection with $u_0 = 1$ and $v_0 = 2$ so that the general solution is

$$u = u_0 + tx_3^* = 1$$
$$v = v_0 + tx_2^* = 2 + t$$

Thus,

$$R = \begin{bmatrix} x_2^* & v \\ x_3^* & u \end{bmatrix} = \begin{bmatrix} 1 & v \\ 0 & u \end{bmatrix}$$

Next,

$$\begin{aligned} D_{12} &= v_0 [W_{11}x_2^* + W_{12}x_3^*] + u_0 [W_{12}x_2^* + W_{22}x_3^*] + ta \\ &= 2[1 \cdot 1 + (-5) \cdot 0] + 1[(-5) \cdot 1 + 26 \cdot 0] + t \\ &= -3 + t \end{aligned}$$

Choose t so that $2 \cdot |D_{12}| = 2 \cdot |-3 + t| \leq 1 = a$. In this case, $t = 3$. Thus,

$$R = \begin{bmatrix} 1 & v \\ 0 & u \end{bmatrix} = \begin{bmatrix} 1 & 2 + t \\ 0 & 1 \end{bmatrix} = \begin{bmatrix} 1 & 5 \\ 0 & 1 \end{bmatrix}$$

Clearly, $D = R^t WR = I_2$ is the 2×2 identity matrix. This result gives

$$N = \begin{bmatrix} 1 & T & Z \\ 0 & 1 & 5 \\ 0 & 0 & 1 \end{bmatrix}$$

Continuing with the notation of Lemma 7.31, identify A with M:

$$B_{11} = A_{11}$$
$$B_{12} = TA_{11} + R_{11}A_{12} + A_{13}R_{21} = T \cdot 1 + 1 \cdot 5 + 1 \cdot 0 = T + 5$$
$$B_{13} = ZA_{11} + A_{12}R_{12} + A_{13}R_{22} = Z \cdot 1 + 5 \cdot 5 + 1 \cdot 1 = Z + 26$$

Select T and Z such that

$$2 \cdot |B_{12}| = 2 \cdot |T + 5| < A_{11} = 1$$

and

$$2 \cdot |B_{13}| = 2 \cdot |Z + 26| < A_{11} = 1$$

Thus, $T = -5$ and $Z = -26$ giving

$$N = \begin{bmatrix} 1 & -5 & -26 \\ 0 & 1 & 5 \\ 0 & 0 & 1 \end{bmatrix}$$

It is straightforward to verify by direct calculation that $N^t A N = I_3$ so that if we let $x = Ny$ then $Q_2(y_1, y_2, y_3) = y^t B y = y_1^2 + y_2^2 + y_3^2$ with $B = N^t A N$ is equivalent to $Q(x_1, x_2, x_3)$.

Example 7.34 *Let the quadratic form $Q(x_1, x_2, x_3)$ have the corresponding matrix*

$$A = \begin{bmatrix} 5 & 34 & 1 \\ 34 & 233 & 0 \\ 1 & 0 & 26 \end{bmatrix}$$

so that $Q(x_1, x_2, x_3) = 5x_1^2 + 68x_1x_2 + 2x_1x_3 + 233x_2^2 + 26x_3^2$. Since $d(Q) = \det(A) = 1$, we want an equivalent quadratic form that is the sum of three squared variables.

We know that since $\det(A) = 1$ we can find $Q_1(y_1, y_2, y_3)$ with corresponding matrix $B = M^t A M$ so that $B_{11} = 1$ and if

$$x = \begin{bmatrix} x_1 \\ x_2 \\ x_3 \end{bmatrix} \quad and \quad y = \begin{bmatrix} y_1 \\ y_2 \\ y_3 \end{bmatrix}$$

then $x = My$. We now proceed to construct M.

Since $Q(x_1, x_2, x_3) = 5x_1^2 + 68x_1x_2 + 2x_1x_3 + 233x_2^2 + 26x_3^2$, we can use the equation

$$A_{11}Q(x_1, x_2, x_3) = V^2 + Q'(x_2, x_3)$$

where $V = A_{11}x_1 + A_{12}x_2 + A_{13}x_3$ and $Q'(x_2, x_3)$ has corresponding matrix

$$\begin{bmatrix} A_{11}A_{22} - A_{12}^2 & A_{11}A_{23} - A_{12}A_{13} \\ A_{11}A_{23} - A_{12}A_{13} & A_{11}A_{33} - A_{13}^2 \end{bmatrix}$$

to obtain

$$5Q(x_1, x_2, x_3) = (5x_1 + 34x_2 + x_3)^2 + Q'(x_2, x_3)$$

where $Q'(x_2, x_3) = 9x_2^2 - 68x_2x_3 + 129x_3^2$ has a minimum positive value of 1 at $(4, 1)$. The minimum for $V = |5x_1 + 34x_2 + x_3| = |5x_1 + 34(4) + 1|$ occurs when $x_1 = -27$. We let -27, 4, and 1 form the first column for M and fill in the rest of M with integers so that $\det(M) = 1$. We will use

$$M = \begin{bmatrix} -27 & 0 & 26 \\ 4 & 1 & 0 \\ 1 & 0 & -1 \end{bmatrix}$$

Hence

$$B = \begin{bmatrix} -27 & 4 & 1 \\ 0 & 1 & 0 \\ 26 & 0 & -1 \end{bmatrix} \begin{bmatrix} 5 & 34 & 1 \\ 34 & 233 & 0 \\ 1 & 0 & 26 \end{bmatrix} \begin{bmatrix} -27 & 0 & 26 \\ 4 & 1 & 0 \\ 1 & 0 & -1 \end{bmatrix}$$

$$= \begin{bmatrix} 1 & 14 & 53 \\ 14 & 233 & 884 \\ 53 & 884 & 3354 \end{bmatrix}$$

and

$$Q_1(y_1, y_2, y_3) = y_1^2 + 28y_1y_2 + 106y_1y_3 + 233y_2^2 + 1768y_2y_3 + 3354y_3^2$$

$$= (y_1 + 14y_2 + 53y_3)^2 + (37y_2^2 + 284y_2y_3 + 545y_3^2)$$

$Q_1'(y_2, y_3) = 37y_2^2 + 284y_2y_3 + 545y_3^2$ has a minimum positive value 1 at $(-4, 1)$.
We now want to form $Q_2(z_1, z_2, z_3)$ with corresponding matrix $C = M_1^t B M_1$ so that if $y = \begin{bmatrix} y_1 & y_2 & y_3 \end{bmatrix}^t$ and $z = \begin{bmatrix} z_1 & z_2 & z_3 \end{bmatrix}^t$, then $y = M_1 z$ and $C = I_3$.
We now proceed to construct M_1. Since $Q_1'(y_2, y_3)$ has the minimum 1 at $(-4, 1)$, M_1 has the form

$$M_1 = \begin{bmatrix} 1 & v & w \\ 0 & -4 & s \\ 0 & 1 & t \end{bmatrix} \quad \text{where} \quad \begin{vmatrix} -4 & s \\ 1 & t \end{vmatrix} = 1$$

Thus,

$$\begin{aligned} y_1 &= z_1 + vz_2 + wz_3 \\ y_2 &= -4z_2 + sz_3 \\ y_3 &= z_2 + tz_3 \end{aligned}$$

Substituting into $Q_1'(y_2, y_3)$, we get $37(-4z_2 + sz_3)^2 + 284(-4z_2 + sz_3)(z_2 + tz_3) + 545(z_2 + tz_3)^2$. When simplified, the z_2z_3 coefficient is $-12s - 46t$. To get rid of the z_2z_3 term we set $-12s - 46t = 0$ and also require that $\begin{vmatrix} -4 & s \\ 1 & t \end{vmatrix} = 1$ so that $-4t - s = 1$. Solving simultaneously the equations

$$\begin{aligned} -12s - 46t &= 0 \\ -s - 4t &= 1 \end{aligned}$$

we get $s = 23$ and $t = -6$. Hence M_1 has the form

$$M_1 = \begin{bmatrix} 1 & v & w \\ 0 & -4 & 23 \\ 0 & 1 & -6 \end{bmatrix}$$

But
$$\begin{aligned} V_1 &= y_1 + 14y_2 + 53y_3 \\ &= z_1 + vz_2 + wz_3 + 14(-4z_2 + 23z_3) + 53(z_2 - 6z_3) \end{aligned}$$

The coefficient of z_2 is $v - 56 + 53$ and the coefficient of z_3 is $w + 322 - 318$. To get rid of the z_2 and z_3 terms in V_1 we set their coefficients equal to 0 and get $v = 3$ and $w = -4$. Hence our matrix M_1 is

$$M_1 = \begin{bmatrix} 1 & 3 & -4 \\ 0 & -4 & 23 \\ 0 & 1 & -6 \end{bmatrix}$$

We now let $Q_2(z_1, z_2, z_3)$ have corresponding matrix $C = M_1^t B M_1$ so that if $y = \begin{bmatrix} y_1 & y_2 & y_3 \end{bmatrix}^t$ and $z = \begin{bmatrix} z_1 & z_2 & z_3 \end{bmatrix}^t$, then $y = M_1 z$ and $C = I_3$. Hence $Q_2(z_1, z_2, z_3) = z_1^2 + z_2^2 + z_3^2$.

Summarizing, we have

$$Q = x^t A x = y^t M^t A M y = z^t M_1^t M^t A M M_1 z = z^t (MM_1)^t A(MM_1)z$$

$$x = My = M(M_1 z) = (MM_1)z \text{ since } y = M_1 z, \text{ and } I_3 = (MM_1)^t A(MM_1).$$

Example 7.35 *In Examples 7.28, 7.29, 7.33, and 7.34 we needed to determine values of the variables in binary positive definite quadratic forms that caused the quadratic form to have its minimum positive value. If the coefficient of a squared term is 1 as in the case of*

$$Q_A(x_2, x_3) = x_2^2 - 10x_2x_3 + 26x_3^2$$

it is obvious that $x_2 = 1$ and $x_3 = 0$ give the value 1, which must be the minimum positive value. For a positive definite quadratic form such as

$$Q_B(x_2, x_3) = 9x_2^2 - 68x_2x_3 + 129x_3^2$$

determining the minimum positive value and finding x_2 and x_3 producing the minimum may not be obvious. We present a general procedure for obtaining the positive minimum of a binary quadratic form.

Suppose that

$$Q(x, y) = ax^2 + bxy + cy^2$$

is a positive definite binary quadratic form and suppose that m is a positive integer. We want integers x and y so that

$$Q(x, y) = ax^2 + bxy + cy^2 = m$$

First, we look for solutions x and y with $y \neq 0$. Dividing by y^2, we have

$$ar^2 + br + c = \frac{m}{y^2} \text{ where } r = \frac{x}{y}$$

If there is a real solution r, then

$$r = \frac{b \pm \sqrt{b^2 - 4a\left(c - m/y^2\right)}}{2a}$$

In order for r to be rational, we need an integer w so that

$$b^2 - 4a\left(c - \frac{m}{y^2}\right) = w^2$$

or

$$\frac{y^2\left(b^2 - 4ac\right) + 4am}{y^2} = w^2$$

Thus, we also need an integer z such that

$$y^2(b^2 - 4ac) + 4am = z^2$$

or

$$z^2 - (b^2 - 4ac)y^2 = 4am$$

Since Q is positive definite, Theorem 7.24 guarantees that $d = b^2 - 4ac < 0$ and we have Pell's equation,

$$z^2 - dy^2 = 4am$$

with $d < 0$. There are at most finitely many integral solutions z and y. Clearly, any solution integers z and y have the property

$$|z| \leq \sqrt{4am} \text{ and } |y| \leq \sqrt{\frac{4am}{d}}$$

There are only finitely many pairs x and z to check for a given m so that a solution with $y \neq 0$ may be found in a finite and bounded number of steps.
    For

$$Q(x,y) = 9x^2 - 68xy + 129y^2$$

we obtain $d = b^2 - 4ac = -20$; and we want to solve

$$z^2 + 20y^2 = 36m$$

where

$$|z| \leq 6\sqrt{m} \text{ and } |y| \leq (9/5)\sqrt{m}$$

    We do not know ahead of time what the minimum positive value of Q will be; however, we do know that $m \leq \min\{a, b\} = 9$ because $Q(1, 0) = 9$. We will try to solve $Q(x, y)$ for $m = 1, 2, \ldots, \min\{a, b\}$ since for each such m, the *solutions, if any, will be found in a finite number of steps depending upon a, m, and d; therefore, the minimum will be found in a finite number of steps.*

  For $m = 1$, *we try* $|z| \leq 6$ *and* $|y| \leq 1$. *Thus,* y *is limited to be* ± 1. *We then have*

$$z^2 + 20(\pm 1)^2 = 36 \ or \ z = \pm 4$$

$$w^2 = \frac{z^2}{y^2} = \frac{16}{1} = 16$$

$$\frac{x}{y} = r = \frac{68 \pm \sqrt{16}}{18} = \frac{4}{1} \ or \ \frac{32}{9}$$

so that

$$x = 4y = \pm 4$$

Therefore, $m = 1$ *is the minimum positive value of* Q *and the minimum is achieved at* $(x, y) = (-4, -1)$ *and* $(4, 1)$.

  Note that a possible minimum corresponding to a solution with $y = 0$ *is quickly compared to the one obtained above since* $Q(x, 0) = ax^2$.

Exercises

1. Determine which of the following are positive definite:

(a) $\begin{bmatrix} 2 & 4 \\ 2 & 3 \end{bmatrix}$
   (b) $\begin{bmatrix} 1 & 0 & 0 \\ 0 & 3 & 0 \\ 0 & 0 & 8 \end{bmatrix}$

(c) $\begin{bmatrix} 2 & -1 & 1 \\ -1 & 4 & 1 \\ 1 & 1 & 4 \end{bmatrix}$
   (d) $\begin{bmatrix} 1 & 3 & 0 \\ 3 & 2 & 1 \\ 0 & 1 & 3 \end{bmatrix}$

2. Determine which of the following are positive definite:

(a) $\begin{bmatrix} 1 & 0 & 0 \\ 0 & 3 & 0 \\ 0 & 0 & -8 \end{bmatrix}$
   (b) $\begin{bmatrix} -1 & 6 & 5 \\ 6 & 3 & 2 \\ 5 & 2 & 8 \end{bmatrix}$

(c) $\begin{bmatrix} 2 & 1 & 1 \\ 1 & 2 & 2 \\ 1 & 2 & 5 \end{bmatrix}$
   (d) $\begin{bmatrix} 2 & -2 \\ 2 & 3 \end{bmatrix}$

3. If

$$Q(x_1, x_2, x_3) = \begin{bmatrix} x_1 & x_2 & x_3 \end{bmatrix} \begin{bmatrix} 2 & -1 & 1 \\ -1 & 4 & 1 \\ 1 & 1 & 4 \end{bmatrix} \begin{bmatrix} x_1 \\ x_2 \\ x_3 \end{bmatrix}$$

expand $Q(x_1, x_2, x_3)$ as a sum.

4. If

$$Q(x_1, x_2, x_3) = \begin{bmatrix} x_1 & x_2 & x_3 \end{bmatrix} \begin{bmatrix} 2 & 5 & -1 \\ 5 & 0 & 3 \\ -1 & 3 & 4 \end{bmatrix} \begin{bmatrix} x_1 \\ x_2 \\ x_3 \end{bmatrix}$$

expand $Q(x_1, x_2, x_3)$ as a sum.

5. Find the following matrices:

(a) If $Q(x, y) = x^2 + 4xy + y^2 = v^t A v$ where A is a symmetric matrix and $v = [x, y]^t$, then find A.

(b) If $Q(x, y) = x^2 - 2xy + y^2 = v^t A v$ where A is a symmetric matrix and $v = [x, y]^t$, then find A.

(c) If $Q(x, y, z) = x^2 + 4xy - 2xz + y^2 + z^2 = v^t A v$ where A is a symmetric matrix and $v = [x, y, z]^t$, then find A.

6. Find the following matrices:

(a) If $Q(x, y) = 2x^2 - 2xy + 2y^2 = v^t A v$ where A is a symmetric matrix and $v = [x, y]^t$, then find A.

(b) If $Q(x, y) = x^2 + 2xy + y^2 = v^t A v$ where A is a symmetric matrix and $v = [x, y]^t$, then find A.

(c) If $Q(x, y, z) = x^2 + 6yz - 2xz + 2y^2 + 4z^2 = v^t A v$ where A is a symmetric matrix and $v = [x, y, z]^t$, then find A.

7. If

$$Q(x_1, x_2, x_3) = \begin{bmatrix} x_1 & x_2 & x_3 \end{bmatrix} \begin{bmatrix} 2 & -1 & 1 \\ -1 & 4 & 1 \\ 1 & 1 & 4 \end{bmatrix} \begin{bmatrix} x_1 \\ x_2 \\ x_3 \end{bmatrix}$$

and

$$\begin{bmatrix} x_1 \\ x_2 \\ x_3 \end{bmatrix} = \begin{bmatrix} 2 & 0 & -1 \\ 0 & 2 & -1 \\ -1 & -1 & 1 \end{bmatrix} \begin{bmatrix} y_1 \\ y_2 \\ y_3 \end{bmatrix}$$

Find $Q_1(y_1, y_2, y_3)$ equivalent to $Q(x_1, x_2, x_3)$.

8. If

$$Q(x_1, x_2, x_3) = \begin{bmatrix} x_1 & x_2 & x_3 \end{bmatrix} \begin{bmatrix} 1 & -1 & 3 \\ -1 & -2 & 1 \\ 3 & 1 & 4 \end{bmatrix} \begin{bmatrix} x_1 \\ x_2 \\ x_3 \end{bmatrix}$$

and

$$\begin{bmatrix} x_1 \\ x_2 \\ x_3 \end{bmatrix} = \begin{bmatrix} 1 & 0 & -1 \\ 0 & 1 & -1 \\ -1 & -1 & 1 \end{bmatrix} \begin{bmatrix} y_1 \\ y_2 \\ y_3 \end{bmatrix}$$

Find $Q_1(y_1, y_2, y_3)$ equivalent to $Q(x_1, x_2, x_3)$.

9. In Example 7.29 we saw that to construct M, we found values x_1^* and x_2^* so that $Q(x_1^*, x_2^*) = 1$. Possible values were $(-2, 1)$, $(2, -1)$, $(-3, 1)$, and $(3, -1)$. Find M using the values:

    (a) $(2, -1)$ **(b)** $(-3, 1)$ **(c)** $(3, -1)$

10. In Example 7.33, in addition to $Q'(x_2, x_3) = x_2^2 - 10x_2x_3 + 26x_3^2$ having minimum 1 at $x_2 = 1$ and $x_3 = 0$, it also has minimum 1 at $x_2 = 5$ and $x_3 = 1$. Using these numbers to construct M, complete the example.

11. In Example 7.34, $Q_1'(y_2, y_3) = 37y_2^2 + 284y_2y_3 + 545y_3^2$ has the minimum 1 at

$$
\begin{aligned}
y_2 &= -4, & y_3 &= 1 \\
y_2 &= 4, & y_3 &= -1 \\
y_2 &= -23, & y_3 &= 6 \\
y_2 &= 23, & y_3 &= -6
\end{aligned}
$$

    (a) Complete the example when $y_2 = 4$, $y_3 = -1$.

    (b) Complete the example when $y_2 = -23$, $y_3 = 6$.

    (c) Complete the example when $y_2 = 23$, $y_3 = -6$.

12. Given the quadratic form $Q(x_1, x_2, x_3)$ with associated matrix A, find an equivalent form $Q_1(y_1, y_2, y_3) = y_1^2 + y_2^2 + y_3^2$ when:

    (a) $A = \begin{bmatrix} 2 & 1 & -1 \\ 1 & 1 & 0 \\ -1 & 0 & 2 \end{bmatrix}$ **(b)** $A = \begin{bmatrix} 10 & 3 & 6 \\ 3 & 1 & 2 \\ 6 & 2 & 5 \end{bmatrix}$

    (c) $A = \begin{bmatrix} 5 & 12 & 1 \\ 12 & 29 & 0 \\ 1 & 0 & 30 \end{bmatrix}$ **(d)** $A = \begin{bmatrix} 1 & 3 & 1 \\ 3 & 10 & 0 \\ 1 & 0 & 11 \end{bmatrix}$

    (e) $A = \begin{bmatrix} 2 & 5 & 1 \\ 5 & 13 & 0 \\ 1 & 0 & 14 \end{bmatrix}$ **(f)** $A = \begin{bmatrix} 5 & 17 & 1 \\ 17 & 58 & 0 \\ 1 & 0 & 59 \end{bmatrix}$

13. Prove Theorem 7.16.

14. Prove Theorem 7.18.

15. Prove Theorem 7.20.

16. Prove Theorem 7.23.

7.5 Integers as Sums of Three Squares

Evidently, 6 is uniquely expressed as the sum of three squares since $6 = 2^2 + 1^2 + 1^2$, while 7 is uniquely expressed as the sum of four squares since $7 = 2^2 + 1^2 + 1^2 + 1^2$. In the following theorem, the conditions under which a positive integer can be represented as the sum of three squares are shown.

Theorem 7.36 *The Diophantine equation $x_1^2 + x_2^2 + x_3^2 = n$ where n is a positive integer has integral solutions if and only if n is not of the form $4^k(8m+7)$.*

Proof. Since $1^2 \equiv 3^2 \equiv 5^2 \equiv 7^2 \equiv 1 \pmod 8$, $0^2 \equiv 4^2 \equiv 0 \pmod 8$, and $2^2 \equiv 6^2 \equiv 4 \pmod 8$, it is impossible to find three squares whose sum is 7 modulo 8. Assume that $4^k(8m + 7)$ can be expressed as the sum of three squares for some k and m. Let k' be the least nonnegative integer such that $4^{k'}(8m + 7)$ is the sum of three positive squares. If $k' \geq 1$, by Theorem 7.6, each of the squares must be divisible by 4 and $4^{k'-1}(8m + 7)$ is the sum of three squares, contradicting the definition of k' being minimal. So $k' = 0$. But $k' = 0$ contradicts the fact that the sum of three squares of positive integers n cannot be congruent to 7.

Conversely, we show that any integer of the form $8k + r$ where $r = 1$, 2, 3, 5, and 6 can be the sum of 3 squares. To do so, we will show that n is representable by some positive definite quadratic form $Q(x_1, x_2, x_3) = x^t A x$; that is, we construct a symmetric A and an x so that $n = x^t A x$ where

$d(Q) = 1, \begin{vmatrix} A_{11} & A_{12} \\ A_{21} & A_{22} \end{vmatrix} > 0$, and $A_{11} > 0$. Thus, we will construct a

reduced matrix of the type required in Theorem 7.30 for Q to be positive definite. Theorem 7.32 will then give the final result.

Let $A_{13} = A_{31} = 1$, $A_{23} = A_{32} = 0$, and $A_{33} = n$, so that

$$A = \begin{bmatrix} A_{11} & A_{12} & 1 \\ A_{21} & A_{22} & 0 \\ 1 & 0 & n \end{bmatrix}$$

and let $x = \begin{bmatrix} 0 & 0 & 1 \end{bmatrix}^t$ so that $x^t A x = n$. It can be shown that

$$d(Q) = \det(A) = n \begin{vmatrix} A_{11} & A_{12} \\ A_{21} & A_{22} \end{vmatrix} - A_{22}$$

by direct computation. Let $q = \begin{vmatrix} A_{11} & A_{12} \\ A_{21} & A_{22} \end{vmatrix}$ so that $A_{22} = nq - 1$ since we want $\det(A) = 1$. Hence A_{22} is positive. Since $A_{11}A_{22} = A_{12}^2 + q$ is positive, A_{11} is positive. Now we need to find q so that there is an integer A_{12} such that A_{11} satisfies the equation $A_{11}A_{22} = A_{12}^2 + q$. This result is true if $A_{12}^2 \equiv -q \pmod{A_{22}} \equiv -q \pmod{(nq - 1)}$.

We first consider the case where $r = 2$ or 6, i.e. $n \equiv 2 \pmod 4$. Since $4n$ and $n - 1$ are relatively prime, by Dirichlet's Theorem 2.18, there exists an m so that $4mn + (n - 1)$ is a prime, say p. Let $q = 4m + 1$ so that

$p = n(4m + 1) - 1 = nq - 1$. Thus, $p = 4mn + (4k + 2) - 1$ for some integer k since $n \equiv 2 \pmod 4$. Since $p = nq - 1 = A_{22}$, we want to find a solution to $x^2 \equiv -q \pmod p$ and let A_{12} be the solution; that is, we want to show using the Jacobi and Legendre symbols that $\left(\dfrac{-q}{p}\right) = 1$. Thus,

$$
\begin{aligned}
\left(\frac{-q}{p}\right) &= \left(\frac{-1}{p}\right)\left(\frac{q}{p}\right) = (-1)^{(p-1)/2}\left(\frac{q}{p}\right) \\
&= \left(\frac{q}{p}\right) \quad \text{since } p \equiv 1 \pmod 4 \\
&= \left(\frac{p}{q}\right) \quad \text{since } p \equiv 1 \pmod 4 \\
&= \left(\frac{nq - 1}{q}\right) = \left(\frac{-1}{q}\right) \quad \text{since } nq - 1 \equiv -1 \pmod q
\end{aligned}
$$

Hence $A_{12}^2 \equiv -q \pmod{A_{22}}$ is solvable since

$$
\left(\frac{-q}{A_{22}}\right) = \left(\frac{-q}{p}\right) = 1
$$

Thus, A_{11} exists. $A_{22} = nq - 1$, and A_{12} is a solution of $A_{12}^2 \equiv -q \pmod{A_{22}}$ so that A may be constructed when $n \equiv 2$ or $6 \pmod 8$.

Assume that $r = 1$, 3, or 5 where $n \equiv r \pmod 8$. We choose c so that $cn - 1 \equiv 2 \pmod 4$. For the case of $r = 3$, we could let $c = 5$ or 1, say. The numbers $4n$ and $(cn - 1)/2$ are relatively prime, so by Dirichlet's Theorem, there exists an s so that $4sn + (cn - 1)/2$ is a prime, say p'. Let $q = 8s + c$ so that $2p' = 8sn + nc - 1 = nq - 1$. Since we are seeking a solution to $A_{12}^2 \equiv -q \pmod{A_{22}}$, we need to show that $-q$ is a quadratic residue modulo $2p' = nq - 1 = A_{22}$; but this result is true if $-q$ is a quadratic residue modulo 2 and $-q$ is a quadratic residue modulo p'. But $q \equiv 1 \pmod 4$ implies that $q \equiv 1 \pmod 2$ and $-q \equiv -1 \equiv 1 \pmod 2$ so that $-q$ is a quadratic residue modulo 2.

We next show that $-q$ is a quadratic residue modulo p'. There are three cases: $r = 1$, $r = 3$, and $r = 5$. We will argue here only the case of $r = 3$. In this case, we will let $c = 5$. Using the Legendre symbol, Theorem 3.96 implies that

$$
\left(\frac{-q}{p'}\right) = \left(\frac{-1}{p'}\right)\left(\frac{q}{p'}\right)
$$

But since n has the form $8k + r = 8k + 3$ and $c = 5$, $2p' = 8sn + 5(8k + 3) - 1 = 8sn + 40k + 14$ so that $p' = 4sn + 20k + 7$, $p' \equiv 3 \pmod 4$, and $\left(\dfrac{-1}{p'}\right) = -1$ using Theorem 3.96.

Because $q \equiv 1 \pmod 4$, we obtain by the Law of Quadratic Reciprocity

$$
\left(\frac{q}{p'}\right) = \left(\frac{p'}{q}\right)
$$

$$= -\left(\frac{p'}{q}\right)\left(\frac{2}{q}\right) \text{ since } \left(\frac{2}{q}\right) = -1$$

$$= -\left(\frac{2p'}{q}\right) = -\left(\frac{nq-1}{q}\right)$$

$$= -\left(\frac{-1}{q}\right) = -1$$

Hence, $\left(\frac{-q}{p'}\right) = 1$ and $-q$ is a quadratic residue modulo p'. Since, from above, $-q$ is also a quadratic residue modulo 2, $-q$ is a quadratic residue modulo $2p'$ also. Hence there is a solution to $A_{12}^2 \equiv -q \pmod{2p'}$ or $A_{12}^2 \equiv -q \pmod{A_{22}}$ when $r = 3$. The cases of $r = 1$ or 5 may be shown similarly.

As before, we have shown that there exist values for A_{11}, A_{12}, A_{21}, and A_{22} so that n can be represented as, and is hence a sum of, three squares. ∎

Example 7.37 *In this example we find 27 as the sum of three squares. We begin by constructing the matrix A corresponding to $Q(x_1, x_2, x_3)$ so that $Q(0,0,1) = 27$. As in the construction used in the proof of Theorem 7.36, we let $A_{13} = 1$, $A_{23} = 0$, and $A_{33} = 27$. Hence our matrix is*

$$A = \begin{bmatrix} A_{11} & A_{12} & 1 \\ A_{21} & A_{22} & 0 \\ 1 & 0 & 27 \end{bmatrix}$$

Since $27 \equiv 3 \pmod 8$ we find c so that $c(27) - 1 \equiv 2 \pmod 4$. We select $c = 1$ making $(cn - 1)/2 = 13$. Choose s so that $4sn + (cn-1)/2 = 4(27)s + 13$ is prime. We select $s = 0$ and $p = 13$ and obtain $q = 8s + c = 1$ and $A_{22} = qn - 1 = 27 - 1 = 26$. We need A_{12} so that $A_{12}^2 \equiv -q \pmod{26}$, that is, $A_{12}^2 \equiv -1 \pmod{26}$. We select $A_{12} = 5$ giving $A_{11} = (q + A_{12}^2)/A_{22} = (1 + 25)/26 = 1$. Now our matrix A is

$$A = \begin{bmatrix} 1 & 5 & 1 \\ 5 & 26 & 0 \\ 1 & 0 & 27 \end{bmatrix}$$

Thus, $Q(x_1, x_2, x_3) = x_1^2 + 10x_1x_2 + 2x_1x_3 + 26x_2^2 + 27x_3^2$. We wish to find $Q_1(y_1, y_2, y_3)$ with corresponding matrix $B = M^t A M$ so that $x = My$ and $Q_1(y_1, y_2, y_3) = y_1^2 + y_2^2 + y_3^2$. By Example 7.33, we know that

$$M = \begin{bmatrix} 1 & -5 & -26 \\ 0 & 1 & 5 \\ 0 & 0 & 1 \end{bmatrix}$$

If the quadratic form $Q_1(y_1, y_2, y_3)$ has corresponding matrix $M^t A M$, then $Q_1(y_1, y_2, y_3) = y_1^2 + y_2^2 + y_3^2$.

Since $Q(0, 0, 1) = 27$, we have, from $x = My$, that

$$\begin{aligned} 0 &= x_1 &= y_1 - 5y_2 - 26y_3 \\ 0 &= x_2 &= y_2 + 5y_3 \\ 1 &= x_3 &= y_3 \end{aligned}$$

Solving these simultaneous equations, we have $y_1 = 1$, $y_2 = -5$, $y_3 = 1$, and $Q_1(1, -5, 1) = Q(0, 0, 1) = 27$. Thus, $1^2 + (-5)^2 + 1^2 = 27$; or using only positive integers, we have $27 = 1^2 + 5^2 + 1^2$ as the sum of three squares. This representation is not unique since also $27 = 3^2 + 3^2 + 3^2$.

Example 7.38 *As in Example 7.37, we seek to express an integer as the sum of three squares. This time we select $n = 26$ and again seek a matrix A corresponding to a quadratic form $Q(x_1, x_2, x_3)$ such that $Q(0, 0, 1) = n = 26$. Let $A_{13} = 1$, $A_{23} = 0$, and $A_{33} = 26$. Hence our matrix is*

$$A = \begin{bmatrix} A_{11} & A_{12} & 1 \\ A_{21} & A_{22} & 0 \\ 1 & 0 & 26 \end{bmatrix}$$

Since $26 \equiv 2 \pmod 8$, we proceed as follows, noting that

$$\det(A) = \begin{vmatrix} A_{11} & A_{12} \\ A_{12} & A_{22} \end{vmatrix} n - A_{22}$$

Letting $q = \begin{vmatrix} A_{11} & A_{12} \\ A_{12} & A_{22} \end{vmatrix}$, we have $A_{22} = nq-1$. Using Dirichlet's theorem we seek s so that $4sn+(n-1)$ is a prime, say p. When $s = 2$, $4(2)(26)+25 = 233$, which is prime, and $A_{22} = 233 = p$ so that $q = (p+1)/n = 234/26 = 9$. We choose A_{12} such that $A_{12}^2 \equiv -q \pmod{A_{22}}$, that is, $A_{12}^2 \equiv -9 \pmod{233}$. The smallest solution is $A_{12} = 34$. Since $q = \begin{vmatrix} A_{11} & A_{12} \\ A_{12} & A_{22} \end{vmatrix}$, $9 = A_{11}(233) - (34)^2$ so that $A_{11} = 5$ and our matrix A is

$$A = \begin{bmatrix} 5 & 34 & 1 \\ 34 & 233 & 0 \\ 1 & 0 & 26 \end{bmatrix}$$

We need $A_{11} = 1$. We know that since $\det(A) = 1$, we can find $Q_1(y_1, y_2, y_3)$ with corresponding matrix $B = M^t A M$ so that $B_{11} = 1$. From Example 7.34 we find that

$$M = \begin{bmatrix} -27 & 0 & 26 \\ 4 & 1 & 0 \\ 1 & 0 & -1 \end{bmatrix}$$

Hence

$$B = \begin{bmatrix} -27 & 4 & 1 \\ 0 & 1 & 0 \\ 26 & 0 & -1 \end{bmatrix} \begin{bmatrix} 5 & 34 & 1 \\ 34 & 233 & 0 \\ 1 & 0 & 26 \end{bmatrix} \begin{bmatrix} -27 & 0 & 26 \\ 4 & 1 & 0 \\ 1 & 0 & -1 \end{bmatrix}$$

$$= \begin{bmatrix} 1 & 14 & 53 \\ 14 & 233 & 884 \\ 53 & 884 & 3354 \end{bmatrix}$$

and

$$Q_1(y_1, y_2, y_3) = y_1^2 + 28y_1y_2 + 106y_1y_3 + 233y_2^2 + 1768y_2y_3 + 3354y_3^2$$

We now want to form $Q_2(z_1, z_2, z_3)$ with corresponding matrix $C = M_1^t B M_1$ so if $y = \begin{bmatrix} y_1 & y_2 & y_3 \end{bmatrix}^t$ and $z = \begin{bmatrix} z_1 & z_2 & z_3 \end{bmatrix}^t$, then $y = M_1 z$ and $C = I_3$. In Example 7.34 we found a matrix M_1 such that $x = My$ and $y = M_1 z$. Further, the quadratic form

$$Q_2(z_1, z_2, z_3) \quad = \quad z^t M_1^t M^t A M M_1 z = z_1^2 + z_2^2 + z_3^2$$

$$= \quad z^t (M M_1)^t A (M M_1) \quad \text{since } B = M^t A M$$

is equivalent to Q provided that

$$M = \begin{bmatrix} -27 & 0 & 26 \\ 4 & 1 & 0 \\ 1 & 0 & -1 \end{bmatrix} \quad \text{and } M_1 = \begin{bmatrix} 1 & 3 & -4 \\ 0 & -4 & 23 \\ 0 & 1 & -6 \end{bmatrix}$$

Therefore, $x = My = MM_1 z$. Since

$$MM_1 = \begin{bmatrix} -27 & 0 & 26 \\ 4 & 1 & 0 \\ 1 & 0 & -1 \end{bmatrix} \begin{bmatrix} 1 & 3 & -4 \\ 0 & -4 & 23 \\ 0 & 1 & -6 \end{bmatrix} = \begin{bmatrix} -27 & -55 & -48 \\ 4 & 8 & 7 \\ 1 & 2 & 2 \end{bmatrix}$$

and since we want $Q(0, 0, 1) = 26$, we need to solve

$$\begin{aligned} 0 &= x_1 = & -27z_1 &- & 55z_2 &- & 48z_3 \\ 0 &= x_2 = & 4z_1 &+ & 8z_2 &+ & 7z_3 \\ 1 &= x_3 = & z_1 &+ & 2z_2 &+ & 2z_3 \end{aligned}$$

The solution is $z_1 = -1$, $z_2 = -3$, and $z_3 = 4$; thus, $(-1)^2 + (-3)^2 + (4)^2 = 26$. Using only positive integers we have $1^2 + 3^2 + 4^2 = 26$.

Theorem 7.39 *Every nonnegative integer can be represented as a sum of four squares.*

Proof. Since elements of the form $8k+3$ can be represented as a sum of three squares, every element of the form $8k + 7 = 8k + 3 + 2^2$ can be represented as the sum of four squares. ∎

Exercises

1. Using the procedures and examples above express the following as the sum of three squares:

   (a) 11 (see Exercise 12(d) in Section 7.4)

   (b) 30 (see Exercise 12(c) in Section 7.4)

   (c) 59 (see Exercise 12(f) in Section 7.4)

(d) 153

(e) 35

(f) 37

(g) 14 (see Exercise 12(e) in Section 7.4)

2. Express as the sum of three squares:

   (a) 32   **(b)** 12

3. In Exercise 10 of Section 7.4, Example 7.33 is completed in an alternative way. Find the effect (if any) this has on the expression of 27 as the sum of three squares.

4. In Exercise 11 of Section 7.4, Example 7.34 is completed in several alternative ways. Find the effect (if any) this has on the expression of 26 as the sum of three squares.

5. Prove or disprove the following: The product of two integers each of which is expressible as the sum of three squares of integers is expressible as the sum of three squares of integers.

6. Prove or disprove the following: The sum of two integers each of which is expressible as the sum of three squares of integers is expressible as the sum of three squares of integers.

7. Prove that if $4m$ is the sum of three squares, then m is the sum of three squares.

7.6   Integers as Sums of Four Squares

In Section 7.3 we showed that primes not of the form $4m + 3$ are the sum of two squares. In Section 7.5 we showed that every nonnegative integer of the form $8m + 3$ can be expressed as the sum of three squares; and therefore, every nonnegative integer of the form $8m + 7$ can be expressed as the sum of four squares. The current section provides an easier and more direct proof that any nonnegative integer can be expressed as the sum of four squares. It is first proved that any prime of the form $4m + 3$ can be expressed as the sum of four squares. We then use the fact that any prime can be expressed as the sum of four squares to prove that any nonnegative integer can be expressed as the sum of four squares.

Theorem 7.40 *The set of integers which can be expressed as the sum of four squares is closed under multiplication and hence forms a subsemigroup of the semigroup of integers.*

Proof.

$$(a_1^2 + b_1^2 + c_1^2 + d_1^2)(a_2^2 + b_2^2 + c_2^2 + d_2^2) = (a_1a_2 + b_1b_2 + c_1c_2 + d_1d_2)^2$$
$$+ (a_1b_2 - a_2b_1 + c_1d_2 - c_2d_1)^2$$
$$+ (a_1c_2 - b_1d_2 - a_2c_1 + b_2d_1)^2$$
$$+ (a_1d_2 + b_1c_2 - b_2c_1 - a_2d_1)^2 \blacksquare$$

Lemma 7.41 *Let p be a prime of the form $4m + 3$. Then there exists an integer k with $0 < k < p$ such that kp is the sum of four squares.*

Proof. We actually show that kp is of the form $a^2 + b^2 + 1^2 + 0^2$ and hence is the sum of three squares, one of which is 1. To show this result, we find an integer t such that there exist $a^2 \equiv -t \pmod{p}$ and $b^2 \equiv t - 1 \pmod{p}$. Since every prime p has $(p-1)/2$ residues and $(p-1)/2$ nonresidues, let t be the smallest positive nonresidue of p. Certainly $t > 1$. Using the Legendre symbol,

$$\left(\frac{-t}{p}\right) = \left(\frac{-1}{p}\right)\left(\frac{t}{p}\right) = (-1)(-1) = 1$$

because p has the form $4m + 3$ (Theorem 3.96). Hence $-t$ is a quadratic residue and we let $a^2 \equiv -t \pmod{p}$. Since t is the smallest nonresidue of p, then $t - 1$ is a quadratic residue so let $b^2 \equiv t - 1 \pmod{p}$ and $a^2 + b^2 + 1 = kp$ for some k. Both a and b may be chosen to be less than $p/2$; for if not, use $p - a$ or $p - b$ instead. Hence $4kp = 4(a^2 + b^2 + 1) < p^2 + p^2 + 4 = 2p^2 + 4 < 4p^2$ and $k < p$. $\blacksquare$

Theorem 7.42 *Any prime p of the form $4m + 3$ can be expressed as the sum of four squares.*

Proof. Let g be the smallest positive integer such that gp is the sum of four squares, say $gp = a^2 + b^2 + c^2 + d^2$. Then g is odd since if g is even, there are either 0, 2, or 4 even squares. In all cases they can be separated into pairs of odd squares and pairs of even integers. If there is a pair of odd integers or a pair of even integers, say x and y, then $(x - y)^2/2$ and $(x + y)^2/2$ are even since $x - y$ and $x + y$ are even and $x^2 + y^2 = (x - y)^2/2 + (x + y)^2/2$. Hence by replacing pairs of odd squares and even squares, we have $gp/2 = (a_1^2 + b_1^2 + c_1^2 + d_1^2)/4$, which contradicts the fact that g is minimal. Hence we assume that $gp = a^2 + b^2 + c^2 + d^2$ where g is odd. There exist a', b', c', and d' such that $a' \equiv a \pmod{g}$, $b' \equiv b \pmod{g}$, $c' \equiv c \pmod{g}$ and $d' \equiv d \pmod{g}$ where $2|a'|$, $2|b'|$, $2|c'|$, and $2|d'|$ are all less than g. This choice is possible since, for example, if $2|a'|$ is not less than g, use $g - a'$ for a', since $(g - a')^2 \equiv (a')^2 \pmod{g}$. Hence

$$(a')^2 + (b')^2 + (c')^2 + (d')^2 \equiv a^2 + b^2 + c^2 + d^2 \equiv 0 \pmod{g}$$

and

$$(a')^2 + (b')^2 + (c')^2 + (d')^2 = ng$$

for some n. Thus,

$$4\left((a')^2 + (b')^2 + (c')^2 + (d')^2\right) < 4g^2$$

so

$$(a')^2 + (b')^2 + (c')^2 + (d')^2 < g^2$$

and $n < g$. Further, $n > 0$; for otherwise $a' = b' = c' = d' = 0$ and a, b, c, and d are multiples of g, which contradicts the fact that g is minimal.

If we multiply $\left(a^2 + b^2 + c^2 + d^2\right)$ and $\left((a')^2 + (b')^2 + (c')^2 + (d')^2\right)$ we get

$$\begin{aligned}
g^2np &= \left(a^2 + b^2 + c^2 + d^2\right)\left((a')^2 + (b')^2 + (c')^2 + (d')^2\right) \\
&= r^2 + s^2 + t^2 + u^2
\end{aligned}$$

where

$$\begin{aligned}
r &= aa' + bb' + cc' + dd' \equiv a^2 + b^2 + c^2 + d^2 \equiv 0 \ (\text{mod } g) \\
s &= ab' - a'b + cd' - c'd \equiv ab - ab + cd - cd \equiv 0 \ (\text{mod } g) \\
t &= ac' - bd' - a'c + b'd \equiv ac - bd - ac + bd \equiv 0 \ (\text{mod } g) \\
u &= ad' + bc' - b'c - a'd \equiv ad + bc - bc - ad \equiv 0 \ (\text{mod } g)
\end{aligned}$$

Hence r^2, s^2, t^2, and u^2 are all divisible by g^2 and we obtain

$$np = \left(\frac{r}{g}\right)^2 + \left(\frac{s}{g}\right)^2 + \left(\frac{t}{g}\right)^2 + \left(\frac{u}{g}\right)^2$$

but n is less than g, contradicting the fact that g is minimal unless $g = n = 1$. Hence p is the sum of four squares. ■

Theorem 7.43 *Any nonnegative integer may be expressed as the sum of four squares.*

Proof. Since every prime may be expressed as a sum of four squares, by Theorems 7.42 and 7.10, all products of primes can be expressed as a sum of four squares. Since 0 and 1 can be expressed as the sum of four squares and all other nonnegative integers can be expressed as a product of primes, the theorem is proved. ■

Waring's Problem asks whether for each $k > 1$ there is a minimal integer $g(k)$ so that each positive integer can be expressed as the sum of $g(k)$ k-th powers of nonnegative integers. We just showed in Theorem 7.43 that $g(2) = 4$. It is known that $g(3) = 9$ and, more generally, that $g(k) = \lfloor (3/2)^k \rfloor + 2^k - 2$, where $\lfloor x \rfloor$ denotes the greatest integer function of x, is valid for all but possibly a finite number of k. It is also known that the $g(k) = \lfloor (3/2)^k \rfloor + 2^k - 2$ holds for all integers $k \le 471600000$. The fact that $g(k)$ exists for all $k > 1$ was proved by the German mathematician David Hilbert; however, Hilbert's proof was existential and no explicit method of obtaining the value of $g(k)$ has been found.

Exercises

1. Express 15 as the sum of 4 squares.

2. Express 23 as the sum of 4 squares.

3. Express $345 = (23)(15)$ as the sum of 4 squares.

4. Express 60 as the sum of 4 squares.

5. Express as the sum of 4 squares:

   (a) 120   **(b)** 2109

6. Express as the sum of 4 squares:

   (a) 1638   **(b)** 2926

7. Prove that every odd integer is the sum of four squares in which two are consecutive using the fact that $4n + 1$ is the sum of three squares and that if $4n + 1 = (2a)^2 + (2b)^2 + (2c + 1)^2$, then $2n + 1 = (a + b)^2 + (a - b)^2 + c^2 + (c + 1)^2$.

8. Show that there are infinitely many primes of the form $a^2 + b^2 + c^2 + 1$ using the fact that there is an infinite number of primes of the form $8m + 7$.

9. Find the numbers between 1 and 20 which cannot be expressed as four positive squares.

10. Find the numbers between 21 and 40 which cannot be expressed as four positive squares.

11. Show that 169 can be expressed as the sum of n positive squares where $n = 1, 2, 3$, and 4.

12. Prove that every integer greater than 169 is the sum of five nonzero squares by expressing $n - 169$ as the sum of 4 squares and considering cases when some of the squares are zero.

13. Prove that if $a^2 + b^2 + c^2 + d^2$ is a multiple of 8, then a, b, c, and d are all even.

14. Using Exercise 13 and induction, prove that $(2)(4^n)$ is never the sum of 4 positive squares.

7.7   The Equation $ax^2 + by^2 + cz^2 = 0$

In the following discussion we give necessary and sufficient conditions that the equation $ax^2 + by^2 + cz^2 = 0$ have a nontrivial solution. We begin by proving two lemmas.

Lemma 7.44 *Let a, b, and c be integers such that abc is positive. Then any congruence $Ax + By + Cz \equiv 0 \pmod{abc}$ has a nontrivial solution with*

$$x^2 \le |bc|, \quad y^2 \le |ac|, \quad and \quad z^2 \le |ab|$$

Proof. Consider x, y, and z such that $0 \le x^2 \le |bc|$, $0 \le y^2 \le |ac|$, and $0 \le z^2 \le |ab|$. There are $(1+|bc|)(1+|ac|)(1+|ab|)-1$ such triples (x^2, y^2, z^2) where x, y, and z are nonnegative and not all zero. But

$$(1 + |bc|)(1 + |ac|)(1 + |ab|) - 1 = |ab| + |ac| + |bc| + |a^2bc| + |ab^2c|$$

$$+ |abc^2| + a^2b^2c^2$$

$$> a^2b^2c^2$$

Hence, since there are as many triples of the form (x, y, z) as there are of the form (x^2, y^2, z^2) and since $a^2b^2c^2 \ge abc$, there are more than abc such triples (x, y, z). Therefore, there must be at least two triples (x, y, z) and (x', y', z') such that $Ax + By + Cz \equiv Ax' + By' + Cz' \pmod{abc}$. Hence $A(x-x')+B(y-y')+C(z-z') \equiv 0 \pmod{abc}$, which is the desired solution. ∎

Lemma 7.45 *If*

$$ax^2 + by^2 + cz^2 \equiv (a_1x + b_1y + c_1z)(a_2x + b_2y + c_2z) \pmod{n}$$

and

$$ax^2 + by^2 + cz^2 \equiv (a_1'x + b_1'y + c_1'z)(a_2'x + b_2'y + c_2'z) \pmod{m}$$

where m and n are relatively prime, then there exist integers A_1, B_1, C_1, A_2, B_2, and C_2 such that

$$ax^2 + by^2 + cz^2 \equiv (A_1x + B_1y + C_1z)(A_2x + B_2y + C_2z) \pmod{mn}$$

The last congruence holds modulo m and modulo n also.

Proof. Let

$$Q(x,y,z) = ax^2 + by^2 + cz^2$$
$$\equiv (a_1x + b_1y + c_1z)(a_2x + b_2y + c_2z) \pmod{n}$$

and

$$ax^2 + by^2 + cz^2 \equiv (a_1'x + b_1'y + c_1'z)(a_2'x + b_2'y + c_2'z) \pmod{m}.$$

Since $\gcd(m, n) = 1$, the Chinese Remainder Theorem implies there is an integer A_1 such that both of the following congruences hold:

$$\begin{aligned} A_1 &\equiv a_1 \ (\text{mod } n) \\ A_1 &\equiv a_1' \ (\text{mod } m) \end{aligned}$$

Similarly, there are integers A_2, B_1, B_2, C_1, C_2 such that

$$\begin{aligned} A_2 &\equiv a_2 \ (\text{mod } n) & B_1 &\equiv b_1 \ (\text{mod } n) \\ A_2 &\equiv a_2' \ (\text{mod } m) & B_1 &\equiv b_1' \ (\text{mod } m) \end{aligned}$$

$$\begin{aligned} B_2 &\equiv b_2 \ (\text{mod } n) & C_1 &\equiv c_1 \ (\text{mod } n) \\ B_2 &\equiv b_2'(\text{mod } m) & C_1 &\equiv c_1' \ (\text{mod } m) \end{aligned}$$

$$\begin{aligned} C_2 &\equiv c_2 \ (\text{mod } n) \\ C_2 &\equiv c_2' \ (\text{mod } m) \end{aligned}$$

So

$$\begin{aligned} Q &\equiv (A_1 x + B_1 y + C_1 z)(A_2 x + B_2 y + C_2 z) \ (\text{mod } n) \\ Q &\equiv (A_1 x + B_1 y + C_1 z)(A_2 x + B_2 y + C_2 z) \ (\text{mod } m) \end{aligned}$$

Hence,

$$Q = ax^2 + by^2 + cz^2 \equiv (A_1 x + B_1 y + C_1 z)(A_2 x + B_2 y + C_2 z) \ (\text{mod } mn)$$

∎

Theorem 7.46 *Let a, b, and c be nonzero integers such that abc is square free, or equivalently, a, b, and c are pairwise relatively prime and square free. Then $ax^2 + by^2 + cz^2 = 0$ has a nontrivial integral solution if and only if these four conditions hold:*

1. *a, b, and c do not all have the same sign.*

2. *$-bc$ is a quadratic residue of $|a|$.*

3. *$-ac$ is a quadratic residue of $|b|$.*

4. *$-ab$ is a quadratic residue of $|c|$.*

Proof. Clearly, if $ax^2 + by^2 + cz^2 = 0$ has a nontrivial integral solution, then a, b, and c cannot all have the same sign. We can assume that there is a primitive solution (x_0, y_0, z_0); otherwise, divide the equation by $\gcd(x_0, y_0, z_0)$ and relabel. The integers z_0 and a are relatively prime since otherwise y_0^2 contains $\gcd(z_0, a)$ because a and b are relatively prime. But then since $(\gcd(z_0, a))^2$ is contained in z_0^2 and y_0^2 and a is square free, $\gcd(z_0, a)$ is contained in x_0, which contradicts the fact that $\gcd(x_0, y_0, z_0) = 1$. Since a and z_0 are relatively prime, there exists a u so that $uz_0 \equiv 1 \ (\text{mod } |a|)$. Certainly since

$ax_0^2 + by_0^2 + cz_0^2 = 0$, $by_0^2 + cz_0^2 \equiv 0 \pmod{|a|}$. Hence $u^2 b(by_0^2 + cz_0^2) \equiv 0 \pmod{|a|}$. But $uz_0 \equiv 1 \pmod{a}$. Thus, $u^2 b^2 y_0^2 + bc \equiv 0 \pmod{|a|}$, $u^2 b^2 y_0^2 \equiv -bc \pmod{|a|}$, and $-bc$ is a quadratic residue of $|a|$. Similarly, we show $-ac$ is a quadratic residue of $|b|$, and $-ab$ is a quadratic residue of $|c|$.

Conversely, assume that $-bc$ is a quadratic residue of $|a|$, $-ac$ is a quadratic residue of $|b|$, and $-ab$ is a quadratic residue of $|c|$. Assume also that a, b, and c do not all have the same sign. By reorganizing, relabeling, and perhaps multiplying both sides of the equation $ax^2 + by^2 + cz^2 = 0$ by -1, we may assume that a is positive, and b and c are negative.

Let $r^2 \equiv -ab \pmod{|c|}$ and a' be defined by $aa' \equiv 1 \pmod{|c|}$. Then

$$
\begin{aligned}
ax^2 + by^2 &\equiv aa'(ax^2 + by^2) \pmod{|c|} \\
&\equiv a'(a^2 x^2 + aby^2) \pmod{|c|} \\
&\equiv a'(a^2 x^2 - r^2 y^2) \pmod{|c|} \\
&\equiv a'(ax - ry)(ax + ry) \pmod{|c|}
\end{aligned}
$$

and, therefore, $ax^2 + by^2 + cz^2$ factors modulo $|c|$. Similarly, $ax^2 + by^2 + cz^2$ factors modulo $|a|$ and $ax^2 + by^2 + cz^2$ factors modulo $|b|$. Hence by Lemma 7.45, $ax^2 + by^2 + cz^2$ factors modulo $|abc|$. Thus,

$$ax^2 + by^2 + cz^2 \equiv (a_1 x + b_1 y + c_1 z)(a_2 x + b_2 y + c_2 z) \pmod{|abc|}$$

for some integers a_1, a_2, b_1, b_2, c_1, and c_2.

Let's consider a general congruence $Ax + By + Cz \equiv 0 \pmod{|abc|}$. By Lemma 7.44 we know there is a nontrivial solution x_0, y_0, z_0 such that $x_0^2 \leq bc$ and such that $x_0^2 = bc$ only if $b = c = -1$ because bc is square free. Also, $y_0^2 \leq -ac$ and, in fact, $y_0^2 = -ac$ only if $a = 1$. In addition, $c = -1$ and $z_0^2 \leq -bc$; but $z_0^2 = -ab$ only if $a = 1$ and $b = -1$. Since b and c are negative, $ax_0^2 + by_0^2 + cz_0^2 \leq ax_0^2 < abc$. Also, $-2abc < ax_0^2 + by_0^2 + cz_0^2$. We now return to our specific congruence problem.

First, consider the case for which b and c are not -1. By Lemma 7.45, there are coefficients such that

$$ax^2 + by^2 + cz^2 \equiv (a_1 x + b_1 y + c_1 z)(a_2 x + b_2 y + c_2 z) \pmod{|abc|}$$

If we identify A, B, and C of the general congruence above as follows: $A = a_1$, $B = b_1$ and $C = c_1$, then Lemma 7.44 implies that there is a nontrivial solution (x_0, y_0, z_0) such that $a_1 x_0 + b_1 y_0 + c_1 z_0 \equiv 0 \pmod{|abc|}$. We then have that $ax_0^2 + by_0^2 + cz_0^2$ is a multiple of abc and is, therefore, equal to either $-abc$ or 0. If it is equal to 0, we are done. If not, let

$$
\begin{aligned}
x_1 &= -by_0 + x_0 z_0 \\
y_1 &= ax_0 + y_0 z_0 \\
z_1 &= z_0^2 + ab
\end{aligned}
$$

By direct computation, one can show that $ax_1^2 + by_1^2 + cz_1^2 = 0$. If (x_1, y_1, z_1) happens to be a trivial solution with $x_1 = y_1 = z_1 = 0$, then $z_0^2 = -ab$, which implies that $z_0 = 1$ or -1. Since ab is square free and $ab = -1$, we get $a = -1$ and $b = 1$. Since $-by_0 + x_0 z_0 = 0$, then we have $(-1)y_0 + x_0 = 0$ or $(-1)y_0 - x_0 = 0$. In either case $x_0^2 = y_0^2$ and $c = 0$. Hence, $x_0 = y_0 = z_0 = 1$ is a solution.

Second, we consider the case when $b = c = -1$. Then we want to solve $ax^2 - y^2 - z^2 = 0$. But $-bc$ is a quadratic residue of $|a|$ and, therefore, -1 is a quadratic residue of $|a|$. Because a is square free, only primes to the first power are factors of a. If -1 is a quadratic residue of $|a|$, then it is a quadratic residue of every prime factor of a. By Theorem 3.96, every odd prime factor of a must be of the form $4k + 1$ and, by Theorem 7.10, is the sum of two squares. If 2 is one of these factors, $2 = 1^2 + 1^2$ is the sum of two squares. But then a can be expressed as the sum of two squares since it is the product of factors which are the sums of two squares (Theorem 7.9). Let y_0^2 and z_0^2 be those squares and let $x_0 = 1$. ∎

Gauss proved Theorem 7.46, as stated above. Dirichlet showed that if $ax^2 + by^2 + cz^2 = 0$, where a, b, and c are relatively prime in pairs, then if a solution (x_0, y_0, z_0) can be found where x_0, y_0, and z_0 are relatively prime in pairs, then all solutions can be determined.

Exercises

1. Consider the equation $Q(x, y, z) = 13x^2 - 10y^2 - 3z^2 = 0$. Apply the methods of this section to find a nontrivial solution as follows:

   (a) Using that 2, 3, and 3 are solutions to

   $$\begin{aligned} t^2 &\equiv 1 \ (\text{mod } 3) \\ t^2 &\equiv 9 \ (\text{mod } 10) \\ t^2 &\equiv 9 \ (\text{mod } 13) \end{aligned}$$

   making 1, 9, and 9 quadratic residues of 3, 10, and 13, respectively, show that

   (i) $Q \equiv (x + 2y + 0z)(x + y + 0z) \ (\text{mod } 3)$

   (ii) $Q \equiv (0x + y + z)(0x + 3y + 10z) \ (\text{mod } 13)$

   (iii) $Q \equiv (x + 0y + z)(3x + 0y + 7z) \ (\text{mod } 10)$

   (b) Show using (i) and (ii) of part (a) that $Q \equiv (13x + 14y + 27z)(13x + 16y + 36z) \ (\text{mod } 39)$ where $39 = 3 \cdot 13$.

   (c) Use part (a) (iii) and part (b) to show that $Q \equiv (91x + 170y + 261z)(13x + 250y + 387z) \ (\text{mod } 390)$ where $390 = 39 \cdot 10$.

   (d) Find a nontrivial solution (x_0, y_0, z_0) to the congruence $91x + 170y +$

$261z \equiv 0 \pmod{390}$ and use it to construct a nontrivial solution to the original equation $13x^2 - 10y^2 - 3z^2 = 0$.

2. Solve:

   (a) Show that $3x^2 - 10y^2 - 17z^2 = 0$ has a nontrivial solution.

   (b) Use the methods of this section to find a nontrivial solution for the equation in part (a).

3. Using an argument similar to that of Lemma 7.44, prove the following: If a, b, and c are positive integers, then the congruence $Ax + By + Cz \equiv 0 \pmod{abc}$ has a nontrivial solution (x_0, y_0, z_0) such that $|x_0| \leq a$, $|y_0| \leq b$, and $|z_0| \leq c$.

7.8   The Equation $x^4 + y^4 = z^2$

Fermat's Last Theorem has been one of the most famous conjectures in modern mathematics. Fermat's Last Theorem states that for $n > 2$ there exist no positive integers x, y, and z such that $x^n + y^n = z^n$. Of course, for $n = 2$ we saw in Section 7.2 that there are infinitely many solutions. Fermat wrote thousands of mathematical papers and notes; however, only one was published by Fermat. His results were often written in letters to other mathematicians or as comments in the margins of a copy of Diophantus' *Arithmetica*. In the margin of Book II near Exercise 8, which was "to divide a given square number into two squares," Fermat wrote that it was impossible for powers greater than two (i.e., to divide a cube into two cubes, etc.) and that he had found an ingenious proof but there was not enough room in the margin to hold it. The search for a proof has vexed mathematicians ever since. Fermat's Last Theorem had been proved only for special cases of n until 1994, when Andrew Wiles and R. Taylor provided a proof. We will prove Fermat's Last Theorem for the case of $n = 4$.

Theorem 7.47 *There are no positive integers x, y, and z which satisfy the equation $x^4 + y^4 = z^2$.*

Proof. Assume that there is a positive integer solution (x_0, y_0, z_0). Then we may divide the equation $x_0^4 + y_0^4 = z_0^2$ by $\gcd(x_0^4, y_0^4)$ to obtain a solution x_1, y_1, z_1 so that $x_1^4 + y_1^4 = z_1^2$ is true and $\gcd(x_1^4, y_1^4) = 1$. Hence we will restrict our solutions to primitive ones. Let

$$S = \{z : \text{there exist } x \text{ and } y \text{ satisfying } x^4 + y^4 = z^2\}$$

Using the Well-Ordering Principle, we may select the least positive z in S.

The equation $(x^2)^2 + (y^2)^2 = z^2$ has a Pythagorean triple primitive solution; and, by Theorem 7.8, there are relatively prime integers r and s where

$r > s > 0$ and $r + s$ is odd, and $x^2 = 2rs$, $y^2 = r^2 - s^2$, and $z = r^2 + s^2$ so that $(x/2)^2 = rs/2$.

Consider the case of s even. Hence r and $s/2$ are both squares since r and $s/2$ are relatively prime, say $r = u^2$ and $s/2 = v^2$. So $x = 2uv$, u and v are relatively prime, $u > 0$, $v > 0$, and u is odd. $y^2 + s^2 = r^2$ implies that $y^2 + 4v^4 = u^4$. Since we again have a Pythagorean triple $y^2 + (2v^2)^2 = (u^2)^2$, there are relatively prime integers a and b where $a > b > 0$ so that $y = a^2 - b^2$, $2v^2 = 2ab$, and $u^2 = a^2 + b^2$. Since $ab = v^2$ and a and b are relatively prime, a and b are both squares. Say that $a = c^2$ and $b = d^2$. Then $u^2 = (c^2)^2 + (d^2)^2 = c^4 + d^4$, but $u \leq r < z$ and we contradict the fact that z is the least positive element in S.

Consider the case of r even. $r/2$ and s are both squares since r and s are relatively prime. So $r/2 = g^2$ and $s = h^2$. Also, $x = 2gh$, g and h are relatively prime, and h is odd. So $y^2 + s^2 = r^2$ or $y^2 + (h^2)^2 = (2g^2)^2$. But according to the discussion before Theorem 7.8, $2g^2$ cannot be even, a contradiction. ∎

Exercises

1. Prove that there are no integers a, b, and c so that $a^4 + b^4 = c^4$.

2. Let a Pythagorean triangle be a right triangle with sides of integral length. Prove that a Pythagorean triangle can never have area equal to a perfect square.

3. Prove that the equation $x^4 - y^4 = 2z^2$ has no positive integral solutions. (*Hint:* Assume that $\gcd(x, y)$ contains no prime factor but 2. Since x and y are both even or both odd, argue that $x^2 + y^2$ and $x^2 - y^2$ are both equal to twice a square. Continue the argument with $x - y$ and $x + y$.)

4. Prove that the only positive integral solution to $x^4 + y^4 = 2z^2$ where x, y, and z are relatively prime is $x = y = z = 1$. (*Hint:* Show that any solution of $x^4 + y^4 = 2z^2$ is also a solution of $z^4 - (xy)^4 = ((x^4 - y^4)/2)^2$.)

5. Prove that $x^4 - 4y^4 = z^2$ has no solution for positive integers x, y, and z by rewriting as $(2y^2)^2 + z^2 = (x^2)^2$ and using requirements on u, v, and w so that $u^2 + v^2 = w^2$. (See Theorem 7.8.)

Chapter 8
ALGEBRA AND NUMBER THEORY

8.1 Algebraic Development of the Integers

In this section further algebraic structures of the integers are developed and the integers are determined algebraically up to isomorphism. Notice that this development does not imply the existence of the integers but only their algebraic properties if they do exist.

Definition 8.1 *A subset I of a ring R is called an ideal of R if*

(a) *I is a subring of R.*

(b) *If x is in I and r is in R, then $x \cdot r$ and $r \cdot x$ are in I.*

For example, in the ring of integers, the set of all multiples of a fixed integer p form an ideal.

Definition 8.2 *Let R be a commutative ring. An ideal I of R is called a* principal ideal *generated by a if I consists of all products of a by elements of R, that is, $I = \langle a \rangle = \{ar : r \in R\}$.*

Theorem 8.3 *Every nonempty ideal I of the ring of integers is a principal ideal.*

Proof. For $I \neq \{0\}$, let p be the least positive integer in I and let m belong to I. By the division algorithm, $m = pq + r$ where $0 \leq r < p$. Since $r = m - pq$, r is in I. Thus, since $r < p$ and p is the smallest positive integer in I, $r = 0$. Hence every integer in I is a multiple of p and $I = \langle p \rangle$. ∎

Note that the proof of this theorem depends upon the Well-Ordering Principle and the Division Algorithm, both of which are available for integers.

Example 8.4 *Consider the ring Z of integers and the two principal ideals generated by the integers 8 and 12:*

$$\begin{aligned} \langle 8 \rangle &= \{8r : r \in Z\} \\ &= \{\ldots, -24, -16, -8, 0, 8, 16, 24, \ldots\} \end{aligned}$$

$$\begin{aligned} \langle 12 \rangle &= \{12s : s \in Z\} \\ &= \{\ldots, -24, -12, 0, 12, 24, \ldots\} \end{aligned}$$

The intersection $\langle 8 \rangle \cap \langle 12 \rangle$ is the set

$$\{\ldots, -48, -24, 0, 24, 48, \ldots\}$$

which is the principal ideal generated by the integer 24. Note that 24 is the least common multiple of 8 and 12. In general,

$$\langle a \rangle \cap \langle b \rangle = \langle \mathrm{lcm}(a, b) \rangle$$

Results pertinent to this example are given in the next theorem. The proof is left to the reader.

Theorem 8.5 *If s and t are nonzero integers and $\langle s \rangle$ and $\langle t \rangle$ are the corresponding principal ideals in the ring Z, then*

(a) *If $\langle s \rangle \subseteq \langle t \rangle$, then $t \mid s$.*

(b) *$\langle s \rangle \cap \langle t \rangle = \langle u \rangle$ where $u = \mathrm{lcm}(s, t)$.*

Example 8.6 *If $\langle a, b \rangle$ is the smallest ideal containing a and b, then $\langle a, b \rangle = \langle \gcd(a, b) \rangle$. For example,*

$$\begin{aligned} \langle 8, 12 \rangle &= \{-16, -12, -8, -4, 0, 4, 8, 12, 16, \ldots\} \\ &= \langle \gcd(8, 12) \rangle \end{aligned}$$

since $\gcd(8, 12) = 4$ and

$$\langle 4 \rangle = \{\ldots, -8, -4, 0, 4, 8, \ldots\}$$

Let I be the smallest ideal containing integers a and b. Then every element of I is of the form $am + bn$, where m and n are integers. The ideal I is said to be generated by a and b. By Theorem 8.5, I is generated by the smallest positive number of this form. But this number is the greatest common divisor of a and b, as mentioned above. This idea may be extended to an ideal generated by any finite number of positive integers. The ideal is again generated by the greatest common divisor of these numbers.

Definition 8.7 *An ideal I of a commutative ring R is a prime ideal if $ab \in I$ implies that either $a \in I$ or $b \in I$.*

Theorem 8.8 *In the ring of integers an ideal $\langle a \rangle$ is a prime ideal if and only if a is a prime number.*

Proof.   Assume that a is not prime, say $a = rs$ where r and s are both integers greater than 1. Then a is in $\langle a \rangle$ but r and s are not. If a is prime and $rs \in \langle a \rangle$, then a divides rs; hence, by Theorem 2.3, a divides r or a divides s. Hence either r is in $\langle a \rangle$ or s is in $\langle a \rangle$. ■

Definition 8.9 *An integral domain D is a* principal ideal domain *if every ideal in D is a principal ideal.*

We have already shown that the integral domain Z of integers is a principal ideal domain.

Definition 8.10 *If A is a commutative ring with unity, let A^* denote the set $\{a \in A : \text{there exists } b \in A \text{ with } ab = 1\}$. The subset A^* is a group under multiplication called the* group of units *of A. Each element of A^* is called a* unit *of A. In a ring with unity, an element s is* irreducible *if it is nonzero, is not a unit, and cannot be expressed as a product of two nonunits.*

Units are just divisors of 1. In the integral domain Z, $ab = 1$ only if $a = b = 1$ or $a = b = -1$ so that 1 and -1 are units in Z. In a field, every nonzero element is a unit because $a \cdot a^{-1} = 1$ for $a \neq 0$.

Example 8.11   *The set $Z_6 = \{[0], [1], [2], [3], [4], [5]\}$ is a commutative ring with unity $[1]$ and zero $[0]$. The multiplication table in Z_6 is*

$\odot$	[0]	[1]	[2]	[3]	[4]	[5]
[0]	[0]	[0]	[0]	[0]	[0]	[0]
[1]	[0]	[1]	[2]	[3]	[4]	[5]
[2]	[0]	[2]	[4]	[0]	[2]	[4]
[3]	[0]	[3]	[0]	[3]	[0]	[3]
[4]	[0]	[4]	[2]	[0]	[4]	[2]
[5]	[0]	[5]	[4]	[3]	[2]	[1]

From the table we see that $[3] \odot [2] = [0]$ but $[3] \mid [0]$ and $[2] \mid [0]$ so that $[3]$ and $[2]$ are nonzero divisors of $[0]$. Thus, Z_6 is not an integral domain.

In the following example it is seen that the cancellation property does not hold since $[3] \odot [1] = [3] = [15] = [3] \odot [5]$ and, although $[3] \neq [0]$, we have $[1] \neq [5]$. The units of Z_6 correspond to integers relative prime to 6; therefore, as the multiplication above indicates, the units of Z_6 are $[1]$ and $[5]$.

In the ring $Z_5 = \{[0], [1], [2], [3], [4]\}$, with unity $[1]$ and zero $[0]$, every nonzero member of Z_5 is a unit since it is relatively prime to the prime 5. Hence every element has an inverse so that Z_5 is a field. See Example 1.63 for operations tables for Z_5.

Definition 8.12 *An integral domain D is said to be a* unique factorization domain *provided that:*

(a) *If an element of D is not zero and not a unit, then it can be factored into a product of a finite number of irreducibles.*

(b) *If an element of D has factorizations $p_1 \cdots p_r$ and $q_1 \cdots q_s$ as products of irreducibles, then $r = s$ and the q_j can be renumbered so that the p_i and q_i differ only by a unit for all i; that is, $p_i = a_i q_i$ for some unit a_i.*

We already know that the integers are a unique factorization domain. We have defined a prime integer as one which has no nontrivial factors. In other words, a prime integer is irreducible. An alternative definition for a prime integer is that p is prime if and only if $p \mid ab$ implies that $p \mid a$ or $p \mid b$. These definitions are equivalent for the integers but not for arbitrary integral domains. For example, let A be the set of all complex numbers of the form

$$a + b\sqrt{5}i$$

It is easily shown that A is an integral domain and that

$$21 = 3 \cdot 7 = (1 + 2\sqrt{5}i)(1 - 2\sqrt{5}i)$$

All of these factor are irreducible and hence A is not a unique factorization domain. However, none of them is a prime using the alternative definition. Complex numbers will be discussed more fully in Section 8.2.

Definition 8.13 *A commutative ring A with unity is an* ordered ring *if and only if there exists a nonempty subset, A^+, of A, called the subset of positive elements of A, such that:*

(a) *If $a, b \in A^+$, then $a + b \in A^+$.*

(b) *If $a, b \in A^+$, then $a \cdot b \in A^+$.*

(c) *Given any $a \in A$, one and only one of the following alternatives holds for a:*

(i) *$a \in A^+$.*

(ii) *$a = 0$.*

(iii) *$-a \in A^+$.*

A commutative ring with unity which has such a set A^+ is said to satisfy the trichotomy axiom. If $a \in A^+$, then we say that $a > 0$. If $-a \in A^+$, then we say that $a < 0$.

The proof of the following theorem is left to the reader.

Theorem 8.14 *Every ordered ring is an integral domain. Given any $a \neq 0$, $a^2 > 0$. In particular, $1^2 > 0$ and hence $1 > 0$.*

Definition 8.15 *An ordered integral domain A is called* well ordered *if and only if any nonempty subset S of A^+ has a first element; that is, there exists $s \in S$ such that if $t < s$, then $t \notin S$.*

We have already seen that the integers form a well-ordered integral domain.

Theorem 8.16 *If A is a well-ordered integral domain, then there is no element c of A such that $0 < c < 1$.*

Proof. Let S be the subset of all c in A such that $0 < c < 1$. If S is not empty, then there exists a least element s in S. But s^2 is in A; and since $s > 0$, $s^2 > 0$. Since $s < 1$, $0 < s$ implies that $s^2 < s$ and $0 < s^2 < s$. Therefore, $s^2 \in S$ and $s^2 < s$ which contradicts the fact that s is the least element of S. Hence S is empty. ■

Theorem 8.17 *In any ordered integral domain the following are equivalent:*

(1) *First Principle of Induction*

(2) *Well-Ordering Principle*

(3) *Second Principle of Induction*

Proof. For (1) implies (2), see Theorem B.22. For (2) implies (3), see Theorem B.23.

Proof of (3) implies (1). Let S be a subset of A^+. Let $p(1)$ be the statement "$n \in S$ implies that $(n+1) \in S$" and $p(2)$ be the statement "If for all $m < n$ with $m \in S$, then $n \in S$." Obviously, $p(1)$ implies $p(2)$. The Second Principle of Induction may be stated " If $1 \in S$ and $p(2)$, then $S = A^+$" and the First Principle of Induction may be stated "if $1 \in S$ and $p(1)$, then $S = A^+$." Since $1 \in S$ and $p(1)$ implies that $1 \in S$ and $p(2)$, then the statement "$1 \in S$ and $p(2)$ implies that $S = A^+$" implies the statement "$0 \in S$ and $p(1)$ implies that $S = A^+$)". (This result is true because $(p \to q) \implies ((q \to r) \to (p \to r))$ as discussed in Appendix A.) ■

Theorem 8.18 *Any two well-ordered, ordered integral domains are isomorphic and, therefore, are isomorphic to Z.*

Proof. Let A be a well-ordered, ordered integral domain and N the set of positive integers. Let I be the multiplicative identity of A. Define $f : N \to A^+$ by $f(1) = I$ and if $f(k) = K$, then $f(k+1) = K + I$. It is easily shown that $f(N)$ satisfies the first principle of induction and hence $f(N) = A^+$. We need to show that $f(m+n) = f(m) + f(n)$ and $f(m \cdot n) = f(m)f(n)$.

Using induction on n, we first show that $f(m+n) = f(m) + f(n)$. For $n = 1$, $f(m+1) = f(m) + I$ by definition of f, so $f(m+1) = f(m) + f(1)$.

Assume that $f(m + k) = f(m) + f(k)$; then

$$
\begin{aligned}
f(m + (k + 1)) &= f((m + k) + 1) \\
&= f(m + k) + I \\
&= f(m) + f(k) + I \\
&= f(m) + f(k + 1)
\end{aligned}
$$

and

$$f(m + n) = f(m) + f(n)$$

for all $m, n \in N$.

We now show that $f(m \cdot n) = f(m)f(n)$ by induction on n. For $n = 1$,

$$
\begin{aligned}
f(m \cdot 1) &= f(m) \\
&= f(m) \cdot I \\
&= f(m)f(1)
\end{aligned}
$$

Assume that $f(m \cdot k) = f(m)f(k)$; then

$$
\begin{aligned}
f(m \cdot (k + 1)) &= f((m \cdot k) + m) \\
&= f(m \cdot k) + f(m) \\
&= f(m)f(k) + f(m) \\
&= f(m)f(k) + f(m)I \\
&= f(m)(f(k) + I) \\
&= f(m)f(k + 1)
\end{aligned}
$$

and $f(m \cdot n) = f(m)f(n)$ for all $m, n \in N$. Therefore, f is an isomorphism. This isomorphism is easily extended to an isomorphism from Z to A. ∎

Definition 8.19 *An integral domain is called a* minimal domain *if and only if it has no subdomains except itself.*

A minimal domain may be found by taking the intersection of all subdomains of an integral domain. The proof of the following theorem is left to the reader.

Theorem 8.20 *Any two ordered minimal integral domains are isomorphic. They are isomorphic to the integers and hence are well ordered.*

Exercises

1. Prove that if F is a field, then the equation $ax = b$ has a unique solution in the field.

2. An element a of a ring R is a zero divisor if there exists a b in R such that $ab = 0$. Prove or disprove that the sum of two elements which are not zero divisors is not a zero divisor.

3. Prove or disprove that the product of two elements that are not zero divisors is not a zero divisor.

4. What are the zero divisors in Z_{10}?

5. What are the zero divisors in Z_7?

6. Prove that every finite integral domain is a field.

7. Prove that $(Z_n, \odot, \oplus)$ is a field if and only if n is prime.

8. Prove that a field F has contains no ideals except itself and the ideal containing only zero.

9. Prove that the intersection of all subdomains of an integral domain is an integral domain and use this to show that every integral domain contains a minimal integral domain.

10. Prove that any two ordered minimal integral domains are isomorphic. Prove that they are isomorphic to the integers and hence are well ordered.

11. In an ordered integral domain prove that for any nonzero element a, $a^2 > 0$. In particular $1 > 0$ (Theorem 8.14).

12. Let A be a commutative ring with unity. Prove that A is an integral domain if and only if $ab = ac$ implies that $b = c$ for all b, c, and nonzero a in A.

13. Let M_3 be the ring of 3×3 matrices. Prove that $\{A \mid \text{such that } \det(A) = 0\}$ is an ideal of M_3.

14. Let M be the ring of 2×2 matrices of the form $\begin{bmatrix} a & b \\ c & 0 \end{bmatrix}$. Prove that the set of all matrices of the form $\begin{bmatrix} a & b \\ 0 & 0 \end{bmatrix}$ is an ideal of M.

15. Prove that the sum of ideals I and J of a commutative ring R defined by $I + J = \{i + j \mid i \in I \text{ and } j \in J\}$ is an ideal of R.

16. Let $A = \langle 14 \rangle$ and $B = \langle 16 \rangle$. Find $A \cap B$ and $A + B$.

17. Describe the smallest ideal of the integers containing 6, 9, and 12.

18. Prove that if I is an ideal of a ring R and $-1 \in I$, then $I = R$.

19. Prove that if R is the ring of Gaussian integers, I is an ideal contained in R, and $i \in I$, then $R = I$.

20. Let $f : R \to R'$ be a ring homomorphism. Prove that $\{x \in R \mid f(x) = 0\}$ is an ideal of R.

21. Prove that the zero divisors of a commutative ring R form an ideal of R.

8.2   Group and Semigroup Characters

Arithmetic functions, many of which were multiplicative, were the subject of Chapter 4. We now discuss multiplicative arithmetic functions called Dirichlet characters. They are useful in proving Dirichlet's Theorem and the Prime Number Theorem. We begin by introducing complex numbers.

Definition 8.21 *Let R be the set of real numbers and C be the set*

$$\{(r, s) : r, s \in R\} = R \times R$$

so that every element of C is an ordered pair of real numbers. Define addition on C by

$$(a, b) + (c, d) = (a + c, b + d)$$

and multiplication by

$$(a, b) \cdot (c, d) = (ac - bd, ad + bc)$$

Then C is called the set of complex numbers.

It is straightforward to show that C is a field. The mapping $f : R \rightarrow C$ defined by $f(a) = (a, 0)$ is a monomorphism and is an isomorphism into the subfield $\{(a, 0) : a \in R\}$. Therefore, we speak of the real number field R as being contained in the complex number field and we say that R is a *subfield* of C. Further, $(0, 1) \cdot (0, 1) = (-1, 0) = -1$. If we define $i = (0, 1)$, then $i^2 = -1$. Hence, we may write (a, b) as $(a, 0) + (b, 0) \cdot (0, 1) = a + bi$. Thus, the definitions of addition and multiplication become

$$(a + bi) + (c + di) = (a + c) + (b + d)i$$

and

$$(a + bi) \cdot (c + di) = (ac - bd) + (ad + bc)i$$

For a complex number $z = a + bi = (a, b)$, a is called the *real part* and b is called the *imaginary part* and we write $\text{real}(z) = \text{real}(a + bi) = a$ and $\text{im}(z) = \text{im}(a + bi) = b$. Just as we often plot ordered pairs (a, b) of real numbers on a Cartesian coordinate system (in the plane), we plot the complex number $a + bi$ in the plane as (a, b). In this context, the horizontal axis is called the real axis and the vertical axis is called the imaginary axis. Thus, the real part of $a + bi$ (that is, $\text{real}(a + bi) = a$) is measured along the horizontal axis and the imaginary part of $a + bi$ (that is, $\text{im}(a + bi) = b$) is measured along the vertical axis. When the Cartesian coordinate system is used in this manner, it is referred to as the *complex plane.*

Definition 8.22 *Let C be the set of complex numbers. For $c = a + bi$ in C, $|c| = \sqrt{a^2 + b^2}$ is called the* absolute value *or* length *of c.*

On the complex plane $|c|$ is the distance from the point $c = a + bi$ to the origin It is easily shown that $|c| \, |d| = |cd|$ and $|c + d| \le |c| + |d|$ for complex numbers c and d. Let r be a complex number which is an n-th root of 1, that is, $r^n = 1$. Then, $|r|^n = |r^n| = |1| = 1$; so r has length 1 and is on the *unit circle*, the circle of all points on the complex plane whose distance from the origin is 1.

An important property of complex numbers is given by the following theorem which we will state without proof.

Fundamental Theorem of Algebra *For any polynomial p of positive degree with coefficients in R, the set of real numbers, or in C, the set of complex numbers, there is an element b in C such that $p(b) = 0$.*

Thus, in particular, every polynomial over Z has at least one root or zero in the set of complex numbers.

Definition 8.23 *A* group character *on a commutative group G is a homomorphism from G into the nonzero complex numbers with the operation of multiplication.*

Definition 8.24 *The* conjugate *of the complex number $c = a + bi$ is the number $a - bi$, which is denoted by* $\operatorname{conj}(c)$.

For the remainder of this section we will assume that all groups forming the domain of the group characters are commutative and finite. Let χ be a group character. Thus there exists $g \in G$ such that $\chi(g) \neq 0$. Since G is finite, there exists n so that $g^n = 1$. Hence $\chi(g^n) = (\chi(g))^n = 1$, $\chi(g)$ is an n-th root of 1, and the element g is mapped onto the unit circle. Thus, a group character maps G into the unit circle.

The proof of the following theorem is left to the reader.

Theorem 8.25 *If $\chi(g) = a + bi$, then $\chi(g^{-1})$ is the conjugate of $\chi(g)$, that is, $\chi(g^{-1}) = a - bi$.*

Definition 8.26 *A primitive n-th root r of a complex number a is an n-th root of a such that $r^k \neq a$ for $0 < k < n$.*

Definition 8.27 *The group character χ_1 defined by $\chi_1(g) = 1$ for all g in the group G is called the* principal character *of the group G. The* character group *of G consists of the set of group characters of G with the operation $(\chi \cdot \chi')(g) = \chi(g)\chi'(g)$ for group characters χ and χ'.*

Definition 8.28 *Let $G_1, G_2, \ldots, G_m$ be subgroups of a group G. Then G is the* direct sum *of $G_1, G_2, \ldots, G_m$, denoted $G_1 \oplus G_2 \oplus \cdots \oplus G_m$, if every element of G may be uniquely written in the form $a_1 a_2 \cdots a_m$ where $a_k \in G_k$.*

Theorem 8.29 *Every finite Abelian group is the direct sum of cyclic subgroups. Thus, if $G = G_1 \oplus G_2 \oplus G_3 \oplus \cdots \oplus G_m$ for cyclic groups G_i with generators g_i of order $k(i)$, then each $g \in G$ is uniquely expressed in the form*

$$g = g_1^{j(1)} g_2^{j(2)} g_3^{j(3)} \cdots g_m^{j(m)}$$

with $0 \leq j(i) < k(i)$.

Proof. We show that there exists an integer m so that $G = G_1 \oplus G_2 \oplus G_3 \oplus \cdots \oplus G_m$ and each element of G can be written uniquely in the form $g_1^{j(1)} g_2^{j(2)} g_3^{j(3)} \cdots g_m^{j(m)}$ where $0 \leq j(i) < k(i)$. Let $g_1 \in G$ have order $k(1)$ and G_1 be the subgroup generated by g_1. If $G_1 = G$, then $m = 1$ and we are done. If not, select g_2 in $G - G_1$ and let G_2 be the group generated by g_2. Then $g_2^n = 1$ for some n since G is finite. Hence some power of g_2 is in G_1. Let $k(2)$ be the smallest power of g_2 so that $g_2^{k(2)}$ is in G_1. If $G_1 \cup G_2 = G_2$, then relabel with $G_1 = G_2$ and start over. If $G_1 \cup G_2 \neq G_2$, then it is easily shown that every element in $G_1 \cup G_2$ can be uniquely written in the form $g_1^j g_2^s$, where $0 \leq j < k(1), 0 \leq s < k(2)$. If we select g_3 in $G - (G_1 \oplus G_2)$ and continue the process, we obtain $G_1 G_2 G_3 \cdots G_r G_{r+1} \cdots$; and since G is finite, there exists an m so that $G = G = G_1 \oplus G_2 \oplus G_3 \oplus \cdots \oplus G_m$ and each element of G can be written uniquely in the form $g_1^{j(1)} g_2^{j(2)} g_3^{j(3)} \cdots g_m^{j(m)}$, where $0 \leq j(i) < k(i)$.

Theorem 8.30 *The character group of a group G is isomorphic to G.*

Proof. Let $G = G_1 \oplus G_2 \oplus \cdots \oplus G_m$, where g_i generates G_i; and let a_i be a primitive n_i-th root of unity, where G_i has order n_i. For each $g \in G$, define χ_g as by

$$\chi_g(b) = a_1^{j(1) \cdot h(1)} a_2^{j(2) \cdot h(2)} \cdots a_m^{j(m) \cdot h(m)}$$

where members g and b of G have the direct sum representations

$$
\begin{aligned}
g &= g_1^{j(1)} g_2^{j(2)} g_3^{j(3)} \cdots g_m^{j(m)} \\
b &= g_1^{h(1)} g_2^{h(2)} g_3^{h(3)} \cdots g_m^{h(m)}
\end{aligned}
$$

with $0 \leq j(i) < n_i$ and $0 \leq h(i) < n_i$. It is straightforward to show that χ_g is a homomorphism from G into the n-th roots of unity for any $g \in G$. Further, if $g \neq g'$, then $\chi_g \neq \chi_{g'}$ so there are $\prod_{i=1}^{m} n_i = n$ group characters, where G has n elements. Define the mapping η from G into the character group of G as follows:

$$\eta(g) = \eta\left(g_1^{j(1)} g_2^{j(2)} g_3^{j(3)} \cdots g_m^{j(m)}\right) = (\chi_{g_1})^{j(1)} (\chi_{g_2})^{j(2)} \cdots (\chi_{g_m})^{j(m)}$$

η is easily shown to be an isomorphism. ∎

Further, if A is a matrix where $A_{ij} = \chi_{g_i}(g_j)$, then

$$
A = \begin{bmatrix}
a_1 & 1 & 1 & \cdots & 1 \\
1 & a_2 & 1 & \cdots & 1 \\
1 & 1 & a_3 & \cdots & 1 \\
& & \vdots & & \\
1 & 1 & 1 & \cdots & a_k
\end{bmatrix}
$$

Example 8.31 *Consider the set $G_5 = \{[1], [2], [3], [4]\}$, which consists of the nonzero elements of Z_5. Since 5 is prime, the reduced residue classes are just the nonzero classes in Z_5, and the nonzero classes corresponding to the reduced residues form a group under multiplication. Thus, G_5 is a group under multiplication. The multiplication table for G_5 is*

$\odot$	$[1]$	$[2]$	$[3]$	$[4]$
$[1]$	$[1]$	$[2]$	$[3]$	$[4]$
$[2]$	$[2]$	$[4]$	$[1]$	$[3]$
$[3]$	$[3]$	$[1]$	$[4]$	$[2]$
$[4]$	$[4]$	$[3]$	$[2]$	$[1]$

To make the following construction of the group characters clearer, we will use the notation found in the proof of Theorem 8.30.

The complex number

$$e^{2\pi t i/n} = \cos\left(\frac{2\pi t}{n}\right) + i\sin\frac{2\pi t}{n}$$

is on the unit circle because, clearly,

$$\left|e^{2\pi t i/n}\right| = \sqrt{\cos^2\left(\frac{2\pi t}{n}\right) + \sin^2\left(\frac{2\pi t}{n}\right)} = 1$$

For any integer t, $e^{2\pi t i/n}$ is an n-th root of unity since

$$\left(e^{2\pi t i/n}\right)^n = e^{2\pi t i} = \cos(2\pi t) + i\sin(2\pi t) = 1$$

In fact, the set

$$\left\{e^{2\pi t i/n} : t = 0, 1, 2, \ldots, (n-1)\right\}$$

is the set of the n distinct complex n-th roots of unity.

For $n = 4$, the order of G_5, the formula above gives the four fourth roots of unity.

t	$e^{(2\pi t i)/4}$
0	1
1	i
2	-1
3	$-i$

Calculating $1^4 = i^4 = (-1)^4 = (-i)^4 = 1$, we verify that $1, i, -1,$ and $-i$ are fourth roots of unity.

If we let $g_1 = [3]$, then

$$g_1^2 = [3]^2 = [4]$$
$$g_1^3 = [3]^3 = [2]$$
$$g_1^4 = [3]^4 = [1] = g_1^0$$

Hence, $k(1) = 4$ and $G_1 = G$ so that $m = 1$ in the notation of the proof of Theorem 8.30. For $g_1 = [3]$, select the 4-th root $r_1 = i$. Since $r_1^2 = -1$,

$r_1^3 = -i$, and $r_1^4 = 1$, $r_1 = i$ is primitive. Using this primitive root, we can define a character for every element g of G_5. The following table gives the definition of all the group characters of G_5.

$j(1)$	$g = g_1^{j(1)}$	$\chi_g(b) = \chi_g(g_1^{h(1)})$	
0	[1]	$r_1^{0 \cdot h(1)}$	$= i^0$
3	[2]	$r_1^{3 \cdot h(1)}$	$= i^{3h(1)}$
1	[3]	$r_1^{1 \cdot h(1)}$	$= i^{h(1)}$
2	[4]	$r_1^{2 \cdot h(1)}$	$= i^{2h(1)}$

Therefore, the 4 group characters of the character group of G_5 are given in the next table, where $b = g_1^{h(1)}$:

g	$\chi_g(b)$		$b =$ [1]	[2]	[3]	[4]
			$h(1) =$ 0	3	1	2
[1]	$\chi_{[1]}(b)$	$=$	1	1	1	1
[2]	$\chi_{[2]}(b)$	$=$	1	i	$-i$	-1
[3]	$\chi_{[3]}(b)$	$=$	1	$-i$	i	-1
[4]	$\chi_{[4]}(b)$	$=$	1	-1	-1	1

Clearly, $\chi_{[1]} = 1$ is the principal character.

To illustrate the homomorphic nature of group characters, consider $\chi_{[2]}$:

$$\chi_{[2]}([3][4]) = \chi_{[2]}([2]) = i$$

$$\chi_{[2]}([3]) = -i$$

$$\chi_{[2]}([4]) = -1$$

$$\chi_{[2]}([3])\chi_{[2]}([4]) = (-i)(-1) = i$$

The other 15 products may be checked for $\chi_{[2]}$ similarly.

Using the table of group character definitions above, it is easy to construct the multiplication table for the character group:

$\cdot$	$\chi_{[1]}$	$\chi_{[2]}$	$\chi_{[3]}$	$\chi_{[4]}$
$\chi_{[1]}$	$\chi_{[1]}$	$\chi_{[2]}$	$\chi_{[3]}$	$\chi_{[4]}$
$\chi_{[2]}$	$\chi_{[2]}$	$\chi_{[4]}$	$\chi_{[1]}$	$\chi_{[3]}$
$\chi_{[3]}$	$\chi_{[3]}$	$\chi_{[1]}$	$\chi_{[4]}$	$\chi_{[2]}$
$\chi_{[4]}$	$\chi_{[4]}$	$\chi_{[3]}$	$\chi_{[2]}$	$\chi_{[1]}$

Definition 8.32 Let S be a finite commutative semigroup. A semigroup character on S is a homomorphism from S into the complex numbers with the operation multiplication. As in the case with groups, we will assume that in this section all semigroups used are commutative and finite.

For a finite commutative semigroup S, let $s \in S$. If $|\chi(s)| > 1$, then

$$|\chi(s^2)| > |\chi(s)|$$

and

$$|\chi(s^3)| > |\chi(s^2)|, \ldots, |\chi(s^{k+1})| > |\chi(s^k)|, \ldots$$

which is impossible since S is finite. Similarly, if

$$0 < |\chi(s)| < 1$$

then

$$0 < |\chi(s^2)| < |\chi(s)|$$

and

$$0 < |\chi(s^3)| < |\chi(s^2)| \cdots$$
$$0 < |\chi(s^{k+1})| < |\chi(s^k)| < \cdots < |\chi(s^2)| < |\chi(s)| < 1$$

which is again impossible. Hence either $|\chi(s)| = 0$ or $|\chi(s)| = 1$ for each s in S.

Definition 8.33 *A subsemigroup I of a commutative semigroup S is an* ideal *of S if $i \in I$ and $s \in S$ implies that $s \cdot i \in I$.*

The proof of the following theorem is left to the reader.

Theorem 8.34 *If S is a finite commutative semigroup and χ is a semigroup character on S, then $I = \{s : |\chi(s)| = 0\}$ is an ideal. Conversely, if I is any ideal in S and there is a character χ such that $\chi(s) \neq 0$ for some s in $S - I$, then there exists a nonzero character χ_I such that $\chi_I(i) = 0$ if and only i is in I.*

At this point we consider the set Z_n of residue classes modulo n. Recall that the set of elements of Z_n consisting of equivalence classes containing integers which are relatively prime to n is called the set of reduced residue classes. The set of reduced residue classes forms a group under class multiplication. We denote this group by G_n. The set Z_n is a semigroup under class multiplication. The set $Z_n - G_n$ is a semigroup ideal since the product of any integer and an integer not relatively prime to n is not relatively prime to n. Hence we can form characters on Z_n which are group characters on G_n and map all of the elements of $Z_n - G_n$ to 0. These characters are called *Dirichlet characters*. We know that the number of Dirichlet characters modulo n is equal to the number of group characters on G_n. Since the set of group characters of G_n is isomorphic to G_n and there are $\phi(n)$ elements in G_n, there are $\phi(n)$ Dirichlet characters modulo n and we have the following theorem.

Theorem 8.35 *There exist $\phi(n)$ distinct Dirichlet characters modulo the positive integer n.*

Lemma 8.36 *Let G be a finite group, g' an arbitrary element of G, and $G' = \{x : x = g'g$ for g in $G\}$. Then $G = G'$.*

Proof. Assume that G' is generated by the element g' as in Lemma 8.36. Certainly $G' \subseteq G$. Let g_1 and g_2 be elements of G. Then $g'g_1$ and $g'g_2$ are in G'. If $g'g_1 = g'g_2$, then $g_1 = (g')^{-1}g'g_1 = g_2$. Hence there are as many elements in G' as in G; and since G is finite, $G = G'$. ∎

Theorem 8.37 *Let G be a finite commutative group of order n. If χ is a group character, then*

$$\sum_{g \in G} \chi(g) = \begin{cases} n & \text{if } \chi = \chi_1 \\ 0 & \text{if } \chi \neq \chi_1 \end{cases}$$

Proof. Since $\chi_1(g) = 1$ for all g in G, the first part is obvious. If $\chi \neq \chi_1$ and g' is an element of G, such that $\chi(g') \neq 1$, then

$$\begin{aligned} \sum_{g \in G} \chi(g) &= \sum_{g \in G} \chi(g'g) && \text{by Lemma 8.36} \\ &= \sum_{g \in G} \chi(g')\chi(g) \\ &= \chi(g') \sum_{g \in G} \chi(g) \end{aligned}$$

so that

$$[1 - \chi(g')] \sum_{g \in G} \chi(g) = 0$$

and since $1 - \chi(g') \neq 0$, $\sum_{g \in G} \chi(g) = 0$. ∎

Theorem 8.38 *If G is a finite commutative group of order n and $g \in G$, then*

$$\sum_{\chi} \chi(g) = \begin{cases} n & \text{if } g = e \\ 0 & \text{if } g \neq e \end{cases}$$

where e is the group identity element and where the sum is over all group characters, χ.

Proof. If $g = e$, then $\chi(e) = 1$ for all χ since every χ is a homomorphism. Thus, the sum equals n. Assume that $g \neq e$ and χ_a is a character such that $\chi_a(g) \neq 1$. Such a character exists since $g = g_1^{j(1)} g_2^{j(2)} g_3^{j(3)} \cdots g_m^{j(m)}$ with $0 \leq j(i) < k(i)$, as shown in the proof of Theorem 8.30. Further, since $g \neq e$, then $j(s) \neq 0$ for some s. We can then let $\chi_a(g)$ be $\chi_{g_s}(g)$ since $\chi_{g_s}(g) = r_s^{j_s} \neq 1$, where r_s is a $k(s)$-th primitive root of unity. Then

$$\begin{aligned} \sum_{\chi} \chi(g) &= \sum_{\chi} (\chi_a \chi)(g) && \text{by Lemma 8.36 applied to character groups} \\ &= \sum_{\chi} \chi_a(g)\chi(g) \\ &= \chi_a(g) \sum_{\chi} \chi(g) \end{aligned}$$

and

$$(1 - \chi_a(g)) \sum_{\chi} \chi(g) = 0$$

Since $\chi_a(g) \neq 1$, we get $\sum_{\chi} \chi(g) = 0$. ∎

The next theorem gives two useful orthogonality properties of group characters. The proof is left to the reader.

Theorem 8.39 *If G is a finite commutative group of order n, $g_i, g_j \in G$, and both χ_a and χ_b are group characters of G, then*

(a)

$$\sum_{\chi} \chi(g_i^{-1})\chi(g_j) = \begin{cases} n & \text{if } g_i = g_j \\ 0 & \text{if } g_i \neq g_j \end{cases}$$

(b)

$$\sum_{g \in G} \chi_a(g)\chi_b^{-1}(g) = \begin{cases} n & \text{if } \chi_a = \chi_b \\ 0 & \text{if } \chi_a \neq \chi_b \end{cases}$$

Theorem 8.40 *Assume that the group G has order n and σ is the $n \times n$ matrix with $\sigma_{ij} = \chi_i(g_j)$, where $\chi_1, \chi_2, \chi_3, \ldots, \chi_n$ are the characters of G. If σ' is the matrix with $\sigma'_{ij} = \text{conj}(\sigma_{ji})$, then*

$$\sigma'\sigma = \begin{bmatrix} n & 0 & 0 & \cdots & 0 \\ 0 & n & 0 & \cdots & 0 \\ 0 & 0 & n & \cdots & 0 \\ & & \vdots & & \\ 0 & 0 & 0 & \cdots & n \end{bmatrix} = \sigma\sigma'$$

Proof. Let $C = [C_{ij}] = \sigma'\sigma$. Then

$$\begin{aligned} C_{ij} &= \sum_{k=1}^{n} \sigma'_{ik}\sigma_{kj} \\ &= \sum_{k=1}^{n} \text{conj}(\sigma'_{ki})\sigma_{kj} \\ &= \sum_{k=1}^{n} \text{conj}(\chi_k(g_i))\chi_k(g_j) \\ &= \sum_{k=1}^{n} \chi_k(g_i^{-1})\chi_k(g_j) \end{aligned}$$

and the first equality of the theorem follows from Theorem 8.39. The proof of the second equality of the theorem is left to the reader. ∎

Theorem 8.41 *There are $\phi(m)$ distinct Dirichlet characters modulo the positive integer m. Each Dirichlet character is totally multiplicative and has the* *property that* $\chi([a+m]) = \chi([a])$, *where* $[a]$ *is the equivalence class of a modulo* m, *or, equivalently,* $\chi(a+m) = \chi(a)$.

Note Since addition and multiplication in Z_m are the same for any member of an equivalence class in Z_m, we often write $\chi(a+m)$ instead of $\chi([a+m])$. Thus χ is extended to have domain Z. It is in this extended sense that we speak of each character as being totally multiplicative.

Proof. Since each Dirichlet character is a unique extension of a group character from the group of reduced residue classes modulo m, then the first part follows from the fact that the number of group characters of a group is equal to the number of elements in the group. Since each "integer" in the image of a Dirichlet character really represents a congruence class modulo m, $[a] = [a+m]$ and $\chi([a+m]) = \chi([a])$.

The character χ is totally multiplicative since if a and b are relatively prime to m, then $[a]$ and $[b]$ are in the reduced residue group and χ is a homomorphism on this group. If either a or b is not relatively prime to m, then neither is the product and so $\chi([a])\chi([b])$ and $\chi([ab])$ are both 0 or, equivalently, $\chi(a)\chi(b)$ and $\chi(ab)$ are both 0. ■

Exercises

1. Prove Theorem 8.25.

2. Prove that if b is an element of a commutative group G and $b \neq e$, the identity element of G, then there is a group character χ such that $\chi(b) \neq 1$.

3. If g and g' are elements of a finite commutative group G such that $g \neq g'$, then prove that $\chi_g \neq \chi_{g'}$.

4. Complete the proof of Theorem 8.30.

5. Prove Theorem 8.34.

6. Prove Theorem 8.39.

7. Complete the proof of Theorem 8.40.

8.3   Polynomials

We introduced polynomial functions briefly in Section 3.4; however, in this chapter we formally define polynomial forms. Polynomial functions $f(x)$ and $g(x)$ are equal provided that $f(b) = g(b)$ for all b in the common domain. Polynomial forms $f(x)$ and $g(x)$ are equal provided that corresponding powers of x have equal coefficients. For polynomials over the integers, the two concepts are the same; however, such is not the case over all commutative rings with unity. We will reconcile the two concepts of equality at the end of this section. Most of the following discussions are in terms of commutative rings with unity and integral domains; but usually we will be using the ring of integers or the rational numbers. To aid in understanding, one may wish to think of A as being the integers in much of the following development. For details on algebraic terminology see Section 2.4.

Definition 8.42 *Let A be a commutative ring with unity and let S be the set of all sequences $(a_0, a_1, a_2, \ldots)$ of elements of A such that if $f \in S$, then there exists an integer N_f so that $a_j = 0$ for all $j > N_f$. If $f \in S$, then f is said to be a* polynomial *or a* polynomial form *over A.*

For example,

$$(1, 0, 0, 0, 0, \ldots)$$

and

$$(1, 1, 1, 0, 0, 0, \ldots)$$

are in S; but

$$(1, 1, 1, 1, 1, 1, 1, 1, \ldots)$$

and

$$(1, 0, 1, 0, 1, 0, 1, 0, 1, \ldots)$$

are not in S. If $A = Z$, then

$$(0, \, -5, 2, 0, 0, \ldots)$$

and

$$(3, 7, 5, 8, 0, 0, \ldots)$$

are polynomials in S.

Definition 8.43 *Let A be a commutative ring and let $f = (a_i)^* = (a_0, a_1, a_2, \ldots)$ and $g = (b_i)^* = (b_0, b_1, b_2, \ldots)$ be in S, the set of polynomials over A. Define the* sum *of f and g to be the sequence $f + g = (a_i + b_i)^* = (a_0 + b_0, a_1 + b_1, \ldots)$ and the* product *of f and g to be the sequence $fg = (c_k)^*$, where $c_k = \sum\limits_{i+j=k} a_i b_j$.*

Theorem 8.44 *For $f, g \in S$ over a commutative ring A with unity, $f + g \in S$ and $fg \in S$.*

Proof. The proof for $f + g$ is left to the reader. To show that $fg \in S$, note that for any $k > N_f + N_g + 1$, if $i + j = k$ and $i \le N_f$, then $j > N_g + 1$, and if $j \le N_g$, then $i > N_f + 1$. Hence $c_k = 0$ for any $k > N_f + N_g + 1$ since either $a_i = 0$ or $b_j = 0$ in the sum c_k. ∎

The proof of the following theorem is left to the reader:

Theorem 8.45 *If S is the set of polynomials over a commutative ring A with unity, then S is also a commutative ring with unity. Its unit is $(1,0,0,0,\ldots)$ and its zero element is $(0,0,0,\ldots)$.*

Definition 8.46

(a) *Let A be a commutative ring with unity, and let $f \in S$ and $f = (a_i)^*$. If $f \neq 0$, let $\deg(f)$ equal the largest integer k, such that $a_k \neq 0$. The function $\deg(f)$ is called the* degree *of f. S is called the* ring of polynomials *over A. Any element of S is called a* polynomial *over A. Any polynomial that is of degree 0 or is equal to zero is called a* constant.

(b) *Let $f = (a_0, a_1, a_2, \ldots)$ belong to S. The terms a_i of the sequence are called the* coefficients *of f. If $f \neq 0$ and $n = \deg(f)$, then a_n is called the* leading coefficient *of f. If $a_n = 1$, then f is called a* monic polynomial. *If $f \neq 0$ has the property that any greatest common divisor of all of its nonzero coefficients is a unit, then f is called a* primitive polynomial.

(c) *Two elements f and g of S are equal, written $f = g$, if their respective coefficients are equal; that is, if $f = (a_0, a_1, a_2, \ldots)$ and $g = (b_0, b_1, b_2, \ldots)$, then $f = g$ if and only if $a_i = b_i$ for every nonnegative integer i.*

(d) *The polynomial f* divides *the polynomial g provided that there is a polynomial h such that $fh = g$. In this case, we say that f and h are* factors *of g.*

For example, if

$$f = (0, 1, 1, 0, 1, 0, 0, \ldots)$$

then $\deg(f) = 4$; if

$$g = (1, 0, 0, 0, \ldots)$$

then $\deg(g) = 0$ and g is a constant polynomial. The zero polynomial

$$(0, 0, 0, \ldots)$$

has no degree. It is left to the reader to show that equality in S is an equivalence relation on S.

Theorem 8.47 *Let A be a commutative ring with unity, let S be the ring of polynomials over A, and let f and g be in S.*

(a) *If $f, g \neq 0$, then $\deg(f + g) \leq \max(\deg(f), \deg(g))$.*

(b) *Either $fg = 0$ or $\deg(fg) \leq \deg(f) + \deg(g)$.*

(c) *If A is an integral domain, then either $fg = 0$ or $\deg(fg) = \deg(f) + \deg(g)$.*

(d) *If A is an integral domain, then so is S.*

Proof. (a) Let $f = (a_0, a_1, a_2, \ldots)$ and $g = (b_0, b_1, b_2, \ldots)$, where $n = \deg(f)$ and $m = \deg(g)$. Since $a_i = 0$ for $i > n$ and $b_j = 0$ for $j > m$, if $k > \max(m, n)$, then $a_k + b_k = 0$ and $\deg(f + g) \leq \max(\deg(f), \deg(g))$.

(b) Let $n = \deg(f)$, $m = \deg(g)$ and $fg = (c_0, c_1, c_2, \ldots)$ so that $c_k = \sum_{i+j=k} a_i b_j$. The product $a_i b_j = 0$ when $i > n$ since $a_i = 0$. If $i \leq n$, and $i + j > m + n$, then $j > m + (n - i)$ so $b_j = 0$ and, in either case, $c_k = 0$. Hence $\deg(fg) \leq \deg(f) + \deg(g)$.

(c) By part(b), $\deg(fg) \leq \deg(f) + \deg(g)$. However since a_n and b_m are not 0, then $c_{n+m} = a_n b_m$ is not 0. Hence $\deg(fg) \geq \deg(f) + \deg(g)$ and we have equality.

(d) Let f and g be nonzero elements of S with degree n and m, respectively. Since A is an integral domain, $\deg(fg) = \deg(f) + \deg(g)$ and fg is not 0. ∎

Theorem 8.48 *There is a monomorphism from A into S, the ring of polynomials over A, which makes the image of A a subring of S. If A is an integral domain, then any unit of S corresponds to a unit of A via the monomorphism.*

Proof.   Define the function $\phi : A \to S$ as follows: If $a \in A$, let $\phi(a) = (a, 0, 0, 0, \ldots)$. If a and b are in A, then

$$
\begin{aligned}
\phi(a + b) &= (a + b, 0, 0, 0, 0, \ldots) \\
&= (a, 0, 0, 0, 0, \ldots) + (b, 0, 0, 0, 0, \ldots) \\
&= \phi(a) + \phi(b)
\end{aligned}
$$

$$
\begin{aligned}
\phi(a \cdot b) &= (a \cdot b, 0, 0, 0, \ldots) \\
&= (a, 0, 0, 0, \ldots) \cdot (b, 0, 0, 0, \ldots) \\
&= \phi(a) \cdot \phi(b)
\end{aligned}
$$

Also, $\phi(1) = (1, 0, 0, 0, \ldots) = 1$. Hence ϕ is a homomorphism from A into S.

If $(c, 0, 0, 0, \ldots) = \phi(c) = 0 = (0, 0, 0, 0, \ldots)$, then $c = 0$. If $\phi(a) = \phi(b)$, then $\phi(a) - \phi(b) = 0$; that is, $\phi(a - b) = 0$ and $a - b = 0$. Therefore, $a = b$ and ϕ is one to one.

Recall that a unit of a ring is an element which has an inverse; that is, if b is a unit there exists an element b^{-1} so that $b \cdot b^{-1} = 1$.

Assume that A is an integral domain and f is a unit in S. Let g be its inverse, so $fg = 1 = (1, 0, 0, 0, \ldots)$. But $0 = \deg(fg) = \deg(f) + \deg(g)$, which implies that $\deg(f)$ and $\deg(g)$ are both 0. Thus, $f = (a, 0, 0, 0, \ldots)$ and $g = (b, 0, 0, 0, \ldots)$ for some $a, b \in A$. Since

$$\phi(a \cdot b) = fg = (1, 0, 0, 0, \ldots) = \phi(1)$$

and since ϕ is one to one, $ab = 1$ and a and b are units in A.

Because the monomorphic image $\phi(A)$ in S is a ring and is isomorphic with A, we identify A with $\phi(A)$ as a subring of S and may refer to A as a subring of S. ∎

Definition 8.49 *The Kronecker δ, δ_{ij}, for integers i and j, is defined by*

$$\delta_{ij} = \begin{cases} 1 & \text{if } i = j \\ 0 & \text{if } i \neq j \end{cases}$$

For example, if $f = (a_0, a_1, a_2, \ldots)$ where $a_i = \delta_{i3}$, then

$$f = (0, 0, 0, 1, 0, 0, \ldots)$$

If $g = \sum_{j=1}^{5} (\delta_{ij})^*$, then

$$g = (0, 1, 1, 1, 1, 1, 0, 0, \ldots)$$

Theorem 8.50 *If $x = (0, 1, 0, 0, 0, \ldots) = (c_i)^*$ where $c_i = \delta_{i1}$, then for each $k > 0$, $x^k = (a_0, a_1, a_2, \ldots)$, where $a_i = \delta_{ik}$.*

Proof. The theorem is clearly true for $k = 1$. Assume that the theorem holds for $k = n$. Since $x^{n+1} = x \cdot x^n$, simple multiplication shows the i-th term in the sequence x^{n+1} to be $\delta_{i,n+1}$. ∎

If $a \in A$ and $f = (a_0, a_1, a_2, \ldots)$, then $af = (aa_0, aa_1, aa_2, \ldots)$. In particular, if $a \in A$, then $ax^k = (0, 0, 0, \ldots, a, 0, 0, 0, \ldots)$, where a occurs in the $(k + 1)$-th place.

Let $f \in S$ be of degree n. Then

$$\begin{aligned} f &= (a_0, a_1, a_2, \ldots, a_n, 0, 0, \cdots) \\ &= (a_0, 0, 0, \ldots) + (0, a_1, 0, \ldots) + \cdots + (0, 0, \ldots, 0, a_n, 0, 0, \ldots) \\ &= a_0(1, 0, 0, \ldots) + a_1(0, 1, 0, 0, \ldots) + \cdots + a_n(0, 0, \ldots, 0, 1, 0, 0, \ldots) \\ &= a_0 + a_1 x + a_2 x^2 + \cdots + a_n x^n \end{aligned}$$

Definition 8.51 *Let A be a commutative ring with unity and let S be the set of polynomials over A. The symbol x is called an* indeterminate *over A. For every*

$$f = a_0 + a_1 x + a_2 x^2 + \cdots + a_n x^n \in S$$

there is associated a function from A to A

$$f(x) = a_0 + a_1 x + a_2 x^2 + \cdots + a_n x^n$$

or

$$f(x) = a_n x^n + \cdots + a_2 x^2 + a_1 x + a_0$$

called a polynomial function. *Let $A(x) = \{f(x) : f \in S\}$. The degree of $f(x)$ is the same as the degree of the corresponding polynomial $f \in S$. The elements f of S, are called* polynomial forms *to distinguish them more easily from the polynomial functions.*

It is easily shown that the map θ indicated in Definition 8.51, which maps elements of S into the corresponding polynomial functions in $A(x)$, is a homomorphism. Since θ is onto, it is an epimorphism as well. Thus, all the properties of S preserved by homomorphisms are imbued upon the commutative ring of polynomial functions $A(x)$.

Definition 8.52 *Let A be a commutative ring with unity. If*

$$f(x) = a_n x^n + \cdots + a_2 x^2 + a_1 x + a_0$$

and

$$g(x) = b_n x^n + \cdots + b_2 x^2 + b_1 x + b$$

where some of the a_i and b_i in A may be 0 including a_n or b_n, then

$$f(x) + g(x) = (a_n + b_n)x^n + \cdots + (a_1 + b_1)x + (a_0 + b_0)$$

and

$$f(x)g(x) = c_m x^m + \cdots + c_2 x^2 + c_1 x + c_0$$

where $c_k = \sum_{i+j=k} a_i b_j$. Define $f(x) = g(x)$ if and only if $f(b) = g(b)$ for all $b \in A$. A solution of the equation $f(x) = 0$ is an element $a \in A$ such that $f(a) = 0$.

From the discussion above, we have the following theorem for polynomial functions over an integral domain A.

Theorem 8.53 *If $f(x)$ and $g(x)$ are polynomial functions over an integral domain A, $f(x)$ has degree n, and $g(x)$ has degree m, then*

(a) *$f(x) + g(x)$ has degree less than or equal to $\max\{m, n\}$.*

(b) *$f(x)g(x)$ has degree $m + n$.*

Two polynomial forms f and g were defined to be equal if and only if their coefficients were equal. Two polynomial functions $f(x)$ and $g(x)$ were defined to be equal provided that $f(b) = g(b)$ for all $b \in A$. The next theorem establishes that both concepts of equality are equivalent if A is an infinite integral domain. In Section 3.4 we discussed polynomial functions and showed that if $f(x)$ is a polynomial function of degree n over an integral domain A, then $f(x) = 0$ has at most n solutions. This result will be used in the proof of the following theorem.

Theorem 8.54 *Let f and g be a polynomials over an infinite integral domain A. Then $f = g$ if and only if the corresponding polynomial functions $f(x)$ and $g(x)$ have the property that $f(b) = g(b)$ for every $b \in A$.*

Proof. If f and g are polynomials over the integral domain A such that $f = g$, then clearly $f(b) = g(b)$ for all $b \in A$.
　　If

$$f(x) = a_n x^n + \cdots + a_2 x^2 + a_1 x + a_0$$

and

$$g(x) = b_n x^n + \cdots + b_2 x^2 + b_1 x + b$$

are polynomial functions of degrees n and m, respectively, and $f(c) = g(c)$ for every $c \in A$. Let $f = (a_i)^*$ and $g = (b_i)^*$ be the respective corresponding polynomials. Then the polynomial function $h(x) = c_s x^s + \cdots + c_2 x^2 + c_1 x + c$, where $s = \max(n, m)$ and $c_i = a_i - b_i$, has the property $h(c) = f(c) - g(c) = 0$ for every $c \in A$. But $h(c) = 0$ for all c in A, which is infinite, and since any nonzero polynomial has a finite number of solutions, $h(x)$ must be the zero polynomial. Therefore, $c_i = a_i - b_i = 0$ or $a_i = b_i$ for $i \geq 1$, and $f = g$. ∎

Corollary 8.55 *If $f = (a_i)^*$ is a polynomial over an infinite integral domain A and $f(x)$ is the corresponding polynomial function with the property that $f(b) = 0$ for all $b \in A$, then $a_i = 0$ for all i.*

Example 8.56 *The set $Z_5 = \{[0], [1], [2], [3], [4]\}$ of integers modulo 5 is a field and, therefore, an integral domain. By Fermat's Theorem, if $a \not\equiv 0 \pmod{p}$ for p prime, then $a^{p-1} \equiv 1 \pmod{p}$. Then $a^p \equiv a \pmod{p}$ for any integer a, even if $a \equiv 0 \pmod{p}$. For $p = 5$, we have $a^5 \equiv a \pmod{5}$ for all $a \in Z$. This congruence is equivalent to $[a]^5 = [a]$ or to $[a]^5 - [a] = [0]$ for any integer a. Thus, if $g(x) = x^5 - x$, then $g(x) = [0]$ holds for any x in Z_5. But if h is the zero polynomial, then $h(x) = [0]$ for any x in Z_5. The polynomials g and h are not equal, but $g(b) = h(b)$ for every b in the integral domain Z. Thus, over a finite integral domain, equality of polynomial forms is not equivalent to equality of polynomial functions. On the other hand, since the set of integers is infinite, the concepts of equality are the same for polynomial forms and polynomial functions over Z.*

Exercises

1. Prove Theorem 8.45

2. Show that equality of S in Definition 8.46 is an equivalence relation.

3. Prove or disprove: If $f, g \neq 0$, then

$$\deg(f(x) + g(x)) = \max(\deg(f(x)), \deg(g(x)))$$

4. Find two polynomials from the set of polynomials over Z_{12} (i.e., the coefficients of the polynomials are from Z_{12}) such that $\deg(f(x)g(x)) < \deg(f(x)) + \deg(g(x))$.

5. Given a function $f : A \to B$, prove that the relation on A defined by aRa' if and only $f(a) = f(a')$ is an equivalence relation. Relate this to cosets of a group homomorphism.

6. Prove or disprove: In the set of polynomial functions over Z_6, if $f(x) = g(x)$ for all x, then the polynomials are equal.

8.4   Algebra and Polynomials

We began our study of number theory with the integers which we used in most of the basic core of the text. We eventually developed the rational and real numbers. Finally, in this chapter, we defined the complex numbers. We now discuss special subrings and subfields of the complex numbers. These consist of the Gaussian integers, algebraic integers, and algebraic numbers.

In this section we will discuss further properties of polynomials and also find extension fields of a given field F so that polynomials which cannot be factored in F can be factored in the extension field. Recall that a unit is a divisor of 1; that is, u is a unit provided that there is an element u' such that $uu' = 1$. Recall also that, in a field, every nonzero element is a unit.

In the following material we build up algebraic structure which we will then apply to polynomials. All rings will be assumed to be commutative.

Definition 8.57 *A field F is called a* subfield *of F' if it is contained in F' and has the same operations. The field F' is called an* extension field *of F. More generally, a field F' is an extension field of F if there is an isomorphism $\eta : F \to E$ where E is a subfield of F'.*

Definition 8.58 *If A is an integral domain, a nonzero element a of A is said to be* irreducible *if when $a = bc$, then either b or c is a unit, and is said to be* reducible *otherwise. Let $A[x]$ denote the set of polynomial functions with coefficients in A. A polynomial $f(x)$ in $A[x]$ with positive degree is called* irreducible *if whenever $f(x) = g(x)h(x)$, then either $g(x)$ or $h(x)$ is an element of A, and is said to be* reducible *otherwise.*

Thus, if a polynomial $f(x)$ is reducible, then there are polynomials $g(x)$ and $h(x)$ of positive degree such that $f(x) = g(x)h(x)$.

We shall see that the definitions of irreducible for integral domains and for polynomials are the same if A is a field or if the polynomials are restricted to monic polynomials.

Definition 8.59 *A polynomial $f(x)$ in $A[x]$ is called* prime *if it has degree greater than 0 and if whenever $f(x) \mid g(x)h(x)$, then either $f(x) \mid g(x)$ or $f(x) \mid h(x)$. More generally, in an integral domain, an element p is prime if $p \mid (ab)$ implies that $p \mid a$ or $p \mid b$.*

In the set of integers, prime integers and irreducible integers are the same; however, such is not true in some rings of polynomials.

Theorem 8.60 *In an integral domain, a prime is always irreducible.*

Proof. Assume that p is a prime and $p = ab$; then by definition, $p \mid a$ or $p \mid b$. Assume that $p \mid a$; then $a = mp$ for some m. Then $p = mpb = mbp$, so $mb = 1$ and b is a unit. Hence p is irreducible. ∎

For a review of ideals, principal ideals, prime ideals, and unique factorization domains, see Section 8.1.

Definition 8.61 *Let $A^{\emptyset}$ denote the nonzero elements of the integral domain A, then A is a Euclidean domain if there exists a function $\phi : A^{\emptyset} \to N$, where N is the set of positive integers, such that:*

(a) *If $a, b \in A^{\emptyset}$, then $\phi(a) \leq \phi(ab)$ or equivalently, if $a \mid c$, then $\phi(a) \leq \phi(c)$.*

(b) *If $a, b \in A^{\emptyset}$, then there exists $q, r \in A^{\emptyset}$ such that $a = bq + r$ and either $r = 0$ or $\phi(r) < \phi(b)$.*

In a Euclidean domain A, $\phi(a)$, for $a \in A^{\emptyset}$, is called the norm *of a.*

Theorem 8.62 *Let F be a field; then $F[x]$ is a Euclidean domain using the norm $\phi(f) = \deg(f)$ for $f \in F[x]$.*

Proof. Let $A^{\emptyset} = F[x] - \{(0)^*\}$ and define $\phi : A^{\emptyset} \to N$ by $\phi(f(x)) = deg(f)$ when $f \neq 0$. Theorem 8.47 establishes part (a). For part (b), let $a(x) \in F[x]$ be a polynomial of degree n and $b(x)$ a polynomial of degree k. Let S be the set of polynomials of the form $a(x) - q(x)b(x)$. If any element of S is 0, we are done; otherwise, select one, say $r(x)$, with least degree and $q(x)$ the corresponding quotient.. This element has degree less than n; for if the leading coefficient of $b(x)$ is v, select a polynomial of degree $n-k$ with leading coefficient c so that vc is the leading coefficient of $a(x)$. Then $a(x) - q(x)b(x)$ has degree less than n. Similarly, if $\deg(r)$ is not less than $\deg(b)$, then

$$r(x) = q'(x)b(x) + r'(x)$$

where $\deg(r')$ is less than $\deg(r)$ so that

$$a(x) - q(x)b(x) = r(x) = q'(x)b(x) + r'(x)$$

and

$$a(x) - (q(x) + q'(x))b(x) = r'(x)$$

where $\deg(r') < \deg(r)$. But this contradicts the definition of $r(x)$. Hence $\deg(r) < \deg(b)$. ∎

Theorem 8.63 *Every Euclidean domain A is a principal ideal domain.*

Proof. Let I be a nonzero ideal of A. Let d be any nonzero element of I with minimal norm value, and let $p \in I$. But $p = dq + r$, where the norm of r is less than the norm of d. Since $r \in I$, d has minimal norm, and the norm of r is less than the norm of d, $r = 0$, and $p = dq$. Hence d generates I. ∎

Corollary 8.64 *Let F be a field; then F[x] is a principal ideal domain.*

Definition 8.65 *An ideal I is maximal in a ring R if for any ideal J, $I \subseteq J \subseteq R$ implies that $I = J$ or $I = R$.*

Theorem 8.66 *An ideal I of a principal ideal domain is maximal if and only if p, the generator of I, is irreducible over F.*

Proof. Assume that I is maximal and p is not irreducible, say $p = rs$. Then p is in the ideal J generated by r(or s) and $I \subset J$, which is a contradiction.

Conversely, if p is irreducible and I is not maximal, say $I \subset J$ and r generates J, then $p = rs$ for some s and this is impossible since p is irreducible. ∎

Theorem 8.67 *Let I be an ideal contained in a ring R. Define a relation $\sim$ on R by $a \sim b$ if and only if $a - b \in I$. The relation $\sim$ is an equivalence relation.*

Proof. The relation $\sim$ has the following properties:

(a) Reflexivity: $a \sim a$ since $0 \in I$.

(b) Symmetry: If $a \sim b$, then $a - b \in I$; hence, $b - a = -(a - b) \in I$ since I is a group under addition. So $b \sim a$.

(c) Transitivity: Let $a \sim b$ and $b \sim c$ so that $a - b \in I$ and $b - c \in I$. Hence

$$(a - b) + (b - c) = a - c \in I$$

and $a \sim c$. ∎

Theorem 8.68 *If $a \sim b$ and $c \sim d$, then $a + c \sim b + d$ and $ac \sim bd$.*

Proof. If $a \sim b$ and $c \sim d$, then $a - b \in I$ and $c - d \in I$, then

$$(a - b) + (c - d) = (a + c) - (b + d) \in I$$

and $a + c \sim b + d$. Also, $c(a - b) = ac - bc \in I$ and $b(c - d) = bc - ad \in I$ since I is an ideal. Hence $(ac - bc) + (bc - bd) = ac - bd \in I$ and $ac \sim bd$. ∎

We can now define addition and multiplication between elements of the equivalence relation. If $a + I$ denotes the equivalence class containing a and if $b + I$ the equivalence class containing b, then we define

$$(a + I) + (b + I) = (a + b) + I$$

and

$$(a + I) \cdot (b + I) = a \cdot b + I$$

When we can multiply and add equivalence classes, then the relation is called a *congruence relation*. The ring of equivalence classes is denoted by R/I. Note that $0 + I$ is the zero element of the ring and $1 + I$ is the multiplicative identity.

The equivalence class $a + I$ is called a *left coset* of I, in particular, the one generated by a. Hence, the distinct left cosets of I are just the equivalence classes in R/I.

Theorem 8.69 *Let R be a commutative ring with unity and I be an ideal of R. Then R/I is a field if and only if I is a maximal ideal.*

Proof. Assume that I is maximal and let $a + I \in R/I$, where $a \notin I$. Let $M = \{ra + b \mid r \in R \text{ and } b \in I\}$. It is straightforward to show that M is an ideal. Since $a \notin I$, one obtains $M \subset I$ and $M = R$. Hence $1 \in M$ and there exists b in R and c in I so that $1 = ab + c$. Thus, $1 + I = ab + I = (a + I)(b + I)$ and $a + I$ has an inverse. Hence R/I is a field.

Conversely, assume that R/I is a field. Assume that there exists an ideal J such that $I \subset J \subseteq R$. Consider $a + I$ where a is in J but not I. Then there exists a' in R so that $a' + I$ is the inverse of $a + I$, making $aa' + I = 1 + I$. Furthermore, since J is an ideal, $aa' \in J$. Also, $1 = aa' + c$ for some c in I. Since aa' and c are in J, 1 is in J. But if 1 is in J, then $r1$ is in J for all r and $J = R$. ∎

Theorem 8.70 *Let R be a commutative ring with identity. Then R/I is an integral domain if and only if I is a prime ideal.*

Proof. If R/I is an integral domain and ij is an element of I, then $ij + I = 0 + I$. Hence $(i + I)(j + I) = 0 + I$. But since R/I is an integral domain, either $i + I$ or $j + I$ equals $0 + I$. Hence either $i \in I$ or $j \in I$ and I is a prime ideal.

Conversely, if I is a prime ideal and $(i + I)(j + I) = 0$, then $ij + I = 0$ and $ij \in I$. But since I is a prime ideal, $i \in I$ or $j \in I$; therefore, $i + I = 0$ or $j + I = 0$ and I is a integral domain. ∎

Theorem 8.71 *Every maximal ideal is a prime ideal.*

Proof. The proof follows immediately from the results of Theorems 8.70 and 8.71. ∎

Theorem 8.72 *In a principal ideal domain A, every irreducible element is a prime.*

Proof. Let p be an irreducible element of A and let $\langle p \rangle$ be the principal ideal generated by p. Since p is irreducible, by Theorem 8.66, $\langle p \rangle$ is maximal; and hence by Theorem 8.71, is a prime ideal. Therefore, if $p \mid (ab)$, then $ab \in \langle p \rangle$. Thus, $a \in \langle p \rangle$ or $b \in \langle p \rangle$ so that $p \mid a$ or $p \mid b$ and p is prime. ∎

Combining Corollary 8.64 and Theorem 8.72, we obtain the following result.

Corollary 8.73 *If F is a field, then in $F[x]$ every irreducible element is a prime.*

In the next theorem, we have an analogy to the unique factorization of integers into products of primes.

Theorem 8.74 *If F is a field, then every nonunit polynomial in $F[x]$ can be uniquely factored into a product of irreducible polynomials in the sense that factors are unique up to multiplication by a unit.*

Proof. Using the second form of induction on the degree of the polynomial, we show that every nonconstant polynomial of degree n can be factored into irreducible elements. For $n = 1$ we have a polynomial of the form $ax+b$, which is already irreducible because if not, then $ax+b = f(x)g(x)$ but $\deg(ax+b) = \deg(f) + \deg(g) \geq 2$, an obvious contradiction. Assume that $k > 1$ and the theorem holds for true for all $n < k$ and let $f(x)$ have degree k. If $f(x)$ is irreducible, we are done. If not, then $f(x)$ factors into polynomials $g(x)$ and $h(x)$, each with positive degree less than $f(x)$. Hence $g(x)$ and $h(x)$ both factor into irreducible elements; therefore, $f(x)$ also factors into irreducible factors.

To show that the factors are unique we again use induction on the number of factors in one reduction. For $n = 1$, the result is obvious. Assume that the uniqueness holds for $n = k$ factors. Assume that $f(x)$ factors into $k + 1$ irreducible elements $g_1(x), g_2(x), \ldots, g_{k+1}(x)$ and into m irreducible elements $g_1'(x), g_2'(x), \ldots, g_m'(x)$. Select the factor $g_1(x)$ from the $k + 1$ irreducible elements. Since an irreducible element is a prime by Corollary 8.73 and $g_1(x)$ divides $f(x)$, $g_1(x)$ must divide one of the m irreducible elements in the second factorization, say $g_j'(x)$. But since $g_j'(x)$ is irreducible, $g_1(x)$ and $g_j'(x)$ differ by a unit. Hence divide both factorizations by $g_1(x)$, leaving the first factorization of $f(x)$ with k irreducible factors $g_2(x), \ldots, g_{k+1}(x)$ and the second factorization with $m-1$ irreducible factors $g_1'(x), \ldots, g_{j-1}'(x), g_{j+1}, \ldots, g_m'(x)$ and a possible unit factor. By induction, the factorization of $f(x)/g_1(x)$ is unique and so the factorization of $f(x)$ is unique. ∎

Recall that a polynomial $f(x) = a_0 + a_1x + \cdots + a_nx^n$ is primitive if $\gcd(a_0, a_1, \ldots, a_n)$ is a unit.

Theorem 8.75 (Gauss) *Let A be an integral domain. The product of two primitive polynomials in $A[x]$ is a primitive polynomial.*

Proof. Let $f(x) = a_0+a_1x+\cdots+a_nx^n$ and $g(x) = b_0+b_1x+\cdots+b_mx^m$ and let p be an irreducible element of A. Since $f(x)$ is primitive, there exists an a_i such that p does not divide a_i. Let a_r be the first such coefficient. Similarly, there is a b_j such that p does not divide b_j. Let b_s be the first such coefficient. Let $h(x) = c_0 + c_1x + \cdots + c_{m+n}x^{m+n}$ be the product of $f(x)$ and $g(x)$; then

$$c_{r+s} = a_0b_{r+s} + \cdots + a_{r-1}b_{s+1} + a_rb_s + a_{r+1}b_{s-1} + \cdots + a_{r+s}b_0$$

But p divides $a_0, \ldots, a_{r-1}$, so p divides $a_0b_{r+s} + \cdots + a_{r-1}b_{s+1}$; and p divides $b_0, \ldots, b_{s-1}$, so p divides $a_{r+1}b_{s-1} + \cdots + a_{r+s}b_0$. But p does not divide a_rb_s so that p does not divide $h(x)$. Hence, $h(x)$ is primitive. ∎

It might seem that a polynomial which is irreducible in the integers might not be irreducible if we allow factorization into polynomials with rational coefficients; however, the following theorem shows that this limitation is not true.

Theorem 8.76 *Let A be a unique factorization domain, that is, an integral domain having unique factorization into primes, and let F be its field of fractions. If $p(x) \in A[x]$ and if $p(x)$ is reducible in $F[x]$, then it is reducible in $A[x]$. If $p(x)$ is primitive in $A[x]$, then it is reducible in $F[x]$ if and only if it is reducible in $A[x]$.*

Proof. Let $p(x) \in A(x)$ be reducible in $F[x]$, say $p(x) = f(x)g(x)$. Let u be the least common multiple of the denominators of the coefficients of $f(x)$ and $g(x)$; then $u \cdot p(x) = f'(x)g'(x)$, where $f'(x)$ and $g'(x)$ are in $A[x]$. But $f'(x) = af''(x)$ and $g'(x) = bg''(x)$, where $f''(x)$ and $g''(x)$ are primitive; and $p(x) = cp'(x)$, where c is primitive. Hence $uc \cdot p'(x) = ab \cdot f''(x)g''(x)$. But since $f''(x)g''(x)$ is also primitive, it is easily shown that two expressions for a primitive polynomial differ only by a unit. Thus $ucv = ab$, where v is a unit. Hence $ucp'(x) = ucvf''(x)g''(x)$; so

$$up(x) = ucvf''(x)g''(x)$$

or

$$p(x) = cvf''(x)g''(x)$$

and $p(x)$ has been factored in $A[x]$.

Obviously, if a polynomial factors in $A[x]$, then it factors in $F[x]$, but if $p(x)$ is not primitive in $A[x]$, then it is reducible in $A[x]$ because it is the product of an integer and a polynomial; however, in $F[x]$ it may be irreducible since an integer is a unit. ■

Theorem 8.77 *An integral domain A is a unique factorization domain if and only if $A[x]$ is.*

Proof. Let F be the field of fractions of A and $p(x)$ be a polynomial in $A[x]$. Then $p(x)$ factors into irreducibles in $F[x]$; and hence, $p(x)$ factors into irreducibles in $A[x]$, where each irreducible of the factorization in $F[x]$ is equal to a corresponding irreducible of the factorization in $A[x]$ multiplied by an element of F.

Let d equal the greatest common divisor of the coefficients of $p(x)$ so that $p(x) = dp'(x)$, where $p'(x)$ is a primitive polynomial. Since d factors uniquely, we need only show $p'(x)$ factors uniquely; hence we will assume that $p(x)$ is primitive. Since, by Theorem 8.74, $F[x]$ is a unique factorization domain, let

$$p'(x) = r_1(x)r_2(x) \cdots r_n(x) = s_1(x)s_2(x) \cdots s_n(x)$$

Assume that by rearrangement we have $r_i(x)$ and $s_i(x)$ differing only by a nonzero element of F. Say that $r_i(x) = (a/b)s_i(x)$ so $br_i(x) = as_i(x)$; but since $r_i(x)$ and $s_i(x)$ are primitive, a and b differ by a unit in $A(x)$ and the factorization is unique.

Conversely, since A is a subset of $A[x]$, if $A[x]$ is a unique factorization domain, then every element of A must factor uniquely and A is a unique factorization domain. ∎

Theorem 8.78 *Let A be a unique factorization domain; then two polynomials in $A[x]$ have a greatest common divisor.*

Proof. Let $p(x)$ and $q(x)$ be polynomials in $A[x]$. Since $A[x]$ is a unique factorization domain, factor $p(x)$ and $q(x)$ into irreducibles. Then $\gcd(p(x), q(x))$ is the product of the least powers of irreducibles occurring as factors in both $p(x)$ and $q(x)$. ∎

Theorem 8.79 *Let A be a principal ideal domain and I be an ideal generated by p and q. If d generates I, then $d = \gcd(p, q)$ and d may be written in the form $up + vq$.*

Proof. Certainly, d divides p and q. Since $d \in I$, d may be written in the form $up + vq$. If c divides p and c divides q, then c divides $up + vq$ and c divides d. Thus, $d = \gcd(p, q)$. ∎

Beginning with an integral domain A, such as the integers, we may of course find the fraction field of A, say F, and form $F[x]$. Since $A[x]$ is an integral domain, we may also form the fraction field of $A[x]$, essentially generating elements of the form $p(x)/q(x)$, where $p(x)$ and $q(x)$ belong to $A[x]$. Since $F[x]$ is also an integral domain, we may form the fraction field of $F[x]$, which essentially consists of elements of the form $p(x)/q(x)$ where $p(x)$ and $q(x)$ belong to $F[x]$. This field is called the *field of rational functions over the field F.*

Definition 8.80 *An element $c \in C$, the set of complex numbers, is an* algebraic integer *if it is a zero of some monic polynomial $p(x)$ in $Z[x]$, that is, $p(c) = 0$.*

Definition 8.81 *An element a of an extension field E of a field F is* algebraic over F *if $f(a) = 0$ for some nonzero $f(x) \in F[x]$.*

Definition 8.82 *An element $c \in C$ that is algebraic over Q, the field of rational numbers, is an* algebraic number,*; that is, there is a polynomial $p(x) \in Q[x]$ such that $p(c) = 0$.*

In the following we show abstractly how, given a field and an irreducible polynomial over that field, we can find an extension field with a root of the irreducible polynomial. We then compare this extension field to one which would be formed by a field that has been extended by a root of the irreducible polynomial in the complex numbers.

Theorem 8.83 *Let F be a field and $p(x)$ be an irreducible polynomial over F. Then F is isomorphic to a subfield of $F[x]/\langle p(x)\rangle$ and there is an element f of $F[x]/\langle p(x)\rangle$ such that $p(f) = 0$.*

Proof. By Theorems 8.66 and 8.69, $F[x]/\langle p(x)\rangle$ is a field. The field F can be identified with a subfield F' of $F[x]/\langle p(x)\rangle$ as follows: Let $g : F \to F'$ be defined by $g(a) = a + \langle p(x)\rangle$. The function g is a monomorphism (i.e., one to one) for if $g(a) = g(b)$, then $a + \langle p(x)\rangle = b + \langle p(x)\rangle$ and $p(x)$ divides $a - b$. Since $p(x)$ is a polynomial and a and b are in F, $a - b = 0$ and $a = b$.

Let $f = x + \langle p(x)\rangle$ and $p(x) = a_0 + a_1 x + \cdots + a_n x^n$. Then

$$
\begin{aligned}
p(f) &= a_0 + a_1(x + \langle p(x)\rangle) + \cdots + a_n(x + \langle p(x)\rangle)^n \\
&= a_0 + a_1 x + \cdots + a_n x^n + \langle p(x)\rangle \\
&= p(x) + \langle p(x)\rangle \\
&= 0 + \langle p(x)\rangle \\
&= 0
\end{aligned}
$$

in $F[x]/\langle p(x)\rangle$. ∎

Theorem 8.84 *Let $p(x)$ be an irreducible polynomial in $F[x]$ and $p(a) = 0$ in an extension field of F. If $g(a) = 0$ for $g(x)$ in $F[x]$, then $p(x)$ divides $g(x)$.*

Proof. It is easily shown that $I = \{g(x) : g(x) \in F[x]$ and $g(a) = 0\}$ is an ideal. This ideal is obviously generated by $p(x)$ since $p(x)$ is irreducible. Hence if $g(x) \in I$, i.e., $g(a) = 0$, then $g(x)$ must be a multiple of $p(x)$. ∎

Theorem 8.85 *Let F be a field and $p(x) \in F[x]$ be an irreducible polynomial. If K is an extension field of F containing a root a of $p(x)$, i.e., $p(a) = 0$, then $F[a]$ is isomorphic to $F[x]/\langle p(x)\rangle$.*

Proof. Define $h : F[x]/\langle p(x)\rangle \to F[a]$ by $h(f(x) + \langle p(x)\rangle) = f(a)$. It is obvious that h is a homomorphism. If $h(f(x) + \langle p(x)\rangle) = h(g(x) + \langle p(x)\rangle)$, then $f(a) = g(a)$ and $(f - g)(a) = 0$ Hence, by the Theorem 8.84, $p(x)$ divides $f(x) - g(x)$ and $f(x) + \langle p(x)\rangle = g(x) + \langle p(x)\rangle$ and h is one to one. Let $f(a) \in F[a]$; then $h(f(x) + \langle p(x)\rangle) = f(a)$, h is onto, and consequently, h is an isomorphism. ∎

Let F be the field of real numbers and $p(x) = x^2 + 1$; and let K be an extension of F containing i which is a root of $p(x)$, that is, $i^2 + 1 = 0$. By the Theorem 8.84, $F[x]/\langle p(x)\rangle$ is isomorphic to $F[i]$. Since $p(x)$ has degree 2, we need only look at $f(i) \in F[i]$ of degree 1. Hence every element of $F[i]$ will be of the form $a + bi$ for real numbers a and b and we have created the complex numbers.

The subring consisting of elements of $F[i]$ where a and b are both integers is called the set of *Gaussian integers*. This may also be considered as the smallest subring of the complex numbers containing the integers and i. It is obvious that 5 is not an irreducible in this ring since $5 = (1 - 2i)(1 + 2i)$. Let the norm n_G on the set of Gaussian integers be defined by $n_G(a+bi) = a^2+b^2$. Then

$$
\begin{aligned}
n_G((a+bi)(c+di)) &= n_G((ac-bd)+(ad+bc)i) \\
&= (ac-bd)^2 + (ad+bc)^2 \\
&= a^2c^2 - 2abcd + b^2d^2 + a^2d^2 + 2abcd + b^2c^2 \\
&= (a^2+b^2)(c^2+d^2) \\
&= n_G(a+bi)n_G(c+di)
\end{aligned}
$$

so that n_G is a multiplicative homomorphism from the Gaussian integers to the integers. Since $n_G(1-2i) = n_G(1+2i) = 5$, then $1-2i$ and $1+2i$ are irreducibles in the ring of Gaussian integers. The product of irreducible Gaussian integers $2+i$ and $2-i$ is also 5. Hence we do not have unique factorization into irreducibles. Note that the irreducibles in this example are not primes.

As another example, let F be the field of rational numbers and $p(x) = x^2 - 2$; then $F[x]/\langle p(x)\rangle = F(\sqrt{2})$. Again only polynomials of degree 1 need be considered, and $F[x]/\langle p(x)\rangle$ is the field of all elements of the rationals of the form $a + b\sqrt{2}$ where a and b are rational numbers. We may consider the subring consisting only of elements of $F[x]/\langle p(x)\rangle$ of the form $a+b\sqrt{2}$ where a and b are integers. $(3+\sqrt{2})(3-\sqrt{2}) = 7$ so that 7 is not an irreducible in this set.

We have assumed that, beginning with the rational numbers, all of the algebraic integers are a subset of the complex numbers. Under this assumption, we prove the following theorem:

Theorem 8.86 *The set A of all algebraic numbers over the rational numbers Q is a subfield of the complex numbers.*

Proof. Let a and b be nonzero algebraic numbers. If a is not rational, then there exists an irreducible polynomial $p(x)$ with degree 2 or greater such that a is a root of $p(x)$. Let $F = Q[x]/\langle p(x)\rangle$. If b is not in F, then there exists an irreducible polynomial in F, say $q(x)$, of degree greater than 2 so that b is a root of q. Let $G = F[x]/\langle q(x)\rangle$; then G is a field containing a and b and so contains $a+b$, $a-b$, ab, and a/b. Since G is contained in A, these elements are in A and A is a field. ∎

Exercises

1. Prove that the set of Gaussian integers form a subring of the set of complex numbers.

2. Prove or disprove that the set of Gaussian integers is an integral domain.

3. Prove that $\sin(x)$ cannot be expressed as a polynomial.

4. Prove or disprove: The product of two prime polynomials is a prime polynomial.

5. Prove or disprove: The sum of two prime polynomials is a prime polynomial.

6. Find the generator of the smallest principal ideal containing the polynomials $x^2 - 4$ and $x^2 + 4x + 4$ where the polynomials are over the integers.

7. Prove or disprove that every Gaussian integer is an algebraic number.

8. Find an ideal of the integers that is not a maximal ideal.

9. Describe the smallest subfield of the real numbers containing the roots of $x^2 - 5 = 0$.

10. Describe the smallest subfield of the real numbers containing the roots of $x^2 + 5 = 0$.

11. Describe the smallest subfield of the real numbers containing the roots of $x^3 - 1 = 0$.

12. Describe the smallest subfield of the real numbers containing the roots of $x^4 - 1 = 0$.

Appendix A
LOGIC AND PROOFS

A.1 Axiomatic Systems

Much of mathematics deals with theorems and proofs of theorems. Theorems are "true" statements about the mathematical system being considered. For example, in Euclidean geometry the statement

> The hypotenuse of a right triangle is longer
> than either of the other two sides.

is a theorem about points, line segments, and angles. The statement is considered to be true because it is "derivable" or "deducible" from previously accepted, or derived truths in the discipline of Euclidean geometry. We will discuss how this deriving takes place and what the procedures are for doing it.

In many areas of mathematics, it is common to begin with undefined terms and state precisely the fundamental characteristics or truths about these terms from which the mathematician attempts to explore a system with such characteristics. The statements specifying what is discovered (proved) using only the fundamental properties (axioms and postulates), previously proved statements, and the rules of logic are called *theorems*. The theorems and their proofs constitute the subject matter of the mathematical system.

A reason for being so particular about a mathematical system is to avoid having false statements or inconsistencies appear as "theorems." Hence the only theorems accepted are those derived as described above.

In a mathematical system all of the information necessary to support a theorem is contained in axioms and previously proven theorems. The complete list of axioms and previously proven theorems may not be written out. Instead, we may accept some of the previously proven theorems as axioms and begin at that point or we may simply assume that certain mathematical concepts are known to be true from other sources or are obvious and hence are used often without comment. For example, the Integer Axioms and the Peano Axioms discussed elsewhere implicitly assume that the axioms of set theory; but since the emphasis in number theory is on the properties of integers, it would be distracting to attempt a full development of set theory as well. So we may start in "midstream" with the properties of either the natural numbers (Peano Axioms) or the integers (Chapter 1).

We will describe in some detail how rules of logical inference are used and how proofs are constructed, but before doing so we need to build up some tools from propositional calculus.

A.2   Propositional Calculus

A *proposition* is a statement or declarative sentence that may be assigned a true or false value. The context of the statement must be such that it makes sense to speak of it as being true or false. We will use $p, q, r, \ldots$ to represent propositions. Examples of propositions are

$$p : \text{The integer 2 is greater than 1.}$$
$$q : \text{Bob has red hair.}$$

Sentences that are not propositions would be

Who are you?

(a question)

This sentence is not true.

(can be neither true nor false).

In English, sentences are combined using connectives and clauses to form more complex compound sentences. Common connectives are "and," "or," "not," "if ... then," "but," "only if," and the like. We will be very specific about the meanings that we will ascribe to these connectives. The main point to grasp is that the truth of the compound proposition depends only upon the truth or falsity of the component parts and not upon what the propositions are about. We will introduce some notation to help highlight the structure of complex propositions.

Let p and q refer to the propositions:

$$p : \text{Jane drives a Ford.}$$
$$q : \text{Bob has red hair.}$$

The compound statement

Jane drives a Ford, and Bob has red hair.

has two parts joined with the connective "and." This statement may be rendered symbolically as

$$p \text{ and } q$$

or, more simply, as

$$p \wedge q$$

where the symbol $\wedge$ is called a "wedge" and the connective it refers to is the conjunction (i.e., "and"). Clearly, when someone says, "Jane drives a Ford, and Bob has red hair," we expect both a red-headed Bob and a Ford that is driven by Jane. If any other situation occurs, we would declare that person to be lying.

There are four possible cases to consider. The proposition p could be true (T) or false (F) and regardless of which value is assigned to p, the proposition q could also be true (T) or false (F). The four possibilities are conveniently arranged in a tableau form called a *truth table*. A truth table exhaustively lists all the combinations of truth and falsity of the component propositions.

Case	p	q	$p \wedge q$
1	T	T	T
2	T	F	F
3	F	T	F
4	F	F	F

It was decided earlier that only for case 1 would $p \wedge q$ be true. The other cases yield a false truth value for $p \wedge q$. For example, case 3 describes how to obtain the truth value of $p \wedge q$ when Jane does not drive a Ford but Bob has red hair.

Similarly, the statement

Jane drives a Ford, or Bob has red hair.

is rendered symbolically by

p or q

or

$p \vee q$

where $\vee$ is called the "vee" symbol and the corresponding connective is called a disjunction (that is, "or"). A person saying "Jane drives a Ford or Bob has red hair" would be lying only when Jane fails to drive a Ford and also Bob is not red headed. Only one of the two component propositions "Jane drives a Ford" and "Bob has red hair" needs to be true to keep this person from lying.

Since the truth value of $p \vee q$ depends only on the truth of p and of q separately, again a truth table will enable us to specify completely when we want $p \vee q$ to be true or to be false.

Case	p	q	$p \vee q$
1	T	T	T
2	T	F	T
3	F	T	T
4	F	F	F

Thus, only in case 4, when both p and q are false, will $p \vee q$ be false. The proposition

It is not true that Jane drives a Ford.

or, equivalently,

Jane does not drive a Ford.

is closely related to the proposition

$$p : \text{Jane drives a Ford.}$$

In fact, it is the negation or denial of p and is indicated by

$$\sim p$$

using the tilde to indicate negation. So when "Jane drives a Ford" is true, then "Jane does not drive a Ford" is false. When "Jane drives a Ford" is false, then "Jane does not drive a Ford" is true. The truth table description (or definition) is

Case	p	$\sim p$
1	T	F
2	F	T

The truth value of p is just the opposite of the truth value of $\sim p$.

The symbols $\wedge$ and $\vee$ are binary connectives since they connect two propositions as in $p \wedge q$ and in $p \vee q$. The symbol $\sim$ is a unary connective because it uses only one proposition.

Next consider the statement

Either Jane drives a Ford or else both Bob doesn't have red hair and Sam walks to work.

If we let r represent the proposition

Sam walks to work.

then the symbolic representation of the compound proposition is

$$p \vee ((\sim q) \wedge r)$$

where we have used parentheses to indicate clearly what propositions are the first and second components of each binary connective.

A truth table will enable us to know exactly those situations when the statement $p \vee ((\sim q) \wedge r)$ is true; however, we must be sure to list all possibilities. Since there are three fundamental propositions p, q, and r in the compound proposition, there will be 8 cases.

Case	p	q	r	$\sim q$	$(\sim q) \wedge r$	$p \vee ((\sim q) \wedge r)$
1	T	T	T	F	F	T
2	T	T	F	F	F	T
3	T	F	T	T	T	T
4	T	F	F	T	F	T
5	F	T	T	F	F	F
6	F	T	F	F	F	F
7	F	F	T	T	T	T
8	F	F	F	T	F	F

In constructing the truth table for the $(\sim q) \wedge r$ column, we refer to the columns for $(\sim q)$ and r and the truth table definition for $\wedge$. The truth table for $\wedge$ indicates that only when both parts, namely, $(\sim q)$ and r, are true will the statement $(\sim q) \wedge r$ be true. This happens in cases 3 and 7 only.

To determine what to do for the $p \vee ((\sim q) \wedge r)$ column, we note that only the truth of p and of $(\sim q) \wedge r$ matters (the truth of the binary statement depends only upon the truth of its first and second components). The truth table defining $\vee$ indicates that the only case in which an "or" statement is false is when both parts are false. Such a combination occurs only for cases 5, 6, and 8. In the other five cases, at least one component of the statement is true.

If Jane does not drive a Ford (i.e., p is false or has the truth value F), Bob doesn't have red hair (q is F), and Sam walks to work (r is T), then we have case 7 and a person saying

Either Jane drives a Ford or both Bob doesn't have red hair and Sam walks to work.

would be speaking the truth.

At this point we know the following:

1. The truth or falsity of compound propositions depends upon only the truth or falsity of its component propositions.

2. The truth behavior of a connective such as or can be specified by a truth table that gives the appropriate truth value in every possible case. This truth table, in fact, defines the connective.

3. The truth of a complex proposition formed of many connectives and many propositions can be determined in all situations that would ever arise by generating a truth table for the proposition.

Two other binary connectives are in common use. They are the conditional ($\rightarrow$) and the biconditional ($\longleftrightarrow$).

If Jane drives a Ford, then Bob has red hair.

is translated as

$$\text{if } p, \text{ then } q$$

or

$$p \rightarrow q$$

We immediately see that probably Jane driving a Ford has no causal relationship with whether Bob has red hair; however, we need to remember that the truth or falsity of a binary compound proposition depends only upon the truth of its component proposition parts and not upon any other relationship or nonrelationship between the parts.

The truth table definition is

Case	p	q	$p \rightarrow q$
1	T	T	T
2	T	F	F
3	F	T	T
4	F	F	T

If someone states

If Jane drives a Ford, then Bob has red hair.

and, if Jane does drive a Ford and Bob has red hair, the statement is true (case 1). On the other hand, if someone makes the if...then statement above, and if Jane did drive a Ford, but Bob had brown hair, then we would feel that we were not told the truth (case 2). Cases 3 and 4 are the "don't care" cases. If Jane drives a Jeep, then we do not expect anything concerning Bob's hair color; so we declare the conditional to be true in cases 3 and 4.

Another example of the use of the connective is the statement:

If an integer is 3, then its square is 9.

This statement should be recognized as a theorem, meaning that it is "always" true. We observe in the definition of $\rightarrow$ that the only possibility of the statement being false is when "an integer is 3" is true but "its square is 9" is false. This case does not actually occur in arithmetic because 3^2 equals 9. Thus, the only situation for which "if an integer is 3, then its square is 9" could be false is prevented from occurring by the axioms of the arithmetic of integers. Notice that case 3 occurs if the integer in question is -3 and case 4 occurs when it is 2.

The conditional is expressed in English in several ways, all of which are written symbolically as $p \rightarrow q$. Thus, the conditional $p \rightarrow q$ has these equivalent ways of being written:

> If p then q.

> p is sufficient for q.
> p is a sufficient conditionfor q.

> q is necessary for p.
> q is a necessary condition for p.

> p only if q.

The meaning of these varieties may become clearer by supposing that we can assert, for particular instances of p and q, that $p \rightarrow q$ is true. Therefore, asserting only p guarantees that q is true because we have case 1 of the definition and we say that knowing p is true is sufficient for knowing that q is true also. On the other hand (still having asserted that $p \rightarrow q$ is true), q must be true when p is (case 1 again); consequently, we state that the truth of q is necessary in order for p to be true. The meaning of "p only if q" is similarly reasoned since (in the context of $p \rightarrow q$ being true) p is allowed to be true only when q is.

The biconditional connective $\longleftrightarrow$ derives from a statement such as

> Jane drives a Ford if and only if Bob has red hair.

This last statement is translated as

> p if and only if q

> or

$$p \longleftrightarrow q$$

"p if and only if q" is an English contraction of

> $(p$ if $q)$ and $(p$ only if $q)$

where parentheses are used to emphasize the grouping. But since the "if" clause pertains to q in "p if q" and since this subordinate "if" clause remains subordinate no matter where it occurs, we have equivalently the form

> (if q then p) and (if p then q)

This last form can be translated directly as

$$(q \rightarrow p) \wedge (p \rightarrow q)$$

The truth table of $(q \rightarrow p) \wedge (p \rightarrow q)$ will determine the truth table definition of $p \longleftrightarrow q$:

Case	p	q	$(q \to p)$	$\wedge$	$(p \to q)$	$p \longleftrightarrow q$
1	T	T	T	T	T	T
2	T	F	T	F	F	F
3	F	T	F	F	T	F
4	F	F	T	T	T	T

From this definition we see that the biconditional, $p \longleftrightarrow q$, is true only when p and q agree in truth value.

The biconditional is expressed in English in several ways, all of which are written symbolically as $p \longleftrightarrow q$. So, using the meanings of p and of q above, we have these equivalent ways of writing the biconditional $p \longleftrightarrow q$:

> p if and only if q.

> p is necessary and sufficient for q.
> p is a necessary and sufficient condition for q.

We are particularly interested in compound statements that are expressed differently but are, in fact, true in exactly the same cases. Such propositions are said to be logically equivalent. Equivalence of two propositions can easily be established by constructing truth tables for both propositions and then comparing the two.

As an example, let p and q be these propositions:

> p: It rained today.
> q: It snowed today.

and consider these compound propositions:

> It is not the case that it rained or snowed today.

or

$$\sim (p \vee q)$$

and

> It didn't rain today and it didn't snow today.

or

$$\sim p \wedge \sim q$$

since, by convention, the negation applies to the proposition that immediately follows it. We can construct truth tables for both:

Case	p	q	$\sim$	$(p \vee q)$	$\sim p$	$\wedge$	$\sim q$
1	T	T	F	T	F	F	F
2	T	F	F	T	F	F	T
3	F	T	F	T	T	F	F
4	F	F	T	F	T	T	T
			$*$			$\#$	

where for convenience we have placed the truth values underneath the various connectives being evaluated rather than creating a new column for each connective truth evaluation. Thus, we see that in all four cases, the truth values for $\sim (p \vee q)$ (indicated by *) are, respectively, the same as the truth values for $\sim p \wedge \sim q$ (indicated by #). This result means that the two propositions in question are logically equivalent; that is,

$$\sim (p \vee q) \equiv \sim p \wedge \sim q$$

This equivalence is a very useful fact; namely, the negation of an "or" statement is the conjunction of the negations of the parts. In other words, to negate an "or" statement, just negate each part and change the "or" to "and."

We now give a list of useful logically equivalent propositions. All can be established using truth tables.

Double Negation

$$\sim (\sim p) \equiv p$$

De Morgan's Laws

$$\sim (p \vee q) \equiv \sim p \wedge \sim q$$

$$\sim (p \wedge q) \equiv \sim p \vee \sim q$$

Commutative Properties

$$p \wedge q \equiv q \wedge p$$

$$p \vee q \equiv q \vee p$$

Associative Properties

$$p \wedge (q \wedge r) \equiv (p \wedge q) \wedge r$$

$$p \vee (q \vee r) \equiv (p \vee q) \vee r$$

Distributive Properties

$$p \wedge (q \vee r) \equiv (p \wedge q) \vee (p \wedge r)$$

$$p \vee (q \wedge r) \equiv (p \vee q) \wedge (p \vee r)$$

Equivalence of Contrapositive

$$p \rightarrow q \equiv \sim q \rightarrow \sim p$$

Other Useful Properties

$$p \rightarrow q \equiv \sim p \vee q$$

$$p \longleftrightarrow q \equiv (p \rightarrow q) \wedge (q \rightarrow p)$$

We can construct propositions that are always true no matter what case occurs. A proposition that is true in every case is said to be *logically true* (*LT*) or to be a *tautology*; and a proposition that is constructed to be false in every case is said to be *logically false* (*LF*) or to be a *contradiction.* Consider a proposition of the form

$$(p \land (p \rightarrow q)) \rightarrow q$$

The truth table for this proposition is

Case	p	q	$(p$	$\land$	$(p \rightarrow q))$	$\rightarrow$	q
1	T	T	T	T	T	T	T
2	T	F	T	F	F	T	F
3	F	T	F	F	T	T	T
4	F	F	F	F	T	T	F
						*	

where we have placed columns under both propositions and connectives for ease of reading. The column marked with an asterisk gives the truth value of the entire compound proposition. This proposition is true in all four possible cases. Recall that a proposition that is logically true is also called a *tautology;* hence, we have constructed a tautology.

Once we have a proposition that is always true, it is easy to construct one that is always false — just negate the logically true proposition. The statement

$$\sim ((p \land (p \rightarrow q)) \rightarrow q)$$

is logically false.

We now give several properties dealing with logically true and logically false statements.

$$p \land LT \equiv p$$
$$p \land LF \equiv LF$$
$$p \lor LT \equiv LT$$
$$p \lor LF \equiv p$$
$$p \land \sim p \equiv LF$$
$$p \lor \sim p \equiv LT$$

Since the truth or falsity of any proposition is dependent only upon the truth behavior of its component parts and not upon the name or form of the component parts, we may substitute a logically equivalent proposition for any component part of a statement without changing the truth value of the whole. For example,

$$
\begin{aligned}
(q \lor r) \lor (p \land \sim r) &\equiv q \lor (r \lor (p \land \sim r)) && \text{associative prop.} \\
&\equiv q \lor ((r \lor p) \land (r \lor \sim r)) && \text{distributive prop.} \\
&\equiv q \lor ((r \lor p) \land LT) && \text{equivalence} \\
&\equiv q \lor (r \lor p) && \text{equivalence} \\
&\equiv q \lor (p \lor r) && \text{commutative prop.} \\
&\equiv (q \lor p) \lor r && \text{associative prop.} \\
&\equiv (p \lor q) \lor r && \text{commutative prop.}
\end{aligned}
$$

Exercises

1. Construct truth tables for the following propositions:

   (a) $\sim (p \rightarrow \sim q) \vee (\sim p \wedge q)$

   (b) $(\sim p \rightarrow (q \vee \sim r)) \longleftrightarrow \sim (p \wedge \sim q)$

2. Use the truth table method to establish the following equivalences:

   (a) De Morgan's Law
   $$\sim (p \wedge q) \equiv \sim p \vee \sim q$$

   (b) associative property for $\vee$
   $$p \vee (q \vee r) \equiv (p \vee q) \vee r$$

   (c) Distributive property for "or" over "and"
   $$p \vee (q \wedge r) \equiv (p \vee q) \wedge (p \vee r)$$

   (d) Conversion of conditional to an "or" statement
   $$p \rightarrow q \equiv \sim p \vee q$$

3. There are several related conditional statements involving two component propositions p and q. They are

$p \rightarrow q$	the conditional
$q \rightarrow p$	the converse (of $p \rightarrow q$)
$\sim q \rightarrow \sim p$	the contrapositive (of $p \rightarrow q$)
$\sim p \rightarrow \sim q$	the inverse (of $p \rightarrow q$)

   Using truth tables, prove that
   $$p \rightarrow q \equiv \sim q \rightarrow \sim p$$

   and that
   $$p \rightarrow q \not\equiv q \rightarrow p$$

   Thus, the conditional is equivalent to its contrapositive; however, the conditional is not equivalent to its converse. Often we use the phrasing "if p, then q; and conversely." This statement really means "if p, then q; and, if q, then p" or
   $$(p \rightarrow q) \wedge (q \rightarrow p)$$

which is equivalent to $p \longleftrightarrow q$ or "p if and only q." Prove, without direct use of truth tables, that $p \rightarrow q \equiv \sim q \rightarrow \sim p$.

4. Using logically equivalent statements and without the direct use of truth tables, show that:

   (a) $\sim (p \rightarrow q) \equiv (p \wedge \sim q)$

   (b) $\sim (p \longleftrightarrow q) \equiv (p \wedge \sim q) \vee (q \wedge \sim p)$

   (c) $p \equiv \sim (p \wedge s) \rightarrow (\sim s \wedge p)$

A.3   Arguments

The primary objective of discussing logic is to be able to understand the structure of statements and the rules of correct inference. Inference deals with being presented with evidence or known facts from our axioms and, based on that evidence, coming to some conclusion not explicitly declared in the evidence. The process of going from evidence to conclusion is called an *argument* in logic.

A correct argument is one whose conclusion is true whenever all the evidence is true. A good argument is referred to as a *valid argument*. An *invalid argument* is one whose conclusion could be false even though all the evidence is true. Correct logical inference uses valid arguments to reach its conclusions.

The general form of an argument is to have several items of evidence, say, H_1, H_2, and H_3 and a conclusion, say C. The argument is often displayed as

$$
\begin{array}{ll}
H_1 & \\
H_2 & \text{evidence} \\
\underline{H_3} & \\
\therefore \ C & \text{conclusion}
\end{array}
$$

in order to highlight the parts. The symbol $\therefore$ means "therefore." The evidence is a list of one or more propositions called hypotheses or premises. In this illustration there are three premises. Hence the argument is valid if

whenever $H_1, H_2,$ and H_3 are true, then C is true

or

whenever $H_1 \wedge H_2 \wedge H_3$ is true, then C is true

It would be useful to have a truth table criterion for whether an argument is valid because then we could quickly determine whether any proposed argument or inference rule is valid. The truth table criterion for a valid argument is to show that the proposition

$$(H_1 \wedge H_2 \wedge H_3) \rightarrow C$$

is logically true, or is a tautology. In fact, we have already shown one argument to be valid:

$$
\begin{array}{cc}
H_1 & p \\
\underline{H_2} & \underline{p \to q} \\
\therefore\ C & \therefore\ q
\end{array}
$$

We earlier constructed the truth table for

$$(H_1 \wedge H_2) \to C$$

or

$$(p \wedge (p \to q)) \to q$$

and showed that it was logically true. This particular valid argument is called the *law of detachment* or *modus ponens*. Note that the order of listing the premises is unimportant since $H_1 \wedge H_2 \equiv H_2 \wedge H_1$.

The value of a valid argument is that if all of the hypotheses are true, then the conclusion is guaranteed to be true. This property is exactly what we need to justify theorems in an axiomatic system. The conclusion, of course, can now be accepted as a theorem.

As an example of the law of detachment, suppose that b is a particular integer. Let p and q be given by

$$
\begin{array}{ll}
p: & b \text{ is even} \\
q: & b \text{ is divisible by 2}
\end{array}
$$

so that

$$p \to q: \quad \text{if } b \text{ is even, then } b \text{ is divisible by 2}$$

The law of detachment gives:

$$
\begin{array}{ll}
p \to q & \text{if } b \text{ is even, then } b \text{ is divisible by 2} \\
\underline{p} & \underline{b \text{ is even}} \\
\therefore\ q & \therefore\ b \text{ is divisible by 2}
\end{array}
$$

If the statement "if b is even, then b is divisible by 2" is derived as a property of the integers and if $b = 12$, then both premises are true; so we know for sure that 12 is divisible by 2. On the other hand, if $b = 13$, then p is false; and even though the argument is valid, we cannot be sure whether $b = 13$ is divisible by 2. Valid arguments say nothing about what happens to the conclusion if one of the premises is false.

The law of detachment is so named because it allows the "then" part of an "if p then q" conditional to be "detached" and written as a conclusion.

For future reference, we list several valid arguments:

Law of Detachment (Modus Ponens)

$$
\begin{array}{c}
p \to q \\
\underline{p} \\
\therefore\ q
\end{array}
$$

Syllogism

$$p \rightarrow q$$
$$\underline{q \rightarrow r}$$
$$\therefore \quad p \rightarrow r$$

Modus Tollens

$$p \rightarrow q$$
$$\underline{\sim q}$$
$$\therefore \quad \sim p$$

Addition

$$\underline{p}$$
$$\therefore \quad p \vee q$$

Specialization

$$\underline{p \wedge q}$$
$$\therefore \quad p$$

Conjunction

$$p$$
$$\underline{q}$$
$$\therefore \quad p \wedge q$$

Cases

$$p$$
$$p \rightarrow (r \vee s)$$
$$r \rightarrow q$$
$$\underline{s \rightarrow q}$$
$$\therefore \quad q$$

Case Elimination

$$p \vee q$$
$$\underline{p \rightarrow (r \wedge \sim r)}$$
$$\therefore \quad q$$

Reductio ad Absurdum

$$\underline{\sim (p \rightarrow q) \rightarrow (r \wedge \sim r)}$$
$$\therefore \quad p \rightarrow q$$

or

$$\underline{(p \wedge \sim q) \rightarrow (r \wedge \sim r)}$$
$$\therefore \quad p \rightarrow q$$

or

$$\underline{\sim w \rightarrow (r \wedge \sim r)}$$
$$\therefore \quad w$$

All of the arguments above can be shown to be valid by the truth table method.

Of course, all arguments are not valid. Consider the following one:

If the apple is red, then it is ripe.
The apple is ripe.
∴   The apple is red.

Using the notation

$$p:\quad \text{the apple is red}$$
$$q:\quad \text{the apple is ripe}$$

the form of the argument is

$$p \rightarrow q$$
$$\underline{q\quad\quad}$$
$$\therefore\ \ p$$

For this argument to be valid, whenever the premises are true, then the conclusion must be also. We can use the truth table test, namely, determine whether

$$((p \rightarrow q) \wedge q) \rightarrow p$$

is logically true.

Case	p	q	$((p \rightarrow q)$	$\wedge$	$q)$	$\rightarrow$	p
1	T	T	T	T	T	T	T
2	T	F	F	F	F	T	T
3	F	T	T	T	T	F	F
4	F	F	T	F	F	T	F
						$*$	

From the truth table it is evident that the argument is not valid. Case 3 presents the situation "the apple is not red" (p is false) and "the apple is ripe" (q is true). For this case both the premise $p \rightarrow q$ and the premise q are true; however, the conclusion p is not. So all of the premises being true do not guarantee the conclusion to be true. This characteristic determines the argument to be invalid. This particular invalid argument is called a *non sequitur*.

In most axioms systems, the theorems and axioms can be very complex; and the weaving together of theorems, axioms, and rules of inference can be complicated. The objective of the following discussion is to make clear what is happening in proofs and how proofs are constructed. In a complex proof, it is usually necessary to use many of these valid argumentative forms over and over again. This application to proofs can be summarized by describing a proof to be a sequence of statements each of which is

(a) True by hypothesis

(b) An axiom or definition

(c) A previously proved theorem or proposition

(d) A statement implied by previous statements as a conclusion of a valid argument

(e) Logically equivalent to a previous statement

As an example using logic symbolism, we will show that

$$p \rightarrow q$$
$$\sim r \rightarrow \sim q$$
$$\underline{\sim r}$$
$$\therefore \quad \sim p$$

is a valid argument. We will discuss examples more directly related to mathematical proofs later.

Number	Statements	Reason
1	$p \rightarrow q$	given
2	$\sim r \rightarrow \sim q$	given
3	$\sim r$	given
4	$\sim q$	2, 3, and law of detachment
5	$\sim q \rightarrow \sim p$	1 and equivalence $p \rightarrow q \equiv \sim q \rightarrow \sim p$
6	$\sim p$	4, 5, and law of detachment

Therefore, the three premises imply the conclusion $\sim p$ and we have proved or derived $\sim p$.

Many theorems are stated in the if ... then form, that is,

$$\text{if } p, \text{ then } q$$

or

$$p \rightarrow q$$

If we can show that in the context of the axiom system and all the previously derived theorems, whenever p is true (note that since p often represents a complex statement itself, we mean under all situations that cause p to be true), then q is true.

This procedure is carried out in mathematical proofs by considering any way that p is true and then showing, using logic, axioms, and previously proved theorems, that in each such circumstance q is true.

Exercises

1. Prove, using the truth table criterion, that the following arguments are valid:

(a) The syllogism

$$p \rightarrow q$$
$$\underline{q \rightarrow r}$$
$$\therefore \quad p \rightarrow r$$

(b) The cases argument

$$p$$
$$p \rightarrow (r \vee s)$$
$$r \rightarrow q$$
$$\underline{s \rightarrow q}$$
$$\therefore \quad q$$

(c) The reductio ad absurdum argument $\qquad \dfrac{\sim w \to (r \wedge \sim r)}{\therefore \quad w}$

2. Show that the stronger conclusion r cannot be validly inferred with the same premises as the syllogism has; thus, show that the following argument is not valid:
$$\begin{array}{c} p \to q \\ \dfrac{q \to r}{\therefore \quad r} \end{array}$$

3. The non sequitur argument

$$\begin{array}{c} p \to q \\ \dfrac{q}{\therefore \quad p} \end{array}$$

was shown to be invalid. The following two arguments have the form of the non sequitur:

$$\begin{array}{c} (\sim s \vee w) \to w \\ \dfrac{w}{\therefore \quad \sim s \vee w} \end{array} \qquad\qquad \begin{array}{c} (s \wedge \sim w) \to w \\ \dfrac{w}{\therefore \quad s \wedge \sim w} \end{array}$$

provided that we identify p with $\sim s \vee w$ in the first argument and with $s \wedge \sim w$ in the second. Show that one is valid and the other is not. These examples illustrate that for an invalid argumentative form, some substitutions for the components statements yield valid arguments and some do not. For a valid argumentative form, all substitutions for the components will always yield valid arguments. The question is, why the difference?

4. Without the direct use of a truth table, show that these arguments are valid:

(a)
$$\begin{array}{c} \sim (s \wedge t) \\ \dfrac{\sim w \to t}{\therefore \quad s \to w} \end{array}$$

(b)
$$\begin{array}{c} p \to q \\ \sim r \to \sim q \\ \dfrac{\sim r}{\therefore \quad \sim p} \end{array}$$

(c)
$$\begin{array}{c} \sim (\sim p \vee q) \\ \sim z \to \sim s \\ (p \wedge \sim q) \to s \\ \dfrac{\sim z \vee r}{\therefore \quad r} \end{array}$$

(d)
$$\begin{array}{c} \sim x \to \sim w \\ (x \vee \sim w) \to z \\ \sim p \to \sim z \\ \dfrac{p \to (\sim r \vee \sim s)}{\therefore \quad (\sim r \vee \sim s)} \end{array}$$

5. Prove that stating that two propositions are logically equivalent means the same as stating that the biconditional of the two propositions is a tautology. That is, for propositions p and q, prove that

$$p \equiv q$$

means the same as

$$p \longleftrightarrow q \text{ is a tautology}$$

6. Further, prove that this last statement means the same as

$$\text{both } p \rightarrow q \text{ and } q \rightarrow p \text{ are tautologies}$$

When the conditional $p \rightarrow q$ is a tautology, we say that p implies q and often write $p \Rightarrow q$. Therefore, $p \longleftrightarrow q$ means the same as having both $p \rightarrow q$ and $q \rightarrow p$.

A.4 Predicate Calculus

Many statements which appear to be propositions are really not because they involve variables whose values are not specified. For example, consider these statements:

$$
\begin{array}{ll}
P(x): & 3 + x \text{ is } 5 \\
Q(x, y, z): & x^2 + y^2 \geq z^2 \\
R(x, y): & x^2 + y^2 \geq 0 \\
S(x): & -1 \leq \sin(x) \leq 1
\end{array}
$$

Such statements are called *predicates*. The predicates $P(x)$, $Q(x, y, z)$, $R(x, y)$, and $S(x)$ include one or more variable names as arguments to emphasize which variables are unspecified. Hence, $P(x)$ is a one-place predicate, $Q(x, y, z)$ is a three-place predicate, and so on. We will use these particular predicates in the illustrations that follow.

Predicates become propositions when their variables are specified. For instance,

$$P(2): \qquad 3 + 2 \text{ is } 5$$

is a proposition and is true, but

$$P(7): \qquad 3 + 7 \text{ is } 5$$

is a proposition and is false. The predicate $P(x)$ is neither true nor false because the value to be substituted is not indicated. Combinations of variable values causing the predicate to be true are said to *satisfy* the *predicate*. Thus, 2 satisfies $P(x)$ because $P(2)$ is true but 7 does not satisfy $P(x)$ because $P(7)$ is false.

Some predicates are true for every possible choice of values of the variables. As above, $S(x)$ is the predicate $-1 \leq \sin(x) \leq 1$, where x is allowed to be from the set of real numbers. Since

$$-1 \leq \sin(x) \leq 1$$

for any choice of x, then we can say that

$$\text{for every } x, \ S(x) \ \text{(is true)}$$

or

$$\forall x S(x)$$

$$\forall x(-1 \le \sin(x) \le 1)$$

In this case, we could equivalently say that

$$\forall y(-1 \le \sin(y) \le 1)$$

where we have replaced every related occurrence of x with y.

The symbol $\forall x$ is called a *quantifier* — a *universal quantifier* in this case — and is read "for every x" or "for each x." The set of values from which x may be taken is called the *universe* or *context* or *domain of discourse* under consideration. Frequently, the universe is the set of all real numbers, the set of all integers, the set of all functions on the closed interval $[0, 1]$, and so on.

The quantified predicate

$$\forall x P(x)$$

mentioned above ($P(x) : 3 + x$ is 5) is not valid or true because there is an instance of x in the universe such that $P(x)$ is false (assuming that the universe is something reasonable such as the integers or the real numbers). The fact that $P(2)$ is true does not help because $P(7)$ is false. For $\forall x P(x)$ to be true, $P(x)$ must be true for each instance of x in the universe.

The predicate $\forall x \, \forall y \, R(x, y)$ is read as "for every x for every y, $R(x, y)$" or "for every x and for every y, $R(x, y)$." Of course, for $\forall x \, \forall y \, R(x, y)$ to be true, for any choice of x and of y from the universe, $x^2 + y^2 \ge 0$ must be true. For the universe of all real numbers, this inequality is true no matter what values of x and y are chosen. Thus

$$\forall x \, \forall y \, R(x, y)$$

is true. Considering the predicate $Q(x, y, z)$ given above ($Q(x, y, z) : x^2 + y^2 \ge z^2$), we see that

$$\forall x \, \forall y \, \forall z \, Q(x, y, z)$$

is not valid because

$$Q(1, 2, 3)$$

is false.

For clarity, we will sometimes surround the quantifier with parentheses, writing $(\forall x) P(x)$ instead of $\forall x P(x)$ and $(\forall x)(\forall y) R(x, y)$ instead of $\forall x \forall y R(x, y)$.

Even though $\forall x P(x)$ is not true, $P(x)$ is true for at least one choice of x, namely, $x = 2$. We express this by using the *existential quantifier* $\exists x$ and writing

$$\exists x P(x)$$

or

$$\text{there is an } x \text{ so that } P(x) \ \text{(is true)}$$

Here, as with $\forall x$, $\exists x$ refers to a value of x within the *universe of discourse under consideration.*

For $P(x)$, there is only one choice of x making $P(x)$ true. For other predicates, say $Q(x, y, z)$, there may be many choices of values for the variables making $Q(x, y, z)$ true. Since $Q(1, 2, 0)$ and $Q(-3, 4, 5)$ are true, we write

$$\exists x\, \exists y\, \exists z\, Q(x, y, z)$$

and declare the statement to be true.

To be a proposition, all variables appearing in the predicates must be bound, that is, addressed by some quantifier. For example, $(\exists x)(\forall z)Q(x, y, z)$ is not a statement because the variable y is not associated with any quantifier.

An example that illustrates how quantifiers can be combined and used is in the definition of limit in calculus.

$$\lim_{x \to 1}(x^2 + 3) = 4$$

can be stated as

For each $\epsilon > 0$ there is a $\delta > 0$ such that
if $0 < |x - 1| < \delta$, then $\left|(x^2 + 3) - 4\right| < \epsilon$.

and which can be rendered symbolically as

$$\forall \epsilon\, \exists \delta\, \forall x\, \left(0 < |x - 1| < \delta \to \left|x^2 + 3) - 4\right| < \epsilon\right)$$

By letting,

$$\begin{array}{ll} A(x, \delta) & \text{be} \quad 0 < x - 1 < \delta \\ B(x, \epsilon) & \text{be} \quad \left|(x^2 + 3) - 4\right| < \epsilon \end{array}$$

we can write

$$\forall \epsilon\, \exists \delta\, \forall x\, (A(x, \delta) \to B(x, \epsilon))$$

where the domain of discourse for x is the set of real numbers and the domain for ϵ and δ is the set of positive real numbers.

If $D(x)$ is a predicate, we saw that $\forall x D(x)$ is true or valid only if, for every instance of x, $D(x)$ is true. The question now is: What does it mean to say that $\forall x D(x)$ is not true? We can express the denial or negation of $\forall x D(x)$ by

$$\sim \forall x D(x)$$

For $\forall x D(x)$ to fail to be true, we need to find only one instance of x so that $D(x)$ is false (or, equivalently, so that $\sim D(x)$ is true). This condition is expressed using the existential quantifier as

$$\exists x(\sim D(x))$$

or

there exists an x (within the domain of discourse) so that
$\sim D(x)$ is true (or, so that $D(x)$ is false).

Similarly, if $G(x)$ is a predicate, we would deny that there exists an x so that $G(x)$ is true by writing

$$\sim (\exists x G(x))$$

Clearly, if there is no instance of x making $G(x)$ true, then all values of x must make $G(x)$ false. In this case, $\sim G(x)$ will be true for all values of x, or equivalently,

$$\forall x(\sim G(x))$$

The results of the discussion are expressed succinctly as equivalences:

$$\sim \forall x(D(x)) \equiv \exists x(\sim D(x))$$

$$\sim \exists x(G(x)) \equiv \forall x(\sim G(x))$$

In other words, to form the negation or denial of a universally quantified predicate, change "$\forall x$" to "$\exists x$" and negate the predicate that follows. To form the negation of an existentially quantified predicate, change "$\exists x$" to "$\forall x$" and negate the predicate that follows.

There is a relationship between quantified predicates and predicates using particular instances of the variables. For example, consider

$$\forall x S(x)$$

or

for every x it is true that $-1 \le \sin(x) \le 1$.

Assume that this proposition is valid or true. We wish to apply it to the particular value 8 of x and conclude:

$$S(8)$$

If $\forall x S(x)$ is true and a is any particular instance of x (in the domain of discourse), then

$$S(a)$$

is true. So for the instance 8 for x, we have as true

$$-1 \le \sin(8) \le 1$$

On the other hand, if b is an arbitrary instance of the universe (that is, the only property of b known or assumed is that b is in the domain of discourse) and we show that $S(b)$ is true, then

$$\text{for every } x, S(x)$$

or

$$\forall x S(x)$$

These two rules allow us to go back and forth between quantified predicates and predicates as applied to constants in the universe.

The statement $\forall x S(x)$ implies that $S(a)$ is true for arbitrary a in the universe (i.e., any one you pick).

The statement $S(b)$ for an arbitrary constant b in the universe implies that $\forall x S(x)$.

As an example, consider the following definition applied to the integers:

if n is even, then $n = 2k$ for some integer k
$$\forall n(n \text{ is even } \rightarrow (\exists k)(n = 2k))$$

Since 10 is an integer instance of n, we can say that

$$10 \text{ is even } \rightarrow (\exists k)(10 = 2k)$$

is true. Since 13 is an integer instance of n, we also have that

$$13 \text{ is even } \rightarrow (\exists k)(13 = 2k)$$

is true. However, only 10 (and not 13) is even, so we have that $(\exists k)(10 = 2k)$ is true using the valid argument:

$$
\begin{array}{ll}
p \rightarrow q & 10 \text{ is even } \rightarrow (\exists k)(10 = 2k) \\
\underline{p} & \underline{10 \text{ is even}} \\
\therefore \quad q & \therefore \quad (\exists k)(10 = 2k)
\end{array}
$$

This last transformation is one in which we moved from the existentially quantified proposition $(\exists k)(10 = 2k)$ to a proposition about constants or particular instances of the universe.

From $(\exists k)(10 = 2k)$ we concluded that there is at least one instance of k in the universe for which it is true that $10 = 2k$. In other words, from $(\exists k)W(k)$ we know that there is a particular instance of k, say L, so that the proposition $W(L)$ is true.

Conversely, if we can find just one instance or constant b in the universe so that $Z(b)$ is true, then we can conclude that $(\exists x)Z(x)$ is true. Therefore, since $10 = 2 \cdot 5$ we can declare as true $(\exists k)(10 = 2k)$.

These two rules allow us to pass back and forth between predicates with particular values to existentially quantified predicates.

We summarize these four relationships.

Universal Instantiation

From $\forall x P(x)$ we can infer $P(a)$ for arbitrary a in the universe.

Universal Generalization

From having arbitrary a in the universe making $P(a)$ true
we can infer that $\forall x(P(x)$ is true.

Existential Instantiation

From $\exists x P(x)$ we can infer that there is an instance b with $P(b)$ true.

Existential Generalization

From an instance c in the universe with $P(c)$ true we can infer that $\exists x P(x)$.

Properties associated with the distribution of quantifiers over "and" and "or" statements are sometimes useful. The following inferences can easily be proved using the four inference relationships above:

$$\forall x(P(x) \wedge Q(x)) \equiv \forall x P(x) \wedge \forall x Q(x)$$

$$\exists x(P(x) \vee Q(x)) \equiv \exists x P(x) \vee \exists x Q(x)$$

From $\forall x P(x) \vee \forall x Q(x)$ we can infer that $\forall x(P(x) \vee Q(x))$

From $\exists x(P(x) \wedge Q(x))$ we can infer that $\exists x P(x) \wedge \exists x Q(x)$

Exercises

1. For $Q(x, y, z) : x^2 + y^2 \geq z^2$ we showed that

$$\forall x \forall y \forall z Q(x, y, z)$$

is false. If the universe is the positive integers, then what can be said about the truth of:

   (a) $\forall x \forall y \exists z Q(x, y, z)$?

   (b) $\forall x \exists y \exists z Q(x, y, z)$?

   (c) $\exists x \forall y \exists z Q(x, y, z)$?

   (d) $\exists x \exists y \exists z Q(x, y, z)$?

   (e) $\exists x \exists y \forall z Q(x, y, z)$?

2. If we know that

$$\forall n(n \text{ is odd } \rightarrow (\exists k)(n = 2k + 1)$$

   (a) Which of these statements is true?

$$7 \text{ is odd } \rightarrow (\exists k)(7 = 2k + 1)$$

$$6 \text{ is odd } \rightarrow (\exists k)(6 = 2k + 1)$$

(b) Which of these statements is true, and why?

$$(\exists k)(7 = 2k + 1)$$

$$(\exists k)(6 = 2k + 1)$$

3. A proof is given below for the equivalence

$$\forall x(P(x) \land Q(x)) \equiv \forall x P(x) \land \forall x Q(x)$$

by assuming that $\forall x(P(x) \land Q(x))$ and deriving $\forall x P(x) \land \forall x Q(x)$ and then by assuming that $\forall x P(x) \land \forall x Q(x)$ and deriving $\forall x(P(x) \land Q(x))$. Supply the reasons for each step in these arguments.

(a) Assume that $\forall x(P(x) \land Q(x))$ and derive $\forall x P(x) \land \forall x Q(x)$.

Number	Statement	Reason
1	$\forall x(P(x) \land Q(x))$	given
2	a	arbitrary instance in universe
3	b	arbitrary instance in universe
4	$P(a) \land Q(a)$	1, 2, and _____
5	$P(a)$	
6	$\forall x P(x)$	2, 5, and _____
7	$P(b) \land Q(b)$	
8	$Q(b)$	
9	$\forall x Q(x)$	
10	$\forall x P(x) \land \forall x Q(x)$	6, 9, and _____

(b) Assume that $\forall x P(x) \land \forall x Q(x)$ and derive $\forall x(P(x) \land Q(x))$.

Number	Statement	Reason
1	$\forall x P(x) \land \forall x Q(x)$	given
2	a	arbitrary instance of universe
3	$\forall x P(x)$	
4	$\forall x Q(x)$	
5	$P(a)$	
6	$Q(a)$	
7	$P(a) \land Q(a)$	
8	$\forall x(P(x) \land Q(x))$	

4. Prove that:

(a) $p \vee q$ is valid using truth tables.

$$p \to r$$
$$q \to r$$
$$\therefore \quad r$$

(b) $\exists x(P(x) \vee Q(x)) \equiv \exists x P(x) \vee \exists x Q(x)$ using part (a).

A.5   Mathematical Proofs

With the ideas of the previous sections in mind, we will show how they are put together to generate a mathematical proof of:

if n is even, then n^2 is even

It is common in mathematics both to state a theorem informally and to prove it informally; however, to illustrate what has just been discussed, we will first give a more formal proof. Since the context of the theorem has already been established to be the integers (the domain of discourse is the set of integers), a more formal statement of the theorem would be

$\forall n(n$ is even $\to n^2$ is even$)$

The proof will formally proceed as follows, where we will ultimately use the principle of universal generalization (for arbitrary a in the universe, we will show that $P(a)$ is true allowing us to infer $\forall n P(n)$) to show that the universally quantified statement is true.

1	a is an integer	given instance in universe
2	a is even	given (assumption)
3	$\forall n(n$ is even $\to (\exists k)(n = 2k))$	definition of even integer
4	a is even $\to (\exists k)(a = 2k)$	1, 3, and universal instantiation
5	$(\exists k)(a = 2k)$	2, 4, and law of detachment
6	$a = 2L$ for some L	5 and existential instantiation
7	$\forall r \forall s(r = s \to r^2 = s^2)$	theorem about integers is assumed known
8	$(a = 2L) \to a^2 = (2L)^2$	7 and universal instantiation
9	$a^2 = (2L)^2$	6, 8, and law of detachment
10	$(2L)^2 = 2(2L^2)$	9 and associative prop. of integers (steps to here omitted for brevity)
11	$\forall x \forall y \forall z((x = y$ and $y = z)$ $\to x = z)$	transitive property of equality assumed
12	$a^2 = (2L)^2$ and $(2L)^2 = 2(2L^2)$ $\to (2L)^2 = 2(2L^2)$	9, 11, and universal instantiation

13	$a^2 = (2L)^2$ and $(2L)^2 = 2(2L^2)$	9, 10, and conjunction valid argumentative form
14	$a^2 = 2(2L^2)$	12, 13, and law of detachment
15	$(\exists k)(a^2 = 2k)$	14 and existential generalization for the instance $(2L^2)$
16	$\forall x((\exists k)(x = 2k) \rightarrow x$ is even$)$	definition of even integer
17	$(\exists k)(a^2 = 2k) \rightarrow x$ is even	16 and universal instantiation with the instance a^2
18	a^2 is even	15, 17, and law of detachment
19	a is even $\rightarrow a^2$ is even	2, 18, and $((p \wedge q) \rightarrow (p \rightarrow q)$ is a tautology)
20	$\forall n(n$ is even $\rightarrow n^2$ is even$)$	1, 19, and universal generalization

The argument above has been written out more formally than is customary except in courses in mathematical logic. In fact, we even shortcut the process somewhat by assuming that some properties as known in order to avoid making the proof too long. We just wanted to illustrate how logic, quantifiers, valid arguments, and inference interact with a rather specific example. The more typical mathematical proof that you will see in this book will appear as follows.

Theorem *If n is even, then n^2 is even.*

Proof. Assume that n is an (arbitrary) even integer. By definition of even integer there is an integer L so that

$$n = 2L$$

If two integers are equal, then their squares are equal so that

$$n^2 = (2L)^2$$

But

$$(2L)^2 = (2L)(2L) = 2(2L^2)$$

using the associative and commutative properties of integer multiplication. So $n^2 = 2(2L^2)$ and $n^2 = 2J$ for some integer J (namely $J = 2L^2$). By the definition of even integer, n^2 is even. ∎

For knowledgeable readers, the proof would be shortened even more. Readers are supposed to fill in the logical details for themselves. As another example, we consider a proof of

Theorem *If n^2 is even, then n is even.*

This theorem has the schematic form

$$p \rightarrow q$$

In trying to construct a proof of this statement by supposing that n^2 is even and obtaining $n^2 = 2L$ for some integer L requires several theorems of number theory to prove that $n = 2K$ for some integer K. Another approach, perhaps easier, is the indirect proof using the contrapositive $\sim q \rightarrow \sim p$ since

$$p \rightarrow q \equiv \sim q \rightarrow \sim p$$

That is, proving $\sim q \rightarrow \sim p$ is equivalent to proving $p \rightarrow q$. The statement of $\sim q \rightarrow \sim p$ is

Theorem *If n is not even, then n^2 is not even.*

Proof. Let n be an integer that is not even (we note that even integers have the form $2K$ and odd integers have the form $2L + 1$ and an integer is either even or odd, but not both). We need to show that n^2 is odd. Since n is odd, there exists an integer L so that

$$n = 2L + 1$$

Squaring both sides, we have

$$
\begin{aligned}
n^2 &= (2L+1)^2 \\
&= 4L^2 + 4L + 1 \\
&= 2(2L^2 + 2L) + 1
\end{aligned}
$$

So if $J = 2L^2 + 2L$, we have

$$n^2 = 2J + 1$$

By definition of odd integers, n^2 is odd and hence is not even; and the theorem (the contrapositive version) is proved. ∎

The next type of proof is that of a biconditional statement, $p \longleftrightarrow q$.

Theorem *n^2 is even if and only if n is even.*

If we let

$$p : n^2 \text{ is even}$$
$$q : n \text{ is even}$$

then we want to prove

$$p \longleftrightarrow q$$

Recalling the equivalence

$$(p \longleftrightarrow q) \equiv (p \rightarrow q) \wedge (q \rightarrow p)$$

we have

Theorem (Equivalent Form) *If n^2 is even, then n is even; and if n is even, then n^2 is even.*

The proof of a biconditional is most often accomplished by proving $p \rightarrow q$ and separately proving $q \rightarrow p$. Then, using the conjunction valid argumentative form

$$
\begin{array}{l}
(p \rightarrow q) \\
(q \rightarrow p) \\
\hline
\therefore \quad (p \rightarrow q) \wedge (q \rightarrow p)
\end{array}
$$

we can obtain the desired result. Note that we have already proved both of the hypotheses $p \to q$ and $q \to p$ in the discussion above.

Another kind of argument that is frequently used in mathematical proofs is the reductio ad absurdum. For this kind of argument, one is usually interested in proving an "if...then" theorem, namely, proving that

$$p \to q$$

is true.

Informally, the rationale of the reductio ad absurdum goes somewhat as follows. Either $p \to q$ is true or it is not, but not both. To prove $p \to q$ we assume that $p \to q$ is false or $\sim (p \to q)$ is true. This simply means there is a circumstance when p is true but q is not. From p being true and q not, we then derive a logically false statement. The usual logically false statement obtained is of the form $r \wedge \sim r$. We then say that we have a contradiction since we should never be able to derive a false statement from true ones ($r \wedge \sim r \equiv LF$, where LF refers to a logically false statement). Since the only statement we assumed true was $\sim (p \to q)$, then $\sim (p \to q)$ must be false; and consequently, $(p \to q)$ must be true. In short, in order to prove $p \to q$, we can just show that the alternative $\sim (p \to q)$ leads to a logically false statement and therefore cannot be derivable. The only alternative is for $(p \to q)$ to be true or derivable.

More formally, we note that

$$
\begin{aligned}
\sim (p \to q) \to LF \ & \equiv \ \sim (\sim p \vee q) \to LF \\
& \equiv \ (\sim\sim p \wedge \sim q) \to LF \quad \text{de Morgan's Law} \\
& \equiv \ (p \wedge \sim q) \to LF \\
& \equiv \ \sim (p \wedge \sim q) \vee LF \\
& \equiv \ (\sim p \vee \sim\sim q) \vee LF \\
& \equiv \ (\sim p \vee q) \vee LF \\
& \equiv \ (\sim p \vee q) \\
& \equiv \ (p \to q)
\end{aligned}
$$

So $\sim (p \to q) \to LF$ is logically equivalent to $(p \to q)$. Any logically false statement will do for LF.

Thus, all we have to do is to derive or deduce the theorem $\sim (p \to q) \to r \wedge \sim r$ and we will have proven $(p \to q)$. We derive $\sim (p \to q) \to r \wedge \sim r$ by showing that it is logically true; that is, whenever $\sim (p \to q)$ is true, then $r \wedge \sim r$ is true.

We will illustrate reductio ad absurdum by reproving

Theorem *If n^2 is even, then n is even.*

Proof. Identify p with "n^2 is even" and q with "n is even." Assume that n^2 is even and n is not even (suppose that $\sim (p \to q)$). Since n is not even, then n is odd and there is an integer j so that

$$n = 2j + 1$$

This last equation gives

$$n^2 = (2j + 1)^2 = 2(2j^2 + 2j) + 1$$

so that $n^2 = 2k + 1$ for the integer $k = 2j^2 + 2j$. This characterization of n^2 means that n^2 is odd. But now we have both that n^2 is even and that n^2 is odd. So letting r be "n^2 is even," we have established $r \wedge \sim r$. Therefore, n^2 cannot be even and n be odd. So, by reductio ad absurdum, the theorem is proved. ∎

Note in this case that the statement r was the same as the statement p; but this redundancy is not required and r may be any statement.

We now consider a second example:

Theorem *If $n = m$, then $n^2 = m^2$.*

Proof. Assume that $n = m$ and $n^2 \neq m^2$. Since $n = m$, we have $n - m = 0$. If $n^2 \neq m^2$, then $n^2 - m^2 \neq 0$. So

$$0 \neq n^2 - m^2 = (n - m)(n + m)$$

But this result means that $n - m$ cannot be zero; for if it were, then $(n-m)(n+m) = 0 \cdot (m + n) = 0$, contradicting $(n - m)(n + m) \neq 0$. Thus $n - m \neq 0$. We now have both $n - m = 0$ and $n - m \neq 0$, which is a contradiction. Thus, by reductio ad absurdum, we have proved that if $n = m$, then $n^2 = m^2$. ∎

Some parts of the argument just given could also be interpreted from the viewpoint of having several alternatives, say,

$$x \vee y \vee z$$

If we can show that all but one of the alternatives lead to a contradiction, then the remaining one will have been derived. In logic argument form the reasoning would be

$$x \vee y \vee z$$
$$x \rightarrow LF$$
$$\underline{y \rightarrow LF}$$
$$\therefore \quad z$$

This argument can easily be shown to be valid since

$$[(x \vee y \vee z) \wedge (x \rightarrow LF) \wedge (y \rightarrow LF] \equiv z$$

Usually, $x \rightarrow LF$ is derived by supposing that x is true and then obtaining a logically false statement. Then by reductio ad absurdum, we declare $\sim x$ to be true. The same reasoning is also applied to y. So, instead of showing that all cases lead to the conclusion, the valid argument just discussed allows one to eliminate various cases from needing any further consideration. We will apply the case elimination argument in the proof of

Theorem *If $n > 0$ and $n^2 > 1$, then $n > 1$.*

Proof. Assume that $n > 0$ and $n^2 > 1$. But $n = 1$ or $n < 1$ or $n > 1$.

Case 1. If $n = 1$. Then

$$n^2 = n \cdot n = 1 \cdot n = n = 1$$

which contradicts $n^2 > 1$. Thus $n = 1$ is false.

Case 2. If $n < 1$, then, since $n > 0$ was assumed above,

$$n^2 = n \cdot n < 1 \cdot n = n < 1$$

We then have $n^2 < 1$, which contradicts $n^2 > 1$. Thus $n < 1$ is false. Since two of the three alternatives lead to contradictions or logically false statements, only the case $n > 1$ remains. ∎

In the discussion of structure of proofs above, sometimes the same theorem was proved several times using different strategies. It should, therefore, be evident that usually there are many ways to prove a particular theorem. The method employed depends upon experience, what theorems are already assumed to be known, and even taste. It should also be the case that some proofs appear awkward and some are very streamlined. Proving theorems requires practice and is an art.

Exercises

In each of the following proofs, outline the structure of the arguments. You should identify the overall argument type as well as any "subarguments" that may appear.

1. Theorem: $\sqrt{2}$ is irrational.

    Proof: Either $\sqrt{2}$ is irrational or $\sqrt{2}$ is rational. If $\sqrt{2}$ is rational, then there are positive integers p and q with no common factors (i.e., p/q is in lowest terms) such that $\sqrt{2} = p/q$. But, by definition of "square root," $(p/q)^2 = 2$ or $p^2 = 2q^2$. Thus 2 must divide p^2 and must consequently divide p. So there is an integer k with $p = 2k$. We can then write $2q^2 = p^2 = (2k)^2 = 4k^2$ or equivalently, $q^2 = 2k^2$. Using the same reasoning, q is also divisible by 2. So both p and q contain the factor 2 giving a contradiction. Thus $\sqrt{2}$ cannot be rational and must be irrational.

2. Theorem: If $n \geq 0$ and $n^2 > 9$, then $n > 3$.

    Proof: Assume that $n \leq 3$. We will show that $n \leq 0$ or $n \leq 9$ as follows.

    Case 1. If $n \leq 0$, then it is not true that $n > 0$ or that both $n > 0$ and $n^2 > 9$.

    Case 2. Assume that $n > 0$. Then since $n \leq 3$ we have

    $$n \cdot n \leq 3 \cdot n \leq 3 \cdot 3 = 9$$

    This means that $n^2 \leq 9$. Thus it is not true that both $n > 0$ and $n^2 > 9$. Since in either case we have obtained that both $n > 0$ and $n^2 > 9$ cannot be true, the proof is complete.

3. Theorem: If $a \cdot b = 0$, then $a = 0$ or $b = 0$.

Proof: Assume that $a \cdot b = 0$ and consider a. Either $a = 0$ or $a \neq 0$. If $a = 0$, we are done, so suppose that $a \neq 0$. Then $a \cdot b = 0 = a \cdot 0$. Dividing both sides by the nonzero a, we obtain $b = 0$. Hence either $a = 0$ or $b = 0$.

Appendix B
PEANO AXIOMS

B.1 Postulates

We now formally define the positive integers using a set of axioms called Peano's Postulates. From these axioms we can develop the necessary properties of the integers given in Chapter 1.

Peano's Postulates Let N be a set and S a relation on $N \times N$. If $n = S(m)$, then n is said to be a *successor* of m. If N satisfies the following axioms, then N is called the set of *positive integers*:

Axiom 1 The relation S is a function; that is, if $m \in N$, then there exists a unique $n \in N$ such that $n = S(m)$.

Axiom 2 There exists an element 1 in N such that for all n in N, $1 \neq S(n)$; that is, 1 is not in the range of S.

Axiom 3 S is one to one; that is, if $n = S(m)$ and $n = S(m')$, then $m = m'$.

Axiom 4 If M is a subset of N which has the following properties:

    (a) $1 \in M$,

    (b) if $m \in M$ and $n = S(m)$, then $n \in M$,

then $M = N$.

Definition B.1 *Let N and S satisfy Peano's Postulates.*

(a) *Define addition of positive integers as follows:*

$$
\begin{aligned}
n + 1 &= S(n) \text{ for all } n \in N \\
n + S(m) &= S(m + n) \text{ for all } n, m \in N
\end{aligned}
$$

(b) *Define multiplication of positive integers as follows:*

$$1 \cdot n = n \text{ for all } n \in N$$
$$S(n) \cdot m = n \cdot m + m$$

B.2   Development

Theorem B.2 *Let n be a positive integer. Either $n = 1$ or n is in the range of S; that is, $n = S(m)$ for some positive integer m.*

Proof.   Let $M = \{x : x$ is a positive integer and either $x = 1$ or $x = S(k)$ for some integer $k\}$. We want to show that $M = N$. By the definition of M, $M \subseteq N$; and, certainly, $1 \in M$. Assume that $m \in M$ and $n = S(m)$. By definition of M, $n \in M$. By Axiom 4, $M = N$. ■

Theorem B.3 *The positive integers are closed under addition and multiplication; that is, the sum of two positive integers is a positive integer, and the product of two positive integers is a positive integer.*

Proof. The proof of closure follows directly from the definitions of addition and multiplication. ■

Theorem B.4 *For all n, m, and $r \in N$, $n + (m + r) = (n + m) + r$.*

Proof. The proof uses induction on r. Assume that m and n are positive integers and $M = \{r : r$ is an integer and $n + (m + r) = (n + m) + r\}$. The integer $1 \in M$ because $n + (m + 1) = n + S(m) = S(n + m) = (n + m) + 1$. Assume that $r = k$ and $k \in M$ so that $n + (m + k) = (n + m) + k$. We need to show that $r = k + 1 \in M$. But

$$
\begin{aligned}
n + (m + S(k)) &= n + S(m + k) \\
&= n + ((m + k) + 1) \\
&= (n + (m + k)) + 1 \\
&= ((n + m) + k) + 1 \\
&= (n + m) + (k + 1) \\
&= (n + m) + S(k)
\end{aligned}
$$

Thus, $k + 1 \in M$. By Axiom 4, $M = N$. ■

Theorem B.5 *For all n, $m \in N$, $n + m = m + n$.*

Proof. First using induction on n, we prove that $n + 1 = 1 + n$ for all $n \in N$. It is trivially true when $n = 1$. Assume that the equality is true for $n = k$

and prove for $n = S(k)$.

$$
\begin{aligned}
1 + S(k) &= 1 + (k+1) \\
&= (1+k) + 1 \\
&= (k+1) + 1 \\
&= S(k) + 1
\end{aligned}
$$

Hence by induction, $n + 1 = 1 + n$ for all n.

Assume that $m \in N$. We prove using induction on n that $m + n = n + m$ for all $n, m \in N$. The equality has already been shown to be true for $n = 1$. Assume that the equality is true for $n = k$ and prove for $n = S(k)$.

$$
\begin{aligned}
m + S(k) &= S(m + k) \\
&= S(k + m) \\
&= k + S(m) \\
&= k + (m + 1) \\
&= k + (1 + m) \\
&= (k + 1) + m \\
&= S(k) + m \qquad \blacksquare
\end{aligned}
$$

Theorem B.6 *For all n, m, and $r \in N$, $n \cdot (m + r) = (n \cdot m) + (n \cdot r)$.*

Proof. Let m and r be positive integers, and use induction on n. When $n = 1$,

$$
\begin{aligned}
1 \cdot (m + r) &= m + r \\
&= (1 \cdot m) + (1 \cdot r)
\end{aligned}
$$

and the theorem is true. Assume that the theorem is true when $n = k$, that is, $k \cdot (m + r) = (k \cdot m) + (k \cdot r)$. Then prove that the equality holds when $n = k+1$ or $n = S(k)$; that is, prove that $S(k) \cdot (m+r) = (S(k) \cdot m) + (S(k) \cdot r)$.

$$
\begin{aligned}
S(k) \cdot (m + r) &= (k \cdot (m + r)) + (m + r) \\
&= ((k \cdot m) + (k \cdot r)) + (m + r) \\
&= (k \cdot m) + ((k \cdot r) + (m + r)) \\
&= (k \cdot m) + (((k \cdot r) + m) + r) \\
&= (k \cdot m) + ((m + (k \cdot r)) + r) \\
&= ((k \cdot m) + (m + (k \cdot r))) + r \\
&= (((k \cdot m) + m) + (k \cdot r)) + r \\
&= ((k \cdot m) + m) + ((k \cdot r) + r) \\
&= (S(k) \cdot m) + (S(k) \cdot r) \qquad \blacksquare
\end{aligned}
$$

Theorem B.7 *For all n, $m, \in N$, $n \cdot m = m \cdot n$.*

Proof. First we prove using induction on n that $n \cdot 1 = n$. It is certainly true for $n = 1$. Assume that the equality of the theorem holds for $n = k$ so that $k \cdot 1 = k$, and prove it true for $n = S(k)$. That is, prove $S(k) \cdot 1 = S(k)$. By definition

$$
S(k) \cdot 1 = (k \cdot 1) + 1 = k + 1 = S(k)
$$

Thus $n \cdot 1 = n$ for every positive integer n.

For a given n, we show using induction on m that $n \cdot m = m \cdot n$ for every m. It was just proven that the equality is true for $m = 1$. Assume that the equality is true for $m = k$; that is, assume that $n \cdot k = k \cdot n$. We will now prove the equality for $m = k+1 = S(k)$; that is, we prove $n \cdot S(k) = S(k) \cdot n$.

$$
\begin{aligned}
n \cdot S(k) &= n \cdot (k+1) \\
&= (n \cdot k) + (n \cdot 1) \\
&= (k \cdot n) + n \\
&= S(k) \cdot n \qquad \blacksquare
\end{aligned}
$$

Corollary B.8 *For all n, m, and $r \in N$, $(n + m) \cdot r = (n \cdot r) + (m \cdot r)$.*

Theorem B.9 *For all n, m, and $r \in N$, $(n \cdot m) \cdot r = n \cdot (m \cdot r)$.*

Proof. The proof proceeds by induction on n. When $n = 1$, we have

$$
\begin{aligned}
(1 \cdot m) \cdot r &= m \cdot r \\
&= 1 \cdot (m \cdot r)
\end{aligned}
$$

Assume that the theorem is true for $n = k$; that is, assume that $(k \cdot m) \cdot r = k \cdot (m \cdot r)$. We now prove that $(S(k) \cdot m) \cdot r = S(k) \cdot (m \cdot r)$.

$$
\begin{aligned}
S(k) \cdot (m \cdot r) &= (k \cdot (m \cdot r)) + (m \cdot r) \\
&= ((k \cdot m) \cdot r) + (m \cdot r) \\
&= ((k \cdot m) + m) \cdot r \\
&= (S(k) \cdot m) \cdot r \qquad \blacksquare
\end{aligned}
$$

Theorem B.10 (Uniqueness of Addition) *For all n, m, and $r \in N$, if $n = m$, then $n + r = m + r$.*

Proof. We use induction on r. Hence we show that if $n = m$, then $n + 1 = m + 1$. But if $n = m$, then $S(n) = S(m)$ because S is a function; and therefore, $n + 1 = m + 1$. Now assume that if $n = m$, then $n + k = m + k$. We now want to prove that if $n = m$, then $n + S(k) = m + S(k)$. But if $n = m$, by assumption $n + k = m + k$; and one obtains $S(n + k) = S(m + k)$ since S is a function. So

$$
\begin{aligned}
n + S(k) &= S(n + k) \\
&= S(m + k) \\
&= m + S(k) \qquad \blacksquare
\end{aligned}
$$

Theorem B.11 (Cancellation Law) *For all n, m, and $r \in N$, if $n + r = m + r$ for some positive integer r, then $n = m$.*

Proof. We proceed by induction on r. If $r = 1$, then show that $n + 1 = m + 1$ implies that $n = m$. But if $n + 1 = m + 1$, then $S(n) = S(m)$; and therefore, $n = m$ since S is one to one. Thus, the theorem is true when $r = 1$. Assume that the equality holds for $r = k$; that is, assume that if $n + k = m + k$, then $n = m$. We now wish to prove it true for $r = S(k)$; that is, prove that if $n + S(k) = m + S(k)$, then $n = m$. If $n + S(k) = m + S(k)$, then

$S(n + k) = S(m + k)$ by definition of addition. Hence $n + k = m + k$ since S is one to one. But then by our assumption, $n = m$. ∎

Theorem B.12 *For all n, m, and $r \in N$, if $n = m$, then $n \cdot r = m \cdot r$.*

Proof. Using induction on r, we first show that the theorem is true if $r = 1$. But if $n \cdot 1 = m \cdot 1$, then $n = m$ by Definition B.1(b) and Theorem B.7. Assume that the theorem is true for $r = k$; that is, assume that if $n = m$, then $n \cdot k = m \cdot k$. We now want to show that if $n = m$, then $n \cdot S(k) = m \cdot S(k)$. If $n \cdot k = m \cdot k$, then $k \cdot n = k \cdot m$ by commutativity. Thus, by uniqueness of addition, $k \cdot n + n = k \cdot m + m$ and $S(k) \cdot n = S(k) \cdot m$. Consequently, $n \cdot S(k) = m \cdot S(k)$. ∎

Theorem B.13 (Multiplicative Cancellation) *For any positive integers m, n, and r, if $n \cdot r = m \cdot r$, then $n = m$.*

Proof. Using induction on n, let $M = \{n : \text{if } m \text{ and } r \text{ are positive integers}$ such that $n \cdot r = m \cdot r$ then $n = m\}$. Assume that $n = 1$ and $1 \cdot r = m \cdot r$. If $m = 1$, then we have $n = m$ and $1 \in M$. But if $m \neq 1$, then m is in the range of S and hence $m = S(p)$ for some positive integer p. Hence

$$
\begin{aligned}
1 \cdot r &= S(p) \cdot r \\
&= p \cdot r + r \\
&= (p \cdot r) + (1 \cdot r)
\end{aligned}
$$

and

$$
1 + (1 \cdot r) = 1 + (p \cdot r) + (1 \cdot r)
$$

By cancellation of addition,

$$
\begin{aligned}
1 &= 1 + (p \cdot r) \\
&= (p \cdot r) + 1 \\
&= S(p \cdot r)
\end{aligned}
$$

which contradicts the fact that 1 is not in the range of S. Hence $m = 1$.

Assume that $n = k \in M$; that is, for any positive integers m and r, if $k \cdot r = m \cdot r$, then $k = m$. Assume that t and s are positive integers such that $S(k) \cdot s = t \cdot s$. We show that $S(k) = t$. If $t = 1$, then since $1 \in M$, $S(k) = 1$, which is impossible because $S(k)$ is in the range of S (Axiom 2). We must, therefore, have $t \neq 1$, making $t = S(m)$ for some m. Hence $S(k) \cdot s = S(m) \cdot s$, $k \cdot s + s = m \cdot s + s$, and $k \cdot s = m \cdot s$ by cancellation. By the induction hypothesis, $k = m$ and so $S(k) = S(m) = t$. Thus $n = k + 1 \in M$ and so $M = N$. ∎

At this point we wish to develop the integers. To do this we first define the symbol 0 to have the properties $0 \cdot n = n \cdot 0 = 0$ and $0 + n = n + 0 = n$ if $n = 0$ or $n \in N$. It is straightforward to verify that all but one of the preceding theorems (associative, commutative, etc.) of this section hold also for $N \cup \{0\}$. The exception is multiplicative cancellation when $r = 0$. We then define the set P of ordered pairs $\{(m, n) : m, n \in N \cup \{0\}\}$. We define a relation R on P by $(m,n)\,R\,(r,s)$ if $m+s=n+r$. The relation R is an equivalence relation and, therefore, partitions P. To justify this claim, we must show that R is reflexive, symmetric, and transitive. It is obvious that R is reflexive and symmetric. To show that R is transitive, assume that $(m,n)\,R\,(r,s)$ and $(r,s)\,R\,(u,v)$. We need to show that $(m,n)\,R\,(u,v)$. Since $(m,n)\,R\,(r,s)$, $m+s=n+r$. Since $(r,s)\,R\,(u,v)$, $r+v=s+u$. Hence

$$
\begin{aligned}
m+s+r+v &= n+r+s+u \\
m+v+s+r &= n+u+s+r
\end{aligned}
$$

and $m+v=n+u$ so that $(m,n)\,R\,(u,v)$. Thus, R is an equivalence relation which partitions P. Let I be the set of equivalence classes of R.

Let $[(m,n)] \in I$ denote the equivalence class containing (m,n). Define addition and multiplication between equivalence classes as follows:

$$
\begin{aligned}
[(m,n)] + [(p,q)] &= [(m+p,n+q)] \\
[(m,n)] \cdot [(p,q)] &= [(mp+nq,mq+np)]
\end{aligned}
$$

We need to show that addition and multiplication are well defined; that is if $[(m,n)] = [(m',n')]$ and $[(p,q)] = [(p',q')]$, then

$$
\begin{aligned}
[(m,n)] + [(p,q)] &= [(m',n')] + [(p',q')] \\
[(m,n)] \cdot [(p,q)] &= [(m',n')] \cdot [(p',q')]
\end{aligned}
$$

First, consider addition. If $[(m,n)] = [(m',n')]$ and $[(p,q)] = [(p',q')]$, then $m+n' = m'+n$ and $p+q' = p'+q$. Hence

$$
\begin{aligned}
m+p+n'+q' &= m+n'+p+q' \\
&= m'+n+p'+q \\
&= m'+p'+n+q
\end{aligned}
$$

Therefore,

$$
[(m+p,n+q)] = [(m'+p',n'+q')]
$$

and

$$
[(m,n)] + [(p,q)] = [(m',n')] + [(p',q')]
$$

Next consider multiplication. By definition of multiplication,

$$
[(m',n')] \cdot [(p',q')] = [(m'p'+n'q',m'q'+n'p')]
$$

If $[(m,n)] = [(m',n')]$ and $[(p,q)] = [(p',q')]$, then $m+n' = m'+n$ and $p+q' = p'+q$. Since $m+n' = n+m'$, multiplying on the right by p and q, respectively, and using the distributive law, we have

(1) $mp+n'p = np+m'p$

and

(2) $nq+m'q = mq+n'q$.

Since $p+q' = p'+q$, multiplying on the left by m' and n', respectively, and using the distributive law, we have

(3) $m'p + m'q' = m'p' + m'q$

and

(4) $n'p' + n'q = n'p + n'q'$.

Adding (1), (2), (3), and (4) and using associativity and commutativity of addition of positive integers, we have

$$mp + nq + m'q' + n'p' + (n'p + m'q + m'p + n'q)$$
$$= m'p' + n'q' + mq + np + (n'p + m'q + m'p + n'q)$$

Hence

$$mp + nq + m'q' + n'p' = m'p' + n'q' + mq + np$$

and

$$[mp + nq, mq + np] = [m'p' + n'q', m'q' + n'p']$$

so that

$$[(m, n)] \cdot [(p, q)] = [(m', n')] \cdot [(p', q')]$$

The proof of the following theorem is straightforward and is left to the reader.

Theorem B.14 *The set I, with the definitions of addition and multiplication given above, has the following properties:*

(a) $a + b = b + a$, $a \cdot b = b \cdot a$ *for all $a, b \in I$ (commutativity of addition and multiplication).*

(b) $a + (b + c) = (a + b) + c$, $a \cdot (b \cdot c) = (a \cdot b) \cdot c$ *for all $a, b, c \in I$ (associativity of addition and multiplication).*

(c) $a \cdot (b + c) = a \cdot b + a \cdot c$ *for all $a, b, c \in I$ (distributive law).*

By definition, $[(n, 0)] = [(n', 0)]$ if and only if $n + 0 = n' + 0$ if and only if $n = n'$. Hence there is a one-to-one correspondence between the positive integers and equivalence classes of the form $[(n, 0)]$ for n in N.

Further,

$$[(n, 0)] + [(n', 0)] = [(n + n', 0)]$$

and

$$[(n, 0)] \cdot [(n', 0)] = [(nn' + 0, n \cdot 0 + 0 \cdot n')] = [(nn', 0)]$$

so we may identify the positive integer n with the equivalence class $[(n, 0)]$. The equivalence class $[(0, 0)]$, as expected, has the properties that

$$[(0, 0)] + [(m, n)] = [(m, n)]$$

and

$$[(0, 0)] \cdot [(m, n)] = [(0, 0)]$$

so that $[(0,0)]$ may be identified with the integer 0. Further,

$$[(1,0)] \cdot [(m,n)] = [(1 \cdot m + 0 \cdot n, m \cdot 0 + n \cdot 1)] = [(m,n)]$$

for all $m, n \in N$. Therefore, the integer 1 is identified with $[(1,0)]$.

Also, $[(n,0)] + [(0,n)] = [(n,n)]$. But $[(n,n)] = [(0,0)]$ because of the identity $n + 0 = n + 0$. We shall denote the element $[(0,n)]$ by $-n$. Hence $n + (-n) = (-n) + n = 0$. Further,

$$
\begin{aligned}
(-n)(-m) &= [(0,n)] \cdot [(0,m)] \\
&= [(0 \cdot 0) + (n \cdot m), (0 \cdot m) + (n \cdot 0)] \\
&= [(mn, 0)] \\
&= mn
\end{aligned}
$$

and

$$
\begin{aligned}
m \cdot (-n) &= [(m,0)] \cdot [(0,n)] \\
&= [(m \cdot 0 + n \cdot 0, m \cdot n + 0 \cdot 0)] \\
&= [(0, mn)] \\
&= -(mn)
\end{aligned}
$$

Note that if $[(n,0)] = [(0,m)]$ for positive integers n and m, then $n + m = 0$, which contradicts the closure of the positive integers under addition.

Define $[(m,n)] > [(p,q)]$ if there exists $[(r,0)]$ for positive integer r such that

$$[(p,q)] + [(r,0)] = [(m,n)]$$

so that, for positive integers a and b, one has $a > b$ if there exists a positive integer c so that $b + c = a$. Hence $p > q$ if and only if $p = q + r$ for some positive integer r if and only if

$$[(p,q)] = [(r,0)] = r$$

Also, $q > p$ if and only if $q = p + r$ for some positive integer r if and only if

$$[(p,q)] = [(0,r)] = -r$$

for some positive integer r. We write $a \geq b$ if and only if $a = b$ or $a > b$, whether a and b are integers or positive integers.

Theorem B.15 *If n and m are positive integers, then either $n \geq m$ or $m \geq n$.*

Proof. The proof will be by induction on m; that is, given a positive integer n, we will show that for every positive integer m, either $n \geq m$ or $m \geq n$. First we consider the case $m = 1$. If $n = 1$ also, then trivially $n \geq 1$. If $n \neq 1$, then $n = S(p)$ for some p so that $n = p + 1$, $n > 1$, and, consequently, $n \geq 1$.

Assume that for $m = k$, either $k \geq n$ or $n \geq k$. We want to show that for $m = S(k) = k + 1$, we have either $n \geq k + 1$ or $k + 1 \geq n$. If $k = n$, then $k + 1 = n + 1$ and $k + 1 \geq n$. If $k > n$, then $k = n + p$ for some positive integer p and $k + 1 = n + (p + 1)$; so $k + 1 > n$. If $n > k$, then $n = k + p$ for some positive integer p. If $p = 1$, then $n = k + 1$ and we are done. If $p > 1$, then $p = S(q)$ for some positive integer q. Therefore,

$$n = k + S(q) = k + q + 1 = k + 1 + q$$

and $n \geq k + 1$. Hence, the theorem is true. ∎

Theorem B.16 *Let n and m be positive integers. If $n \geq m$ and $m \geq n$, then $n = m$.*

Proof. If $n > m$ and $m > n$, then $n = m + r$ and $m = n + t$ for some positive integers r and t. Hence

$$\begin{aligned} n &= n + t + r \\ n + 1 &= n + t + r + 1 \end{aligned}$$

and

$$1 = t + r + 1 = S(t + r)$$

which is a contradiction because 1 is not in the range of S. ∎

Definition B.17 *Let $p = [(n, m)]$ be an integer. Define $-p$ to be the integer $[(m, n)]$.*

This definition is obviously consistent with the definition of $-p$ when p is a positive integer. Further, it is obvious that $-(-p) = p$ for every integer p.

Theorem B.18 (Law of Trichotomy) *Let p be an integer. Then one and only one of the following is true:*

(a) $p = 0$.

(b) p *is a positive integer.*

(c) $-p$ *is a positive integer.*

Proof. If p is an integer, then $p = [(m, n)]$ for some positive integers m and n. If $m = n$, then $p = 0$. If $m \neq n$, then either $n > m$ or $m > n$. If $m > n$, then p is a positive integer. If $n > m$ then $-p$ is a positive integer. ∎

Definition B.19 *If m and n are integers, then $m - n = m + (-n)$.*

Theorem B.20 *Every positive integer is greater than or equal to 1.*

Proof. Let $S = \{n : n \in N \text{ and } n \geq 1\}$. We want to show that $S = N$. Obviously, $1 \in S$. Assume that $k \in N$ and $k \neq 1$; then $k = S(r)$ for some r. Hence $k = r + 1$, $k > 1$, and $N = S$. ∎

The following theorem is easily proven. The proof is left to the reader.

Theorem B.21 *For all integers n, m, and p:*

(a) *If $n > m$ and $m > p$, then $n > p$.*

(b) *If $n > m$, then $n + p > m + p$.*

(c) *If $n > m$ and p is positive, then $np > mp$.*

(d) *If $n > m$ and $-p$ is positive, then $mp > np$.*

We now introduce two alternatives to the principle of induction

Theorem B.22 (Well-Ordering Principle) *Every nonempty subset of the positive integers has a least element.*

Proof.   Let $S = \{n \in N$: if $n \in T$ where T is a subset of N, then T has a least element$\}$. We want to show that $S = N$. Obviously, $1 \in S$; for if $1 \in T$, then 1 is the least element of T. Assume that $k \in S$ and let $k + 1 \in T$. We wish to show that T has a least element. If $1 \in T$, then T has a least element and we are done. Assume that $1 \notin T$. Let $T' = \{s : s + 1 \in T\}$. Since $k \in T'$, T' has a least element, say r. By definition, $r + 1 \in T$. Also, $r + 1$ is the least element in T; for, if $u \in T$ and $u < r + 1$, then since $u = v + 1$ for some $v \in T'$, we obtain $v < r$, which is a contradiction. Hence T has a least element, $k + 1 \in S$ and $S = N$. ■

Theorem B.23 *Assume that S is a subset of N with the properties that:*

(a) $1 \in S$,

(b) *if $k \in N$ and $s \in S$ for all $s < k$, then $k \in S$,*

then $S = N$.

Proof.   Choose S with the properties above. We want to show that $S = N$. Let $T = \{n : n \in N$ and $n \notin S\}$. We want to show that T is empty. If T is not empty, by the Well-Ordering Principle, T contains a least element, say t, and $t \neq 1$ because 1 is in S and t is not. Hence $t = S(r)$ for some r. The set of all $x \in N$ such that $x < t$ is nonempty since r is in it. All of these elements belong to S since t is the least positive integer not belonging to S. Hence by part (b), $t \in S$ which is a contradiction; therefore, T is empty. ■

Exercises

1.   Let $N^0 = N \cup \{0\}$ be the set formed from adding the new symbol $0 \notin N$ to the set of positive integers N. Define the function $S^0 : N^0 \to N^0$ as follows:

$$S^0(n) = \begin{cases} S(n) & \text{if } n \in N \\ 1 & \text{if } n = 0 \end{cases}$$

where S is the successor function given in Peano's Axioms. Prove the following extended Peano Axioms:

(a) S^0 is a function from N^0 into N^0.

(b) There exists an element $0 \in N^0$ such that for all $n \in N^0$, $(n, 0) \notin S^0$.

(c) S^0 is one to one.

(d) If M is a subset of N^0 which has the following properties:

    (i) $0 \in M$,

    (ii) if $m \in M$ and $m = S^0(n)$, then $m \in M$,

    then $M = N^0$.

Next define the operations $+$ and $\cdot$ on N^0 as follows:
$$
\begin{aligned}
n + 1 &= S^0(n) && \text{for all } n \in N^0 \\
n + S^0(m) &= S^0(m + n) && \text{for all } m, n \in N^0 \\
0 \cdot n &= 0 && \text{for all } n \in N^0 \\
S^0(n) \cdot m &= n \cdot m + m && \text{for all } m, n \in N^0
\end{aligned}
$$

Prove analogs of Theorems B.2, B.4, and B.5 for the extension of the Peano Axioms to N^0 as described in Exercise 1, namely,

(e) If $n \in N^0$, then either $n = 0$ or n is in the range of S^0.

(f) For all $n, m, r \in N^0$, $n + (m + r) = (n + m) + r$.

(g) For all $n, m \in N^0$, $n + m = m + n$.

2. An integer in I was defined to be an equivalence class $[(m, n)]$ for the equivalence relation R defined on $(N \cup \{0\}) \times (N \cup \{0\})$ by $(m, n)R(r, s)$ if and only if $m + s = n + r$. Consider the integers $-5_I, -2_I, 0_I, 2_I, 3_I, 5_I$, and 6_I as members of I. (The subscript I is to indicate that the notation refers to members of I and not to members of N so that $3 \in N$ but $3_I \in I$.) For example,

$$- 5_I = [(m, n)] = \{(r, s) : (r, s)R(m, n)\}$$

for appropriate m and n in $N \cup \{0\}$.

(a) Specify a choice for (m, n) and two other elements (r, s) and (r', s') in $-5_I = [(m, n)]$ _for_ -5.

(b) Repeat part (a) for the other integers $-2_I, 0_I, 2_I, 3_I, 5_I, 6_I$ in I.

(c) Compute $5_I + (-2_I)$ explicitly using one representative for each integer and verify the following:

$$5_I + (-2_I) = [(a,b)] \text{ for } 5 + [(c,d)] \text{ for } -2 = [(e,f)] \text{ for } 3 = 3_I$$

(d) Repeat part (c) with another set of representatives for 5_I and -2_I.

(e) Repeat parts (c) and (d) for the sum $(-5_I) + 3_I$.

(f) In the same manner as parts (c) and (d), compute explicitly $2_I \cdot 3_I$ and show that the definition yields 6_I.

3. Prove Theorem B.14.

Appendix C
TABLES

Table C.1. Factors

The least prime factor f of odd n with $5 \nmid n$ is given unless n is prime.

n	f	n	f	n	f	n	f	n	f	n	f
2	.	93	3	189	3	283	.	379	.	473	11
3	.	97	.	191	.	287	7	381	3	477	3
5	.	99	3	193	.	289	17	383	.	479	.
7	.	101	.	197	.	291	3	387	3	481	13
9	3	103	.	199	.	293	.	389	.	483	3
11	.	107	.	201	3	297	3	391	17	487	.
13	.	109	.	203	7	299	13	393	3	489	3
17	.	111	3	207	3	301	7	397	.	491	.
19	.	113	.	209	11	303	3	399	3	493	17
21	3	117	3	211	.	307	.	401	.	497	7
23	.	119	7	213	3	309	3	403	13	499	.
27	3	121	11	217	7	311	.	407	11	501	3
29	.	123	3	219	3	313	.	409	.	503	.
31	.	127	.	221	13	317	.	411	3	507	3
33	3	129	3	223	.	319	11	413	7	509	.
37	.	131	.	227	.	321	3	417	3	511	7
39	3	133	7	229	.	323	17	419	.	513	3
41	.	137	.	231	3	327	3	421	.	517	11
43	.	139	.	233	.	329	7	423	3	519	3
47	.	141	3	237	3	331	.	427	7	521	.
49	7	143	11	239	.	333	3	429	3	523	.
51	3	147	3	241	.	337	.	431	.	527	17
53	.	149	.	243	3	339	3	433	.	529	23
57	3	151	.	247	13	341	11	437	19	531	3
59	.	153	3	249	3	343	7	439	.	533	13
61	.	157	.	251	.	347	.	441	3	537	3
63	3	159	3	253	11	349	.	443	.	539	7
67	.	161	7	257	.	351	3	447	3	541	.
69	3	163	.	259	7	353	.	449	.	543	3
71	.	167	.	261	3	357	3	451	11	547	.
73	.	169	13	263	.	359	.	453	3	549	3
77	7	171	3	267	3	361	19	457	.	551	19
79	.	173	.	269	.	363	3	459	3	553	7
81	3	177	3	271	.	367	.	461	.	557	.
83	.	179	.	273	3	369	3	463	.	559	13
87	3	181	.	277	.	371	7	467	.	561	3
89	.	183	3	279	3	373	.	469	7	563	.
91	7	187	11	281	.	377	13	471	3	567	3

Table C.1. Factors (Continued)

n	f	n	f	n	f	n	f	n	f	n	f
569	.	663	3	759	3	853	.	949	13	1043	7
571	.	667	23	761	.	857	.	951	3	1047	3
573	3	669	3	763	7	859	.	953	.	1049	.
577	.	671	11	767	13	861	3	957	3	1051	.
579	3	673	.	769	.	863	.	959	7	1053	3
581	7	677	.	771	3	867	3	961	31	1057	7
583	11	679	7	773	.	869	11	963	3	1059	3
587	.	681	3	777	3	871	13	967	.	1061	.
589	19	683	.	779	19	873	3	969	3	1063	.
591	3	687	3	781	11	877	.	971	.	1067	11
593	.	689	13	783	3	879	3	973	7	1069	.
597	3	691	.	787	.	881	.	977	.	1071	3
599	.	693	3	789	3	883	.	979	11	1073	29
601	.	697	17	791	7	887	.	981	3	1077	3
603	3	699	3	793	13	889	7	983	.	1079	13
607	.	701	.	797	.	891	3	987	3	1081	23
609	3	703	19	799	17	893	19	989	23	1083	3
611	13	707	7	801	3	897	3	991	.	1087	.
613	.	709	.	803	11	899	29	993	3	1089	3
617	.	711	3	807	3	901	17	997	.	1091	.
619	.	713	23	809	.	903	3	999	3	1093	.
621	3	717	3	811	.	907	.	1001	7	1097	.
623	7	719	.	813	3	909	3	1003	17	1099	7
627	3	721	7	817	19	911	.	1007	19	1101	3
629	17	723	3	819	3	913	11	1009	.	1103	.
631	.	727	.	821	.	917	7	1011	3	1107	3
633	3	729	3	823	.	919	.	1013	.	1109	.
637	7	731	17	827	.	921	3	1017	3	1111	11
639	3	733	.	829	.	923	13	1019	.	1113	3
641	.	737	11	831	3	927	3	1021	.	1117	.
643	.	739	.	833	7	929	.	1023	3	1119	3
647	.	741	3	837	3	931	7	1027	13	1121	19
649	11	743	.	839	.	933	3	1029	3	1123	.
651	3	747	3	841	29	937	.	1031	.	1127	7
653	.	749	7	843	3	939	3	1033	.	1129	.
657	3	751	.	847	7	941	.	1037	17	1131	3
659	.	753	3	849	3	943	23	1039	.	1133	11
661	.	757	.	851	23	947	.	1041	3	1137	3

Table C.1. Factors (Continued)

n	f	n	f	n	f	n	f	n	f	n	f
1139	17	1233	3	1329	3	1423	.	1519	7	1613	.
1141	7	1237	.	1331	11	1427	.	1521	3	1617	3
1143	3	1239	3	1333	31	1429	.	1523	.	1619	.
1147	31	1241	17	1337	7	1431	3	1527	3	1621	.
1149	3	1243	11	1339	13	1433	.	1529	11	1623	3
1151	.	1247	29	1341	3	1437	3	1531	.	1627	.
1153	.	1249	.	1343	17	1439	.	1533	3	1629	3
1157	13	1251	3	1347	3	1441	11	1537	29	1631	7
1159	19	1253	7	1349	19	1443	3	1539	3	1633	23
1161	3	1257	3	1351	7	1447	.	1541	23	1637	.
1163	.	1259	.	1353	3	1449	3	1543	.	1639	11
1167	3	1261	13	1357	23	1451	.	1547	7	1641	3
1169	7	1263	3	1359	3	1453	.	1549	.	1643	31
1171	.	1267	7	1361	.	1457	31	1551	3	1647	3
1173	3	1269	3	1363	29	1459	.	1553	.	1649	17
1177	11	1271	31	1367	.	1461	3	1557	3	1651	13
1179	3	1273	19	1369	37	1463	7	1559	.	1653	3
1181	.	1277	.	1371	3	1467	3	1561	7	1657	.
1183	7	1279	.	1373	.	1469	13	1563	3	1659	3
1187	.	1281	3	1377	3	1471	.	1567	.	1661	11
1189	29	1283	.	1379	7	1473	3	1569	3	1663	.
1191	3	1287	3	1381	.	1477	7	1571	.	1667	.
1193	.	1289	.	1383	3	1479	3	1573	11	1669	.
1197	3	1291	.	1387	19	1481	.	1577	19	1671	3
1199	11	1293	3	1389	3	1483	.	1579	.	1673	7
1201	.	1297	.	1391	13	1487	.	1581	3	1677	3
1203	3	1299	3	1393	7	1489	.	1583	.	1679	23
1207	17	1301	.	1397	11	1491	3	1587	3	1681	41
1209	3	1303	.	1399	.	1493	.	1589	7	1683	3
1211	7	1307	.	1401	3	1497	3	1591	37	1687	7
1213	.	1309	7	1403	23	1499	.	1593	3	1689	3
1217	.	1311	3	1407	3	1501	19	1597	.	1691	19
1219	23	1313	13	1409	.	1503	3	1599	3	1693	.
1221	3	1317	3	1411	17	1507	11	1601	.	1697	.
1223	.	1319	.	1413	3	1509	3	1603	7	1699	.
1227	3	1321	.	1417	13	1511	.	1607	.	1701	3
1229	.	1323	3	1419	3	1513	17	1609	.	1703	13
1231	.	1327	.	1421	7	1517	37	1611	3	1707	3

Table C.1. Factors (Continued)

n	f	n	f	n	f	n	f	n	f	n	f
1709	.	1803	3	1899	3	1993	.	2089	.	2183	37
1711	29	1807	13	1901	.	1997	.	2091	3	2187	3
1713	3	1809	3	1903	11	1999	.	2093	7	2189	11
1717	17	1811	.	1907	.	2001	3	2097	3	2191	7
1719	3	1813	7	1909	23	2003	.	2099	.	2193	3
1721	.	1817	23	1911	3	2007	3	2101	11	2197	13
1723	.	1819	17	1913	.	2009	7	2103	3	2199	3
1727	11	1821	3	1917	3	2011	.	2107	7	2201	31
1729	7	1823	.	1919	19	2013	3	2109	3	2203	.
1731	3	1827	3	1921	17	2017	.	2111	.	2207	.
1733	.	1829	31	1923	3	2019	3	2113	.	2209	47
1737	3	1831	.	1927	41	2021	43	2117	29	2211	3
1739	37	1833	3	1929	3	2023	7	2119	13	2213	.
1741	.	1837	11	1931	.	2027	.	2121	3	2217	3
1743	3	1839	3	1933	.	2029	.	2123	11	2219	7
1747	.	1841	7	1937	13	2031	3	2127	3	2221	.
1749	3	1843	19	1939	7	2033	19	2129	.	2223	3
1751	17	1847	.	1941	3	2037	3	2131	.	2227	17
1753	.	1849	43	1943	29	2039	.	2133	3	2229	3
1757	7	1851	3	1947	3	2041	13	2137	.	2231	23
1759	.	1853	17	1949	.	2043	3	2139	3	2233	7
1761	3	1857	3	1951	.	2047	23	2141	.	2237	.
1763	41	1859	11	1953	3	2049	3	2143	.	2239	.
1767	3	1861	.	1957	19	2051	7	2147	19	2241	3
1769	29	1863	3	1959	3	2053	.	2149	7	2243	.
1771	7	1867	.	1961	37	2057	11	2151	3	2247	3
1773	3	1869	3	1963	13	2059	29	2153	.	2249	13
1777	.	1871	.	1967	7	2061	3	2157	3	2251	.
1779	3	1873	.	1969	11	2063	.	2159	17	2253	3
1781	13	1877	.	1971	3	2067	3	2161	.	2257	37
1783	.	1879	.	1973	.	2069	.	2163	3	2259	3
1787	.	1881	3	1977	3	2071	19	2167	11	2261	7
1789	.	1883	7	1979	.	2073	3	2169	3	2263	31
1791	3	1887	3	1981	7	2077	31	2171	13	2267	.
1793	11	1889	.	1983	3	2079	3	2173	41	2269	.
1797	3	1891	31	1987	.	2081	.	2177	7	2271	3
1799	7	1893	3	1989	3	2083	.	2179	.	2273	.
1801	.	1897	7	1991	11	2087	.	2181	3	2277	3

Table C.1. Factors (Continued)

n	f	n	f	n	f	n	f	n	f	n	f
2279	43	2373	3	2469	3	2563	11	2659	.	2753	.
2281	.	2377	.	2471	7	2567	17	2661	3	2757	3
2283	3	2379	3	2473	.	2569	7	2663	.	2759	31
2287	.	2381	.	2477	.	2571	3	2667	3	2761	11
2289	3	2383	.	2479	37	2573	31	2669	17	2763	3
2291	29	2387	7	2481	3	2577	3	2671	.	2767	.
2293	.	2389	.	2483	13	2579	.	2673	3	2769	3
2297	.	2391	3	2487	3	2581	29	2677	.	2771	17
2299	11	2393	.	2489	19	2583	3	2679	3	2773	47
2301	3	2397	3	2491	47	2587	13	2681	7	2777	.
2303	7	2399	.	2493	3	2589	3	2683	.	2779	7
2307	3	2401	7	2497	11	2591	.	2687	.	2781	3
2309	.	2403	3	2499	3	2593	.	2689	.	2783	11
2311	.	2407	29	2501	41	2597	7	2691	3	2787	3
2313	3	2409	3	2503	.	2599	23	2693	.	2789	.
2317	7	2411	.	2507	23	2601	3	2697	3	2791	.
2319	3	2413	19	2509	13	2603	19	2699	.	2793	3
2321	11	2417	.	2511	3	2607	3	2701	37	2797	.
2323	23	2419	41	2513	7	2609	.	2703	3	2799	3
2327	13	2421	3	2517	3	2611	7	2707	.	2801	.
2329	17	2423	.	2519	11	2613	3	2709	3	2803	.
2331	3	2427	3	2521	.	2617	.	2711	.	2807	7
2333	.	2429	7	2523	3	2619	3	2713	.	2809	53
2337	3	2431	11	2527	7	2621	.	2717	11	2811	3
2339	.	2433	3	2529	3	2623	43	2719	.	2813	29
2341	.	2437	.	2531	.	2627	37	2721	3	2817	3
2343	3	2439	3	2533	17	2629	11	2723	7	2819	.
2347	.	2441	.	2537	43	2631	3	2727	3	2821	7
2349	3	2443	7	2539	.	2633	.	2729	.	2823	3
2351	.	2447	.	2541	3	2637	3	2731	.	2827	11
2353	13	2449	31	2543	.	2639	7	2733	3	2829	3
2357	.	2451	3	2547	3	2641	19	2737	7	2831	19
2359	7	2453	11	2549	.	2643	3	2739	3	2833	.
2361	3	2457	3	2551	.	2647	.	2741	.	2837	.
2363	17	2459	.	2553	3	2649	3	2743	13	2839	17
2367	3	2461	23	2557	.	2651	11	2747	41	2841	3
2369	23	2463	3	2559	3	2653	7	2749	.	2843	.
2371	.	2467	.	2561	13	2657	.	2751	3	2847	3

Table C.1. Factors (Continued)

n	f	n	f	n	f	n	f	n	f	n	f
2849	7	2943	3	3039	3	3133	13	3229	.	3323	.
2851	.	2947	7	3041	.	3137	.	3231	3	3327	3
2853	3	2949	3	3043	17	3139	43	3233	53	3329	.
2857	.	2951	13	3047	11	3141	3	3237	3	3331	.
2859	3	2953	.	3049	.	3143	7	3239	41	3333	3
2861	.	2957	.	3051	3	3147	3	3241	7	3337	47
2863	7	2959	11	3053	43	3149	47	3243	3	3339	3
2867	47	2961	3	3057	3	3151	23	3247	17	3341	13
2869	19	2963	.	3059	7	3153	3	3249	3	3343	.
2871	3	2967	3	3061	.	3157	7	3251	.	3347	.
2873	13	2969	.	3063	3	3159	3	3253	.	3349	17
2877	3	2971	.	3067	.	3161	29	3257	.	3351	3
2879	.	2973	3	3069	3	3163	.	3259	.	3353	7
2881	43	2977	13	3071	37	3167	.	3261	3	3357	3
2883	3	2979	3	3073	7	3169	.	3263	13	3359	.
2887	.	2981	11	3077	17	3171	3	3267	3	3361	.
2889	3	2983	19	3079	.	3173	19	3269	7	3363	3
2891	7	2987	29	3081	3	3177	3	3271	.	3367	7
2893	11	2989	7	3083	.	3179	11	3273	3	3369	3
2897	.	2991	3	3087	3	3181	.	3277	29	3371	.
2899	13	2993	41	3089	.	3183	3	3279	3	3373	.
2901	3	2997	3	3091	11	3187	.	3281	17	3377	11
2903	.	2999	.	3093	3	3189	3	3283	7	3379	31
2907	3	3001	.	3097	19	3191	.	3287	19	3381	3
2909	.	3003	3	3099	3	3193	31	3289	11	3383	17
2911	41	3007	31	3101	7	3197	23	3291	3	3387	3
2913	3	3009	3	3103	29	3199	7	3293	37	3389	.
2917	.	3011	.	3107	13	3201	3	3297	3	3391	.
2919	3	3013	23	3109	.	3203	.	3299	.	3393	3
2921	23	3017	7	3111	3	3207	3	3301	.	3397	43
2923	37	3019	.	3113	11	3209	.	3303	3	3399	3
2927	.	3021	3	3117	3	3211	13	3307	.	3401	19
2929	29	3023	.	3119	.	3213	3	3309	3	3403	41
2931	3	3027	3	3121	.	3217	.	3311	7	3407	.
2933	7	3029	13	3123	3	3219	3	3313	.	3409	7
2937	3	3031	7	3127	53	3221	.	3317	31	3411	3
2939	.	3033	3	3129	3	3223	11	3319	.	3413	.
2941	17	3037	.	3131	31	3227	7	3321	3	3417	3

Table C.1. Factors (Continued)

n	f	n	f	n	f	n	f	n	f	n	f
3419	13	3513	3	3609	3	3703	7	3799	29	3893	17
3421	11	3517	.	3611	23	3707	11	3801	3	3897	3
3423	3	3519	3	3613	.	3709	.	3803	.	3899	7
3427	23	3521	7	3617	.	3711	3	3807	3	3901	47
3429	3	3523	13	3619	7	3713	47	3809	13	3903	3
3431	47	3527	.	3621	3	3717	3	3811	37	3907	.
3433	.	3529	.	3623	.	3719	.	3813	3	3909	3
3437	7	3531	3	3627	3	3721	61	3817	11	3911	.
3439	19	3533	.	3629	19	3723	3	3819	3	3913	7
3441	3	3537	3	3631	.	3727	.	3821	.	3917	.
3443	11	3539	.	3633	3	3729	3	3823	.	3919	.
3447	3	3541	.	3637	.	3731	7	3827	43	3921	3
3449	.	3543	3	3639	3	3733	.	3829	7	3923	.
3451	7	3547	.	3641	11	3737	37	3831	3	3927	3
3453	3	3549	3	3643	.	3739	.	3833	.	3929	.
3457	.	3551	53	3647	7	3741	3	3837	3	3931	.
3459	3	3553	11	3649	41	3743	19	3839	11	3933	3
3461	.	3557	.	3651	3	3747	3	3841	23	3937	31
3463	.	3559	.	3653	13	3749	23	3843	3	3939	3
3467	.	3561	3	3657	3	3751	11	3847	.	3941	7
3469	.	3563	7	3659	.	3753	3	3849	3	3943	.
3471	3	3567	3	3661	7	3757	13	3851	.	3947	.
3473	23	3569	43	3663	3	3759	3	3853	.	3949	11
3477	3	3571	.	3667	19	3761	.	3857	7	3951	3
3479	7	3573	3	3669	3	3763	53	3859	17	3953	59
3481	59	3577	7	3671	.	3767	.	3861	3	3957	3
3483	3	3579	3	3673	.	3769	.	3863	.	3959	37
3487	11	3581	.	3677	.	3771	3	3867	3	3961	17
3489	3	3583	.	3679	13	3773	7	3869	53	3963	3
3491	.	3587	17	3681	3	3777	3	3871	7	3967	.
3493	7	3589	37	3683	29	3779	.	3873	3	3969	3
3497	13	3591	3	3687	3	3781	19	3877	.	3971	11
3499	.	3593	.	3689	7	3783	3	3879	3	3973	29
3501	3	3597	3	3691	.	3787	7	3881	.	3977	41
3503	31	3599	59	3693	3	3789	3	3883	11	3979	23
3507	3	3601	13	3697	.	3791	17	3887	13	3981	3
3509	11	3603	3	3699	3	3793	.	3889	.	3983	7
3511	.	3607	.	3701	.	3797	.	3891	3	3987	3

Table C.1. Factors (Continued)

n	f	n	f	n	f	n	f	n	f	n	f
3989	.	4083	3	4179	3	4273	.	4369	17	4463	.
3991	13	4087	61	4181	37	4277	7	4371	3	4467	3
3993	3	4089	3	4183	47	4279	11	4373	.	4469	41
3997	7	4091	.	4187	53	4281	3	4377	3	4471	17
3999	3	4093	.	4189	59	4283	.	4379	29	4473	3
4001	.	4097	17	4191	3	4287	3	4381	13	4477	11
4003	.	4099	.	4193	7	4289	.	4383	3	4479	3
4007	.	4101	3	4197	3	4291	7	4387	41	4481	.
4009	19	4103	11	4199	13	4293	3	4389	3	4483	.
4011	3	4107	3	4201	.	4297	.	4391	.	4487	7
4013	.	4109	7	4203	3	4299	3	4393	23	4489	67
4017	3	4111	.	4207	7	4301	11	4397	.	4491	3
4019	.	4113	3	4209	3	4303	13	4399	53	4493	.
4021	.	4117	23	4211	.	4307	59	4401	3	4497	3
4023	3	4119	3	4213	11	4309	31	4403	7	4499	11
4027	.	4121	13	4217	.	4311	3	4407	3	4501	7
4029	3	4123	7	4219	.	4313	19	4409	.	4503	3
4031	29	4127	.	4221	3	4317	3	4411	11	4507	.
4033	37	4129	.	4223	41	4319	7	4413	3	4509	3
4037	11	4131	3	4227	3	4321	29	4417	7	4511	13
4039	7	4133	.	4229	.	4323	3	4419	3	4513	.
4041	3	4137	3	4231	.	4327	.	4421	.	4517	.
4043	13	4139	.	4233	3	4329	3	4423	.	4519	.
4047	3	4141	41	4237	19	4331	61	4427	19	4521	3
4049	.	4143	3	4239	3	4333	7	4429	43	4523	.
4051	.	4147	11	4241	.	4337	.	4431	3	4527	3
4053	3	4149	3	4243	.	4339	.	4433	11	4529	7
4057	.	4151	7	4247	31	4341	3	4437	3	4531	23
4059	3	4153	.	4249	7	4343	43	4439	23	4533	3
4061	31	4157	.	4251	3	4347	3	4441	.	4537	13
4063	17	4159	.	4253	.	4349	.	4443	3	4539	3
4067	7	4161	3	4257	3	4351	19	4447	.	4541	19
4069	13	4163	23	4259	.	4353	3	4449	3	4543	7
4071	3	4167	3	4261	.	4357	.	4451	.	4547	.
4073	.	4169	11	4263	3	4359	3	4453	61	4549	.
4077	3	4171	43	4267	17	4361	7	4457	.	4551	3
4079	.	4173	3	4269	3	4363	.	4459	7	4553	29
4081	7	4177	.	4271	.	4367	11	4461	3	4557	3

Table C.1. Factors (Continued)

n	f	n	f	n	f	n	f	n	f	n	f
4559	47	4653	3	4749	3	4843	29	4939	11	5033	7
4561	.	4657	.	4751	.	4847	37	4941	3	5037	3
4563	3	4659	3	4753	7	4849	13	4943	.	5039	.
4567	.	4661	59	4757	67	4851	3	4947	3	5041	71
4569	3	4663	.	4759	.	4853	23	4949	7	5043	3
4571	7	4667	13	4761	3	4857	3	4951	.	5047	7
4573	17	4669	7	4763	11	4859	43	4953	3	5049	3
4577	23	4671	3	4767	3	4861	.	4957	.	5051	.
4579	19	4673	.	4769	19	4863	3	4959	3	5053	31
4581	3	4677	3	4771	13	4867	31	4961	11	5057	13
4583	.	4679	.	4773	3	4869	3	4963	7	5059	.
4587	3	4681	31	4777	17	4871	.	4967	.	5061	3
4589	13	4683	3	4779	3	4873	11	4969	.	5063	61
4591	.	4687	43	4781	7	4877	.	4971	3	5067	3
4593	3	4689	3	4783	.	4879	7	4973	.	5069	37
4597	.	4691	.	4787	.	4881	3	4977	3	5071	11
4599	3	4693	13	4789	.	4883	19	4979	13	5073	3
4601	43	4697	7	4791	3	4887	3	4981	17	5077	.
4603	.	4699	37	4793	.	4889	.	4983	3	5079	3
4607	17	4701	3	4797	3	4891	67	4987	.	5081	.
4609	11	4703	.	4799	.	4893	3	4989	3	5083	13
4611	3	4707	3	4801	.	4897	59	4991	7	5087	.
4613	7	4709	17	4803	3	4899	3	4993	.	5089	7
4617	3	4711	7	4807	11	4901	13	4997	19	5091	3
4619	31	4713	3	4809	3	4903	.	4999	.	5093	11
4621	.	4717	53	4811	17	4907	7	5001	3	5097	3
4623	3	4719	3	4813	.	4909	.	5003	.	5099	.
4627	7	4721	.	4817	.	4911	3	5007	3	5101	.
4629	3	4723	.	4819	61	4913	17	5009	.	5103	3
4631	11	4727	29	4821	3	4917	3	5011	.	5107	.
4633	41	4729	.	4823	7	4919	.	5013	3	5109	3
4637	.	4731	3	4827	3	4921	7	5017	29	5111	19
4639	.	4733	.	4829	11	4923	3	5019	3	5113	.
4641	3	4737	3	4831	.	4927	13	5021	.	5117	7
4643	.	4739	7	4833	3	4929	3	5023	.	5119	.
4647	3	4741	11	4837	7	4931	.	5027	11	5121	3
4649	.	4743	3	4839	3	4933	.	5029	47	5123	47
4651	.	4747	47	4841	47	4937	.	5031	3	5127	3

Table C.2. Primitive Roots

The least primitive root g is given for each prime p with $p \leq 1439$.

p	g	p	g	p	g	p	g	p	g	p	g
2	1	167	5	389	2	631	3	883	2	1153	5
3	2	173	2	397	5	641	3	887	5	1163	5
5	2	179	2	401	3	643	11	907	2	1171	2
7	3	181	2	409	21	647	5	911	17	1181	7
11	2	191	19	419	2	653	2	919	7	1187	2
13	2	193	5	421	2	659	2	929	3	1193	3
17	3	197	2	431	7	661	2	937	5	1201	11
19	2	199	3	433	5	673	5	941	2	1213	2
23	5	211	2	439	15	677	2	947	2	1217	3
29	2	223	3	443	2	683	5	953	3	1223	5
31	3	227	2	449	3	691	3	967	5	1229	2
37	2	229	6	457	13	701	2	971	6	1231	3
41	6	233	3	461	2	709	2	977	3	1237	2
43	3	239	7	463	3	719	11	983	5	1249	7
47	5	241	7	467	2	727	5	991	6	1259	2
53	2	251	6	479	13	733	6	997	7	1277	2
59	2	257	3	487	3	739	3	1009	11	1279	3
61	2	263	5	491	2	743	5	1013	3	1283	2
67	2	269	2	499	7	751	3	1019	2	1289	6
71	7	271	6	503	5	757	2	1021	10	1291	2
73	5	277	5	509	2	761	6	1031	14	1297	10
79	3	281	3	521	3	769	11	1033	5	1301	2
83	2	283	3	523	2	773	2	1039	3	1303	6
89	3	293	2	541	2	787	2	1049	3	1307	2
97	5	307	5	547	2	797	2	1051	7	1319	13
101	2	311	17	557	2	809	3	1061	2	1321	13
103	5	313	10	563	2	811	3	1063	3	1327	3
107	2	317	2	569	3	821	2	1069	6	1361	3
109	6	331	3	571	3	823	3	1087	3	1367	5
113	3	337	10	577	5	827	2	1091	2	1373	2
127	3	347	2	587	2	829	2	1093	5	1381	2
131	2	349	2	593	3	839	11	1097	3	1399	13
137	3	353	3	599	7	853	2	1103	5	1409	3
139	2	359	7	601	7	857	3	1109	2	1423	3
149	2	367	6	607	3	859	2	1117	2	1427	2
151	6	373	2	613	2	863	5	1123	2	1429	6
157	5	379	2	617	3	877	2	1129	11	1433	3
163	2	383	5	619	2	881	3	1151	17	1439	7

Table C.3. Indices

$n = \left[\left[g^{index}\right]\right]_p$, where g is the least primitive root of p.

	Prime p											
n	2	3	5	7	11	13	17	19	23	29	31	37
1	1	2	4	6	10	12	16	18	22	28	30	36
2	.	1	1	2	1	1	14	1	2	1	24	1
3	.	.	3	1	8	4	1	13	16	5	1	26
4	.	.	2	4	2	2	12	2	4	2	18	2
5	.	.	.	5	4	9	5	16	1	22	20	23
6	.	.	.	3	9	5	15	14	18	6	25	27
7	.	.	.	.	7	11	11	6	19	12	28	32
8	.	.	.	.	3	3	10	3	6	3	12	3
9	.	.	.	.	6	8	2	8	10	10	2	16
10	.	.	.	.	5	10	3	17	3	23	14	24
11	.	.	.	.	.	7	7	12	9	25	23	30
12	.	.	.	.	.	6	13	15	20	7	19	28
13	.	.	.	.	.	.	4	5	14	18	11	11
14	.	.	.	.	.	.	9	7	21	13	22	33
15	.	.	.	.	.	.	6	11	17	27	21	13
16	.	.	.	.	.	.	8	4	8	4	6	4
17	.	.	.	.	.	.	.	10	7	21	7	7
18	.	.	.	.	.	.	.	9	12	11	26	17
19	.	.	.	.	.	.	.	.	15	9	4	35
20	.	.	.	.	.	.	.	.	5	24	8	25
21	.	.	.	.	.	.	.	.	13	17	29	22
22	.	.	.	.	.	.	.	.	11	26	17	31
23	.	.	.	.	.	.	.	.	.	20	27	15
24	.	.	.	.	.	.	.	.	.	8	13	29
25	.	.	.	.	.	.	.	.	.	16	10	10
26	.	.	.	.	.	.	.	.	.	19	5	12
27	.	.	.	.	.	.	.	.	.	15	3	6
28	.	.	.	.	.	.	.	.	.	14	16	34
29	.	.	.	.	.	.	.	.	.	.	9	21
30	.	.	.	.	.	.	.	.	.	.	15	14
31	.	.	.	.	.	.	.	.	.	.	.	9
32	.	.	.	.	.	.	.	.	.	.	.	5
33	.	.	.	.	.	.	.	.	.	.	.	20
34	.	.	.	.	.	.	.	.	.	.	.	8
35	.	.	.	.	.	.	.	.	.	.	.	19
36	.	.	.	.	.	.	.	.	.	.	.	18

Table C.3. Indices (Continued)

n	\multicolumn{13}{c}{Prime p}
	41	43	47	53	59	61	67	71	73	79	83	89	97
1	40	42	46	52	58	60	66	70	72	78	82	88	96
2	26	27	18	1	1	1	1	6	8	4	1	16	34
3	15	1	20	17	50	6	39	26	6	1	72	1	70
4	12	12	36	2	2	2	2	12	16	8	2	32	68
5	22	25	1	47	6	22	15	28	1	62	27	70	1
6	1	28	38	18	51	7	40	32	14	5	73	17	8
7	39	35	32	14	18	49	23	1	33	53	8	81	31
8	38	39	8	3	3	3	3	18	24	12	3	48	6
9	30	2	40	34	42	12	12	52	12	2	62	2	44
10	8	10	19	48	7	23	16	34	9	66	28	86	35
11	3	30	7	6	25	15	59	31	55	68	24	84	86
12	27	13	10	19	52	8	41	38	22	9	74	33	42
13	31	32	11	24	45	40	19	39	59	34	77	23	25
14	25	20	4	15	19	50	24	7	41	57	9	9	65
15	37	26	21	12	56	28	54	54	7	63	17	71	71
16	24	24	26	4	4	4	4	24	32	16	4	64	40
17	33	38	16	10	40	47	64	49	21	21	56	6	89
18	16	29	12	35	43	13	13	58	20	6	63	18	78
19	9	19	45	37	38	26	10	16	62	32	47	35	81
20	34	37	37	49	8	24	17	40	17	70	29	14	69
21	14	36	6	31	10	55	62	27	39	54	80	82	5
22	29	15	25	7	26	16	60	37	63	72	25	12	24
23	36	16	5	39	15	57	28	15	46	26	60	57	77
24	13	40	28	20	53	9	42	44	30	13	75	49	76
25	4	8	2	42	12	44	30	56	2	46	54	52	2
26	17	17	29	25	46	41	20	45	67	38	78	39	59
27	5	3	14	51	34	18	51	8	18	3	52	3	18
28	11	5	22	16	20	51	25	13	49	61	10	25	3
29	7	41	35	46	28	35	44	68	35	11	12	59	13
30	23	11	39	13	57	29	55	60	15	67	18	87	9
31	28	34	3	33	49	59	47	11	11	56	38	31	46
32	10	9	44	5	5	5	5	30	40	20	5	80	74

Table C.3. Indices (Continued)

n	41	43	47	53	59	61	67	71	73	79	83	89	97
						Prime p							
33	18	31	27	23	17	21	32	57	61	69	14	85	60
34	19	23	34	11	41	48	65	55	29	25	57	22	27
35	21	18	33	9	24	11	38	29	34	37	35	63	32
36	2	14	30	36	44	14	14	64	28	10	64	34	16
37	32	7	42	30	55	39	22	20	64	19	20	11	91
38	35	4	17	38	39	27	11	22	70	36	48	51	19
39	6	33	31	41	37	46	58	65	65	35	67	24	95
40	20	22	9	50	9	25	18	46	25	74	30	30	7
41	.	6	15	45	14	54	53	25	4	75	40	21	85
42	.	21	24	32	11	56	63	33	47	58	81	10	39
43	.	.	13	22	33	43	9	48	51	49	71	29	4
44	.	.	43	8	27	17	61	43	71	76	26	28	58
45	.	.	41	29	48	34	27	10	13	64	7	72	45
46	.	.	23	40	16	58	29	21	54	30	61	73	15
47	.	.	.	44	23	20	50	9	31	59	23	54	84
48	.	.	.	21	54	10	43	50	38	17	76	65	14
49	.	.	.	28	36	38	46	2	66	28	16	74	62
50	.	.	.	43	13	45	31	62	10	50	55	68	36
51	.	.	.	27	32	53	37	5	27	22	46	7	63
52	.	.	.	26	47	42	21	51	3	42	79	55	93
53	.	.	.	.	22	33	57	23	53	77	59	78	10
54	.	.	.	.	35	19	52	14	26	7	53	19	52
55	.	.	.	.	31	37	8	59	56	52	51	66	87
56	.	.	.	.	21	52	26	19	57	65	11	41	37
57	.	.	.	.	30	32	49	42	68	33	37	36	55
58	.	.	.	.	29	36	45	4	43	15	13	75	47
59	.	.	.	.	.	31	36	3	5	31	34	43	67
60	.	.	.	.	.	30	56	66	23	71	19	15	43
61	.	.	.	.	.	.	7	69	58	45	66	69	64
62	.	.	.	.	.	.	48	17	19	60	39	47	80
63	.	.	.	.	.	.	35	53	45	55	70	83	75
64	.	.	.	.	.	.	6	36	48	24	6	8	12

Table C.3. Indices (Continued)

n	\multicolumn{7}{c}{Prime p}						
	67	71	73	79	83	89	97
65	34	67	60	18	22	5	26
66	33	63	69	73	15	13	94
67	.	47	50	48	45	56	57
68	.	61	37	29	58	38	61
69	.	41	52	27	50	58	51
70	.	35	42	41	36	79	66
71	.	.	44	51	33	62	11
72	.	.	36	14	65	50	50
73	.	.	.	44	69	20	28
74	.	.	.	23	21	27	29
75	.	.	.	47	44	53	72
76	.	.	.	40	49	67	53
77	.	.	.	43	32	77	21
78	.	.	.	39	68	40	33
79	.	.	.	.	43	42	30
80	.	.	.	.	31	46	41
81	.	.	.	.	42	4	88
82	.	.	.	.	41	37	23
83	.	.	.	.	.	61	17
84	.	.	.	.	.	26	73
85	.	.	.	.	.	76	90
86	.	.	.	.	.	45	38
87	.	.	.	.	.	60	83
88	.	.	.	.	.	44	92
89	.	.	.	.	.	.	54
90	.	.	.	.	.	.	79
91	.	.	.	.	.	.	56
92	.	.	.	.	.	.	49
93	.	.	.	.	.	.	20
94	.	.	.	.	.	.	22
95	.	.	.	.	.	.	82
96	.	.	.	.	.	.	48

Table C.3. Indices (Continued)

Index	Prime p											
	2	3	5	7	11	13	17	19	23	29	31	37
1	1	2	2	3	2	2	3	2	5	2	3	2
2	.	1	4	2	4	4	9	4	2	4	9	4
3	.	.	3	6	8	8	10	8	10	8	27	8
4	.	.	1	4	5	3	13	16	4	16	19	16
5	.	.	.	5	10	6	5	13	20	3	26	32
6	.	.	.	1	9	12	15	7	8	6	16	27
7	.	.	.	.	7	11	11	14	17	12	17	17
8	.	.	.	.	3	9	16	9	16	24	20	34
9	.	.	.	.	6	5	14	18	11	19	29	31
10	.	.	.	.	1	10	8	17	9	9	25	25
11	.	.	.	.	.	7	7	15	22	18	13	13
12	.	.	.	.	.	1	4	11	18	7	8	26
13	.	.	.	.	.	.	12	3	21	14	24	15
14	.	.	.	.	.	.	2	6	13	28	10	30
15	.	.	.	.	.	.	6	12	19	27	30	23
16	.	.	.	.	.	.	1	5	3	25	28	9
17	.	.	.	.	.	.	.	10	15	21	22	18
18	.	.	.	.	.	.	.	1	6	13	4	36
19	.	.	.	.	.	.	.	.	7	26	12	35
20	.	.	.	.	.	.	.	.	12	23	5	33
21	.	.	.	.	.	.	.	.	14	17	15	29
22	.	.	.	.	.	.	.	.	1	5	14	21
23	.	.	.	.	.	.	.	.	.	10	11	5
24	.	.	.	.	.	.	.	.	.	20	2	10
25	.	.	.	.	.	.	.	.	.	11	6	20
26	.	.	.	.	.	.	.	.	.	22	18	3
27	.	.	.	.	.	.	.	.	.	15	23	6
28	.	.	.	.	.	.	.	.	.	1	7	12
29	.	.	.	.	.	.	.	.	.	.	21	24
30	.	.	.	.	.	.	.	.	.	.	1	11
31	.	.	.	.	.	.	.	.	.	.	.	22
32	.	.	.	.	.	.	.	.	.	.	.	7
33	.	.	.	.	.	.	.	.	.	.	.	14
34	.	.	.	.	.	.	.	.	.	.	.	28
35	.	.	.	.	.	.	.	.	.	.	.	19
36	.	.	.	.	.	.	.	.	.	.	.	1

Table C.3. Indices (Continued)

Index	41	43	47	53	59	61	67	71	73	79	83	89	97
							Prime p						
1	6	3	5	2	2	2	2	7	5	3	2	3	5
2	36	9	25	4	4	4	4	49	25	9	4	9	25
3	11	27	31	8	8	8	8	59	52	27	8	27	28
4	25	38	14	16	16	16	16	58	41	2	16	81	43
5	27	28	23	32	32	32	32	51	59	6	32	65	21
6	39	41	21	11	5	3	64	2	3	18	64	17	8
7	29	37	11	22	10	6	61	14	15	54	45	51	40
8	10	25	8	44	20	12	55	27	2	4	7	64	6
9	19	32	40	35	40	24	43	47	10	12	14	14	30
10	32	10	12	17	21	48	19	45	50	36	28	42	53
11	28	30	13	34	42	35	38	31	31	29	56	37	71
12	4	4	18	15	25	9	9	4	9	8	29	22	64
13	24	12	43	30	50	18	18	28	45	24	58	66	29
14	21	36	27	7	41	36	36	54	6	72	33	20	48
15	3	22	41	14	23	11	5	23	30	58	66	60	46
16	18	23	17	28	46	22	10	19	4	16	49	2	36
17	26	26	38	3	33	44	20	62	20	48	15	6	83
18	33	35	2	6	7	27	40	8	27	65	30	18	27
19	34	19	10	12	14	54	13	56	62	37	60	54	38
20	40	14	3	24	28	47	26	37	18	32	37	73	93
21	35	42	15	48	56	33	52	46	17	17	74	41	77
22	5	40	28	43	53	5	37	38	12	51	65	34	94
23	30	34	46	33	47	10	7	53	60	74	47	13	82
24	16	16	42	13	35	20	14	16	8	64	11	39	22
25	14	5	22	26	11	40	28	41	40	34	22	28	13
26	2	15	16	52	22	19	56	3	54	23	44	84	65
27	12	2	33	51	44	38	45	21	51	69	5	74	34
28	31	6	24	49	29	15	23	5	36	49	10	44	73
29	22	18	26	45	58	30	46	35	34	68	20	43	74
30	9	11	36	37	57	60	25	32	24	46	40	40	79
31	13	33	39	21	55	59	50	11	47	59	80	31	7
32	37	13	7	42	51	57	33	6	16	19	77	4	35

Table C.3. Indices (Continued)

Index	Prime p												
	41	43	47	53	59	61	67	71	73	79	83	89	97
33	17	39	35	31	43	53	66	42	7	57	71	12	78
34	20	31	34	9	27	45	65	10	35	13	59	36	2
35	38	7	29	18	54	29	63	70	29	39	35	19	10
36	23	21	4	36	49	58	59	64	72	38	70	57	50
37	15	20	20	19	39	55	51	22	68	35	57	82	56
38	8	17	6	38	19	49	35	12	48	26	31	68	86
39	7	8	30	23	38	37	3	13	21	78	62	26	42
40	1	24	9	46	17	13	6	20	32	76	41	78	16
41	.	29	45	39	34	26	12	69	14	70	82	56	80
42	.	1	37	25	9	52	24	57	70	52	81	79	12
43	.	.	44	50	18	43	48	44	58	77	79	59	60
44	.	.	32	47	36	25	29	24	71	73	75	88	9
45	.	.	19	41	13	50	58	26	63	61	67	86	45
46	.	.	1	29	26	39	49	40	23	25	51	80	31
47	.	.	.	5	52	17	31	67	42	75	19	62	58
48	.	.	.	10	45	34	62	43	64	67	38	8	96
49	.	.	.	20	31	7	57	17	28	43	76	24	92
50	.	.	.	40	3	14	47	48	67	50	69	72	72
51	.	.	.	27	6	28	27	52	43	71	55	38	69
52	.	.	.	1	12	56	54	9	69	55	27	25	54
53	.	.	.	.	24	51	41	63	53	7	54	75	76
54	.	.	.	.	48	41	15	15	46	21	25	47	89
55	.	.	.	.	37	21	30	34	11	63	50	52	57
56	.	.	.	.	15	42	60	25	55	31	17	67	91
57	.	.	.	.	30	23	53	33	56	14	34	23	67
58	.	.	.	.	1	46	39	18	61	42	68	69	44
59	.	.	.	.	.	31	11	55	13	47	53	29	26
60	.	.	.	.	.	1	22	30	65	62	23	87	33
61	.	.	.	.	.	.	44	68	33	28	46	83	68
62	.	.	.	.	.	.	21	50	19	5	9	71	49
63	.	.	.	.	.	.	42	66	22	15	18	35	51
64	.	.	.	.	.	.	17	36	37	45	36	16	61

Table C.3. Indices (Continued)

Index	Prime p						
	67	71	73	79	83	89	97
65	34	39	39	56	72	48	14
66	1	60	49	10	61	55	70
67	.	65	26	30	39	76	59
68	.	29	57	11	78	50	4
69	.	61	66	33	73	61	20
70	.	1	38	20	63	5	3
71	.	.	44	60	43	15	15
72	.	.	1	22	3	45	75
73	.	.	.	66	6	46	84
74	.	.	.	40	12	49	32
75	.	.	.	41	24	58	63
76	.	.	.	44	48	85	24
77	.	.	.	53	13	77	23
78	.	.	.	1	26	53	18
79	.	.	.	.	52	70	90
80	.	.	.	.	21	32	62
81	.	.	.	.	42	7	19
82	.	.	.	.	1	21	95
83	.	.	.	.	.	63	87
84	.	.	.	.	.	11	47
85	.	.	.	.	.	33	41
86	.	.	.	.	.	10	11
87	.	.	.	.	.	30	55
88	.	.	.	.	.	1	81
89	.	.	.	.	.	.	17
90	.	.	.	.	.	.	85
91	.	.	.	.	.	.	37
92	.	.	.	.	.	.	88
93	.	.	.	.	.	.	52
94	.	.	.	.	.	.	66
95	.	.	.	.	.	.	39
96	.	.	.	.	.	.	1

Appendix D
ANSWERS

Answers are given for odd-numbered exercises requiring a numerical result.

Chapter 0 - Section 0.1

1. (a) $R^{-1} = \{(7,1),(6,4),(6,5),(8,2)\}$, $S^{-1} = \{(10,6),(11,6),(10,7),(13,8)\}$

(b) $S \circ R = \{(1,10),(4,10),(4,11),(5,10),(5,11),(2,13)\}$

(c) $S \circ S^{-1} = \{(10,10),(10,11),(11,10),(11,11),(10,10),(13,13)\}$, $S^{-1} \circ S = \{(6,6),(6,7),(6,6),(7,6),(7,7),(8,8)\}$

(d) $R^{-1} \circ S^{-1} = \{(10,4),(10,5),(11,4),(11,5),(10,1),(13,2)\}$

(e) $T \circ (S \circ R) = \{(1,\triangle),(4,\triangle),(5,\triangle)(2,\bigstar),(2,\bigcirc)\}$

(f) $T \circ S = \{(6,\triangle),(7,\triangle),(8,\bigstar),(8,\bigcirc)\}$

(g) $(T \circ S) \circ R = \{(1,\triangle),(4,\triangle),(5,\triangle)(2,\bigstar),(2,\bigcirc)\}$

5. (a) symmetric, antisymmetric, transitive

(b) reflexive, symmetric, antisymmetric, transitive

(c) reflexive, symmetric

(d) reflexive

(e) reflexive, symmetric, transitive

Chapter 0 - Section 0.3

1. (a) $(0,3)$, (b) $[1,2]$

3. (a) $(2,8)$, (b) $(3,6)$, (c) $(-\infty,3] \cup [6,\infty)$, (d) $(-\infty,2] \cup [8,\infty)$

Chapter 1 - Section 1.4

3. (a) $q = 9$, $r = 3$, (b) $q = 20$, $r = 2$,

(c) $q = 9$, $r = 0$, (d) $q = 0$, $r = 16$

5. (a) 75, (b) 54, (c) 11799, (d) 108953, (e) 179196

7. 7

21. (a) $83(8) + 17(-39) = 1$

(b) $361(7) + 418(-6) = 19$

(c) $25(-1) + 15(2) = 5$

(d) $81(1) + 9(-8) = 9$

(e) $216(-1) + 324(1) = 108$

23. (a) $x = -20 + (81/3)u = -20 + 27u$
$y = 6 - (24/3)u = 6 - 8u$

(b) $x = 15 + 14u$
$y = -78 - 73u$

(c) $x = 180 + 151u$
$y = -87 - 73u$

(d) $x = -55 + 38u$
$y = 22 - 15u$

(e) $x = 12 + 26u$
$y = -4 - 9u$

Chapter 1 - Section 1.5

1. 102400_5, 110110010011_{two}, 11202201_{three}, 6623_{eight}

3.

ten	three	five	eight	ten	three	five	eight
1	1	1	1	9	100	14	11
2	2	2	2	10	101	20	12
3	10	3	3	11	102	21	13
4	11	4	4	12	110	22	14
5	12	10	5	13	111	23	15
6	20	11	6	14	112	24	16
7	21	12	7	$\vdots$	$\vdots$	$\vdots$	$\vdots$
8	22	13	10				

5. (a) $256_{ten} = 100_{hex}$, $32767_{ten} = 7FFF_{hex}$, $7203_{ten} = 1C23_{hex}$,

(b) $7A6F_{hex} = 31343_{ten}$, $FF_{hex} = 255_{ten}$, $ABCD_{hex} = 43981_{ten}$

7. $[1 \cdot 2 \cdot 3 \cdot 4 \cdots (n-4)(n-3)(n-1)]_n \cdot (n-1)_n = 1111 \cdots 1$ $(n-1$ ones$)$

Chapter 1 - Section 1.6

1. $n = 1, 2, 4, 5, 8, 10, 20, 40$

3. $13x \equiv 39 \pmod{169}$

5. (a) 1, (b) 0, (c) 4, (d) 5, (e) 10

7. $5 + 6k$ $k = 0, 1, 2, 3, \ldots$

21. $5, 13, 21, 29, 37, 45, 53$

Chapter 1 - Section 1.7

3. $f^{-1}(n) = [[1976663449 \cdot n]]_{10^{10}}$

Chapter 1 - Section 1.8

3. $\begin{aligned} x_1 &= 181636770 & U_1 &= 0.0845812 \cdots \\ x_2 &= 1194931003 & U_2 &= 0.5564331 \cdots \\ x_3 &= 2085784324 & U_3 &= 0.9712690 \cdots \end{aligned}$

Chapter 1 - Section 1.9

3. (a) 256, (b) $-128 \le i \le 127$

5.

(a)	(b)	(c)	(a)	(b)	(c)
−128	−10000000	10000000	1	00000001	00000001
−127	−01111111	10000001	2	00000010	00000010
−126	−01111110	10000010	3	00000011	00000011
−3	−00000011	11111101	125	01111101	01111101
−2	−00000010	11111110	126	01111110	01111110
−1	−00000001	11111111	127	01111111	01111111
0	00000000	00000000			

9. $B(c) = [1, 0, 1, 0, 1, 0, 1, 1]$, $c = 10101011_{two}$

$T(B(a)) = -01011010_{two} = -90_{ten}$

$T(B(b)) = 00000101_{two} = 5_{ten}$

$T(B(c)) = -01010101_{two} = -85_{ten}$

$-85 \equiv -90 + 5 \pmod{256}$

Chapter 2 - Section 2.2

1. (a) $2^3 \cdot 3^3 \cdot 5$, (b) $7^2 \cdot 11$, (c) $5 \cdot 191$, (d) $11 \cdot 53$, (e) 349, (f) $71 \cdot 127$, (g) $2^3 \cdot 3^4 \cdot 7^2$

13. $32^2 = 45^2 - 1001$ not a prime.

Chapter 2 - Section 2.3

1. $35! + 1$ and $35! + 36$

Chapter 2 - Section 2.4

1.

$\oplus$	[0]	[1]	[2]	[3]
[0]	[0]	[1]	[2]	[3]
[1]	[1]	[2]	[3]	[0]
[2]	[2]	[3]	[0]	[1]
[3]	[3]	[0]	[1]	[2]

$\odot$	[0]	[1]	[2]	[3]
[0]	[0]	[0]	[0]	[0]
[1]	[0]	[1]	[2]	[3]
[2]	[0]	[2]	[0]	[2]
[3]	[0]	[3]	[2]	[1]

3. [2]

5. [3]

7. yes

11. no, [2] has no inverse.

Chapter 2 - Section 2.5

3. $7, 8, 11, 12, 13, 14, 15, 16$

5. K in order to hold K primes p.

K in order to hold $N_p(x)$ for K primes p.

K in order to hold $N_p(Y(j))$ for K primes p.

K in order to hold $[[2^{n-1}]]_{10}$ for K primes p.

4 counters for "looping" over primes and Y.

number of order $\approx 4K$.

Chapter 2 - Section 2.6

1. Using $x_0 = 2$ and $c = 1$, for $k = 5$ and $2k = 10$, $x_{10} = 30263$, $x_5 = 24827$. Thus $\gcd(x_5 - x_{10}, 35183) = 151$ and $35183 = 151 \cdot 233$.

3. Using $x_0 = 2$ and $c = 1$, for $k = 4$ and $2k = 8$, $x_8 = 6627$, $x_4 = 13201$. Thus $\gcd(x_4 - x_8, 14359) = 173$ and $14359 = 83 \cdot 173$.

Chapter 3 - Section 3.2

1. (a) $x \equiv 81 \pmod{300}$, (b) $x \equiv 815 \pmod{1260}$,

(c) $x \equiv 16a - 15b \pmod{240}$, (d) $x \equiv 1272 \pmod{2310}$

5. 1459 marbles; 97 rows of 15, 182 rows of 8, and 63 rows of 23.

7. (a) $x \equiv 21 \pmod{72}$, (b) $x \equiv 104 \pmod{180}$, (c) no solution

Chapter 3 - Section 3.3

1. (a) -41, (b) 8, (c) 113, (d) -112

3. (a) $\begin{bmatrix} -2 & 6 & 2 \\ 4 & 1 & 3 \\ 7 & 5 & -4 \end{bmatrix}$, (b) $\begin{bmatrix} 69 & 27 & -20 \\ 27 & 62 & -5 \\ -20 & -5 & 29 \end{bmatrix}$, (c) $\begin{bmatrix} 44 & 4 & 8 \\ 4 & 26 & 21 \\ 8 & 21 & 90 \end{bmatrix}$

5. (a) $\begin{bmatrix} 49 & 2 \\ 46 & -32 \end{bmatrix}$, (b) $\begin{bmatrix} 3 & -8 & 7 \\ 21 & -36 & 69 \\ 10 & 0 & 50 \end{bmatrix}$, (c) $\begin{bmatrix} 1 & 2 \\ 0 & -4 \\ 5 & 6 \end{bmatrix}$,

(d) $\begin{bmatrix} -1 & 3 & 10 \\ 2 & 9 & 0 \end{bmatrix}$, (e) $\begin{bmatrix} 3 & 21 & 10 \\ -8 & -36 & 0 \\ 7 & 69 & 50 \end{bmatrix}$, (f) $\begin{bmatrix} 3 & 21 & 10 \\ -8 & -36 & 0 \\ 7 & 69 & 50 \end{bmatrix}$

13. (a) $\begin{bmatrix} 3 & 2 \\ 2 & 1 \end{bmatrix}$, (b) $\begin{bmatrix} 2 & 0 & 1 \\ 0 & 3 & 4 \\ 1 & 3 & 3 \end{bmatrix}$

15. (a) $x \equiv 6 \pmod{15}$, $y \equiv 11 \pmod{15}$, (b) no solution,

(c)

x_1	x_2	x_3
7	1	0
13	1	3
10	1	9

(d)

x_1	x_2
11	2
29	11
11	20
29	29

Chapter 3 - Section 3.4

5. (a) $1, 3$, (b) 6, (c) 6, (d) $2, 4, 6$, (e) $20, 37, 39$,

(f) $37, 137, 167$, (g) $3, 13, 18, 33$

Chapter 3 - Section 3.5

1. $m = 4,\ n = 2$

7. 1080

Chapter 3 - Section 3.6

9.

(a)

a	1	2	4	5	7	8
$ord_n a$	1	6	3	6	3	2

(b)

a	1	3	7	9	11	13	17	19
$ord_n a$	1	4	4	2	2	4	4	2

(c)

a	1	2	4	5	7	8	10	11	13
$ord_n a$	1	18	9	18	9	6	3	18	9

a	14	16	17	19	20	22	23	25	26
$ord_n a$	18	9	6	3	18	9	18	9	2

11. 1

13. (a) 7, (b) 376, (c) 68, (d) 961, (e) 535044134

Chapter 3 - Section 3.7

3. (a) 3 and 5 for 7^2 and for $2 \cdot 7^2$

(b) 2 and 3 for 19^k, $2 + 19^k$ and 3 for $2 \cdot 19^k$

(c) 2 and 5 for 3^2, 5 and 11 for 3^3, 5 and 11 for $2 \cdot 3^2$

(d) 2 and 3 for 5^4, and 627 and 3 for $2 \cdot 5^4$

Chapter 3 - Section 3.8

3. (a) $x = 5$, (b) $x = 1, 5, 9$, (c) no solution

5. see Tables

7. (a) 12, (b) 16, 35, 54, (c) no solution, (d) 12, 13, (e) 9, 14, (f) 4, 15

Chapter 3 - Section 3.9

1. Equation $x^2 \equiv 0 \pmod 2$ has solution $x \equiv 0 \pmod 2$.

Equation $x^2 + 1 \equiv 0 \pmod 2$ has solution $x \equiv 1 \pmod 2$.

Equation $x^2 + x \equiv 0 \pmod{2}$ has solutions $x \equiv 1 \pmod{2}$ and $x \equiv 0 \pmod{2}$.
Equation $x^2 + x + 1 \equiv 0 \pmod{2}$ has no solution.

3. Quadratic residues are $1, 4, 5, 6, 7, 9, 11, 16$, and 17.

Quadratic nonresidues are $2, 3, 8, 10, 12, 13, 14, 15$, and 18.

7. $(a) - 1$, $(b)\ 1$, $(c)\ 1$, $(d)\ 1$

15. 3 and 2

17. $p \equiv \pm(21 + 8t) \pmod{28}$ where t is a quadratic residue of 7, i.e., $t = 1$, 2, or 4.

Chapter 3 - Section 3.10

1. $(a)\ 1$, $(b) - 1$, $(c)\ 1$

7. $(a) - 1$, $(b) - 1$, $(c)\ 1$

Chapter 3 - Section 3.11

1.

$$
\begin{bmatrix}
1 & 1 & 1 & 1 \\
1 & -1 & 1 & -1 \\
1 & -1 & -1 & 1 \\
1 & 1 & -1 & -1
\end{bmatrix}
\qquad
\begin{bmatrix}
1 & 1 & 1 & 1 & 1 & 1 & 1 & 1 \\
1 & -1 & 1 & 1 & -1 & 1 & -1 & -1 \\
1 & -1 & -1 & 1 & 1 & -1 & 1 & -1 \\
1 & -1 & -1 & -1 & 1 & 1 & -1 & 1 \\
1 & 1 & -1 & -1 & -1 & 1 & 1 & -1 \\
1 & -1 & 1 & -1 & -1 & -1 & 1 & 1 \\
1 & 1 & -1 & 1 & -1 & -1 & -1 & 1 \\
1 & 1 & 1 & -1 & 1 & -1 & -1 & -1
\end{bmatrix}
$$

5. If U_m is an $n \times n$ U matrix, then the desired matrices are

$(1/2)U_4$, $(1/\sqrt{8})U_8$, and $(1/\sqrt{12})U_{12}$.

9. $\{b, d, e, f, j\}$, $\{c, e, f, g, k\}$, $\{a, d, f, g, h\}$, $\{b, e, g, h, i\}$, $\{c, f, h, i, j\}$, $\{d, g, i, j, k\}$, $\{a, e, h, j, k\}$, $\{a, b, f, i, k\}$, $\{a, b, c, g, j\}$, $\{b, c, d, h, k\}$, $\{a, c, d, e, i\}$

11. $(a)\ \{a, c, e, g\}$, $\{b, c, f, g\}$, $\{a, b, e, f\}$, $\{d, e, f, g\}$, $\{a, c, d, f\}$, $\{b, c, d, e\}$, $\{a, b, d, g\}$

Chapter 3 - Section 3.12

1. 534600000

3.

p	$\left[\left[7^{\frac{m-1}{p}}\right]\right]_m$	p	$\left[\left[7^{\frac{m-1}{p}}\right]\right]_m$
2	2147483646	31	512
3	1513477735	151	535044134
7	1205362885	331	1761855083
11	1969212174		

5. $\left[\left[2^{\frac{m-1}{331}}\right]\right]_m = 1,\ \left[\left[3^{\frac{m-1}{3}}\right]\right]_m = 1,\ \left[\left[5^{\frac{m-1}{11}}\right]\right]_m = 1,\ \left[\left[6^{\frac{m-1}{3}}\right]\right]_m = 1$

Chapter 3 - Section 3.13

5. $(a)\ D = 12,\ C = 2231403840,\ (b)\ D + 60,\ C = 1607985850884$

Chapter 3 - Section 3.14

1. 11

3. 112

Chapter 4 - Section 4.2

1. (a) 2, (b) 42, (c) 1682, (d) 40

3. (a) 12, (b) 2028, (c) 1481116, (d) 600

9. $n = 2^j q_1 q_2 \cdots q_s$, where the $q_i = 2^{r_i} + 1$ are distinct primes and $k = (j - 1) + \sum_{i=1}^{s} r_i$.

17. n such that if the prime p is a factor of n then so is $p - 1$.

23. (a) 53, (b) 5003

Chapter 4 - Section 4.3

1. $(a)\ -1,\ (b)\ 0,\ (c)\ 1,\ (d)\ -1,\ (e)\ -1$

3. $F(1) = 5,\ F(2) = 2,\ F(3) = 12,\ F(4) = 7,\ F(5) = 15,$
$F(6) = 7,\ F(7) = 8,\ F(8) = 7,\ F(9) = 12,\ F(10) = 12$

Chapter 4 - Section 4.4

7. 499222

9. (*c*)

box 1	box 2	box 3
w_1	w_2	$w_3 w_4$
w_1	w_3	$w_2 w_4$
w_1	w_4	$w_2 w_3$
w_2	w_3	$w_1 w_4$
w_2	w_4	$w_1 w_3$
w_3	w_4	$w_1 w_2$

(*d*)

box 1	box 2
w_1	$w_2 w_3 w_4$
w_2	$w_1 w_3 w_4$
w_3	$w_1 w_2 w_4$
w_4	$w_1 w_2 w_3$
$w_1 w_2$	$w_3 w_4$
$w_1 w_3$	$w_2 w_4$
$w_1 w_4$	$w_2 w_3$

Chapter 5 - Section 5.1

1. (*a*) $[3; 2, 1, 3]$, $[3; 2, 1, 2, 1]$, (*b*) $[0; 20, 1, 8, 1, 1, 2]$, $[0; 20, 1, 8, 1, 1, 1, 1]$,

(*c*) $[-1; 1, 6, 1, 3, 1, 5, 1, 1, 1, 2]$, $[-1; 1, 6, 1, 3, 1, 5, 1, 1, 1, 1, 1]$,

(*d*) $[0; 3, 2, 1, 3]$, $[0; 3, 2, 1, 2, 1]$, (*e*) $[-1; 1, 7, 1, 4]$, $[-1; 1, 7, 1, 3, 1]$

3. (*a*) $[2; 4, 4, 4, 4, 4, 2 + \sqrt{5}]$, (*b*) $[1; 2, 2, 2, 2, 2, 1 + \sqrt{2}]$,

(*c*) $[3; 7, 15, 1, 292, 1, 1.736 \cdots]$, (*d*) $[1; 1, 1, 1, 1, 1, (1 + \sqrt{5})/2]$

Chapter 5 - Section 5.2

3. (*a*)

k	p_k	q_k	$x - (p_k/q_k)$
0	2	1	$0.236 \cdots$
1	9	4	$-0.0139 \cdots$
2	38	17	$7.73 \cdots \times 10^{-4}$
3	161	72	$-4.31 \cdots \times 10^{-5}$
4	682	305	$2.40 \cdots \times 10^{-6}$
5	2889	1292	$-1.33 \cdots \times 10^{-7}$

(*b*)

k	p_k	q_k	$x - (p_k/q_k)$
0	1	1	$0.414 \cdots$
1	3	2	$-8.57 \cdots \times 10^{-2}$
2	7	5	$1.42 \cdots \times 10^{-2}$
3	17	12	$-2.45 \cdots \times 10^{-3}$
4	41	29	$4.20 \cdots \times 10^{-4}$
5	99	70	$-7.21 \cdots \times 10^{-5}$

(*c*)

k	p_k	q_k	$x - (p_k/q_k)$
0	3	1	$0.141 \cdots$
1	22	7	$-1.26 \cdots \times 10^{-3}$
2	333	106	$8.32 \cdots \times 10^{-5}$
3	355	113	$-2.66 \cdots \times 10^{-7}$
4	103993	33102	$5.77 \cdots \times 10^{-10}$
5	104348	33215	$-3.31 \cdots \times 10^{-10}$

(d)

k	p_k	q_k	$x - (p_k/q_k)$
0	1	1	$0.618\cdots$
1	2	1	$-0.381\cdots$
2	3	2	$0.118\cdots$
3	5	3	$-0.0486\cdots$
4	8	5	$0.0180\cdots$
5	13	8	$-0.00696\cdots$

7. $p_{-2} = 0$, $q_{-2} = 1$

Chapter 5 - Section 5.3

5. (a)

k	t_k	p_k	q_k
0	3	3	1
1	1	4	1
2	1	7	2
3	12	88	25

(b)

k	t_k	p_k	q_k
0	7	7	1
1	4	29	4
2	6	181	25
3	1	210	29
4	2	601	83
5	3	2013	278
6	1	2614	361
7	2	7241	1000

(c)

k	t_k	p_k	q_k
0	-9	-9	1
1	1	-8	1
2	8	-73	9
3	11	-811	100

Chapter 5 - Section 5.4

13. (a) $\dfrac{15 + \sqrt{257}}{8}$, (b) $\dfrac{-1 + \sqrt{77}}{2}$, (c) $\dfrac{2 + \sqrt{10}}{2}$, (d) $\dfrac{10 + \sqrt{37}}{3}$, (e) $\sqrt{47}$

15. (e) $x = \sqrt{3}$. For $k = 2 : 1, 3/2, 5/3$. For $k = 3 : 2, 7/4$. For $k = 4 : 5/3$, $12/7, 19/11$.

$x = \sqrt{5}$. For $k = 2 : 2, 11/5, 20/9, 29/13, 38/17$. For $k = 3 : 9/4, 47/21$, $85/38, 123/55, 161/72$. For $k = 4 : 38/17, 199/89, 360/161, 521/233, 682/305$.

(f) $179/57, 201/64, 223/71, 245/78, 267/85, 289/92, 311/99$. $333/106$ is the next convergent after $22/7$.

17. (a) $\overline{[12; 1, 5, 1]}$, (b) $\overline{[1; 5, 1, 12]}$

Chapter 5 - Section 5.5

1. (a) none, (b) $(\pm 2, \pm 3)$, (c) none, (d) none, (e) $(\pm 7, \pm 3)$, (f) $(\pm 13, \pm 4)$,

$(\pm 5, 0)$, (g) none, (h) $(\pm 5, \pm 2)$

3. $(5, 1)$, $(5585, 1549)$, $(7249325, 2010601)$

Chapter 5 - Section 5.6

1. $p_3/q_3 = 17/12$, $p_4/q_4 = 41/29$, $p_9/q_9 = 3363/2378$

3. $p_3/q_3 = 135/16 = (45/20) \cdot (90/24)$, so 90, 24, 45, and 20.

5.

k	actual $\lvert x - p_k/q_k \rvert$	at most $1/(q_k q_{k+1})$
0	$0.0027\cdots$	$0.0027\cdots$
1	$1.8\cdots \times 10^{-6}$	$1.8\cdots \times 10^{-6}$
2	$5.8\cdots \times 10^{-8}$	$6.4\cdots \times 10^{-8}$
3	$6.0\cdots \times 10^{-9}$	$7.8\cdots \times 10^{-9}$
4	$1.7\cdots \times 10^{-9}$	$1.7\cdots \times 10^{-9}$
5	$9.5\cdots \times 10^{-12}$	$9.6\cdots \times 10^{-12}$
6	$1.4\cdots \times 10^{-13}$	$2.0\cdots \times 10^{-13}$

Chapter 6 - Section 6.2

1. (a) 3, (b) 4, (c) -5, (d) -4

11. 2499

13. (a) $1, 4, 6, 4, 1$ (b) $1, 5, 10, 10, 5, 1$

15. (a) $x^6 + 6x^5y + 15x^4y^2 + 20x^3y^3 + 15x^2y^4 + 6xy^5 + y^6$

(b) $64a^6 + 576a^5b + 2160a^4b^2 + 4320a^3b^3 + 4860a^2b^4 + 2916ab^5 + 729b^6$

(c) $729a^6 - 2916a^5b + 4860a^4b^2 - 4320a^3b^3 + 2160a^2b^4 - 576ab^5 + 64b^6$

(d) $65536 + 655360w + 2867200w^2 + 7168000w^3 + 11200000w^4 + 11200000w^5 + 7000000w^6 + 2500000w^7 + 390625w^8$

23. $\begin{cases} 0 \text{ if } p = 2 \\ \sum_{j \geq 1} \left(\left\lfloor \dfrac{2n}{p^j} \right\rfloor - \left\lfloor \dfrac{n}{p^j} \right\rfloor \right) \text{ if } p > 2 \end{cases}$

Chapter 7 - Section 7.1

1. (a) $(13, -26, 13)$, (b) $(-2, 2, 1)$, (c) $(-300, 600, -600)$, (d) $(25, 25, -25, 25)$

5. (a) $z = u_1$, $y = 2 - 2u_1 + 3u_2$, $x = -1 + u_1 - 2u_2$

(b) $z = u_1$, $y = 2 - 2u_1 + 3u_2$, $x = 3 + u_1 - 2u_2$

(c) $z = u_1$, $y = u_2$, $x = 14 - 7u_1 - 5u_2$

(d) $z = u_1$, $y = u_2$, $x = u_1 + 2u_3$, $w = 8 + 2u_1 + u_2 - 3u_3$

Chapter 7 - Section 7.2

1. (a) $(112, 15, 113)$, (b) $(40, 9, 41)$

2. $(20, 21, 29)$, $(24, 7, 25)$, $(8, 15, 17)$, $(12, 5, 13)$, $(4, 3, 5)$

Chapter 7 - Section 7.3

1. (a) not possible, (b) $221 = 14^2 + 5^2 = 10^2 + 11^2$, (c) $1073 = 32^2 + 7^2$

3. (a) $400 = 101^2 - 99^2$, (b) $401 = 201^2 - 200^2$, (c) $75 = 38^2 - 37^2$

5. (a) not possible, (b) not possible

Chapter 7 - Section 7.4

1. (a) not positive definite, (b) positive definite, (c) positive definite
(d) not positive definite

3. $2x_1^2 + 4x_2^2 + 4x_3^2 - 2x_1x_2 + 2x_1x_3 + 2x_2x_3$

5.

(a) $\begin{bmatrix} 1 & 2 \\ 2 & 1 \end{bmatrix}$, (b) $\begin{bmatrix} 1 & -1 \\ -1 & 1 \end{bmatrix}$, (c) $\begin{bmatrix} 1 & 2 & -1 \\ 2 & 1 & 0 \\ -1 & 0 & 1 \end{bmatrix}$

7.

$Q_1(y_1, y_2, y_3) = \begin{bmatrix} y_1 & y_2 & y_3 \end{bmatrix} \begin{bmatrix} 8 & -4 & -2 \\ -4 & 16 & -6 \\ -2 & -6 & 4 \end{bmatrix} \begin{bmatrix} y_1 \\ y_2 \\ y_3 \end{bmatrix}$

9.

(a) $\begin{bmatrix} 2 & 3 \\ -1 & -1 \end{bmatrix}$, (b) $\begin{bmatrix} -3 & 2 \\ 1 & -1 \end{bmatrix}$, (c) $\begin{bmatrix} 3 & -2 \\ -1 & 1 \end{bmatrix}$

11.

$$(a) \begin{bmatrix} 1 & -3 & 4 \\ 0 & 4 & -23 \\ 0 & -1 & 6 \end{bmatrix}, \quad (b) \begin{bmatrix} 1 & 4 & 3 \\ 0 & -23 & -4 \\ 0 & 6 & 1 \end{bmatrix}, \quad (c) \begin{bmatrix} 1 & -4 & -3 \\ 0 & 23 & 4 \\ 0 & -6 & -1 \end{bmatrix}$$

Chapter 7 - Section 7.5

1.

$$(a)\ Q(x_1, x_2, x_3) = \begin{bmatrix} x_1 & x_2 & x_3 \end{bmatrix} \begin{bmatrix} 1 & 3 & 1 \\ 3 & 10 & 0 \\ 1 & 0 & 11 \end{bmatrix} \begin{bmatrix} x_1 \\ x_2 \\ x_3 \end{bmatrix}$$

$$\begin{bmatrix} x_1 \\ x_2 \\ x_3 \end{bmatrix} = \begin{bmatrix} 1 & -3 & -10 \\ 0 & 1 & 3 \\ 0 & 0 & 1 \end{bmatrix} \begin{bmatrix} z_1 \\ z_2 \\ z_3 \end{bmatrix}$$

For $\begin{bmatrix} x_1 & x_2 & x_3 \end{bmatrix} = \begin{bmatrix} 0 & 0 & 1 \end{bmatrix}$, $\begin{bmatrix} z_1 & z_2 & z_3 \end{bmatrix} = \begin{bmatrix} 1 & -3 & 1 \end{bmatrix}$ so that $11 = 3^2 + 1^2 + 1^2$.

$$(b)\ Q(x_1, x_2, x_3) = \begin{bmatrix} x_1 & x_2 & x_3 \end{bmatrix} \begin{bmatrix} 5 & 12 & 1 \\ 12 & 29 & 0 \\ 1 & 0 & 30 \end{bmatrix} \begin{bmatrix} x_1 \\ x_2 \\ x_3 \end{bmatrix}$$

$$\begin{bmatrix} x_1 \\ x_2 \\ x_3 \end{bmatrix} = \begin{bmatrix} -2 & 0 & 1 \\ 1 & 0 & 0 \\ 0 & 1 & 0 \end{bmatrix} \begin{bmatrix} y_1 \\ y_2 \\ y_3 \end{bmatrix}$$

$$\begin{bmatrix} y_1 \\ y_2 \\ y_3 \end{bmatrix} = \begin{bmatrix} 1 & -2 & -12 \\ 0 & 0 & -1 \\ 0 & 1 & 5 \end{bmatrix} \begin{bmatrix} z_1 \\ z_2 \\ z_3 \end{bmatrix}$$

For $\begin{bmatrix} x_1 & x_2 & x_3 \end{bmatrix} = \begin{bmatrix} 0 & 0 & 1 \end{bmatrix}$, $\begin{bmatrix} y_1 & y_2 & y_3 \end{bmatrix} = \begin{bmatrix} 0 & 1 & 0 \end{bmatrix}$ and $\begin{bmatrix} z_1 & z_2 & z_3 \end{bmatrix} = \begin{bmatrix} -2 & 5 & -1 \end{bmatrix}$ so that $30 = 2^2 + 1^2 + 5^2$.

$$(c)\ Q(x_1, x_2, x_3) = [x_1, x_2, x_3] \begin{bmatrix} 5 & 17 & 1 \\ 17 & 58 & 0 \\ 1 & 0 & 59 \end{bmatrix} \begin{bmatrix} x_1 \\ x_2 \\ x_3 \end{bmatrix}$$

$$\begin{bmatrix} x_1 \\ x_2 \\ x_3 \end{bmatrix} = \begin{bmatrix} -3 & 1 & 0 \\ 1 & 0 & 0 \\ 0 & 0 & -1 \end{bmatrix} \begin{bmatrix} y_1 \\ y_2 \\ y_3 \end{bmatrix}$$

$$\begin{bmatrix} y_1 \\ y_2 \\ y_3 \end{bmatrix} = \begin{bmatrix} 1 & -2 & -17 \\ 0 & 1 & 7 \\ 0 & 0 & 1 \end{bmatrix} \begin{bmatrix} z_1 \\ z_2 \\ z_3 \end{bmatrix}$$

For $\begin{bmatrix} x_1 & x_2 & x_3 \end{bmatrix} = \begin{bmatrix} 0 & 0 & 1 \end{bmatrix}$, $\begin{bmatrix} y_1 & y_2 & y_3 \end{bmatrix} = \begin{bmatrix} 0 & 0 & -1 \end{bmatrix}$ and $\begin{bmatrix} z_1 & z_2 & z_3 \end{bmatrix} = \begin{bmatrix} 3 & 7 & -1 \end{bmatrix}$ so that $59 = 3^2 + 7^2 + 1^2$.

(d) $Q(x_1, x_2, x_3) = [x_1, x_2, x_3] \begin{bmatrix} 78 & 189 & 1 \\ 189 & 458 & 0 \\ 1 & 0 & 153 \end{bmatrix} \begin{bmatrix} x_1 \\ x_2 \\ x_3 \end{bmatrix}$

$\begin{bmatrix} x_1 \\ x_2 \\ x_3 \end{bmatrix} = \begin{bmatrix} -160 & -1 & 0 \\ 66 & 0 & 65 \\ 1 & 0 & 1 \end{bmatrix} \begin{bmatrix} y_1 \\ y_2 \\ y_3 \end{bmatrix}$

$\begin{bmatrix} y_1 \\ y_2 \\ y_3 \end{bmatrix} = \begin{bmatrix} 1 & -1 & 6 \\ 0 & 315 & -1103 \\ 0 & 2 & -7 \end{bmatrix} \begin{bmatrix} z_1 \\ z_2 \\ z_3 \end{bmatrix}$

For $\begin{bmatrix} x_1 & x_2 & x_3 \end{bmatrix} = \begin{bmatrix} 0 & 0 & 1 \end{bmatrix}$, $\begin{bmatrix} y_1 & y_2 & y_3 \end{bmatrix} = \begin{bmatrix} -65 & 10400 & 66 \end{bmatrix}$

and $\begin{bmatrix} z_1 & z_2 & z_3 \end{bmatrix} = \begin{bmatrix} -7 & -2 & -10 \end{bmatrix}$ so that $153 = 7^2 + 2^2 + 10^2$.

(e) $Q(x_1, x_2, x_3) = [x_1, x_2, x_3] \begin{bmatrix} 5 & 13 & 1 \\ 13 & 34 & 0 \\ 1 & 0 & 35 \end{bmatrix} \begin{bmatrix} x_1 \\ x_2 \\ x_3 \end{bmatrix}$

$\begin{bmatrix} x_1 \\ x_2 \\ x_3 \end{bmatrix} = \begin{bmatrix} -3 & 0 & 1 \\ 1 & 0 & 0 \\ 0 & 1 & 0 \end{bmatrix} \begin{bmatrix} y_1 \\ y_2 \\ y_3 \end{bmatrix}$

$\begin{bmatrix} y_1 \\ y_2 \\ y_3 \end{bmatrix} = \begin{bmatrix} 1 & 2 & -13 \\ 0 & 0 & -1 \\ 0 & 1 & -5 \end{bmatrix} \begin{bmatrix} z_1 \\ z_2 \\ z_3 \end{bmatrix}$

For $\begin{bmatrix} x_1 & x_2 & x_3 \end{bmatrix} = \begin{bmatrix} 0 & 0 & 1 \end{bmatrix}$, $\begin{bmatrix} y_1 & y_2 & y_3 \end{bmatrix} = \begin{bmatrix} 0 & 1 & 0 \end{bmatrix}$ and

$\begin{bmatrix} z_1 & z_2 & z_3 \end{bmatrix} = \begin{bmatrix} -3 & -5 & -1 \end{bmatrix}$ so that $35 = 3^2 + 5^2 + 1^2$.

(f) $Q(x_1, x_2, x_3) = [x_1, x_2, x_3] \begin{bmatrix} 26 & 161 & 1 \\ 161 & 998 & 0 \\ 1 & 0 & 37 \end{bmatrix} \begin{bmatrix} x_1 \\ x_2 \\ x_3 \end{bmatrix}$

$\begin{bmatrix} x_1 \\ x_2 \\ x_3 \end{bmatrix} = \begin{bmatrix} -37 & 1 & 0 \\ 6 & 0 & 1 \\ 1 & 0 & 0 \end{bmatrix} \begin{bmatrix} y_1 \\ y_2 \\ y_3 \end{bmatrix}$

$\begin{bmatrix} y_1 \\ y_2 \\ y_3 \end{bmatrix} = \begin{bmatrix} 1 & -5 & -1 \\ 0 & 1 & -6 \\ 0 & 0 & 1 \end{bmatrix} \begin{bmatrix} z_1 \\ z_2 \\ z_3 \end{bmatrix}$

For $\begin{bmatrix} x_1 & x_2 & x_3 \end{bmatrix} = \begin{bmatrix} 0 & 0 & 1 \end{bmatrix}$, $\begin{bmatrix} y_1 & y_2 & y_3 \end{bmatrix} = \begin{bmatrix} 1 & 37 & -6 \end{bmatrix}$ and

$\begin{bmatrix} z_1 & z_2 & z_3 \end{bmatrix} = \begin{bmatrix} 0 & 1 & -6 \end{bmatrix}$ so that $37 = 0^2 + 1^2 + 6^2$.

(g) $Q(x_1, x_2, x_3) = [x_1, x_2, x_3] \begin{bmatrix} 2 & 5 & 1 \\ 5 & 13 & 0 \\ 1 & 0 & 14 \end{bmatrix} \begin{bmatrix} x_1 \\ x_2 \\ x_3 \end{bmatrix}$

$$\begin{bmatrix} x_1 \\ x_2 \\ x_3 \end{bmatrix} = \begin{bmatrix} -2 & 0 & 1 \\ 1 & 0 & 0 \\ 0 & 1 & 0 \end{bmatrix} \begin{bmatrix} y_1 \\ y_2 \\ y_3 \end{bmatrix}$$

$$\begin{bmatrix} y_1 \\ y_2 \\ y_3 \end{bmatrix} = \begin{bmatrix} 1 & -1 & -5 \\ 0 & 0 & -1 \\ 0 & 1 & 3 \end{bmatrix} \begin{bmatrix} z_1 \\ z_2 \\ z_3 \end{bmatrix}$$

For $\begin{bmatrix} x_1 & x_2 & x_3 \end{bmatrix} = \begin{bmatrix} 0 & 0 & 1 \end{bmatrix}$, $\begin{bmatrix} y_1 & y_2 & y_3 \end{bmatrix} = \begin{bmatrix} 0 & 1 & -6 \end{bmatrix}$ and $\begin{bmatrix} z_1 & z_2 & z_3 \end{bmatrix} = \begin{bmatrix} -2 & 3 & -1 \end{bmatrix}$ so that $14 = 2^2 + 3^2 + 1^2$.

3.

$$Q(x_1, x_2, x_3) = [x_1, x_2, x_3] \begin{bmatrix} 1 & 5 & 1 \\ 5 & 26 & 0 \\ 1 & 0 & 27 \end{bmatrix} \begin{bmatrix} x_1 \\ x_2 \\ x_3 \end{bmatrix}$$

$$\begin{bmatrix} x_1 \\ x_2 \\ x_3 \end{bmatrix} = \begin{bmatrix} 1 & -26 & 5 \\ 0 & 5 & -1 \\ 0 & 1 & 0 \end{bmatrix} \begin{bmatrix} z_1 \\ z_2 \\ z_3 \end{bmatrix}$$

For $\begin{bmatrix} x_1 & x_2 & x_3 \end{bmatrix} = \begin{bmatrix} 0 & 0 & 1 \end{bmatrix}$, $\begin{bmatrix} z_1 & z_2 & z_3 \end{bmatrix} = \begin{bmatrix} 1 & 1 & 5 \end{bmatrix}$ so that $27 = 1^2 + 1^2 + 5^2$. Previous method gives $27 = 1^2 + (-5)^2 + 1^2$, no change.

Chapter 7 - Section 7.6

1. $15 = 3^2 + 2^2 + 1^2 + 1^2$

3. $345 = (23)(15) = (3^2 + 3^2 + 2^2 + 1^2)(3^2 + 2^2 + 1^2 + 1^2) = 18^2 + 2^2 + 4^2 + 1^2$

5. $(a)\ 8^2 + 6^2 + 4^2 + 2^2$, $(b)\ 32^2 + 16^2 + 27^2 + 10^2$

9. $1, 2, 3, 5, 6, 8, 9, 11, 14, 17$

11. $13^2 = 5^2 + 12^2 = 4^2 + 3^2 + 12^2 = 9^2 + 4^2 + 6^2 + 6^2$

Chapter 7 - Section 7.7

1. $(1, 1, -1)$

Chapter 8 - Section 8.1

5. $[0]$

17. $< 3 >$

Chapter 8 - Section 8.4

7. no

11. all elements of the form $a + b\sqrt{5}$ where a and b are rational.

13. all elements of the form $a + b\sqrt{3}i$ where a and b are rational.

Appendix A - Section A.2

1. (a)

p	q	$\sim$	$(p$	$\rightarrow$	$\sim q)$	$\vee$	$(\sim p$	$\wedge$	$q)$
T	T	T	T	F	F	T	F	F	T
T	F	F	T	T	T	F	F	F	F
F	T	F	F	T	F	T	T	T	T
F	F	F	F	T	T	F	T	F	F

(b)

p	q	r	$(\sim p$	$\rightarrow$	$(q$	$\vee$	$\sim r))$	$\leftrightarrow$	$\sim$	$(p$	$\wedge$	$\sim q)$
T	T	T	F	T	T	T	F	T	T	T	F	F
T	T	F	F	T	T	T	T	T	T	T	F	F
T	F	T	F	T	F	F	F	F	F	T	T	T
T	F	F	F	T	F	T	T	F	F	T	T	T
F	T	T	T	T	T	T	F	T	T	F	F	F
F	T	F	T	T	T	T	T	T	T	F	F	F
F	F	T	T	F	F	F	F	F	T	F	F	T
F	F	F	T	T	F	T	T	T	T	F	F	T

Appendix A - Section A.3

1. (a)

p	q	r	$[(p$	$\rightarrow$	$q)$	$\wedge$	$(q$	$\rightarrow$	$r)]$	$\rightarrow$	$(p$	$\rightarrow$	$r)$
T	T	T	T	T	T	T	T	T	T	T	T	T	T
T	T	F	T	T	T	F	T	F	F	T	T	F	F
T	F	T	T	F	F	F	F	T	T	T	T	T	T
T	F	F	T	F	F	F	F	T	F	T	T	F	F
F	T	T	F	T	T	T	T	T	T	T	F	T	T
F	T	F	F	T	T	F	T	F	F	T	F	T	F
F	F	T	F	T	F	T	F	T	T	T	F	T	T
F	F	F	F	T	F	T	F	T	F	T	F	T	F

(b)

p	q	r	s	A: p	B: $(p \to$	$(r \lor s))$	C: $r \to q$	D: $s \to q$	E: $\to$	F: q
T	T	T	T	T	T	T	T	T	T	T
T	T	T	F	T	T	T	T	T	T	T
T	T	F	T	T	T	T	T	T	T	T
T	T	F	F	T	F	F	T	T	T	T
T	F	T	T	T	T	T	F	F	T	F
T	F	T	F	T	T	T	F	T	T	F
T	F	F	T	T	T	T	T	F	T	F
T	F	F	F	T	F	F	T	T	T	F
F	T	T	T	F	T	T	T	T	T	T
F	T	T	F	F	T	T	T	T	T	T
F	T	F	T	F	T	T	T	T	T	T
F	T	F	F	F	T	F	T	T	T	T
F	F	T	T	F	T	T	F	F	T	F
F	F	T	F	F	T	T	F	T	T	F
F	F	F	T	F	T	T	T	F	T	F
F	F	F	F	F	T	F	T	T	T	F

where column E gives the truth values for $(A \land B \land C \land D) \to F$.

(c)

w	r	$[\sim w$	$\to$	$(r \land \sim r)]$	$\to$	w
T	T	F	T	F	T	T
T	F	F	T	F	T	T
F	T	T	F	F	T	F
F	F	T	F	F	T	F

Appendix A - Section A.3

3. For an invalid argument the combination of premises that cause the conjugation of the premises to be false and the conclusion to be true may not occur when a component statement is replaced by a compound statement.

Appendix A - Section A.4

1. Note that the domain of discourse is the positive integers.

(a) true   (b) true   (c) true   (d) true   (e) false

3. (a)

(4)−Universal Instantiation

(5) − (4) and Specialization

(6)−Universal Generalization

(7) − (1), (3), and Universal Instantiation

(8) − (7) and Specialization

(9) − (3), (8), and Universal Generalization

(10)−Conjunction

(b)

(3) − (1) and Specialization

(4) − (1) and Specialization

(5) − (2), (3), and Universal Instantiation

(6) − (2), (4), and Universal Instantiation

(7) − (5), (6) and Conjunction

(8) − (2), (7), and Universal Generalization

Bibliography

[1] L. M. Adleman, "A Subexponential Algorithm for the Discrete Logarithm Problem with Applications to Cryptography," *Proc. 20th IEEE Found. Comput. Sci. Symp.* (1979), 55-60.

[2] W. R. Alford, A. Granville, and C. Pomerance, "There Are Infinitely Many Carmichael Numbers," *Ann. Math.*, Volume 140 (1994), 703-722.

[3] J. A. Anderson and E. F. Wilde, "A Generalization of a Mathematical Curiosity," *J. Recreational Math.*, to appear, 1996.

[4] G. E. Andrews, *Number Theory*, Dover, New York, 1994.

[5] E. T. Bell, *The Development of Mathematics*, 2nd ed., McGraw-Hill, New York, 1945.

[6] E. T. Bell, *Men of Mathematics*, Simon & Schuster, New York, 1965.

[7] C. Berge, *Principles of Combinatorics*, Academic Press, New York, 1971.

[8] R. S. Boyer and J. S. Moore, "A Fast String Searching Algorithm," *Commun. ACM*, Volume 20, No. 10 (1977), 762-772.

[9] R. P. Brent, "Irregularities in the Distribution of Primes and Twin Primes," *Math. Comput.*, Volume 29 (1975), 43-56.

[10] R. P. Brent and J. M. Pollard, "Factorization of the Eighth Fermat Number," *Math. of Comput.*, Volume 36, No. 154 (1981), 627-630.

[11] D. M. Burton, *Elementary Number Theory*, 3rd ed., Wm. C. Brown, Dubuque, Iowa, 1994.

[12] F. Cajori, *A History of Mathematics*, 2nd ed., Macmillan, New York, 1961.

[13] R. D. Carmichael, "On Composite Numbers P Which Satisfy the Fermat Congruence $a^{P-1} \equiv 1 \bmod P$," *Amer. Math. Monthly*, Volume 19, No. 2 (1912), 22-27.

[14] C. C. Chang and J. C. Shieh, "A Fast Algorithm for Constructing Reciprocal Hashing Functions," *Proc. Internat. Symp. New Directions Comput.* (1985), 232-236.

[15] C. C. Chang and J. C. Shieh, "Pairwise Relatively Prime Generating Polynomials and Their Applications," *Proc. Internat. Workshop Discrete Algorithms Complexity*, November (1989), 137-139.

[16] Nan-xian Chen, "Modified Möbius Inversion Formula and Its Application

to Physics," *Phys. Rev. Lett.*, Volume 64, No. 11 (1990), 1193-1195.

[17] L. W. Cohen and G. Ehrlich, *The Structure of the Real Number System*, Van Nostrand Reinhold, New York, 1963.

[18] H. Cohn, *Advanced Number Theory*, Dover, New York, 1980.

[19] D. Coppersmith, "Cryptography," *IBM J. Res. Develop.*, Volume 31, No. 2 (1987), 244-248.

[20] R. Crandall, J. Doenias, C. Norrie, and J. Young, "The Twenty-Second Fermat Number Is Composite," *Math. Comput.*, Volume 64 (1995), 863-868.

[21] T. Dantzig, *Number: The Language of Science*, 4th ed., Macmillan, New York, 1954.

[22] H. Davenport, *The Higher Arithmetic*, 6th ed., Cambridge University Press, New York, 1992.

[23] L. E. Dickson, *Studies in the Theory of Numbers*, Chelsea, New York, 1957.

[24] W. Diffie and M. E. Hellman, "New Directions in Cryptography," *IEEE Trans. Inform. Theory*, Volume IT-22, No. 6 (1976), 644-654.

[25] H. W. Eves, *In Mathematical Circles, Quadrants I and II*, Prindle, Weber & Schmidt, Boston, 1969.

[26] C. V. Eynden, *Elementary Number Theory*, McGraw-Hill, New York, 1987.

[27] G. S. Fishman and L. R. Moore, "A Statistical Evaluation of Multiplicative Congruential Random Number Generators with Modulus $2^{31} - 1$," *J. Amer. Statist. Assoc.*, Volume 77 (1982), 129-136.

[28] G. S. Fishman and L. R. Moore III, "An Exhaustive Analysis of Multiplicative Congruential Random Generators with Modulus $2^{31} - 1$," *SIAM J. Sci. Statist. Comput.*, Volume 7, No. 1 (1986), 24-45.

[29] M. Gardner, "Through the Looking Glass," by Lewis Carroll, from *The Annotated Alice*, Bramhall House, New York, 1960.

[30] C. W. Gear, *Computer Organization and Programming*, 2nd ed., McGraw-Hill, New York, 1974.

[31] J. Gilbert and L. Gilbert, *Elements of Modern Algebra*, 4th ed. PWS-KeNT, Boston, 1995.

[32] E. Grosswald, *Representations of Integers as Sums of Squares*, Springer-Verlag, New York, 1985.

[33] G. H. Hardy and E. M. Wright, *An Introduction to the Theory of Numbers*, 5th ed., Clarendon Press, Oxford, 1979.

[34] M. E. Hellman, "The Mathematics of Public-Key Cryptography," *Sci. Amer.*, Volume 241, August (1979), 146-157.

[35] F. S. Hillier and G. Lieberman, *Operations Research*, 2nd ed., Holden-Day, San Francisco, 1974.

[36] A. E. Ingham, *The Distribution of Prime Numbers*, Cambridge University Press, New York, 1992.

[37] G. Jaeschke, "Reciprocal Hashing: A Method for Generating Minimal Perfect Hashing Functions," *Commun. ACM*, Volume 24, No. 12 (1981), 829-833.

[38] D. Kahn, *The Codebreakers: The Story of Secret Writing*, Macmillan, 1968.

[39] R. M. Karp and M. O. Rabin, "Efficient Randomized Pattern-Matching Algorithms," *IBM J. Res. Develop.*, Volume 31, No. 2, March (1987), 249-260.

[40] A. Ya. Khinchin, *Continued Fractions*, The University of Chicago Press, Chicago, 1964.

[41] H. A. Klein, *The Science of Measurement: A Historical Survey*, Dover, New York, 1988.

[42] M. Kline, *Mathematical Thought from Ancient to Modern Times*, Oxford University Press, New York, 1972.

[43] D. E. Knuth, *The Art of Computer Programming*, 2nd ed., Volumes 2 and 3, Addison-Wesley, Reading, Mass., 1981 and 1973.

[44] D. E. Knuth, J. H. Morris, and V. R. Pratt, "Fast Pattern Matching in Strings," *SIAM J. Comput.*, Volume 6, No. 2, June (1977), 323-350.

[45] J. C. Lagarias, V. S. Miller, and A. M. Odlyzko, "Computing $\pi(x)$: The Meissel-Lehmer Method," *Math. Comput.*, Volume 44 (1985), 537-560.

[46] P. L'Ecuyer, "Efficient and Portable Combined Random Number Generators," *Commun. ACM*, Volume 31 (1988), 742-774.

[47] A. K. Lenstra and H. W. Lenstra, Jr., editors, *The Development of the Number Field Sieve*, Springer-Verlag, New York 1993.

[48] W. J. LeVeque, *Elementary Theory of Numbers*, Dover, New York, 1990.

[49] G. Marsaglia, "Random Numbers Fall Mainly in the Planes," *Proc. Nat. Acad. Sci. U.S.A.*, Volume 61 (1968), 25-28.

[50] C. H. Meyer and S. M. Matyas, *Cryptography: A New Dimension in Computer Data Security*, Wiley, New York, 1982.

[51] R. P. Millane, "A Product Form of the Möbius Transform," *Phys. Lett. A*, Volume 162 (1992), 213-214.

[52] R. P. Millane, "Möbius Transform Pairs," *J. Math. Phys.*, Volume 34 (1993), 875-877.

[53] O. Neugebauer, *The Exact Sciences in Antiquity*, 2nd ed., Dover, New York, 1969.

[54] O. Neugebauer and A. Sachs, *Mathematical Cuneiform Texts*, American Oriental Society, New Haven, Conn. 1945.

[55] J. R. Newman, *The World of Mathematics*, Simon & Schuster, New York, 1956.

[56] B. W. Ninham et al., "Möbius, Mellin, and Mathematical Physics," *Physica A*, Volume 186 (1992), 441-481.

[57] I. Niven and H. S. Zuckerman, *An Introduction to the Theory of Numbers*, 4th ed., Wiley, New York, 1980.

[58] J. M. H. Olmsted, *The Real Number System*, Appleton-Century-Crofts, New York, 1962.

[59] G. Ottewell, *The Astronomical Companion*, Astronomical Workshop, Furman University, Greenville, S.C., 1979.

[60] R. E. A. C. Paley, "On Orthogonal Matrices," *J. of Math. Phys.*, Volume 12 (1933), 311-320.

[61] B. K. Parady, J. R. Smith, and S. Zarantonello, "Largest Known Twin Primes," *Math. Comput.*, Volume 55 (1990), 381-382.

[62] S. K. Park and K. W. Miller, "Random Number Generators: Good Ones Are Hard to Find," *Commun. ACM*, Volume 31 (1988), 1192-1201.

[63] R. L. Plackett and J. P. Burman, "The Design of Optimum Multifactorial Experiments," *Biometrika*, Volume 33 (1946), 305-325.

[64] S. Pohlig and M. Hellman, "An Improved Algorithm for Computing Logarithms over $GF(p)$ and Its Cryptographic Significance," *IEEE Trans. Inform. Theory*, Vol. IT-24, No. 1 (1978), 106-110.

[65] J. M. Pollard, "Theorems on Factorization and Primality Testing," *Proc. Cambridge Philos. Soc.*, Volume 76 (1974), 521-528.

[66] J. M. Pollard. "A Monte Carlo Method for Factorization," *Bit*, (1975), 331-334.

[67] C. Pomerance, *Lecture Notes on Primality Testing and Factoring*, The Mathematical Association of America, Washington, D.C., 1984.

[68] J. Purdum, "Pattern Matching Alternatives: Theory vs. Practice," *Comput. Language*, November (1987), 33-44.

[69] M. Rabin, "Probabilistic Algorithm for Testing Primality," *J. Number Theory*, Volume 12 (1980), 128-138.

[70] F. Reid and K. Honeycutt, "A Digital Clock for Sidereal Time," *Sky & Telescope*, July (1976), 59-63.

[71] S. Y. Ren and J. D. Dow, "Generalized Möbius Transforms for Inverse Problems," *Phys. Lett. A*, Volume 154 (1991), 215-216.

[72] P. Ribenboim, *The Little Book of Big Primes*, Springer-Verlag, New York, 1991.

[73] R. L. Rivest, A. Shamir, and L. Adleman, "A Method for Obtaining Digital Signatures and Public Key Cryptosystems," *Commun. ACM*, Volume 21, No. 2 (1978), 120-126.

[74] N. Robbins, *Beginning Number Theory*, Wm. C. Brown, Dubuque, Iowa, 1993.

[75] K. H. Rosen, *Elementary Number Theory and Its Applications*, 3d ed.,

558

Addison-Wesley, Reading, Mass., 1993.

[76] J. B. Rosser and L. Schoenfeld, "Approximate Formulas for Some Functions of Prime Numbers," *Illinois J. Math.*, Volume 6 (1962), 64-94.

[77] A. Schimmel, *The Mystery of Numbers*, Oxford University Press, New York, 1993.

[78] M. R. Schroeder, *Number Theory in Science and Communication*, 2nd ed., Springer-Verlag, Berlin, 1986.

[79] D. Shanks, *Solved and Unsolved Problems in Number Theory*, Chelsea, New York, 1985.

[80] R. W. Sinnott, *Sky & Telescope*, January (1989), 80-82.

[81] D. E. Smith, *History of Mathematics*, Dover, New York, 1958.

[82] D. E. Smith, *A Source Book in Mathematics*, Dover, New York, 1959.

[83] R. Solovay and V. Strassen, "A Fast Monte-Carlo Test for Primality," *SIAM J. Comput.*, Volume 6, No. 1 (1977), 84-85. "Erratum: A Fast Monte-Carlo Test for Primality," *SIAM J. Comput.*, Volume 7, No. 1 (1978), 118.

[84] B. M. Stewart, *Theory of Numbers*, 2nd ed., Macmillan, New York, 1964.

[85] J. K. Strayer, *Elementary Number Theory*, PWS Publishing Company, Boston, 1993.

[86] D. M. Sunday, "A Very Fast Substring Search Algorithm," *Commun. ACM*, Volume 33, No. 8 (1990), 132-142.

[87] J. A. Todd, "A Combinatorial Problem," *J. Math. Phys.*, Volume 12 (1933), 321-333.

[88] B. L. van der Waerden, *Science Awakening*, Vol. I, 3rd ed., Oxford University Press, New York, 1971.

[89] A. E. Western and J. C. Miller, *Tables of Indices and Primitive Roots*, Royal Society Mathematical Tables, Vol. 9, Cambridge University Press, Cambridge, 1968.

Index